Rapid Sensory Profiling Techniques and Related Methods

Related titles

Sensory analysis for food and beverage quality control
(ISBN 978-1-84569-476-0)

Consumer-driven innovation in food and personal care products
(ISBN 978-1-84569-567-5)

Flavour development, analysis and perception in food and beverages
(ISBN 978-1-78242-103-0)

Woodhead Publishing Series in Food Science,
Technology and Nutrition: Number 274

Rapid Sensory Profiling Techniques and Related Methods

Applications in New Product Development and Consumer Research

Edited by

Julien Delarue, J. Ben Lawlor and Michel Rogeaux

AMSTERDAM • BOSTON • CAMBRIDGE • HEIDELBERG
LONDON • NEW YORK • OXFORD • PARIS • SAN DIEGO
SAN FRANCISCO • SINGAPORE • SYDNEY • TOKYO

Woodhead Publishing is an imprint of Elsevier

Woodhead Publishing is an imprint of Elsevier
80 High Street, Sawston, Cambridge, CB22 3HJ, UK
225 Wyman Street, Waltham, MA 02451, USA
Langford Lane, Kidlington, OX5 1GB, UK

Copyright © 2015 Elsevier Ltd. All rights reserved

No part of this publication may be reproduced, stored in a retrieval system or transmitted in any form or by any means electronic, mechanical, photocopying, recording or otherwise without the prior written permission of the publisher.

Permissions may be sought directly from Elsevier's Science & Technology Rights Department in Oxford, UK: phone (+44) (0) 1865 843830; fax (+44) (0) 1865 853333; email: permissions@elsevier.com. Alternatively, you can submit your request online by visiting the Elsevier website at http://elsevier.com/locate/permissions, and selecting Obtaining permission to use Elsevier material.

Notice
No responsibility is assumed by the publisher for any injury and/or damage to persons or property as a matter of products liability, negligence or otherwise, or from any use or operation of any methods, products, instructions or ideas contained in the material herein. Because of rapid advances in the medical sciences, in particular, independent verification of diagnoses and drug dosages should be made.

British Library Cataloguing-in-Publication Data
A catalogue record for this book is available from the British Library.

Library of Congress Control Number: 2014944404

ISBN 978-0-08-101332-8 (print)
ISBN 978-1-78242-258-7 (online)

For information on all Woodhead Publishing publications visit our website at http://store.elsevier.com/

Typeset by Newgen Knowledge Works Pvt Ltd, India
Printed and bound in the United Kingdom

Contents

List of contributors — xiii
Woodhead Publishing Series in Food Science, Technology and Nutrition — xv

Part One Evolution of the methods used for sensory profiling — 1

1 The use of rapid sensory methods in R&D and research: an introduction — 3
  J. Delarue
  1.1 Introduction and context — 3
  1.2 Methodological evolution — 8
  1.3 Consequences on sensory activities — 14
  1.4 Conclusions — 20
  References — 22

2 Alternative methods of sensory testing: advantages and disadvantages — 27
  H. Stone
  2.1 Introduction — 27
  2.2 The subjects in sensory testing — 28
  2.3 Methods in sensory testing — 30
  2.4 Further important considerations in sensory testing — 37
  2.5 Developing descriptive analysis capability — 39
  2.6 Other descriptive methods — 48
  2.7 Future trends — 49
  2.8 Conclusions — 50
  References — 50

3 Measuring sensory perception in relation to consumer behavior — 53
  J.E. Hayes
  3.1 Introduction — 53
  3.2 Sensation — 54
  3.3 Hedonics — 58
  3.4 Measuring product use and intake — 60
  3.5 Linking sensations, liking, and intake — 61
  3.6 Summary — 63
  References — 63

4	**Insights into measuring emotional response in sensory and consumer research**	**71**
	M. Ng, J. Hort	
	4.1 Introduction	71
	4.2 Defining emotion	72
	4.3 The importance of measuring emotions in sensory and consumer research	73
	4.4 Approaches to measuring emotional response	75
	4.5 Verbal self-report emotion lexicon	77
	4.6 Application of verbal self-report emotion techniques in the sensory and consumer field	78
	4.7 Relating sensory properties to consumers' emotional response	84
	4.8 Unresolved issues and topics for future research in verbal self-report emotion measurement	86
	References	87
5	**Expedited procedures for conceptual profiling of brands, products and packaging**	**91**
	D.M.H. Thomson	
	5.1 Introduction	91
	5.2 Fundamentals of new product success and failure	92
	5.3 Measurement using direct scaling	93
	5.4 Concepts, conceptualisation and conceptual structure	95
	5.5 Emotion profiling versus conceptual profiling – some theoretical considerations	97
	5.6 Conceptual profiling in practice	99
	5.7 Applications and case studies	105
	5.8 Conclusion	116
	Acknowledgements	117
	References	117

Part Two Rapid methods for sensory profiling   119

6	**Flash Profile, its evolution and uses in sensory and consumer science**	**121**
	J. Delarue	
	6.1 The method and its origins	121
	6.2 Flash Profile (FP) methodology through an example: evaluation of dark chocolates	124
	6.3 Further methodological considerations	132
	6.4 Metrological properties of Flash Profile	134
	6.5 Limitations of Flash Profile	137
	6.6 Evolution in the use of Flash Profile	138
	6.7 Conclusions and future trends	148
	References	148

7	**Free sorting as a sensory profiling technique for product development**		**153**
	P. Courcoux, E.M. Qannari, P. Faye		
	7.1	Introduction	153
	7.2	The free sorting task	154
	7.3	Statistical treatment of free sorting data	160
	7.4	A case study in the automotive industry: understanding the consumer perception of car body style	170
	7.5	Conclusion	180
		References	181
8	**Free multiple sorting as a sensory profiling technique**		**187**
	C. Dehlholm		
	8.1	Introduction	187
	8.2	Overview of free multiple sorting (FMS)	187
	8.3	Theoretical framework	188
	8.4	Practical framework and design of experiments	189
	8.5	Implementation and data collection	190
	8.6	Data analysis	191
	8.7	Advantages, disadvantages and applications	194
	8.8	Future trends and further information	195
		References	195
9	**Napping and sorted Napping as a sensory profiling technique**		**197**
	S. Lê, T.M. Lê, M. Cadoret		
	9.1	Introduction	197
	9.2	From projective tests to Napping	197
	9.3	From Napping to sorted Napping	206
	9.4	Analysing Napping and sorted Napping data using the R statistical software	210
	9.5	Conclusion	212
		References	213
10	**Polarized sensory positioning (PSP) as a sensory profiling technique**		**215**
	E. Teillet		
	10.1	Introduction	215
	10.2	Polarized sensory positioning (PSP) methodologies	216
	10.3	Data analyses	217
	10.4	PSP and the taste of water	219
	10.5	Discussion of the choice of the poles	223
	10.6	Conclusion	224
		References	224

11	**Check-all-that-apply (CATA) questions with consumers in practice: experimental considerations and impact on outcome**	**227**
	G. Ares, S.R. Jaeger	
	11.1 Introduction	227
	11.2 Implementation of check-all-that-apply (CATA) questions	229
	11.3 Analysis of data from CATA questions	233
	11.4 Case study: application of CATA questions for sensory characterization of plain yoghurt	235
	11.5 Pros, cons and opportunities of the application of CATA questions	240
	11.6 Conclusions	242
	Acknowledgments	242
	References	242
12	**Open-ended questions in sensory testing practice**	**247**
	B. Piqueras-Fiszman	
	12.1 Introduction	247
	12.2 General pros and cons of open-ended questions	249
	12.3 When open-ended questions are appropriate	251
	12.4 Processing the answers: from raw to clean data	252
	12.5 Analysing the data: getting valuable outcomes from different applications	255
	12.6 Future trends and social media	261
	12.7 Conclusions	264
	References	265
13	**Temporal dominance of sensations (TDS) as a sensory profiling technique**	**269**
	N. Pineau, P. Schilch	
	13.1 Introduction	269
	13.2 Overview of temporal dominance of sensations (TDS)	270
	13.3 TDS experiment and panel training	272
	13.4 Data analysis: representation of the sequence	279
	13.5 Data analysis: representation of the product space	284
	13.6 Data analysis: comparison between products	287
	13.7 Panel performance	291
	13.8 Some applications	295
	13.9 Future trends in TDS	299
	13.10 Conclusion	302
	References	304
14	**Ideal profiling as a sensory profiling technique**	**307**
	T. Worch, P.H. Punter	
	14.1 Introduction	307
	14.2 Principle and properties of the Ideal Profile Method (IPM)	309
	14.3 IPM, a tool for product development and product optimization	313
	14.4 Additional valuable properties of the IPM	316

14.5	Illustration of the Ideal Profile Analysis (IPA)	319
14.6	Conclusions	328
	References	330

Part Three   Applications in new product development and consumer research   333

15 Adoption and use of Flash Profiling in daily new product development: a testimonial   335
C. Petit, E. Vanzeveren

15.1	Introduction	335
15.2	Flash Profile as a starting point	335
15.3	Flash Profile as a reference methodology	338
15.4	Limitations and perspectives in the use of Flash Profile	341
15.5	Conclusion	343
	References	343

16 Improving team tasting in the food industry   345
M. Rogeaux

16.1	Introduction: the ever-increasing importance of new tasting methods within the project teams	345
16.2	Precise analysis of the concrete situations where evaluation by team tasting is appropriate	346
16.3	Analysis of opportunities and constraints linked to project team evaluation	347
16.4	An approach adapted to Danone's needs but integrated with the limits of the team tasting	348
16.5	Implementation examples (common in R&D field)	357
16.6	Analysis and prospects	360
	References	361

17 Alternative methods of sensory testing: working with chefs, culinary professionals and brew masters   363
M.B. Frøst, D. Giacalone, K.K. Rasmussen

17.1	Introduction	363
17.2	Background: fast descriptive methods and persons with no formal sensory training in sensory tests	363
17.3	Data analysis of projective descriptive methods	365
17.4	Case study 1: brewers and novices assessing beer	367
17.5	Results and discussion of partial napping of beer	368
17.6	Case study 2: exploring the world of spice blends and pastes with chefs and other food experts	371
17.7	Results and discussion of spice blends and pastes	373
17.8	General discussion and recommendations	377
	References	379
	Appendix: Projective mapping versus napping (see also Chapter 9)	382

18	Sensory testing with flavourists: challenges and solutions	383
	B. Veinand	
	18.1 Introduction	383
	18.2 Roles and responsibilities	384
	18.3 Different ways of working	386
	18.4 Strategies to complement both types of expertise	390
	18.5 Future trends	397
	References	398
19	Projective Flash Profile from experts to consumers: a way to reveal fragrance language	401
	S. Ballay, E. Loescher, G. Gazano	
	19.1 Introduction: an industrial approach to the assessment of fragrances	401
	19.2 Flash Profile of fragrances: perfumers vs consumers	401
	19.3 An extension to Flash Profile of fragrances with consumers: beyond sensory description	410
	19.4 Discussion and conclusion	423
	References	424
20	Use of rapid sensory methods in the automotive industry	427
	D. Blumenthal, N. Herbeth	
	20.1 Introduction	427
	20.2 Example 1: gearbox sensations and comfort	430
	20.3 Example 2: role and lateral support perception	435
	20.4 Example 3: idle noises of diesel engines	444
	20.5 Conclusion: pros and cons of rapid sensory methods in the automotive context	450
	References	451
21	Testing consumer insight using mobile devices: a case study of a sensory consumer journey conducted with the help of mobile research	455
	D. Lutsch, R. Möslein, M. Strack, S. Kunze	
	21.1 Mobile research: status quo	455
	21.2 Mobile sensory research: a new mobile research method	456
	21.3 Case study: a sensory consumer journey conducted with the help of mobile research	459
	21.4 Summary and discussion	466
	21.5 Conclusion	468
	References	469

Part Four Applications in sensory testing with specific populations and methodological consequences — 471

22	Sensory testing in new product development: working with children	473
	S. Nicklaus	
	22.1 Introduction	473

22.2	Reasons for studying sensory aspects in children	474
22.3	How to organize sensory evaluation testing with children	475
22.4	Application of different sensory evaluation techniques to children of different ages	476
22.5	Conclusion	480
22.6	Future trends	480
22.7	Sources of further information	481
	References	481

23 Sensory testing in new product development: working with older people 485
I. Maitre, R. Symoneaux, C. Sulmont-Rossé

23.1	Introduction	485
23.2	The elderly market: a challenge between needs and pleasure	485
23.3	The heterogeneity of the elderly	487
23.4	Impact of age and dependence on performance at a sensory task: key findings on scale use in a monadic sequential presentation	491
23.5	Running sensory descriptive analysis with an elderly panel: recommendations	497
23.6	Conclusion and future trends	503
	Acknowledgements	504
	References	505

24 Empathy and Experiment™: dealing with the algebra of the mind to understand and change food habits 509
H.R. Moskowitz, M. Reisner, L. Ettinger Lieberman, B. Batalvi, M. Beg

24.1	Introduction	509
24.2	The origins of the study	509
24.3	Background: Golden Rice – the positives	510
24.4	Background: Golden Rice – the negatives	512
24.5	Empathy and Experiment™: the two halves of the approach	512
24.6	The value of experimentation and implementation of Golden Rice evaluations among Pakistanis	516
24.7	Summary of the elements and process of the experiment	518
24.8	The material of the interview and analysis of structured experimental design data	521
24.9	Explicating the results – the total panel versus gender	525
24.10	Culture-mind-set segments	528
24.11	Summary and future trends	535
	Acknowledgment	536
	References	536

Index 539

List of contributors

G. Ares Universidad de la República, Montevideo, Uruguay

S. Ballay LVMH Parfums & Cosmetiques, Paris, France

B. Batalvi SB & B Marketing Research, Pty, Toronto, Canada and Lahore, Pakistan

M. Beg SB & B Marketing Research, Pty, Toronto, Canada and Lahore, Pakistan

D. Blumenthal UMR1145 Ingénierie Procédés Aliments, AgroParisTech, INRA, Cnam, Massy, France

M. Cadoret Kuzulia, Plabennec, France

P. Courcoux L'UNAM University, ONIRIS and INRA, Nantes, France

J. Delarue UMR1145 Ingénierie Procédés Aliments, AgroParisTech, INRA, Cnam, Massy, France

C. Dehlholm Danish Technological Institute, Aarhus, Denmark

L. Ettinger Lieberman Moskowitz Jacobs Inc., White Plains, NY, USA

P. Faye L'UNAM University, ONIRIS and PSA Peugeot Citroen, Nantes, France

M.B. Frøst University of Copenhagen and Nordic Food Lab, Frederiksberg, Denmark

G. Gazano LVMH Parfums & Cosmetiques, Paris, France

D. Giacalone University of Copenhagen, Frederiksberg, Denmark

J.E. Hayes The Pennsylvania State University, University Park, PA, USA

N. Herbeth Renault SAS, Guyancourt, France

J. Hort University of Nottingham, Nottingham, UK

S.R. Jaeger The New Zealand Institute for Plant & Food Research Limited, Auckland, New Zealand

S. Kunze isi GmbH & Co. KG, Niedersachsen, Germany

J.B. Lawlor Danone Nutricia Research, Utrecht, the Netherlands

S. Lê Agrocampus Ouest, Rennes, France

T.M. Lê Agrocampus Ouest, Rennes, France

E. Loescher LVMH Parfums & Cosmetiques, Paris, France

D. Lutsch isi GmbH & Co. KG, Niedersachsen, Germany

I. Maitre UPSP GRAPPE, Groupe ESA, SFR QUASAV 4207, Angers, France

H.R. Moskowitz iNovum LLC, White Plains, NY, USA

R. Möslein isi GmbH & Co. KG, Niedersachsen, Germany

S. Nicklaus Centre des Sciences du Gout et de l'Alimentation, CNRS, INRA and Université de Bourgogne, Dijon, France

M. Ng PepsiCo Europe R&D, Leicester, UK and University of Nottingham, Nottingham, UK

C. Petit Puratos N.V., Groot-Bijgaarden, Belgium

N. Pineau Nestlé Research Center (Nestec Ltd), Lausanne, Switzerland

B. Piqueras-Fiszman Wageningen University and Research Centre, Wageningen, The Netherlands

P.H. Punter OP&P Product Research, Utrecht, the Netherlands

E.M. Qannari L'UNAM University, ONIRIS and INRA, Nantes, France

K.K. Rasmussen University of California, Berkeley, CA, USA

M. Reisner Moskowitz Jacobs Inc., USA, White Plains, NY, USA

M. Rogeaux Danone Nutricia Research, Palaiseau, France

P. Schlich Centre des Sciences du Gout et de l'Alimentation, Dijon, France

H. Stone Sensory Consulting Services, USA

M. Strack isi GmbH & Co. KG, Niedersachsen, Germany

C. Sulmont-Rossé INRA, UMR 1324, CNRS, UMR 6265, Université de Bourgogne, Centre des Sciences du Goût et de l'Alimentation, Dijon, France

R. Symoneaux UPSP GRAPPE, Groupe ESA, SFR QUASAV 4207, Angers France

E. Teillet SensoStat, Dijon, France

D.M.H. Thomson MMR Research Worldwide, Wallingford, UK

E. Vanzeveren Puratos N.V., Groot-Bijgaarden, Belgium

B. Veinand Givaudan International SA, Kemptthal, Switzerland

T. Worch Qi Statistics, Ruscombe, Reading, UK

Woodhead Publishing Series in Food Science, Technology and Nutrition

1 **Chilled foods: A comprehensive guide**
   Edited by C. Dennis and M. Stringer
2 **Yoghurt: Science and technology**
   A. Y. Tamime and R. K. Robinson
3 **Food processing technology: Principles and practice**
   P. J. Fellows
4 **Bender's dictionary of nutrition and food technology Sixth edition**
   D. A. Bender
5 **Determination of veterinary residues in food**
   Edited by N. T. Crosby
6 **Food contaminants: Sources and surveillance**
   Edited by C. Creaser and R. Purchase
7 **Nitrates and nitrites in food and water**
   Edited by M. J. Hill
8 **Pesticide chemistry and bioscience: The food-environment challenge**
   Edited by G. T. Brooks and T. Roberts
9 **Pesticides: Developments, impacts and controls**
   Edited by G. A. Best and A. D. Ruthven
10 **Dietary fibre: Chemical and biological aspects**
    Edited by D. A. T. Southgate, K. W. Waldron, I. T. Johnson and G. R. Fenwick
11 **Vitamins and minerals in health and nutrition**
    M. Tolonen
12 **Technology of biscuits, crackers and cookies Second edition**
    D. Manley
13 **Instrumentation and sensors for the food industry**
    Edited by E. Kress-Rogers
14 **Food and cancer prevention: Chemical and biological aspects**
    Edited by K. W. Waldron, I. T. Johnson and G. R. Fenwick
15 **Food colloids: Proteins, lipids and polysaccharides**
    Edited by E. Dickinson and B. Bergenstahl
16 **Food emulsions and foams**
    Edited by E. Dickinson
17 **Maillard reactions in chemistry, food and health**
    Edited by T. P. Labuza, V. Monnier, J. Baynes and J. O'Brien
18 **The Maillard reaction in foods and medicine**
    Edited by J. O'Brien, H. E. Nursten, M. J. Crabbe and J. M. Ames
19 **Encapsulation and controlled release**
    Edited by D. R. Karsa and R. A. Stephenson

20 **Flavours and fragrances**
   Edited by A. D. Swift
21 **Feta and related cheeses**
   Edited by A. Y. Tamime and R. K. Robinson
22 **Biochemistry of milk products**
   Edited by A. T. Andrews and J. R. Varley
23 **Physical properties of foods and food processing systems**
   M. J. Lewis
24 **Food irradiation: A reference guide**
   V. M. Wilkinson and G. Gould
25 **Kent's technology of cereals: An introduction for students of food science and agriculture Fourth edition**
   N. L. Kent and A. D. Evers
26 **Biosensors for food analysis**
   Edited by A. O. Scott
27 **Separation processes in the food and biotechnology industries: Principles and applications**
   Edited by A. S. Grandison and M. J. Lewis
28 **Handbook of indices of food quality and authenticity**
   R. S. Singhal, P. K. Kulkarni and D. V. Rege
29 **Principles and practices for the safe processing of foods**
   D. A. Shapton and N. F. Shapton
30 **Biscuit, cookie and cracker manufacturing manuals Volume 1: Ingredients**
   D. Manley
31 **Biscuit, cookie and cracker manufacturing manuals Volume 2: Biscuit doughs**
   D. Manley
32 **Biscuit, cookie and cracker manufacturing manuals Volume 3: Biscuit dough piece forming**
   D. Manley
33 **Biscuit, cookie and cracker manufacturing manuals Volume 4: Baking and cooling of biscuits**
   D. Manley
34 **Biscuit, cookie and cracker manufacturing manuals Volume 5: Secondary processing in biscuit manufacturing**
   D. Manley
35 **Biscuit, cookie and cracker manufacturing manuals Volume 6: Biscuit packaging and storage**
   D. Manley
36 **Practical dehydration Second edition**
   M. Greensmith
37 **Lawrie's meat science Sixth edition**
   R. A. Lawrie
38 **Yoghurt: Science and technology Second edition**
   A. Y. Tamime and R. K. Robinson
39 **New ingredients in food processing: Biochemistry and agriculture**
   G. Linden and D. Lorient
40 **Benders' dictionary of nutrition and food technology Seventh edition**
   D. A. Bender and A. E. Bender

41 **Technology of biscuits, crackers and cookies Third edition**
   D. Manley
42 **Food processing technology: Principles and practice Second edition**
   P. J. Fellows
43 **Managing frozen foods**
   Edited by C. J. Kennedy
44 **Handbook of hydrocolloids**
   Edited by G. O. Phillips and P. A. Williams
45 **Food labelling**
   Edited by J. R. Blanchfield
46 **Cereal biotechnology**
   Edited by P. C. Morris and J. H. Bryce
47 **Food intolerance and the food industry**
   Edited by T. Dean
48 **The stability and shelf-life of food**
   Edited by D. Kilcast and P. Subramaniam
49 **Functional foods: Concept to product**
   Edited by G. R. Gibson and C. M. Williams
50 **Chilled foods: A comprehensive guide Second edition**
   Edited by M. Stringer and C. Dennis
51 **HACCP in the meat industry**
   Edited by M. Brown
52 **Biscuit, cracker and cookie recipes for the food industry**
   D. Manley
53 **Cereals processing technology**
   Edited by G. Owens
54 **Baking problems solved**
   S. P. Cauvain and L. S. Young
55 **Thermal technologies in food processing**
   Edited by P. Richardson
56 **Frying: Improving quality**
   Edited by J. B. Rossell
57 **Food chemical safety Volume 1: Contaminants**
   Edited by D. Watson
58 **Making the most of HACCP: Learning from others' experience**
   Edited by T. Mayes and S. Mortimore
59 **Food process modelling**
   Edited by L. M. M. Tijskens, M. L. A. T. M. Hertog and B. M. Nicolaï
60 **EU food law: A practical guide**
   Edited by K. Goodburn
61 **Extrusion cooking: Technologies and applications**
   Edited by R. Guy
62 **Auditing in the food industry: From safety and quality to environmental and other audits**
   Edited by M. Dillon and C. Griffith
63 **Handbook of herbs and spices Volume 1**
   Edited by K. V. Peter

64 **Food product development: Maximising success**
   M. Earle, R. Earle and A. Anderson
65 **Instrumentation and sensors for the food industry Second edition**
   Edited by E. Kress-Rogers and C. J. B. Brimelow
66 **Food chemical safety Volume 2: Additives**
   Edited by D. Watson
67 **Fruit and vegetable biotechnology**
   Edited by V. Valpuesta
68 **Foodborne pathogens: Hazards, risk analysis and control**
   Edited by C. de W. Blackburn and P. J. McClure
69 **Meat refrigeration**
   S. J. James and C. James
70 **Lockhart and Wiseman's crop husbandry Eighth edition**
   H. J. S. Finch, A. M. Samuel and G. P. F. Lane
71 **Safety and quality issues in fish processing**
   Edited by H. A. Bremner
72 **Minimal processing technologies in the food industries**
   Edited by T. Ohlsson and N. Bengtsson
73 **Fruit and vegetable processing: Improving quality**
   Edited by W. Jongen
74 **The nutrition handbook for food processors**
   Edited by C. J. K. Henry and C. Chapman
75 **Colour in food: Improving quality**
   Edited by D. MacDougall
76 **Meat processing: Improving quality**
   Edited by J. P. Kerry, J. F. Kerry and D. A. Ledward
77 **Microbiological risk assessment in food processing**
   Edited by M. Brown and M. Stringer
78 **Performance functional foods**
   Edited by D. Watson
79 **Functional dairy products Volume 1**
   Edited by T. Mattila-Sandholm and M. Saarela
80 **Taints and off-flavours in foods**
   Edited by B. Baigrie
81 **Yeasts in food**
   Edited by T. Boekhout and V. Robert
82 **Phytochemical functional foods**
   Edited by I. T. Johnson and G. Williamson
83 **Novel food packaging techniques**
   Edited by R. Ahvenainen
84 **Detecting pathogens in food**
   Edited by T. A. McMeekin
85 **Natural antimicrobials for the minimal processing of foods**
   Edited by S. Roller
86 **Texture in food Volume 1: Semi-solid foods**
   Edited by B. M. McKenna

87 **Dairy processing: Improving quality**
   Edited by G. Smit
88 **Hygiene in food processing: Principles and practice**
   Edited by H. L. M. Lelieveld, M. A. Mostert, B. White and J. Holah
89 **Rapid and on-line instrumentation for food quality assurance**
   Edited by I. Tothill
90 **Sausage manufacture: Principles and practice**
   E. Essien
91 **Environmentally-friendly food processing**
   Edited by B. Mattsson and U. Sonesson
92 **Bread making: Improving quality**
   Edited by S. P. Cauvain
93 **Food preservation techniques**
   Edited by P. Zeuthen and L. Bøgh-Sørensen
94 **Food authenticity and traceability**
   Edited by M. Lees
95 **Analytical methods for food additives**
   R. Wood, L. Foster, A. Damant and P. Key
96 **Handbook of herbs and spices Volume 2**
   Edited by K. V. Peter
97 **Texture in food Volume 2: Solid foods**
   Edited by D. Kilcast
98 **Proteins in food processing**
   Edited by R. Yada
99 **Detecting foreign bodies in food**
   Edited by M. Edwards
100 **Understanding and measuring the shelf-life of food**
   Edited by R. Steele
101 **Poultry meat processing and quality**
   Edited by G. Mead
102 **Functional foods, ageing and degenerative disease**
   Edited by C. Remacle and B. Reusens
103 **Mycotoxins in food: Detection and control**
   Edited by N. Magan and M. Olsen
104 **Improving the thermal processing of foods**
   Edited by P. Richardson
105 **Pesticide, veterinary and other residues in food**
   Edited by D. Watson
106 **Starch in food: Structure, functions and applications**
   Edited by A.-C. Eliasson
107 **Functional foods, cardiovascular disease and diabetes**
   Edited by A. Arnoldi
108 **Brewing: Science and practice**
   D. E. Briggs, P. A. Brookes, R. Stevens and C. A. Boulton

109 **Using cereal science and technology for the benefit of consumers: Proceedings of the 12th International ICC Cereal and Bread Congress, 24 – 26th May, 2004, Harrogate, UK**
Edited by S. P. Cauvain, L. S. Young and S. Salmon
110 **Improving the safety of fresh meat**
Edited by J. Sofos
111 **Understanding pathogen behaviour: Virulence, stress response and resistance**
Edited by M. Griffiths
112 **The microwave processing of foods**
Edited by H. Schubert and M. Regier
113 **Food safety control in the poultry industry**
Edited by G. Mead
114 **Improving the safety of fresh fruit and vegetables**
Edited by W. Jongen
115 **Food, diet and obesity**
Edited by D. Mela
116 **Handbook of hygiene control in the food industry**
Edited by H. L. M. Lelieveld, M. A. Mostert and J. Holah
117 **Detecting allergens in food**
Edited by S. Koppelman and S. Hefle
118 **Improving the fat content of foods**
Edited by C. Williams and J. Buttriss
119 **Improving traceability in food processing and distribution**
Edited by I. Smith and A. Furness
120 **Flavour in food**
Edited by A. Voilley and P. Etievant
121 **The Chorleywood bread process**
S. P. Cauvain and L. S. Young
122 **Food spoilage microorganisms**
Edited by C. de W. Blackburn
123 **Emerging foodborne pathogens**
Edited by Y. Motarjemi and M. Adams
124 **Benders' dictionary of nutrition and food technology Eighth edition**
D. A. Bender
125 **Optimising sweet taste in foods**
Edited by W. J. Spillane
126 **Brewing: New technologies**
Edited by C. Bamforth
127 **Handbook of herbs and spices Volume 3**
Edited by K. V. Peter
128 **Lawrie's meat science Seventh edition**
R. A. Lawrie in collaboration with D. A. Ledward
129 **Modifying lipids for use in food**
Edited by F. Gunstone

130 **Meat products handbook: Practical science and technology**
G. Feiner
131 **Food consumption and disease risk: Consumer–pathogen interactions**
Edited by M. Potter
132 **Acrylamide and other hazardous compounds in heat-treated foods**
Edited by K. Skog and J. Alexander
133 **Managing allergens in food**
Edited by C. Mills, H. Wichers and K. Hoffman-Sommergruber
134 **Microbiological analysis of red meat, poultry and eggs**
Edited by G. Mead
135 **Maximising the value of marine by-products**
Edited by F. Shahidi
136 **Chemical migration and food contact materials**
Edited by K. Barnes, R. Sinclair and D. Watson
137 **Understanding consumers of food products**
Edited by L. Frewer and H. van Trijp
138 **Reducing salt in foods: Practical strategies**
Edited by D. Kilcast and F. Angus
139 **Modelling microorganisms in food**
Edited by S. Brul, S. Van Gerwen and M. Zwietering
140 **Tamime and Robinson's Yoghurt: Science and technology Third edition**
A. Y. Tamime and R. K. Robinson
141 **Handbook of waste management and co-product recovery in food processing Volume 1**
Edited by K. W. Waldron
142 **Improving the flavour of cheese**
Edited by B. Weimer
143 **Novel food ingredients for weight control**
Edited by C. J. K. Henry
144 **Consumer-led food product development**
Edited by H. MacFie
145 **Functional dairy products Volume 2**
Edited by M. Saarela
146 **Modifying flavour in food**
Edited by A. J. Taylor and J. Hort
147 **Cheese problems solved**
Edited by P. L. H. McSweeney
148 **Handbook of organic food safety and quality**
Edited by J. Cooper, C. Leifert and U. Niggli
149 **Understanding and controlling the microstructure of complex foods**
Edited by D. J. McClements
150 **Novel enzyme technology for food applications**
Edited by R. Rastall
151 **Food preservation by pulsed electric fields: From research to application**
Edited by H. L. M. Lelieveld and S. W. H. de Haan

152 **Technology of functional cereal products**
   Edited by B. R. Hamaker
153 **Case studies in food product development**
   Edited by M. Earle and R. Earle
154 **Delivery and controlled release of bioactives in foods and nutraceuticals**
   Edited by N. Garti
155 **Fruit and vegetable flavour: Recent advances and future prospects**
   Edited by B. Brückner and S. G. Wyllie
156 **Food fortification and supplementation: Technological, safety and regulatory aspects**
   Edited by P. Berry Ottaway
157 **Improving the health-promoting properties of fruit and vegetable products**
   Edited by F. A. Tomás-Barberán and M. I. Gil
158 **Improving seafood products for the consumer**
   Edited by T. Børresen
159 **In-pack processed foods: Improving quality**
   Edited by P. Richardson
160 **Handbook of water and energy management in food processing**
   Edited by J. Klemeš, R. Smith and J.-K. Kim
161 **Environmentally compatible food packaging**
   Edited by E. Chiellini
162 **Improving farmed fish quality and safety**
   Edited by Ø. Lie
163 **Carbohydrate-active enzymes**
   Edited by K.-H. Park
164 **Chilled foods: A comprehensive guide Third edition**
   Edited by M. Brown
165 **Food for the ageing population**
   Edited by M. M. Raats, C. P. G. M. de Groot and W. A Van Staveren
166 **Improving the sensory and nutritional quality of fresh meat**
   Edited by J. P. Kerry and D. A. Ledward
167 **Shellfish safety and quality**
   Edited by S. E. Shumway and G. E. Rodrick
168 **Functional and speciality beverage technology**
   Edited by P. Paquin
169 **Functional foods: Principles and technology**
   M. Guo
170 **Endocrine-disrupting chemicals in food**
   Edited by I. Shaw
171 **Meals in science and practice: Interdisciplinary research and business applications**
   Edited by H. L. Meiselman
172 **Food constituents and oral health: Current status and future prospects**
   Edited by M. Wilson
173 **Handbook of hydrocolloids Second edition**
   Edited by G. O. Phillips and P. A. Williams

174 **Food processing technology: Principles and practice Third edition**
   P. J. Fellows
175 **Science and technology of enrobed and filled chocolate, confectionery and bakery products**
   Edited by G. Talbot
176 **Foodborne pathogens: Hazards, risk analysis and control Second edition**
   Edited by C. de W. Blackburn and P. J. McClure
177 **Designing functional foods: Measuring and controlling food structure breakdown and absorption**
   Edited by D. J. McClements and E. A. Decker
178 **New technologies in aquaculture: Improving production efficiency, quality and environmental management**
   Edited by G. Burnell and G. Allan
179 **More baking problems solved**
   S. P. Cauvain and L. S. Young
180 **Soft drink and fruit juice problems solved**
   P. Ashurst and R. Hargitt
181 **Biofilms in the food and beverage industries**
   Edited by P. M. Fratamico, B. A. Annous and N. W. Gunther
182 **Dairy-derived ingredients: Food and neutraceutical uses**
   Edited by M. Corredig
183 **Handbook of waste management and co-product recovery in food processing Volume 2**
   Edited by K. W. Waldron
184 **Innovations in food labelling**
   Edited by J. Albert
185 **Delivering performance in food supply chains**
   Edited by C. Mena and G. Stevens
186 **Chemical deterioration and physical instability of food and beverages**
   Edited by L. H. Skibsted, J. Risbo and M. L. Andersen
187 **Managing wine quality Volume 1: Viticulture and wine quality**
   Edited by A. G. Reynolds
188 **Improving the safety and quality of milk Volume 1: Milk production and processing**
   Edited by M. Griffiths
189 **Improving the safety and quality of milk Volume 2: Improving quality in milk products**
   Edited by M. Griffiths
190 **Cereal grains: Assessing and managing quality**
   Edited by C. Wrigley and I. Batey
191 **Sensory analysis for food and beverage quality control: A practical guide**
   Edited by D. Kilcast
192 **Managing wine quality Volume 2: Oenology and wine quality**
   Edited by A. G. Reynolds

193 **Winemaking problems solved**
    Edited by C. E. Butzke
194 **Environmental assessment and management in the food industry**
    Edited by U. Sonesson, J. Berlin and F. Ziegler
195 **Consumer-driven innovation in food and personal care products**
    Edited by S. R. Jaeger and H. MacFie
196 **Tracing pathogens in the food chain**
    Edited by S. Brul, P. M. Fratamico and T. A. McMeekin
197 **Case studies in novel food processing technologies: Innovations in processing, packaging, and predictive modelling**
    Edited by C. J. Doona, K. Kustin and F. E. Feeherry
198 **Freeze-drying of pharmaceutical and food products**
    T.-C. Hua, B.-L. Liu and H. Zhang
199 **Oxidation in foods and beverages and antioxidant applications Volume 1: Understanding mechanisms of oxidation and antioxidant activity**
    Edited by E. A. Decker, R. J. Elias and D. J. McClements
200 **Oxidation in foods and beverages and antioxidant applications Volume 2: Management in different industry sectors**
    Edited by E. A. Decker, R. J. Elias and D. J. McClements
201 **Protective cultures, antimicrobial metabolites and bacteriophages for food and beverage biopreservation**
    Edited by C. Lacroix
202 **Separation, extraction and concentration processes in the food, beverage and nutraceutical industries**
    Edited by S. S. H. Rizvi
203 **Determining mycotoxins and mycotoxigenic fungi in food and feed**
    Edited by S. De Saeger
204 **Developing children's food products**
    Edited by D. Kilcast and F. Angus
205 **Functional foods: Concept to product Second edition**
    Edited by M. Saarela
206 **Postharvest biology and technology of tropical and subtropical fruits Volume 1: Fundamental issues**
    Edited by E. M. Yahia
207 **Postharvest biology and technology of tropical and subtropical fruits Volume 2: Açai to citrus**
    Edited by E. M. Yahia
208 **Postharvest biology and technology of tropical and subtropical fruits Volume 3: Cocona to mango**
    Edited by E. M. Yahia
209 **Postharvest biology and technology of tropical and subtropical fruits Volume 4: Mangosteen to white sapote**
    Edited by E. M. Yahia
210 **Food and beverage stability and shelf life**
    Edited by D. Kilcast and P. Subramaniam

211 **Processed Meats: Improving safety, nutrition and quality**
Edited by J. P. Kerry and J. F. Kerry
212 **Food chain integrity: A holistic approach to food traceability, safety, quality and authenticity**
Edited by J. Hoorfar, K. Jordan, F. Butler and R. Prugger
213 **Improving the safety and quality of eggs and egg products Volume 1**
Edited by Y. Nys, M. Bain and F. Van Immerseel
214 **Improving the safety and quality of eggs and egg products Volume 2**
Edited by F. Van Immerseel, Y. Nys and M. Bain
215 **Animal feed contamination: Effects on livestock and food safety**
Edited by J. Fink-Gremmels
216 **Hygienic design of food factories**
Edited by J. Holah and H. L. M. Lelieveld
217 **Manley's technology of biscuits, crackers and cookies Fourth edition**
Edited by D. Manley
218 **Nanotechnology in the food, beverage and nutraceutical industries**
Edited by Q. Huang
219 **Rice quality: A guide to rice properties and analysis**
K. R. Bhattacharya
220 **Advances in meat, poultry and seafood packaging**
Edited by J. P. Kerry
221 **Reducing saturated fats in foods**
Edited by G. Talbot
222 **Handbook of food proteins**
Edited by G. O. Phillips and P. A. Williams
223 **Lifetime nutritional influences on cognition, behaviour and psychiatric illness**
Edited by D. Benton
224 **Food machinery for the production of cereal foods, snack foods and confectionery**
L.-M. Cheng
225 **Alcoholic beverages: Sensory evaluation and consumer research**
Edited by J. Piggott
226 **Extrusion problems solved: Food, pet food and feed**
M. N. Riaz and G. J. Rokey
227 **Handbook of herbs and spices Second edition Volume 1**
Edited by K. V. Peter
228 **Handbook of herbs and spices Second edition Volume 2**
Edited by K. V. Peter
229 **Breadmaking: Improving quality Second edition**
Edited by S. P. Cauvain
230 **Emerging food packaging technologies: Principles and practice**
Edited by K. L. Yam and D. S. Lee
231 **Infectious disease in aquaculture: Prevention and control**
Edited by B. Austin

232 **Diet, immunity and inflammation**
   Edited by P. C. Calder and P. Yaqoob
233 **Natural food additives, ingredients and flavourings**
   Edited by D. Baines and R. Seal
234 **Microbial decontamination in the food industry: Novel methods and applications**
   Edited by A. Demirci and M.O. Ngadi
235 **Chemical contaminants and residues in foods**
   Edited by D. Schrenk
236 **Robotics and automation in the food industry: Current and future technologies**
   Edited by D. G. Caldwell
237 **Fibre-rich and wholegrain foods: Improving quality**
   Edited by J. A. Delcour and K. Poutanen
238 **Computer vision technology in the food and beverage industries**
   Edited by D.-W. Sun
239 **Encapsulation technologies and delivery systems for food ingredients and nutraceuticals**
   Edited by N. Garti and D. J. McClements
240 **Case studies in food safety and authenticity**
   Edited by J. Hoorfar
241 **Heat treatment for insect control: Developments and applications**
   D. Hammond
242 **Advances in aquaculture hatchery technology**
   Edited by G. Allan and G. Burnell
243 **Open innovation in the food and beverage industry**
   Edited by M. Garcia Martinez
244 **Trends in packaging of food, beverages and other fast-moving consumer goods (FMCG)**
   Edited by N. Farmer
245 **New analytical approaches for verifying the origin of food**
   Edited by P. Brereton
246 **Microbial production of food ingredients, enzymes and nutraceuticals**
   Edited by B. McNeil, D. Archer, I. Giavasis and L. Harvey
247 **Persistent organic pollutants and toxic metals in foods**
   Edited by M. Rose and A. Fernandes
248 **Cereal grains for the food and beverage industries**
   E. Arendt and E. Zannini
249 **Viruses in food and water: Risks, surveillance and control**
   Edited by N. Cook
250 **Improving the safety and quality of nuts**
   Edited by L. J. Harris
251 **Metabolomics in food and nutrition**
   Edited by B. C. Weimer and C. Slupsky

252 **Food enrichment with omega-3 fatty acids**
Edited by C. Jacobsen, N. S. Nielsen, A. F. Horn and A.-D. M. Sørensen
253 **Instrumental assessment of food sensory quality: A practical guide**
Edited by D. Kilcast
254 **Food microstructures: Microscopy, measurement and modelling**
Edited by V. J. Morris and K. Groves
255 **Handbook of food powders: Processes and properties**
Edited by B. R. Bhandari, N. Bansal, M. Zhang and P. Schuck
256 **Functional ingredients from algae for foods and nutraceuticals**
Edited by H. Domínguez
257 **Satiation, satiety and the control of food intake: Theory and practice**
Edited by J. E. Blundell and F. Bellisle
258 **Hygiene in food processing: Principles and practice Second edition**
Edited by H. L. M. Lelieveld, J. Holah and D. Napper
259 **Advances in microbial food safety Volume 1**
Edited by J. Sofos
260 **Global safety of fresh produce: A handbook of best practice, innovative commercial solutions and case studies**
Edited by J. Hoorfar
261 **Human milk biochemistry and infant formula manufacturing technology**
Edited by M. Guo
262 **High throughput screening for food safety assessment: Biosensor technologies, hyperspectral imaging and practical applications**
Edited by A. K. Bhunia, M. S. Kim and C. R. Taitt
263 **Foods, nutrients and food ingredients with authorised EU health claims: Volume 1**
Edited by M. J. Sadler
264 **Handbook of food allergen detection and control**
Edited by S. Flanagan
265 **Advances in fermented foods and beverages: Improving quality, technologies and health benefits**
Edited by W. Holzapfel
266 **Metabolomics as a tool in nutrition research**
Edited by J.-L. Sébédio and L. Brennan
267 **Dietary supplements: Safety, efficacy and quality**
Edited by K. Berginc and S. Kreft
268 **Grapevine breeding programs for the wine industry: Traditional and molecular technologies**
Edited by A. G. Reynolds
269 **Handbook of antimicrobials for food safety and quality**
Edited by M. Taylor
270 **Managing and preventing obesity: Behavioural factors and dietary interventions**
Edited by T. P. Gill

271 **Electron beam pasteurization and complementary food processing technologies**
Edited by S. D. Pillai and S. Shayanfar
272 **Advances in food and beverage labelling: Information and regulations**
Edited by P. Berryman
273 **Flavour development, analysis and perception in food and beverages**
Edited by J. K. Parker, S. Elmore and L. Methven
274 **Rapid sensory profiling techniques and related methods: Applications in new product development and consumer research**
Edited by J. Delarue, J. B. Lawlor and M. Rogeaux
275 **Advances in microbial food safety: Volume 2**
Edited by J. Sofos
276 **Handbook of antioxidants in food preservation**
Edited by F. Shahidi
277 **Lockhart and Wiseman's crop husbandry including grassland: Ninth edition**
H. J. S. Finch, A. M. Samuel and G. P. F. Lane
278 **Global legislation for food contact materials: Processing, storage and packaging**
Edited by J. S. Baughan
279 **Colour additives for food and beverages: Development, safety and applications**
Edited by M. Scotter
280 **A complete course in canning and related processes 14th Edition: Volume 1**
Revised by S. Featherstone
281 **A complete course in canning and related processes 14th Edition: Volume 2**
Revised by S. Featherstone
282 **A complete course in canning and related processes 14th Edition: Volume 3**
Revised by S. Featherstone

Part One

Evolution of the methods used for sensory profiling

The use of rapid sensory methods in R&D and research: an introduction

J. Delarue

UMR1145 Ingénierie Procédés Aliments, AgroParisTech, INRA, Cnam, Massy, France

1.1 Introduction and context

1.1.1 Evolution in the use of descriptive analysis (DA)

Over the past decades, the use of sensory techniques has undergone drastic change. Sensory professionals in industry, as well as sensory scientists in academia, have changed the way they use these methods, and the way they use their outcomes in research and development projects. Students in sensory and food science programmes are now familiar with a series of different techniques, and the number of new methods published every year in the literature is accelerating. Among other factors, time and economic constraints that go along with industrial needs have certainly driven this evolution.

Description and quantification of human perception are difficult tasks, and sensory DA techniques are among the most sophisticated tools in the arsenal of the sensory scientist (Lawless and Heymann, 2010). "Sensory profiling" is often used as shorthand for sensory DA, which is in fact a name for a class of methods rather than a unique technique (Dijksterhuis and Byrne, 2005). For a review of these methods, see also Murray *et al.* (2001).

Sensory DA has long been a "must have" tool in the fast moving consumer goods (FMCG) industry, and it has proven to be extremely useful to key industrial functions such as R&D and quality management. However, probably victims of their own success, conventional profiling methods such as quantitative descriptive analysis (QDA®) (Stone *et al.*, 1974) or the Spectrum® method (Meilgaard *et al.*, 1999) are becoming less adaptable to growing and changing demand. In everyday industrial practice, the need for descriptive information is now very diverse, and sensory services are frequently overwhelmed with demands from stakeholders who may have multiple objectives. The cost for accuracy and reliability of conventional profiling methods is time and effort. As a result, sensory teams that rely only on conventional profiling often cannot be fast enough to meet this demand. Also, these methods are not very flexible, and certainly lack adaptability, in the face of an ever-changing market. Some companies thus question the worth of time and investment in heavily trained panels, and look for faster and cheaper methods. Varela and Ares (2012) have pointed out that the constraints in having a trained descriptive panel could also be a problem in academic research, when a short project does not justify the training of a panel from scratch, or when the lack of funding does not allow it. Besides, although conventional profiling is considered as a reference measurement in DA, a more general observation is that it is questionable whether a single method can fulfil multiple different objectives.

1.1.2 Emergence of rapid sensory profiling methods

Following this evolution, sensory practitioners have started using various methods in addition to conventional profiling to capture sensory perception for their research or in the frame of product development. Some of these methods have been known for decades, whether in sensory science or other fields, while others are new or are adaptations of older methods. They can all be seen as rapid alternatives to sensory profiling although, strictly speaking, not all of them can be called sensory profiling methods as they do not rely on description per se, i.e. when products are described in terms of sensory attributes. However, like conventional DA, they may all provide access to the relative sensory positioning of a set of products, which is most often represented in the form of a sensory map. As a result, even if it may be considered improper usage, these methods are commonly referred to as rapid sensory profiling methods. In their review of new descriptive methods, Valentin *et al.* (2012) have categorized these alternative methods into three classes depending on the nature of the evaluation task assigned to the panellists: methods based on verbal descriptions of products (e.g. Flash Profile (Dairou and Sieffermann, 2002), check-all-that-apply (CATA) (Adams *et al.*, 2007)); methods based on the measurements of between-product similarity or differences (e.g. Free Sorting (Lawless, 1989; Lawless *et al.*, 1995) and its variants, projective mapping (Risvik *et al.*, 1994) and napping (Pagès, 2003, 2005)); and methods based on the comparison of individual products with a reference or a set of references, as in polarized sensory positioning (PSP) (Teillet *et al.*, 2010). Dynamic methods, such as temporal dominance of sensations (TDS) (Pineau *et al.*, 2009), could form a fourth category, although TDS is also based on the use of attributes.

Interestingly, the increasing use of these "new" methods has been conducive to enlarging the scope of traditional descriptive sensory analysis, as is discussed later in this chapter and amply illustrated throughout this book. Thus, it is worth underscoring that these rapid sensory profiling techniques are not only rapid alternatives to conventional sensory profiling ones, but that they may also be used in situations where sensory profiling had not been previously possible.

1.1.3 Aims and needs

Today, the circumstances under which DA is needed are very diverse, and range from daily industrial practice to academic research, and from quality control to marketing studies. They differ in terms of both objectives and constraints. As a result, it would be inappropriate to rely on just one technique to fit all situations. Therefore, before going through the various sensory techniques presented in this book, it will be useful to review the different uses of DA with respect to these situations and the related objectives. Accordingly, it is important to analyse the major consequences in terms of requirements (what is needed) and constraints (applicability), as these considerations will generally guide the choice of a DA method. What is more, as noted by Dijksterhuis and Byrne (2005), the objectives (the "research question") may also influence the methodological choices at different steps of a given sensory profiling study.

It should first be noted that quality control is outside the scope of this book, although it is an important field of application for sensory analysis. Rapid profiling methods are indeed not developed for the purpose of monitoring product changes over time, variations in raw materials or production, which require measurement tools that may take time to implement but which can be used with confidence in the long run. That is quite opposite to rapid profiling methods. Of course, one could imagine adapting some of these methods so that they could be used in a quality control context, but conventional sensory profiling or other dedicated methods would seem to be much more adapted. The reader interested in such use of sensory measurements is thus invited to refer, for instance, to Muñoz (2002) and Costell (2002).

1.1.3.1 Use of DA in R&D projects

Most R&D projects in the FMCG industry aim at developing new products, optimizing existing products, or maintaining product properties with respect to other constraints (e.g. supply issues, regulation changes, cost reduction, packaging, sustainability, etc.). In such projects, developers need to take into account the sensory properties of the products they are working on, even when they are not the primary focus of the project. The way the sensory properties are taken into account may, however, take different forms that in fact reflect different objectives. The most frequent objectives of DA are:

- To compare products with existing products on the market, to compare a new or an improved product to previous versions, or to compare several prototypes;
- To improve or to optimize. Most frequently, optimization is understood in terms of liking (but not limited to it). Therefore, it implies that one has the ability to relate the sensory properties of the product to liking, in order to identify drivers of preference and, ultimately, to determine optimal sensory properties;
- To understand how product formulae or process variables affect its sensory properties. An underlying objective is usually to acquire knowledge and develop technical skills and expertise;
- To communicate with others: other team members, other teams (marketing, suppliers and customers, subsidiaries, local teams, etc.).

One could easily imagine that, depending on these objectives, the needs for sensory information will differ. Sometimes, information on how products can be sensorily grouped or positioned relative to each other would be sufficient, while in other cases an accurate quantification with finely tuned attributes would be needed.

1.1.3.2 Use of DA in marketing and consumer research

Obviously, sensory DA aims at measuring product characteristics. However, it is now widely assumed that sensory measurements are subjective, and essentially capture a *product* × *subject* interaction (i.e. the characteristics being perceived by the subjects) rather than absolute properties of the products. Naturally, it does not mean that sensory measurements are not sound and reliable, but a direct consequence of this subjectivity is the acknowledgement of inter-individual differences as the foundation of sensory measurements. Accordingly, many of the descriptive techniques that have been developed over the past 10–15 years allow for taking into account inter-individual differences in perception. This is naturally conducive to seeing the sensory panel not only as a sensory measure instrument, but also as a sample of subjects that may be representative of consumers' perceptions. That approach to sensory measurement has been coined "Sensory Evaluation II" by O'Mahony (1995).

This point becomes crucial in a competitive context where FMCG companies pay more and more attention to the way consumers perceive their products. In this respect, R&D and market research share common objectives. Consequently, there is a very strong trend toward more consumer-orientated approaches to product innovation and development (MacFie, 2007; Jaeger and MacFie, 2010), which implies consumer-orientated measurements, including for descriptive purposes (van Kleef *et al.*, 2005; Tuorila and Monteleone, 2009). Many sensory professionals have thus accepted to somehow compromise on the accuracy of DA in favour of consumer-orientated results.

Dijksterhuis and Byrne (2005), however, have called into question the use of conventional sensory profiling studies in the perspective of studying consumer behaviour. They argue that the cognitive processes involved in (attribute-based) profiling require conscious action by the panellists, and hence they are required to enter into an analytical mode of thinking that may lead to results that are somewhat different from the global, and sometimes unconscious, sensory image that forms in our minds when normally consuming or using a product. An implicit assumption in sensory profiling is, indeed, that perception can be split up into separate attributes, which may sometimes be challenging (Lawless, 1999; Murray *et al.*, 2001).

Another major limitation to performing DA with consumers is the use of a common language and the associated training required to align concepts (O'Mahony, 1991). Seeking a common language may lead to overlooking the inter-individual diversity of consumers in terms of their descriptive language and in their perception. Removing the semantic constraints through free choice of attributes, descriptive words or semantic-free tasks has allowed sensory scientists to conduct DA more easily with consumers (Thomson and McEwan, 1988; Gains and Thomson, 1990; Jack and Piggott, 1992; Veinand *et al.*, 2011). Beside methods that do not rely on a common vocabulary, the emergence of non-verbal and holistic methods has expanded the possibilities of capturing consumers' sensory perceptions (Risvik *et al.*, 1994; Faye *et al.*, 2004; Moussaoui and Varela, 2010).

Overall, more attention is now paid to the variety of consumers' perceptions and judgements. The main challenge for sensory scientists is then to deal with this variety and to provide efficient solutions to stakeholders. In some cases, this implies investigating a possible typology of the consumers in terms of perception, not just in terms of preferences (see Chapter 6 and Chapter 20 for examples of such approaches).

In addition to this, a common feature of rapid sensory profiling methods is the absence of (or very limited) training. This has further heightened interest in the use of these methods in consumer studies. As a result, as Varela and Ares (2012) have underlined, the line is now blurred between sensory and consumer science, and rapid sensory profiling methods are now increasingly used to capture consumers' perceptions.

In the same line of thinking, market researchers have attempted to use rapid sensory methods to evaluate how consumers perceive products, not only in terms of sensory attributes, but also in terms of expectations, emotions, evocations, lifestyles, etc. Ballay *et al.* present a very nice example of how Flash Profile can be adapted to investigate these aspects in Chapter 19 of this book. This evolution has also led researchers to adapt sensory techniques to investigate concept fit (Carr *et al.*, 2001; Lee and O'Mahony, 2005) and conceptual associations (see Chapter 5 of this book).

The expanding use of rapid sensory methods in this direction may also be seen as an aid in better communicating sensory properties to consumers, either in a direct (verbally using claims, cobweb plots, etc.) or indirect manner (using sensory marketing through colours, shapes, etc.).

1.1.3.3 Use of DA in research

Naturally, DA is also used in research for various purposes, ranging from the most applied product-related issues (especially in food science) to the more fundamental psychophysics. Below are the main categories of research objectives for which DA may be involved:

- To understand how product properties affect perception (e.g. study of food-flavour interactions);
- To link sensory properties with instrumental measurements;
- To understand human perception of sensory stimuli and its link to the determinants of consumption behaviour;
- To understand inter-individual differences in perception (genetic, cross-cultural differences, etc.).

From a practical point of view it must be noted that, for all these categories of research objectives, rapid methods may be useful, especially when experimental constraints predominate (e.g. use of custom-made product tailored for research purposes, use of short shelf life samples, short time availability with the subjects, or the need to interview large samples of subjects in a given population). In addition to this, it is worth noting that rapid sensory methods can also be used with subjects with limited cognitive abilities, because these methods usually involve simpler tasks. Examples of such uses are discussed in Chapters 22 and 23.

From a conceptual point of view, it must be noted that thanks to the generalized use of multivariate analysis techniques, researchers in sensory science have started to pay more attention to the relative positioning of the objects rather than to product scores on separate attributes. As a result, rather than measuring the stimulus by conventional physical means as a psychophysicist might do, one might create a multidimensional space, and then use the coordinates of that space as a surrogate set of physical measures made on the same stimuli in the test set (Moskowitz, 2003).

The increasing use of rapid sensory profiling methods in research is an integral part of how that evolution has taken place (Delarue *et al.*, 2004). This has indeed been conducive to creating a more holistic picture of response to stimuli. As described by Moskowitz (2003), this type of thinking represents a psychophysical mindset (functional relations between variables) applied to new types of data (locations of products using a multidimensional coordinate space).

Last but not least is the research in sensory methodology and in sensometrics that is actively contributing to the evolution of sensory science. Perhaps the main issue for this methodological research is to keep contributing to the development of new methods and adaptation of existing methods with due regard to the objectives and constraints for potential users of these methods.

1.2 Methodological evolution

Along with the evolution of sensory techniques, the greater availability of data analysis techniques and statistical software has greatly contributed to the increasingly widespread use of rapid sensory methods. Sophisticated multivariate techniques such as generalized procrustes analysis (GPA), multiple factorial analysis (MFA) and multidimensional scaling (MDS), which allow analysis of more complex datasets provided by rapid techniques, are now available to anyone. As a result, it is possible for any well-trained sensory practitioner to analyse data with appropriate techniques and hence to use rapid profiling methods, without it being a hurdle. This has obviously resulted in the diversification of the toolkit for the sensory practitioner, who may now choose which method to apply depending on the type of information that is needed and on the metrological properties of each method, but also according to criteria of rapidity and flexibility. In addition to this, panel training being a lesser requirement with rapid sensory profiling methods, new types of subjects can be used for conducting DA, which opens additional perspectives.

1.2.1 Rapidity

It must be stressed that not all the methods presented in the book are rapid *per se*. Rapidity here is indeed a relative notion, and is in fact multifaceted. Many of the so-called rapid profiling methods are seen as rapid because they do not require an extensive training phase. As a matter of fact, most of them do not require training at all. As a result, the time necessary to acquire sensory data starting from scratch is indeed much shorter than with any conventional profiling method. However, this may only be true in the absence of prior information or of a trained panel or, as Murray *et al.* (2001) observed, that while, on the one hand, time constraints may be limiting in their profiling methods based on extensive training, on the other hand they may be considered to be *timesaving* in the long run.

Actually, when considering the time needed for subjects to complete the evaluation task, the cards are being reshuffled. Although the simplest comparative tasks, as in Free Sorting, may still be faster, monadic sequential evaluation of the samples in conventional profiling is not much longer than other evaluation tasks. In some cases, this is in fact quite the opposite; this would be the case for instance with flash profiling if we consider the length of the evaluation session.

It may also be worth considering other parameters, such as the time to recruit subjects and to set up a panel, the time spent by the experimenter to collect and organize data, and the time needed for data analysis and interpretation.

Eventually, it should be stressed that "rapid" does not necessarily mean "instantaneous." Rather, one should consider rapid profiling methods as ways to acquire data more rapidly than with their conventional equivalents. In Chapter 20, Blumenthal and Herbeth report that in order to evaluate sensations while driving, each assessor in their study drove a total of more than 25 h and 600 km, even if they used a quicker approach than conventional profiling. The same goes with other methods: TDS is not as rapid, but it is much faster than traditional time–intensity measurements; the Ideal Profile Method may also take some time for the panellists, but it is clearly more rapid than a full external preference mapping study.

1.2.2 Flexibility

A key point in the application of rapid sensory profiling methods is that they do not require an important initial investment, since there is no need to train and to maintain a panel. There is thus no need to mobilize substantial resources to run a sensory profile (except when a large consumer sample is to be recruited). In a way, those methods thus democratize access to sensory information.

Accordingly, practitioners have started to realize that rapidity comes along with flexibility, which is perhaps an even more interesting feature. This has opened ways for the use of DA in situations to which it was previously not adapted, or even not possible to apply. These situations would notably include: sensory profiling of products with supply or sampling constraints (short shelf life, prototypes, etc.); earlier use of sensory profiling in product development or in research projects; sensory profiling in small facilities; sensory profiling with subjects who are usually not available for repeated training sessions. Flexibility would also allow using sensory information in a more interactive way, as for example for the selection of sensory-relevant variables or sensory-relevant product sets in the frame of a project. In general, having such flexible tools at hand may facilitate decision making and project management without spending too much time.

1.2.3 Diversification of the sensory toolkit

The evolution of sensory methods, and more specifically the development of rapid sensory profiling techniques, has resulted in a broad variety of available methods. The "sensory toolkit" is now much more complete than even a few years ago and offers several alternative solutions to the sensory practitioner to address many different objectives. When taking one step back, one become aware of the impressive number of options that underlie DA methods (Fig. 1.1). These methodological options range from the type of subjects that are included in the panel to data handling and data

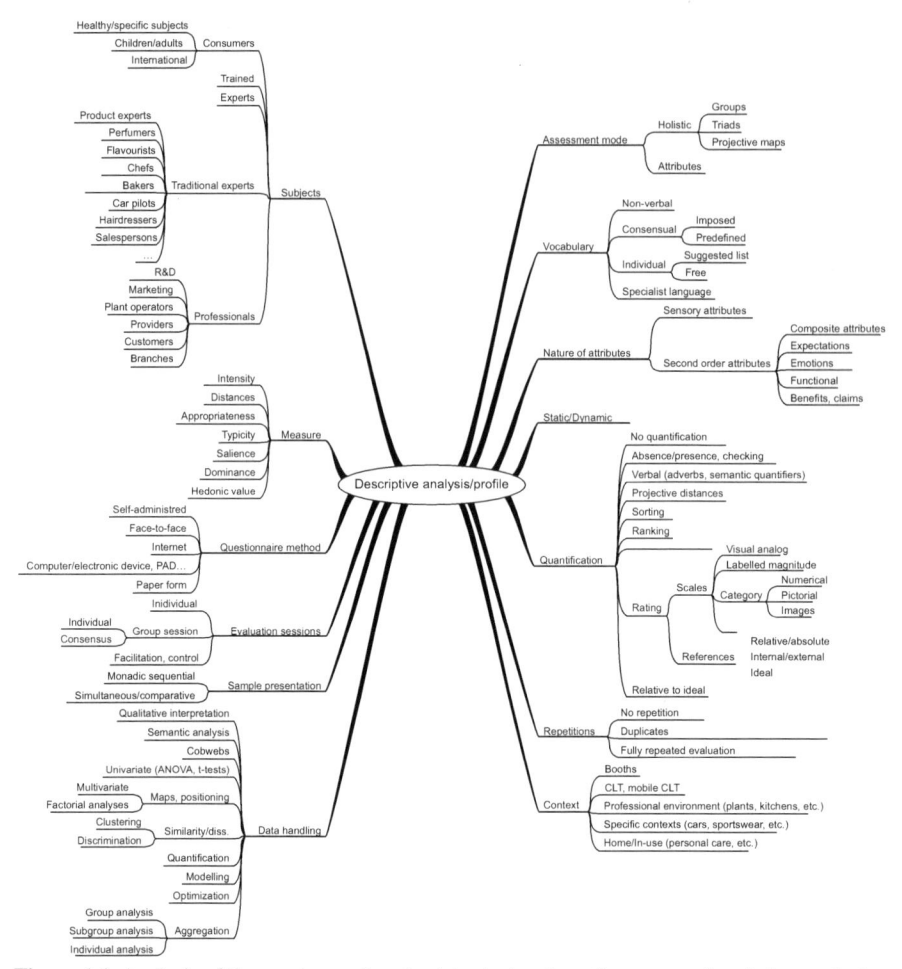

Figure 1.1 Analysis of the spectrum of methodological options for sensory descriptive analysis.

analysis issues, and include the different cognitive tasks, the type of measure and the nature of the attributes, the assessment mode, the type of vocabulary, the means of quantification, the context of evaluation and the design of evaluation sessions. Some of these topics have given rise to literature reviews and to further methodological development, but there are still many unexplored options and combinations that may allow addressing new goals in the future.

1.2.4 Use of DA with different types of subjects

A major consequence of the developments in DA methods is linked to the typology of subjects that are used to perform the descriptive task. In effect, descriptive sensory tasks are traditionally performed with panels of subjects with varying degrees of training or expertise, depending on the method. Although a minimal training is considered to be necessary for a series of reasons (need for a semantic consensus, attribute understanding and quantification, use of scales, repeatability, etc.), sensory scientists have found ways to get round some of these requirements in order to reduce training. The reduction or even the absence of training may imply compromising on the quality of the measurement (in terms of accuracy, precision or repeatability) or may induce more complex data interpretation. Nevertheless, many methodological developments have paved the way for running DA with subjects other than trained panellists. To date, most emphasis has been placed on the "consumers versus experts" paradigm (see for example Scriven (2005) for a discussion of these two types of panels), but there are certainly more options as will be presented in the following sections.

1.2.4.1 DA with consumers

As already discussed, getting descriptive information directly from consumers has long been very tempting to sensory scientists (Guy *et al.*, 1989; Jack and Piggott, 1992). However, the custom in sensory science is to interview consumers for hedonic testing only. Accordingly, it is generally recommended to avoid descriptive or analytical questions when running a hedonic test, in order to limit potential biases (Lawless and Heymann, 2010; Prescott *et al.*, 2011). All the more reason for not training consumers and hence risking biasing their perceptions and responses, not to mention the fact that training consumers would certainly be too time-consuming within the course of a consumer survey. The ability of "untrained consumers" to give reliable DA results could thus be questioned, although this ability seems to be largely underestimated (Husson *et al.*, 2001; Worch *et al.*, 2010). Interestingly, QDA was initially designed as a way to capture consumers' perceptions by eliciting consumer-orientated vocabulary from the panellists under supervision of a panel administrator (Stone *et al.*, 1974). However, recruiting consumers and training them into "no-longer-naïve" subjects would indeed lose much of its interest. In common practice, QDA subjects are therefore most frequently referred to as trained panellists, and consumer panels are dedicated to hedonic testing.

Yet conducting DA with consumers is extremely appealing, as it is a potential way to assess consumers' diversity in their perception of products, while this cannot be achieved using trained panellists (because of the training and of the limited panel size). Alternately, qualitative techniques that are traditionally used in market research would only partially render this diversity, and would fail to provide actionable or quantitative data.

Since then, the number of DA studies with consumers has increased impressively. Recently, several comparison studies have shown that running such evaluations is perfectly feasible, and that the results compare very well with DA conducted with experts (Moussaoui and Varela, 2010; Worch *et al.*, 2010). This tends to indicate that trained panellists yield sensory information that is not distorted compared to that obtained with consumers, hence supporting the use of DA with trained panels with respect to external validity. It should be noted, however, that a meta-analysis of this issue is difficult because usually there is relatively little information regarding the recruitment of participants in most published studies. For instance, the term "consumers" is frequently used as a synonym for "untrained subjects," which may result in biased conclusions (Schutz, 1999; Scriven, 2005). The same applies to trained panellists and experts: the degree of training and/or level expertise is often not specified. Sadly, a very good illustration of this is the use of students as panellists. Using students is not intrinsically a bad thing (the present author has conducted many studies with students), but it is striking to see that they may alternatively be considered either as experts or as consumers, with apparently not much difference in their training or prior sensory knowledge. Nevertheless, conducting DA with consumers draws considerable interest. This way to use DA indeed opens perspectives to sensory science for understanding consumer perceptions, and relates more closely to consumer insights. In particular, the use of this type of analysis for investigating the diversity of consumers' perceptions is very promising.

How many consumers should be recruited for conducting a DA is a frequent question. Although not well defined, the most frequent panel size in published studies is of about 40–50 participants, but this may vary from one study to another, from 20 to several hundred. The appropriate sample size will probably depend on the objectives of the study. Obviously, the more consumers that are interviewed, the better the products will be statistically discriminated. The logic behind hedonic testing may apply. Greater accuracy would be obtained with larger consumer samples. According to most standards, such reasoning would imply running these tests with at least 100 consumers. Besides, increasing the number of consumers would automatically increase the amount of sensory information collected. In this respect, one may refer to qualitative studies, where researchers generally use saturation as a guiding principle during their data collection. Saturation occurs when the collection of new data does not shed any further light on the issue under investigation (Glaser and Strauss, 1967). To our knowledge, no study has addressed this point in the perspective of consumer-based DA.

Another consideration is the possibility to explore consumer' diversity, which may only be possible when a large number of consumers participate in the test or when careful segmented sampling is applied. Eventually, we should also keep in mind that the number of consumers recruited to participate in such tests greatly impacts the time and the cost of the test. This may be a major limitation to the development of such studies.

1.2.4.2 Professional sensory experts

Professionals who work with the product every day develop a sensory expertise that could be the core of their activity (e.g. perfumers), or that may be less conscious and yet extremely useful in their work (e.g. plant operators, hairdressers). In many cases, it would potentially be very interesting to include the input from those professional experts in sensory studies. Recently, researchers have started to apply rapid sensory profiling methods as a way to get sensory input from professionals. Such attempts are as yet very infrequent, but this trend is very promising, with many options remaining to be investigated.

In this category of subjects, we consider professionals who use their senses on a daily basis in their work in order to create, develop or optimize products. Among such experts, flavourists, perfumers and oenologists immediately come to mind because of their remarkable skills in olfaction. However, other talented professionals, such as chefs and bakers, also rely on their senses to create and control the course of their production. They have probably developed a more multisensory expertise. Acute sensory expertise can also be found in non-food sectors. For instance, aestheticians and hairdressers are certainly sensory experts; the same goes for professional car drivers.

Obviously, these experts would be expected to provide a richer, faster and perhaps more discriminating analysis than normal trained panellists, especially when difficult, subtle descriptions are needed. A common trait among these professionals is that they are often able to translate their perceptions into technical variables (raw materials, formula and process parameters, product tuning) and use this associated knowledge to make decisions (selecting an ingredient, changing the formula, adjusting the process, etc.). However, they do not often participate in sensory studies, either because they are not asked to, or because they are not willing to be trained according to traditional descriptive techniques that would be depreciative with respect to their own expertise. Most often, however, they would not have the time to train (as part of a sensory panel) because of their professional constraints.

Besides, a major difficulty in taking such sensory expertise into account relates to language. Professional sensory experts indeed use a very specific, almost personal, descriptive vocabulary. As a result, it is sometimes difficult for them to objectively share their perceptions with other experts (Brochet and Dubourdieu, 2001; Feria-Morales, 2002; Sauvageot *et al.*, 2006) or to communicate with R&D.

Successful attempts to collect product descriptions from professional sensory experts have relied on Flash Profile approaches (Eladan *et al.* (2005) with perfumers, Lassoued *et al.* (2008) with bakers and milling professionals, Dairou *et al.* (2003) with car pilots), on Napping and Projective Mapping (Perrin *et al.* (2008) with wine professionals, Nestrud and Lawless (2008) with culinary professionals), or on Free Sorting (Soufflet *et al.* (2004) with textile experts, Ballester *et al.* (2008) with wine experts). Several other examples are presented throughout this book.

It is, of course, questionable whether the descriptions provided by these experts can be representative of consumers' perceptions (see Mehta *et al.*, 2011 for a discussion of this topic). Nevertheless, such descriptions would probably not be used toward that goal. On the contrary, it would be very much complementary to a consumer-orientated approach.

1.2.4.3 Plant operators

When DA applies to quality control goals, it is frequent to witness sensory panels that are constituted of plant workers. In most cases, the techniques that are implemented are variations of conventional profiling techniques. However, these techniques do not allow fully taking into account the expertise that the plant operators may have developed. Unfortunately, the possibility to rely on such expertise to improve quality control has been barely considered so far (Ioannou *et al.*, 2002).

1.2.4.4 Team tasting

Informal sensory assessment is a very common situation in many FMCG companies. During project meetings, team members often evaluate sets of prototypes, of varied recipes in order to select different options (to set process parameters, to select ingredients, etc.). Important decisions can be made on the basis of these evaluations. Participants may include R&D team members as well as members from other departments (marketing, quality, purchasers and top management). Sometimes, suppliers also participate in such evaluations (e.g. from flavour houses).

In these meetings, sensory characterization of the products is usually only one aspect of the discussion. Thus, trying to set up a full conventional DA in these circumstances would be nonsense, but in some cases making the team tasting more formal using a fast sensory DA technique could be highly beneficial to the project. This is worthwhile as long as immediate feedback can be given to the participants so that the results can help making decisions. Although such applications of DA are hardly publishable because of confidentiality constraints, the authors have had such experiences and believe they present specific methodological challenges that would deserve more research (see Chapter 16).

1.2.4.5 Salespersons

These represent a specific category of professional experts whose job is to communicate sensory information to the consumers to help them make their choice. A very good example of a sensory salesperson is the seller at perfume stores. Another example is the "sommelier".

Is DA useful for them? Perhaps not directly, but a DA study may provide them with more accurate sensory knowledge they can communicate. Having them participate in the DA study is surely the best way to make this process efficient. The same would apply to sales engineers from food ingredient companies in business-to-business situations. To our knowledge, no example of this particular use of DA has been published, although some companies have experienced success with it (see Chapter 15 for an example of the use of Flash Profile to better communicate with customers).

1.3 Consequences on sensory activities

The development of rapid sensory profiling methods may have potential consequences on sensory activities themselves, since it broadens the spectrum of available methods and opens way for measurements that were previously not possible. Besides offering new opportunities in the use of sensory data in R&D and research projects, this development may also have an organizational impact on sensory services and their relationships with stakeholders. As a result of this evolution, the practice of sensory descriptive analysis certainly becomes richer but also more complex and challenging.

1.3.1 Advocacy for earlier and more integrated measurements

Designers and product developers stand to benefit from using sensory evaluation in their projects much earlier than they usually do, and in a more integrated way. Considering sensory responses at the different stages of a product's development provides faster feedback and, indeed, allows driving the project according to sensory objectives. Overall, if sensory profiling is not limited to validation of the final result, there is much to gain in terms of responsiveness and efficiency.

According to this principle, product developers normally use many ordinary measure instruments (weighing scales, pH-meters, refractometers, viscometers, etc.) in their daily work. This is, however, not the case for sensory measurements. The fact that available sensory profiling methods have long remained very restrictive (either time-consuming or expensive) has certainly prevented truly integrating such sensory data. In addition to this, traditional sensory measures are sometimes not completely pertinent in regard to developers' needs and may not really answer their questions.

Besides, in many firms, sensory measurements are usually carried out either by a separate dedicated sensory team or by product developers themselves who informally directly assess their trials, alone or collectively. This type of organization may crystallize and fuel two extreme scenarios that are sometimes unfortunately observed. The first situation is when sensory profiling is considered essential as a corporate standard and demands for "profiling" products are made without prioritization and not much adaptation to the needs and objectives of the stakeholders. The sensory evaluation service is therefore quickly overwhelmed and unable to provide rapid feedback. When the results are communicated to the stakeholders, it could be too late. As a consequence, developers end up not using sensory analysis and simply rely on their own assessments within their team. The second pitfall is met when the sensory team deliver results that are either too simplistic or too complex to be actionable. Developers who often associate a strong sensory expertise with their technical knowledge of the products (see Subsection 1.2.4.2) are more efficient working alone and no longer have confidence in sensory analysis.

It is also noted that, even with good intentions, developers censor themselves in their use of sensory analysis: *"we do not have sufficient resources," "it is too long for this type of project"* are frequently heard comments. However, similar to instrumental measurements, one could imagine using sensory profiling in totally different ways. A number of rapid sensory profiling methods that are currently available could indeed be integrated in the course of R&D projects, and could even be conducted by the developers themselves. All things considered, sending samples to sensory facilities may take more time than just evaluating the product by the means of rapid profiling methods. One solution might thus be to delegate the simplest and quickest methods to developers (with assistance from the sensory team). Meanwhile, more complex and thorough measures would still need to be conducted by dedicated specialized services, just as sophisticated instrumental measurements, such as gas chromatography-mass spectrometry (GC-MS), rheometry and so on, are carried out by specialists.

1.3.2 A richer and more complex job

1.3.2.1 Need for greater expertise

As a result of the development of rapid profiling methods and of the accompanying evolution in the use of sensory measurements, sensory specialists may become both consultants and warrantors of good practice. They may indeed be more involved in advising researchers and product developers and helping them carrying out their own tests. Should such a change occur in good conditions, this would allow more flexibility, provide quicker results and hopefully consume less resource. Sensory teams would thus be in charge of more challenging and more rewarding tasks.

When only conventional profiling methods are available, setting up a descriptive panel implies an important investment (in time and money). It is thus determinative to make the right choice and to determine what the best method is for the company or for research purposes. Once the choice of a conventional profiling method is made, there is indeed a difficult way back. The method may then be applied routinely without wondering for each evaluation if the method best addresses one's needs.

The current diversity in methods available makes sensory science more demanding in terms of methodological skills and theoretical background. Given the broad spectrum of methods and objectives that sensory professionals may face in their daily practice, methodological choices will undoubtedly be frequent. This requires that sensory professionals have a good knowledge of the methods available and that they can select and sometimes adapt these methods depending on objectives, stakes and constraints of the context of each study.

1.3.2.2 Knowledge and dissemination

As has just been pointed out, knowledge of the various sensory profiling methods is a key determinant in the current and future practice of sensory science. This raises the issue of teaching and dissemination of these methods. Until the late 1990s, good sensory professionals were supposed to master the main conventional descriptive methods (usually QDA and/or Spectrum). Reasonable skills in statistics were required for interpreting descriptive results and monitoring panel performances. Most advanced classes in sensory science included preference mapping techniques.

Nowadays, most programmes in sensory science include rapid sensory methods as well. However, it is noted that in many cases, professors naturally tend to teach techniques they have developed, or that they are used to applying in their research projects. As a matter of fact, it is difficult to have a comprehensive view of all existing sensory profiling techniques and related methods. In an online survey that we conducted in 2011 with attendees at the Pangborn Sensory Science Symposium, we observed a lack of knowledge and trust in alternative DA methods (Rogeaux et al., 2011).

In this survey we focused on six methods: conventional sensory profiling, Free Choice Profiling, Flash Profiling, CATA, Repertory Grid and Free Sorting. The questionnaire was built in three dimensions in order to assess, for each method, the level of use and of knowledge, the expected output when a method is used (in terms of quality, innovation, etc.), and the main strengths and weaknesses that were perceived in each case.

Overall, 104 respondents, mainly Europeans and North Americans, completed the survey. More than half were sensory professionals from industry (52%) with various levels of responsibility; other respondents were from sensory research and service institutes (27%) or from academia (21%). In Table 1.1 the three main limitations to the use of alternatives descriptive methods, namely lack of formalization, complexity of data analysis, and lack of trust and knowledge, are highlighted.

Lack of formalization is indeed part of sensory professionals' concerns, as new methods are generally built in the research field. Often, these methods are progressively improved, sometimes hybridized, and there are many ways to apply a given method. Free Sorting is a good example of such an evolution. Many variants of sorting are indeed possible, depending on specific instructions regarding the task (number of groups, full of partial sorting, verbalization, etc.), and data processing. The academic point of view could be that each variant may have an added value, and that the method is thus of great interest for research. For the sensory practitioner, however, the situation is more delicate, and there is always doubt about which option to choose. The lack of methodological formalization and guidelines is therefore seen as a limitation.

Besides, respondents have reported that the statistical techniques that would be needed to analyse new types of data are not always available in standard sensory acquisition software (although this situation is changing rapidly). As a consequence, practitioners need to use some statistical software, sometimes with the assistance of well-trained people. Even when routines are available, the data analysis is often seen as more complex than for conventional profiling and thus prevents the use of rapid methods.

Eventually, we found that alternative descriptive methods were not always well known. From this perspective, the lack of formalization is of course a weakness. When a method is not easy to define and to process, it is more difficult for sensory users to acquire practical knowledge of that method. Given the increasing prevalence of methodological choices, recommendations and guidelines are much needed to help the users to fit methods to objectives.

By providing sound presentation of the methods, together with guidelines and testimonies of their usage in industrial context, this book will hopefully help sensory professionals to be more comfortable with their choices and to apply rapid profiling methods with more confidence.

1.3.2.3 Critical points in method selection

As in any methodological choice, selecting a sensory profiling technique has to be made according to one's objectives and constraints. Consequences and future use of the data that will be provided must also be taken into consideration. As a general guideline, careful attention should be paid to the following points:

1. *Challenge the objectives.* Focusing on the objectives of the measure is decisive. Although it seems obvious, this point is not always carefully considered and, apart from the basic rules of sensory evaluation (i.e. deciding between hedonic, descriptive and difference testing), many sensory professionals tend to apply a method they are used to, not the method that would best fit their goals. Besides, in the view of rapidity and efficiency, it would seem wise to select the simplest method available. That is to say, a method should be proportionate to

Table 1.1 Summary of main outcomes from the 2011 online survey among Pangborn Sensory Science Symposium attendees (104 respondents)

Method	Summary	Strengths/weaknesses
Conventional sensory profile (like QDA® or Spectrum®)	The most known (94%) The most used (77%) The reference	+: the most reliable method; the best method to track product changes and to correlate data. Easy to use and to communicate; well defined in the literature −: time-consuming, expensive, needing dedicated resources, not adapted to reveal inter-individual differences
Free Choice Profiling	Low level of use (18%) Good level of knowledge (72%) A very specific method with low interest	+: no real strengths versus other methods, quick and flexible, feasible and easy for any subject −: data processing is too complex, added value not well perceived (not enough described?, appears less pertinent than Flash Profiling)
Flash Profile	Low level of use (25%) Good level of knowledge (63%) Well adapted to describe and communicate a sensory product range	+: can be executed by different types of subjects (from experts to consumers), produces a fast, quick and easy description of a product range, well adapted to define ways to communicate on products −: lack of software and data processing not easy, method inadequately documented in practical view, consensus on verbatim not easy to find (if no exchange on list of items)
CATA	Low level of use (27%) Level of knowledge (52%)	+: very easy to run, to communicate, to explain to the subjects, can be achieved by different types of subjects (from experts to consumers) −: not well known enough to be implemented, data analysis process not defined and described in the literature
Repertory Grid	Low level of use (21%) Knowledge quite low (51%) A specific method dedicated to elicit consumer verbatim	+: great to generate consumer vocabulary, easy to run and explain to consumers −: data analysis process not easy, not well enough described to be used correctly, specific method
Free Sorting	Low level of use (34%) Good level of knowledge (65%) Well adapted to understand a competitor view	+: task easy to understand, method easy to run and organize, can be achieved by different types of subjects (from experts to consumers), based on theory in psychology −: no dedicated software, data analysis is often an issue, needs a specific number of products (between 6 and 12), not adapted to monitor product quality

the level of information that is actually needed. For example, running a full conventional profile when only Free Sorting is needed would be like running a high-performance liquid chromatography (HPLC) analysis in chemistry to measure the concentration of acids when only a pH measurement is needed. This would clearly be a waste of time and resources. It is good advice therefore to focus on what information is really needed.

2. *Define the available methods.* In view of the objectives, one has to consider the variety of methods available. The first thing to remember is that there is no ideal, all-purpose method. Indeed, no method exists that is simultaneously accurate, discriminant and reproducible, that would give access to quantifiable data with possible translation into technical specifications, that is close to consumers' perception and language, that would provide both hedonic and analytical data, that would be convenient, fast and cheap, and that would easily transfer to different countries and cultures. Sometimes, on the other hand, several methods might possibly fit the same objective. However, each method has its advantages and drawbacks. Knowledge of the various methods is thus a good start, but it is of course important to know what methods are actually available under given circumstances. That is where constraints must be taken into account.

3. *Constraints.* Two types of constraints are to be anticipated: the first are practical constraints (budget, time, sensory resources (panel availability), product availability (supply issues, possibility to make prototypes, shelf life, etc.)); and second are the constraints intrinsic to the product set to be evaluated (i.e. those constraints related to the type of products, the number of products to be evaluated and magnitude of the sensory differences). These two types of constraints are common to all sensory methods, including conventional profiling. However, the limited time and resources may orientate the choice toward rapid methods. On the other hand, issues relating to product availability may turn out to be a serious limitation to all rapid methods that are based on the comparative assessment of a product set. If all samples are not available at the same time, applications of such techniques as Free Sorting, Napping or Flash Profiling might not be feasible. Regarding intrinsic constraints, each case must be carefully considered before selecting a method, as each constraint may rule out a particular method. In this regard, the maximum number of samples to evaluate always comes to mind, but minimum number of products or the size of inter-product differences are often underestimated constraints.

4. *Subjects.* Closely linked to the objectives is the question of the type of subjects that should be or could be involved in the test. This issue is of course central and has been extensively discussed already.

5. *Stakes.* Eventually, method selection also depends on the stakes. In a business context, it would indeed make sense to adapt the choice of a method to the importance of expected results for decision making. Not respecting this rule could lead to overwhelmed sensory resources and misuse of sensory measures. This goes along with prioritization. The key advice is to discipline yourself in order to avoid comfort and routine, and always to challenge the needs. Always go for simpler methods, and use more sophisticated methods only when there is a real need. Additionally, it is important to educate your stakeholders so that they understand that not all demands are addressed with the same method.

1.3.2.4 Toward a broadening range of activities

The evolution that is observed and encouraged in the use of sensory DA may conduct users of sensory measurements to work differently, hopefully in a more integrated way. Improvements may indeed be expected in the way sensory science can contribute to product development and to research.

These future evolutions would imply that sensory practitioners, together with their sponsors, make informed methodological choices. Obviously, they are already selecting the most suitable sensory profiling methods according to objectives and constraints. However, in addition to their current role, they may build upon rapid sensory methods to suggest strategies in project management. Such strategies may include combinations of methods. For example, one could imagine starting to evaluate a large set of products using Free Sorting with team members in order to define a subset of the most relevant or most representative products, and then move to QDA. This conventional profile could be supplemented by a TDS analysis for dynamic aspects, and finally a Flash Profile could be run with consumers should one wish to have a more consumer-orientated description of a product set.

In the same line of thinking, and in order to gain optimum advantage from the use of rapid profiling methods, sensory scientists should develop a "design of experiment" mindset that leads them to pilot their studies in more efficiently than in earlier stages by selecting sensory-relevant product sets and sensory-relevant variables. New methods indeed allow earlier measurements with minimal cost and thus allow possible feedback loops in the course of a study. Ultimately, this working pattern would lead developers to adopt a sensory engineering approach to product development. From an organizational point of view, this means that sensory people become full partners of projects instead of mere service providers.

1.4 Conclusions

To conclude this introductory chapter, we attempt a SWOT analysis of rapid sensory profiling methods and of their use in the context of R&D and research.

1.4.1 Strengths

As it has been amply discussed, it comes as no surprise that the main strengths of the methods presented in this book are rapidity and, perhaps more important, flexibility. Generally speaking, these methods would be more affordable to most potential users as there is no need for important initial investment (in training and panel set-up), which makes them perfect candidates to start acquiring sensory descriptive information from scratch.

Another strength compared to conventional profiling methods is the possible adaptation of rapid sensory profiling techniques so that they can be used with consumers. Sensory and consumer scientists, as well as market researchers, have quickly realized the advantage of these methods. Accordingly, most of these methods allow for better taking into account inter-individual differences in human perceptions.

1.4.2 Weaknesses

One of the main weaknesses of most rapid methods is that their validity is difficult to assess in terms of metrological properties. Even though they compare well with conventional profiling in terms of product positioning, users may be frustrated by their lack of quality indices and panel monitoring tools. This may contribute to the feeling that these methods are not reliable and hence prevent their widespread use.

Besides, the sensory descriptions provided by these methods are usually not as accurate as those obtained by conventional profiling, though this criterion may not always apply (e.g. in the case of Free Sorting with consumers). This lower accuracy is notably linked to the more difficult interpretation when using free vocabulary. A direct consequence of this is the need for more advanced data analysis techniques. Eventually, the translation of sensory properties into technical product variables may be more difficult.

1.4.3 Opportunities

The development in the use of rapid sensory profiling methods opens up particularly interesting prospects for more integrated sensory measurements. Mastering of this new methodological portfolio may change the way researchers and product developers use sensory data in their daily work, potentially with considerable gains in terms of reactivity and interactivity. Eventually, better interaction with stakeholders may be expected.

Rapid methods can be adapted so that team tasting can be optimized and yield more actionable results. Many of these methods can also be considered for integrating traditional sensory expertise. Accordingly, they may allow sensory description with different types of subjects (including children, elder people and patients) or even sensory profiling at home for in-use evaluation of products.

Finally, rapid profiling techniques may be adapted to go beyond sensory description and measure consumer responses linked to expectations, emotions, evocations, images, lifestyles, etc. Most recent developments in this direction are very promising.

1.4.4 Threats

Although the development of new sensory profiling methods has accelerated, it must be stressed that not all of the new methods have been thoroughly tested. Like any measurement technique, rapid sensory profiling techniques may be sensitive to methodological details in their protocol. However, experience of the practical impacts of such details is as yet minimal. To date, the influence of methodological variants on results is indeed not well documented. A major threat is that users of these methods may overlook this subtle methodological consideration and apply the methods in their own way, thus biasing the measurement with no guarantee of validity of the final outcome.

Another threat lies in the way results from rapid sensory profiling methods are communicated to stakeholders, or to anyone else, who are not familiar with these methods. First, it must be noted that multivariate data analysis and sensory maps are not understood by everyone, which may be seriously hazardous if results are misinterpreted. Sensory practitioners must thus be very careful when delivering their results and they should make extra efforts in communication. In addition to this, stakeholders may be puzzled by the fact that apparently similar results (i.e. sensory maps) may sometimes be obtained in few hours, while in other cases they require several months of training. Here too, preparation and the communicating skills of sensory professionals are decisive.

In spite of these reservations, it may be emphasized that when appropriately applied, rapid sensory profiling techniques are powerful tools for clever use of sensory analysis. Many examples and testimonies of successful uses of these methods, both in industry and in academia, are presented throughout this book. This will hopefully spark the interest of students in sensory programmes, as well as that of sensory professionals, sensory scientists and, more generally, all users of sensory data. May this book help them in their daily work, provide them with some solutions and contribute to fostering innovation in sensory science.

References

Adams, J., Williams, A., Lancaster, B. and Foley, M. (2007). Advantages and uses check-all-that-apply response compared to traditional scaling of attributes for salty snacks. *7th Pangborn Senssory Science Symposium.* Mineapolis, MN.

Ballester, J., Patris, B., Symoneaux, R. and Valentin, D. (2008). Conceptual vs. perceptual wine spaces: does expertise matter? *Food Quality and Preference,* **19**, 267–276.

Brochet, F. and Dubourdieu, D. (2001). Wine descriptive language supports cognitive specificity of chemical senses. *Brain and Language,* **77**, 187–196.

Carr, B. T., Craig-Petsinger, D. and Hadlich, S. (2001). A case study in relating sensory descriptive data to product concept fit and consumer vocabulary. *Food Quality and Preference,* **12**, 407–412.

Costell, E. (2002). A comparison of sensory methods in quality control. *Food Quality and Preference,* **13**, 341–353.

Dairou, V. and Sieffermann, J.-M. (2002). A comparison of 14 jams characterized by conventional profile and a quick original Method, the Flash Profile. *Journal of Food Science,* **67**, 826–834.

Dairou, V., Sieffermann, J. M., Priez, A. and Danzart, M. (2003). Sensory evaluation of car brake systems. The use of flash Profile as a preliminary study before a conventional profile. *SAE World Congress,* Detroit, MI.

Delarue, J., Danzart, M. and Sieffermann, J.-M. (2004). Flash profile gives insights into human sensory perception. *5th International Multisensory Research Forum,* Barcelona, Spain.

Dijksterhuis, G. B. and Byrne, D. V. (2005). Does the mind reflect the mouth? Sensory profiling and the future. *Critical Reviews in Food Science and Nutrition,* **45**, 527–34.

Eladan, N., Gazano, G., Ballay, S. and Sieffermann, J.-M. (2005). Flash profile and fragrance research: the world of perfume in the consumer's words. ESOMAR Fragrance Research Conference, 15–17 May 2005 New York, NY.

Faye, P., Brémaud, D., Durand Daubin, M., Courcoux, P., Giboreau, A. and Nicod, H. (2004). Perceptive free sorting and verbalization tasks with naive subjects: an alternative to descriptive mappings. *Food Quality and Preference,* **15**, 781–791.

Feria-Morales, A. M. (2002). Examining the case of green coffee to illustrate the limitations of grading systems/expert tasters in sensory evaluation for quality control. *Food Quality and Preference*, **13**, 355–367.

Gains, N. and Thomson, D. M. H. (1990). Sensory profiling of canned lager beers using consumers in their own homes. *Food Quality and Preference*, **2**, 39–47.

Glaser, B. G. and Strauss, A. L. (1967). *The Discovery of Grounded Theory: Strategies for Qualitative Research*, Chicago, Aldine Publishing.

Guy, C., Piggott, J. R. and Marie, S. (1989). Consumer profiling of Scotch whisky. *Food Quality and Preference*, **1**, 69–73.

Husson, F., Le Dien, S. and Pagès, J. (2001). Which value can be granted to sensory profiles given by consumers? Methodology and results. *Food Quality and Preference*, **12**, 291–296.

Ioannou, I., Perrot, N., Hossenlopp, J., Mauris, G. and Trystram, G. (2002). The fuzzy set theory: a helpful tool for the estimation of sensory properties of crusting sausage appearance by a single expert. *Food Quality and Preference*, **13**, 589–595.

Jack, F. R. and Piggott, J. R. (1992). Free choice profiling in consumer research. *Food Quality and Preference*, **3**, 129–134.

Jaeger, S. R. and Macfie, H. (2010). *Consumer-Driven Innovation in Food and Personal Care Products*, Cambridge, UK, Woodhead Publishing Limited.

Lassoued, N., Delarue, J., Launay, B. and Michon, C. (2008). Baked product texture: correlations between instrumental and sensory characterization using Flash Profile. *Journal of Cereal Science*, **48**, 133–143.

Lawless, H. T. (1989). Exploration of fragrance categories and ambiguous odors using multidimensional scaling and cluster analysis. *Chemical Senses*, **14**, 349–360.

Lawless, H. T. (1999). Descriptive analysis of complex odors: reality, model or illusion? *Food Quality and Preference*, **10**, 325–332.

Lawless, H. T. and Heymann, H. (2010). *Sensory Evaluation of food. Principles and Practices*, New York, Springer.

Lawless, H. T., Sheng, N. and Knoops, S. S. C. P. (1995). Multidimensional scaling of sorting data applied to cheese perception. *Food Quality and Preference*, **6**, 91–98.

Lee, H. S. and O'Mahony, M. (2005). Sensory evaluation and marketing: measurement of a consumer concept. *Food Quality and Preference*, **16**, 227–235.

Macfie, H. (2007). *Consumer-Led Food Product Development*, Cambridge, UK, Woodhead Publishing Limited.

Mehta, R., Hoegg, J. and Chakravarti, A. (2011). Knowing too much: expertise-induced false recall effects in product comparison. *Journal of Consumer Research*, **38**, 535–554.

Meilgaard, M., Civille, G. V. and Carr, B. T. (1999). *Sensory Evaluation Techniques*, Boca Raton, FL, CRC Press.

Moskowitz, H. R. (2003). The intertwining of psychophysics and sensory analysis: historical perspectives and future opportunities – a personal view. *Food Quality and Preference*, **14**, 87–98.

Moussaoui, K. A. and Varela, P. (2010). Exploring consumer product profiling techniques and their linkage to a quantitative descriptive analysis. *Food Quality and Preference*, **21**, 1088–1099.

Muñoz, A. M. (2002). Sensory evaluation in quality control: an overview, new developments and future opportunities. *Food Quality and Preference*, **13**, 329–339.

Murray, J. M., Delahunty, C. M. and Baxter, I. A. (2001). Descriptive sensory analysis: past, present and future. *Food Research International*, **34**, 461–471.

Nestrud, M. A. and Lawless, H. T. (2008). Perceptual mapping of citrus juices using projective mapping and profiling data from culinary professionals and consumers. *Food Quality and Preference,* **19**, 431–438.

O'Mahony, M. (1991). Descriptive analysis and concept alignment. *In:* Lawless, H. T. and Klein, B. P. (eds.) *Sensory Science Theory and Applications in Foods.* New York, Marcel Dekker, 223–267.

O'Mahony, M. (1995). Sensory measurement in food science: fitting methods to goals. *Food Technology,* **49**, 72–82.

Pagès, J. (2003). Recueil direct de distances sensorielles: application à l'évaluation de dix vins blancs du Val-de-Loire. *Science des Aliments,* **23**, 679–688.

Pagès, J. (2005). Collection and analysis of perceived product inter-distances using multiple factor analysis: application to the study of 10 white wines from the Loire Valley. *Food Quality and Preference,* **16**, 642–649.

Perrin, L., Symoneaux, R., Maître, I., Asselin, C., Jourjon, F. and Pagès, J. (2008). Comparison of three sensory methods for use with the Napping® procedure: case of ten wines from Loire valley. *Food Quality and Preference,* **19**, 1–11.

Pineau, N., Schlich, P., Cordelle, S., Mathonnière, C., Issanchou, S., Imbert, A., Rogeaux, M., Etiévant, P. and Köster, E. (2009). Temporal dominance of sensations: construction of the TDS curves and comparison with time–intensity. *Food Quality and Preference,* **20**, 450–455.

Prescott, J., Lee, S. M. and Kim, K.-O. (2011). Analytic approaches to evaluation modify hedonic responses. *Food Quality and Preference,* **22**, 391–393.

Risvik, E., McEwan, J. A., Colwill, J. S., Rogers, R. and Lyon, D. H. (1994). Projective mapping: a tool for sensory analysis and consumer research. *Food Quality and Preference,* **5**, 263–269.

Rogeaux, M., Lawlor, B., Punter, P. and Delarue, J. (2011). Current status and future directions for alternative descriptive sensory methods workshop. *9th Pangborn Sensory Science Symposium,* Toronto.

Sauvageot, F., Urdapilleta, I. and Peyron, D. (2006). Within and between variations of texts elicited from nine wine experts. *Food Quality and Preference,* **17**, 429–444.

Schutz, H. G. (1999). Consumer data – sense and nonsense. *Food Quality and Preference,* **10**, 245–251.

Scriven, F. (2005). Two types of sensory panels or are there more? *Journal of Sensory Studies,* **20**, 526–538.

Soufflet, I., Calonnier, M. and Dacremont, C. (2004). A comparison between industrial experts' and novices' haptic perceptual organization: a tool to identify descriptors of the handle of fabrics. *Food Quality and Preference,* **15**, 689–699.

Stone, H., Sidel, J., Oliver, S., Woosley, A. and Singleton, R. C. (1974). Sensory evaluation by quantitative descriptive analysis. *Food Technology,* **28**, 24–34.

Teillet, E., Schlich, P., Urbano, C., Cordelle, S. and Guichard, E. (2010). Sensory methodologies and the taste of water. *Food Quality and Preference,* **21**, 967–976.

Thomson, D. M. H. and McEwan, J. A. (1988). An application of the repertory grid method to investigate consumer perceptions of foods. *Appetite,* **10**, 181–193.

Tuorila, H. and Monteleone, E. (2009). Sensory food science in the changing society: opportunities, needs, and challenges. *Trends in Food Science and Technology,* **20**, 54–62.

Valentin, D., Chollet, S., Lelièvre, M. and Abdi, H. (2012). Quick and dirty but still pretty good: a review of new descriptive methods in food science. *International Journal of Food Science and Technology,* **47**, 1563–1578.

Van Kleef, E., Van Trijp, H. C. M. and Luning, P. (2005). Consumer research in the early stages of new product development: a critical review of methods and techniques. *Food Quality and Preference*, **16**, 181–201.

Varela, P. and Ares, G. (2012). Sensory Profiling, the blurred line between sensory and consumer science. A review of novel methods for product characterization. *Food Research International*, **48**, 893–908.

Veinand, B., Godefroy, C., Adam, C. and Delarue, J. (2011). Highlight of important product characteristics for consumers. Comparison of three sensory descriptive methods performed by consumers. *Food Quality and Preference*, **22**, 474–485.

Worch, T., Lê, S. and Punter, P. (2010). How reliable are the consumers? Comparison of sensory profiles from consumers and experts. *Food Quality and Preference*, **21**, 309–318.

Alternative methods of sensory testing: advantages and disadvantages

H. Stone
Sensory Consulting Services, USA

2.1 Introduction

Sensory evaluation is a science that measures, analyzes, and interprets the responses of people to products as perceived by the senses. This brief statement serves as an introduction to this chapter; it is the foundation on which sensory evaluation is based. With this in mind, one can better understand development of methods and more confidently modify existing methods to be less time consuming without jeopardizing the quality of the obtained results.

Sensory evaluation has experienced significant growth in the past four decades, for many reasons but primarily because of the value of the information. As competition has expanded well beyond national borders, food and beverage companies have realized the value of product sensory information. The need for actionable product information is always urgent, as is the need to be ahead of competition. With the constant flow of new ingredients, flavoring materials including alternative sweeteners, and health concerns associated with obesity, and fat and salt levels in foods, companies need information that they can use to their best advantage. These developments represent challenges as well as opportunities. The impact of the internet has further complicated product business decisions, either as part of a marketing effort to introduce a new product or when issues such as food safety erupt. In this mix is the consumer, whose food choices are changing. Because sensory tests involve consumers, their responses provide a connection with technology and the market strategy, so it is not surprising that interest in sensory information is so high. Of the different kinds of sensory methods, descriptive are especially useful because they provide product descriptions in quantitative terms based on such variables as ingredient changes and preference differences. Before discussing descriptive analysis and contemporary practices, it is useful to provide a brief summary of what constitutes sensory resources and their applications in the context of a business environment.

Sensory resources are an integral part of most consumer products companies. By resources are meant the tools used by sensory scientists to obtain actionable information. This information may be used by technology, quality control, consumer insights, marketing, and brand managers. The resources include: subjects, methods, facilities, and data capture and analysis capabilities. In Section 2.3, the methodologies are described. Section 2.5 is a more detailed exposition about descriptive analysis. More information about these resources can be found in Lawless and Heymann (2010) and Stone et al. (2012).

2.2 The subjects in sensory testing

Sensory methods are small panel procedures; that is, they use relatively few subjects, usually less than 50, and based on their responses, results are generalized to a larger population. For some testing, the numbers are 25 or fewer. As Pangborn (1981) noted, "...I share the desire of most behavioral scientists to observe and measure sensory responses which can be generalized to populations beyond the confines of a small panel of select individuals." This means that we put considerable reliance on these few subjects and their responses, and we have to be confident that they are qualified to participate. If individuals are simply recruited at random, then the information obtained will be random and not likely to be useful, especially not for a company that markets its products to specific segments of the consuming population. In this section a procedure for qualifying subjects is described.

Recruiting, screening, and qualifying subjects begin with the recognition that all people are different. No two people are alike, and their sensory skills are different (Pangborn, 1981; Stone et al., 2012). The simplest way to demonstrate this is to measure the individual thresholds for various taste and odor stimuli of a group of people. A second example is to obtain intensity measures for an array of stimuli at different concentrations, and a third is to measure the difference thresholds. In each case, no two subjects will have the same results and no amount of "training" will change this simple but critical observation. From these tests it is clear that about 30% of any population is relatively insensitive to differences among products they regularly consume and should be excluded from future testing. This observation is age-, gender-, and culture-independent. Sensory scientists need to take this information into account whenever a sensory test is planned. In its own way, this is not different from the recruiting criteria used by marketing when measuring consumer attitudes about a product or a marketing message. Using unqualified subjects significantly increases the risk of decision errors. Many failed product sensory tests can usually be traced to results from using unqualified subjects. Claims that subjects can be trained to be invariant are simply that; there are no data to support it.

In the early years of sensory testing, the question of who should be selected for a sensory test was discussed at length (Pfaffman et al., 1954; Schlosberg et al., 1954). Initially, it was thought that scientists, technologists, and those involved in the production process were best suited to the process. However, it was soon realized that these individuals were biased because of their product knowledge; i.e., they knew what the product was supposed to be and responded accordingly vs responding based on their perceptions. Selecting subjects based on their absolute thresholds to various stimuli such as sweet, sour, salt, and bitter also was found to be ineffective; research showed that the correlation with product differentiation was poor, about 50% or less. See also Girardot et al. (1952) and Mackey and Jones (1954) for more on this topic.

That sensory thresholds yielded low correlations with actual product evaluations should not be surprising since product characteristics are well above threshold and much more perceptually complex compared with a model system. It is interesting to note that this practice of measuring absolute thresholds continues to be used despite a lack of evidence for its effectiveness as an indicator of future performance.

Research by Sawyer et al. (1962) reported that results from preliminary discrimination tests were a good indicator of how an individual would be expected to perform in future tests. With this in mind, one of the authors developed a system using the duo–trio difference method and test products modified to reflect typical differences. If an individual could not discriminate differences among these products at better than chance, then any higher order task such as scoring would be in question. After examining results from several tests, it was determined that the discrimination model was an effective system for identifying those individuals qualified to participate in descriptive as well as discrimination tests. The procedure involved selecting product pairs that would challenge the subjects in terms of degree of difficulty, include all product modalities (appearance, aroma, etc.), and be time efficient. The system developed into a process requiring no more than three sessions, each lasting about 60–75 min. In each session, a subject will have participated in six or seven trials with a replicate to yield a total of 12 or 14 decisions each session for a total of 36–42 decisions on which to select qualified subjects. To minimize sensory fatigue, a 2 min interval between product pairs was introduced. In addition, appearance pairs were interspersed between taste and mouthfeel differences. It was also learned that about 30–40 trials was a reasonable number on which to assess a subject's skill. Since people's sensory skills develop at different rates, there had to be a sufficient number of pairs ranging from easy to difficult to allow for this skill to develop and also be challenged without losing motivation.

The most challenging part of the process was selecting product pairs that represented a range of differences across modalities and also were easy to prepare. Conversation with product technologists and staff bench screening provided examples. Easy differences could be competitive products, and difficult differences could be samples from a production line 30 s apart. By making use of available materials, preparation of product pairs would be easy and not time consuming. At the end of the process, subjects selected for a panel were those whose percent correct was greater than 50% and were especially successful with the more difficult pairs. As a general rule, subjects selected achieved at least 65% correct. The idea was to have a pool of subjects representative of the discriminating segment of the population; i.e., those most likely to detect a difference.

An added benefit of using this screening system was to identify a subset of the pairings that could be used when screening new subjects for replacing subjects who cannot maintain the appropriate skill level or to develop a new panel. By using fewer product pairs, the time allocated for screening could be reduced from three to two sessions.

The argument that one should recruit subjects on a random basis fails to recognize these differences in skill and the consequences. Testing subjects who cannot discriminate at better than chance increases the risk of concluding that there is no difference when, in fact, there is. In statistical terms, this reduces the power of that test. The argument that one can correct for these differences by using a large number of subjects – for example, 100 or more – not only fails to appreciate the consequences of these sensitivity differences but also introduces other challenges such as production variation. Also, adding more subjects does not reduce error and as the number increases the risk of obtaining statistical but not practical significance increases substantially. Companies do not market products randomly; they are marketed to specific populations and a goal is to minimize risk of detection of a difference by the discriminating segment of their customer base.

From the perspective of rapid methods, the ability to screen and qualify subjects in two or three sessions is a small price to pay for having a pool of qualified subjects. The argument that one can achieve the same level of confidence using a statistical analysis based on various assumptions about behavior is a risky argument; a sensory test is a behavioral test and use of statistics enables one to summarize the information and reach a conclusion in terms of risk associated with using qualified subjects.

2.3 Methods in sensory testing

There are a finite number of methods described in the sensory literature and a very large number of modifications, many of which are based on a need to try a different method, or belief that a modified method will improve results. In most instances, the modifications are driven by a statistical, not a behavioral approach. Increasing the power of a test is a goal and certainly will have a significant impact on the results, but how much more power can be achieved if one is not using qualified subjects. Confidence in results derives from knowing what subjects were used, the chosen design, and the analysis of the data.

Methods can be classified as either analytical or affective. Analytical methods are either discriminative or descriptive; they provide product analytical information. Affective tests are hedonic or paired preference; they provide product liking or preference information. These methods provide different kinds of information and should not be combined, a topic discussed later in this section. These methods are the foundation on which none evaluation has developed. Each provides different kinds of information and none of them is superior to another in terms of sensitivity.

Analytical methods provide product analytical information and there are two types – *discrimination* and *descriptive*.

2.3.1 Discrimination

Discrimination methods enable one to determine whether the difference between products is perceived at some previously established significance or confidence level. Since all products are different from each other, the question is whether that difference is detected by a sufficient number of subjects to be able to conclude that the result is not due to chance and the null hypothesis of no difference can be rejected. Results cannot tell us whether that difference is important, only that a difference is detected. Its importance is determined by other testing; e.g., measure product liking.

There are numerous methods described in the literature but only few are used to any degree of frequency. The most frequently used methods are described here.

1. Paired-comparison method is a two-product test; both products are coded, and the subject's task is to indicate which one has more of a specific characteristic, such as sweetness. There are two orders of presentation – AB and BA. As long as one could specify the attribute and the subjects recognized it, the method is quite useful, because of its simplicity in terms of instructions and implementation. The complexity of technology in today's environment has reduced the popularity of this method.
2. A-not-A method also is a two-product test and differs in terms of the instructions; no attribute or direction is provided, only to indicate whether (or not) the products are similar. This option is useful when the attribute could not be specified as with the paired-comparison. Another possible use is a situation where products are served in a monadic sequential order to minimize a non-test difference influencing the decision. There are four orders of presentation – AB, BA, AA, and BB.
3. Duo–trio method is a three-product test in which one product is identified as "R" or Reference and the other two are coded. The subject's task is to indicate which coded sample is most like the reference. There are four orders of presentation – R_1AB, R_1BA, R_2AB, and R_2BA.
4. Triangle method is a three-product test in which all three products are coded. The subject's task is to identify which one is most different from the other two. There are six orders of presentation – AAB, ABA, ABB, BBA, BAB, and BAB.

Other methods typically include more samples; e.g., four or five, and the subject's task is to identify which product is most like one or more of the others. For example, it could be a 2 of 5 or some such combination. This type of test is like a sorting task but with more sensory fatigue potential. These methods are often developed within an industry and are unique to that industry; however, the sensory fatigue issue is a concern.

In addition to these multi-product tests, considerable attention has been given to expanding the subject's activities during a test, most often associated with the triangle method. Some of these changes have used a 4-product set. The main focus has been to promote various statistical hypotheses claiming significantly improved sensitivity, reduced β risk, and a concomitant reduction in the number of subjects in a test to less than 50. The subjects are asked to indicate the magnitude of the difference, their confidence in their decision, and sort products based on a specified attribute (see Ennis and Bi, 1998 and Ennis and Jesionka, 2011 for more discussion about these methods). The claims remain to be demonstrated as having practical application. No actual data are provided using qualified subjects. The assumption that subjects can perceive a specific attribute also remains to be demonstrated without some kind of training or practice as will be discussed in the descriptive analysis section of this chapter. One additional issue with these alternative methods is the use of responses from those subjects who did not make the correct match. Should one accept information obtained from a subject when the decision about the difference was not correct? For a historical perspective on this issue, the reader is referred to Bradley and Harmon (1964) and Gridgeman (1959).

At various times the question of which method is most sensitive is asked and almost always the reply is the triangle test based on the probability of 1/3 ($p = 0.33$) while it is 1/2 ($p = 0.50$) for the paired and duo–trio methods. This is a surprising claim, as there are no data to support that statement. If this were true, then it would suggest that a 4-product test would be more sensitive, as $p = 0.25$, etc. As the number of products in a test increases from two to three to four, the amount of product exposure increases as does sensory fatigue. In addition, the test becomes a memory game as subjects try to remember their perceptions when responding. If one thinks in terms of product exposure, the triangle test has the most and the paired test, the least. For the triangle test, there are three comparisons to reach a decision: A vs B, A vs C and B vs C. For the duo–trio test there are two comparisons for each decision, R vs A and R vs B. For the paired test there is one comparison for each decision, A vs B. Claims of method sensitivity must begin with qualified subjects and a common product base not based on assumptions about perception that may not be relevant. The choice of method should be based on the objective, how the results will be used, and the product itself; i.e., the degree to which it leads to sensory fatigue, and not based on habit or false assumptions about method sensitivity.

In any conversation about sensory testing, a topic of great interest is deciding on the number of subjects for a test. For all tests, this decision is based on several variables including the magnitude of the difference between products, the sensitivity of the subjects, variability of the products, the power of the test, and experimenter error. Since one cannot know the answers to these questions, one has to create a set of assumptions. For example, the sensitivity of subjects can be estimated from knowledge of the discrimination skill often referred to as the just-noticeable-difference (jnd). For most differences involving aroma and taste, the jnd is about 0.20–0.25 (20–25%); i.e., a 20–25% change in a product ingredient will be detected. The magnitude of the difference between products can be estimated from the variable being tested. But there will be many situations in which the variable being tested has many other effects in a product, so the problem becomes more complicated. Investigators have created a series of tables based on these assumptions and including the α and β risks. These tables, however, do not account for use of repeated trials or use of qualified subjects. Continued study of these problems should be helpful. Until then one can consider the following as a practical option. As background, consider the following. First, discrimination testing is an analytical test, so one starts with qualified subjects and a design that includes replication; i.e., each subject evaluates each pairing more than once but in a different order. If there are a total of 50 trials from a panel of 25 this is sufficient to reach a conclusion of accepting/rejecting the null hypothesis. Having more than a single judgment enables one to examine response patterns and develop greater confidence in the conclusion. The choice of 25 subjects is arbitrary; one could have as few as 20 to as many as 30. This range is based not only on the qualifying and replicate criteria, but also considering the information shown in Table 2.1. As the number of judgments increases from 10 to 40, the percent correct for statistical significance decreases from 90% to 64% at $\alpha = 0.05$. As N increases to 60 and beyond, the percent correct for significance does not decrease in the same proportion. In

Table 2.1 **Number of total judgements, correct matches and percentage correct for statistical significance at $\alpha = 0.05$**

No. of total judgements (N)	No. of correct matches	% correct, $\alpha = 0.05$
10	9	90
20	15	75
40	26	65
60	37	62
80	48	60
100	59	59
120	70	58
200	112	56

With $N = 40$ (20 subjects, each tested twice), 65% correct is needed for statistical significance at $\alpha = 0.05$. With $N = 80$ (40 subjects), 60% correct is needed. Plotting these data shows the impact of increasing N on significance.

effect, screening, qualifying, and testing more subjects does not add significantly to the efficiency of the test considering the cost to add more subjects. This approach is efficient, taking not more than 1 h to field and less time to interpret results and make a recommendation to the requestor. In five decades of testing using this model, the author has never experienced a reversal; i.e., a company stating the results were opposite from what was obtained with this discrimination test model. For more discussion on this topic see Stone *et al.* (2012).

2.3.2 Descriptive

Descriptive methods attract considerable attention because of the information obtained. They are defined as methods that provide *"quantitative descriptions of products, obtained from the perceptions of qualified subjects. It is a complete sensory description, taking into account all sensations that are perceived – visual, auditory, olfactory, kinesthetic, etc – when the product is evaluated. ... The evaluation is defined in part by the product characteristics, and in part by the nature of the problem"* (Stone *et al.*, 2012, p. 234).

Sensory scientists will agree that descriptive analysis has many applications but there is relatively little agreement as to how specific methods are developed and used. For example, subjects may or may not have been screened; the number of subjects can range from as few as 5 to as many as 20; there may be a formal language development process or not; the number of attributes on a scorecard may be limited; references may or may not be used; replication may or may not be part of the design; data analysis can be simple (e.g., summary statistics such as means and variance measures) or can be complex (involving multi-variate analyses) or a combination of both. Other differences include the time to develop a panel and use of subjects for more than one product category. For sensory professionals with limited knowledge about descriptive analysis, developing a capability can be a daunting process. Further challenges develop when a technologist or a project manager requests results within a few days. Knowing that one can obtain consumer information using an internet survey system, there can be additional pressure on sensory staff to take shortcuts to satisfy a request. Of course, the sensory professional can satisfy a request quickly if there is a pool of qualified subjects. However, without qualified subjects and no appreciation for the most appropriate test plan, results will be misleading. A more extensive discussion about descriptive analysis is provided in Section 2.5.

2.3.3 Affective

Affective methods provide measures of liking or preference. There are two kinds of methods, hedonic (providing a measure of liking) and preference (determining which product is preferred). Within a sensory program, these methods were initially intended as a screening system to assess progress during development efforts or to reduce the number of alternatives for consideration in a test market. In recent years some have expanded the activity to collect information about attributes, imagery, and related information. Of the two, the hedonic method is more informative providing a liking measure for each product as well as location on the liking continuum. This enables comparison with past results. For example, frozen desserts generally receive scores above a 7.0 (using a 9-pt hedonic scale) and if test products are significantly lower, then further testing is not justified until such time as formulations have been changed and the scores achieve a minimum score of 7.0.

The sensory hedonic model has advantages and limitations. The advantages are that it provides an early and inexpensive indication of a product's potential. Its disadvantages are the limited number of judgments, usually about 75, and the response of liking is a passive judgment, i.e., liking does not mean it will be purchased. The fact that it is a small scale test does not minimize its usefulness; however, the size of the panel is often forgotten when actions are taken based on the results. Action-oriented decisions, such as purchase intent, require larger numbers of responses (usually more than 100) as they are impacted by context and related issues.

Establishing an affective program needs access to several hundred consumers. Each sensory test will use about 75 consumers, and they should not be used any more frequently than once every 3 or 4 weeks to minimize any bias about the products or the test itself. Consumers are recruited using some criteria obtained from marketing, to be sure that they are a reasonable representation of the actual customer; i.e., they consume the product at a frequency similar to the regular customer. Since these consumers are only asked to provide a measure of liking or preference, this recruiting profile should be sufficient. One of the reasons for keeping the test model simple is to avoid the appearance of being a larger scale test in which many different kinds of questions are asked and especially those associated with imagery.

Name: _____ Code: _____ Date: _____

Circle the statement that best reflects your opinion about this product

Like extremely

Like very much

Like moderately

Like slightly

Neither like nor dislike

Dislike slightly

Dislike moderately

Dislike very much

Dislike extremely

Figure 2.1 The 9-point hedonic scale (Peryam and Pilgrim, 1957). The subject is instructed to place a check adjacent to the statement or circle the words that best reflect their overall opinion for that product. The responses are converted to numerical values from 1 = dislike extremely to 9 = like extremely.

The system enables one to screen as many as six to eight products in a test in a single session of about an hour. Testing usually is done in a central location with booths; however, it can be fielded in other locations depending on the type of product and the product preparation requirements.

There are two basic methods, hedonic and paired preference. The 9-pt hedonic scale (see Fig. 2.1) is a method that has been used for more than half a century and used successfully in many countries by all ages of consumers (see Peryam and Pilgrim, 1957). It is interesting to note that the scale is bipolar with a neutral center of "neither like or dislike"; in use, however, subjects respond as if it were an equal-interval scale, thus enabling use of the analysis of variance (AOV) and related statistical methods.

A recent development, the Labeled Affective Magnitude scale (Schutz and Cardello, 2001; Cardello *et al.*, 2008), has been found to be useful to provide greater separation among products that are well liked but where the differences are thought to be small. A detailed description of the scale and its use is provided in the above cited references.

An important advantage of using scaled responses of liking is the ability to examine response patterns, as shown in Table 2.2. One also can examine responses by serving order to gain a better understanding of the product differences as well as product deficiencies by examining the responses at the lower end of the scale.

Table 2.2 A summary of the results from the 9-pt hedonic scale listing the % responses in each scale category for each product as well as their means and standard deviations

	Product A %	Product B %	Product C %
9 – Like extremely	18	16	5
8 – Like very much	27	31	25
7 – Like moderately	25	24	30
6 – Like slightly	13	10	14
5 – Neither like nor dislike	1	4	3
4 – Dislike slightly	8	8	13
3 – Dislike moderately	2	3	3
2 – Dislike very much	4	2	4
1 – Dislike extremely	0	0	1
Mean	6.88	6.96	6.35
Std deviation	1.89	1.75	1.89

This display provides a quick summary of scale use and enables sensory staff to better understand the results before reaching a conclusion.

The third method, paired preference, is the oldest of the methods and is not often used, in part because of the limited information obtained – which product is preferred. Primary focus today is to obtain a measure of the magnitude of the liking relative to other products in a test. One of the features of any test should be the efficiency of the method. In the hedonic test, each subject provides a judgment for each sample; in the paired-comparison test, two samples are evaluated and one judgment is obtained. Nonetheless, there can be situations where product preference is needed; for example, wanting to know if a product is preferred as a prelude to a larger scale preference claim test. Figure 2.2 shows the paired method along with a more typical way in which the results are displayed.

The availability of a pool of qualified consumers also allows for testing in a typical use situation such as a home-use test. This capability enhances the value of a sensory capability. Here too its goal is to provide support for a product before larger scale testing.

One of the challenges facing sensory staff is the request to add questions to the hedonic test. For example, to have subjects assign liking scores to attributes or to provide agree–disagree responses to attribute questions such as flavor – too weak, too strong, just-about-right (JAR), usually referred to as the JAR scale. Considering what is known about the differences in sensitivity among the population, it is remarkable that such procedures are used. A compelling argument for doing this is the comment that a result is obtained so it must be a good use of subject time in a test. This misuse is further complicated by the application of the AOV and other related statistical procedures based on assumptions about the responses. The JAR scale is a nominal

Alternative methods of sensory testing 37

Name:_____ Code:_____ Date:_____

Evaluate both products starting from the left.
Check the box for the product you prefer.
You must make a choice.

        347 ( )   602 ( )

80 consumers participated.

Product	Served first	Served second
347	25 (62%)	21 (52%)
602	15 (38%)	19 (48%)

Figure 2.2 An example of the paired preference scorecard. In this example 80 subjects participated. Test was a balanced design and results are summarized on the right showing responses by serving order which is pooled for the analysis.

scale and one should not be going beyond summations, percentages, and use of chi square statistics. This topic has been discussed elsewhere, and the interested reader is referred to Lawless and Heymann (2010) and Stone et al. (2012). The use of affective testing enables the sensory staff to provide more product information, and as long as it focuses on product, it serves an essential role in support of product development and other technological efforts. It is unlikely that sensory staff will be able to limit the focus, given the need for actionable information in very short timelines; however, the manner in which results are shared will limit the extent to which such information is projected to the larger market.

2.4 Further important considerations in sensory testing

2.4.1 Facilities

The design of sensory facilities has been well documented in various publications (Kueston and Kruse, 2008; Stone et al., 2012). What is most important from a practical perspective is to be able to work closely with architects and designers to share the unique requirements of a sensory facility. Avoidance of bright colors and daylight in the booth area are to be avoided, while proper ventilation and physical location are essential requirements for a facility to operate efficiently. It should be easy for subjects to access the site and enter and leave the booths without crowding. In today's electronic environment, booth space must be able to accommodate screens and keyboards or a tablet as well as product and rinse water. A common problem observed by the author is booth height, usually set at desk height which is incorrect. Subjects are seated but servers are standing. A complaint from servers is the constant bending as they serve and respond to subjects' questions leading to back-pain complaints. In the past year this author has encountered more than six new facilities with this incorrect arrangement. This appears to be caused by failure to access the aforementioned facilities' design documents.

As more testing moves away from the typical controlled environment, sensory staff have to give additional attention to use of more replicates to take account of greater variability in these other locations. The nature of the products also has to be considered; for example, when testing a product prepared at a fast-food location where transport alters the actual evaluation experience and testing has to be done on site. For products that are inherently variable, for example, an ice cream with fruit or any product with particulate matter, additional replicates may be needed. The decision has to be a sensory one, reflecting prior product testing knowledge.

2.4.2 Date capture and analysis

As noted in previous sections, current practice for most sensory facilities includes direct data entry. There are many systems available from suppliers, and the sensory staff need to determine what system best meets their needs. All systems also provide experimental design and analysis features, as well as the ability to monitor the data collection process and statistical output via cloud computing. These are features that enable one to provide results on a 24/7 basis, thus quelling complaints about the time needed before results are available. Unfortunately, these features come at a price and that price is losing connection with the data and basing one's conclusions on the output from the computer. Many of the systems do not allow one to decide, for example, which AOV model is most appropriate for the problem. As Groopman (2009) noted, "statistics is not a good substitute for thinking." To rely solely on a statistical output and not spending sufficient time thinking about whether the results make sense is a mistake that will happen. All too often one finds it easy/convenient to use a system that provides the test design, serving order, the data collection and the output with tables, plots, multi-variate maps, and significance values. This approach enables one to satisfy the need for a rapid system but it has some disadvantages, not the least of which is the inability to know what was actually being done to the raw data and whether the results made sense based on one's knowledge of the products. It is reasonable to expect that newer systems will allow for more insight to the actual computations and their effects.

Since much of the testing being done today, and for the foreseeable future, will involve scoring of a product characteristic, the AOV becomes an essential resource in support of data analysis and interpretation. Since there are many AOV models, one needs to be familiar with those most appropriate for sensory data; for example, the AOV mixed model (fixed and random effects) with replication is appropriate. Other features should allow for the ability to test the main effect by interaction when interaction is significant, and so forth. Finally, one needs to be cautious when using software that allows for exclusion of some data but without providing the details of what was excluded. Procrustes analysis is one such system. See Huitson (1989) for more discussion on this topic when applied to sensory data. The problem with any computation that removes some data is an assumption that data are an aberration when it may not. How does one know that this does or does not represent a unique population? Failure to examine results in their entirety could potentially mean missing valuable product information. In most product categories there are unique preference segments that are driven by specific sensory attributes and, any time some data are suppressed, information is lost. The larger question has to do with the use of this practice for all tests to where one questions whether any of such results have any external validity. Sensory staff needs to be careful about using default systems without appreciating the consequences.

2.4.3 Products

Products are what we evaluate and it is often forgotten that products are biological material and are a source of variability. All too often, results are criticized as being too variable and the cause is usually focused on the subjects and the method. There is no argument that subjects are variable; however, this variability is accounted for through qualifying the subjects, use of repeated trials as part of the design, and the AOV that allows for partitioning the subject effect from the product effect, etc. Product variability is accounted for, in part, through the use of servings from different containers and familiarity of the product category derived from earlier testing. This is a product management issue, one that has to be taken into account when planning a test. For some products, the variation within a brand can be as large as the differences between brands. Without replication, this problem can be lost within the analysis and expected differences are not obtained. This is another reason why replication in a sensory test is essential.

2.5 Developing descriptive analysis capability

As noted previously a descriptive analysis capability attracts attention because of the results it provides; for example, the ability to describe specific differences among an array of competitive products. This section provides some details on how a descriptive analysis capability can be developed and a panel made ready to evaluate products in a relatively short time period. Starting with newly recruited subjects, one can have a panel available in 2 weeks – three session days for screening, five for language, and one for a pre-test. Once operational, however, a test can be organized in a day, followed by data collection. The duration of a test depends on the number and type of products and, as always, the objective. So any discussion about rapid methods must first begin with what is meant by rapid in the context of existing methods.

Historically, descriptive analysis tests, like other sensory tests, were fielded in a laboratory environment, but with the advent of the electronic age, one could look at the methodology as a mobile resource, i.e., able to be used in any environment based on the kinds of information needed and the specific type of product; e.g., a home care product. This author recalls using the QDA® method (in the 1970s) to evaluate golf clubs, with players actually using the clubs on a golf course. In that time period without a mobile device, it was a paper and pencil exercise requiring lots of paper and time to transfer responses to a data file for the analysis. Despite the scorecard logistics, this approach yielded reliable and actionable results. In today's environment, capturing responses on a real time basis regardless of where a product is tested is more easily accomplished using a tablet, software, and the internet. For a company evaluating home and/or personal care products, this dramatically improves data capture and provides a real-time product usage experience. For these non-food products, the ability to have a rapid method can be realized once the product attributes have been characterized and correlated with other product attributes. The sections that follow provide a commentary on the resources required to develop a descriptive analysis capability with consideration of it being a rapid method.

As mentioned, there is a range of possibilities available to the sensory professional intending to develop a new or upgrade an existing descriptive capability. Reading the literature, participation in conferences, short courses, or in conversation with other sensory professionals are typically the ways in which possibilities are considered and plans made. In some instances, the decision is to select what is believed to be the best features of each method. On the surface this makes sense, but it too can have its own problems and this will be better understood once these first two stages of panel development are described. For additional background reading on this topic, see Stone et al. (1974), Stone and Sidel (1998, 2007, 2009) and Stone et al. (2012).

There are four basic stages in developing a capability:

1. Recruit, screen, and qualify subjects
2. Develop a language, an evaluation, and scoring protocol
3. Test design and data analysis
4. Reporting

All four of these activities are important, and the first two are especially so because the quality of the results is entirely dependent on having qualified subjects, a language, and a testing protocol.

2.5.1 Recruiting, screening, and qualifying subjects

The subject qualifying process follows the plan described in Section 2.2. Descriptive panel size is usually 10–12, so one would start screening with about 20–22 people and assume 30% will not meet the discrimination qualifying criterion. The QDA method as originally developed (Stone et al., 1974) recommended a panel of 12 and that has not changed. While more could be recruited and tested, there is no evidence that it will improve sensitivity; early research indicated that the level of significance would increase; e.g., from 0.00 to 0.001, possibly because of a larger N, but it did not change decisions about specific product differences. When a panel with less than ten subjects is tested, it is not unusual to obtain results where the significance level reaches 0.10 but not 0.00, based on prior expectations. This can occur because of the smaller number of subjects. Product is another source of variability that impacts responses. With fewer subjects, fewer product units are served. To some degree this can be addressed through use of a repeated trials design; however, it is better to add subjects to a panel when it is known that the number available is ten or below. As a general guide, having a panel of 12 is optimal but one could have one more or one less without significantly impacting results. It becomes more a matter of scheduling, not a matter of improving the quality of the results.

2.5.2 Language development

Developing a sensory language is an iterative and interactive process, and there is considerable confusion about the actual process and especially the timing. Like the qualifying process, developing a language can be time consuming, depending on what method is being considered. For example, Flavor Profile® and Spectrum Analysis® require 14 weeks of training (Cairncross and Sjöström, 1950; Meilgaard *et al.*, 2006) or it can be ignored along with subject qualifying, as in the case of Free Choice Profiling (Williams and Arnold, 1985; Williams and Langron, 1984), or in its current version, called Flash Profiling (FP) described by Dairou and Sieffermann (2002) and Delarue and Sieffermann (2004). To save time, techniques such as these rely on statistical software to adjust the responses. This will be discussed later in this chapter. Between these extremes are quantitative descriptive analysis (QDA) (Stone *et al.*, 2102), which takes 8 days – three qualifying sessions and five language sessions for a new panel. Of course, these eight-session days could be reduced by having two sessions per day, assuming subjects are available. For an existing panel, one needs only a single pre-test session. So the question of timing depends on what is meant by a rapid method; it also depends on whether one is considering developing a panel, or using an experienced panel.

As was emphasized in the previous section, all subjects are different from each other in terms of their sensory discrimination skills and will differ in their use of words to represent their perceptions. Stated another way, people will use the same words to represent different sensations, or the converse of using different words to represent the same sensations. Without any time to discuss these differences, how is one supposed to make sense of results regardless of the data analysis system used? With no way of assessing the quality of the information, any sensory scientist should be concerned when reporting results.

Language development is a group effort under the direction of an individual designated as the panel leader (PL). This individual works with the subjects, organizing each session, providing samples, and encourages the dialog among the subjects. In addition to the language, the subjects are advised by the PL that they have to develop explanations/definitions for the words, the order of the words within a modality on the scorecard, and an evaluation protocol. They also practice scoring and, in the process, become familiar with the sensory characteristics of the products. All these activities require five 90 min sessions. Early research showed no advantage to having more sessions, while having fewer sessions yielded data that tended to be more variable, as evidenced by increased standard deviation values of about 10%. Using experienced subjects but a different product category reduces the number of sessions to three. If one is fielding a test of products from the same category, then there is only an orientation session followed by data collection. So the issue of a rapid method is reduced to about an hour, and the number of sessions is determined by the number of products and the number of replicates, topics discussed further in this section.

The language development process, as noted above, can vary based on the method chosen. If one is using a language that is technical in content, or a language that does not allow for changes, then it is likely to add significant amounts of time to the process. Technical words should not be used because they have different meanings if one has a technical background. In effect, subjects have to learn to ignore any language that is not typical of the language they use in an everyday situation when using that product. Not allowing subjects to modify a language also creates problems for subjects for reasons already noted. Subjects may believe that the PL has a set of words in mind and the subjects' task is to state those words, thus gaining approval of the PL as having learned what to say.

In QDA (Stone *et al.*, 1974, 2012), the first session begins with an orientation explaining why the subjects were selected, what they will be doing over the next 5 days, and emphasizing their role in this process: they will sample products, describing their perceptions, discussing the words they use, and as a group coming to agreement, and in the process develop a scorecard and an evaluation procedure. The subjects are served a product and are advised to open their tablets (or given a piece of paper) and instructed to write any words they want that reflect the product's appearance, aroma, etc. They also are informed to not write whether they like the product or whether they consider it good quality, only what they perceive. This initial effort takes about 15 min and the PL lists all the words for everyone to see. A summary of the information is provided and a second product is served with instructions to now think in terms of the order of occurrence; i.e., appearance, odor, etc., as they evaluate the sample. This process continues with a third product and a fourth. The subjects are able to speak more openly as they learn what is expected of them, and the sequence of activities goes easier and the dialog becomes more efficient. By the end of the first session the subjects may have sampled all six products and have developed as many as 70+ words. An example of such a listing is shown in Table 2.3. The PL may ask for comments and/or clarification for some words, but making clear that the language is from the subjects, not from the PL – otherwise some subjects might believe that the PL is expecting to hear certain words. Once the session is completed, the PL summarizes the language and, where appropriate, groups the words; for example, burned, scorched, and overcooked flavors are grouped as burned for this next session. Such a checklist is shown in Table 2.4. For the second session, the subjects are provided with the checklist and served two products, one at a time, and are instructed to evaluate a sample and to add any new words, placing a check for words already listed. After the first product there is another listing of words, followed by the second product. The subjects are encouraged to talk with one another when there is a question about use of a word and/or an explanation for that word. Subjects are reminded that they can make changes but also they need to be prepared to share that information with the other subjects. This effort is intended to remind them that the development of the scorecard is their responsibility. Before serving the fifth product, the PL encourages the subjects to think about the order of the modalities (appearance followed by aroma, etc.) and the order of attributes within a modality. Table 2.5 lists some attributes and their definitions developed during the language sessions. Toward the end of this second session the PL usually has the time to introduce the

Alternative methods of sensory testing

Table 2.3 A compilation of appearance words used by qualified but not experienced subjects in an initial language session

1.	Skin on edges, black/brown edges, dark edges, burned edges, black edges, peel, moldy edges, golden brown edges, spots
2.	Thin/thick
3.	Transparent (translucent)
4.	Butter yellow color, pale, light, orange, yellow, dark brown, uneven color
5.	Broken
6.	Bubbles on surface, uneven surface, air bubbles
7.	Veins, branching; light/dark spots, smooth, bumpy
8.	Folded
9.	Oval – irregular shapes, form, not round, uneven pieces, uniform shape – duck bill, saddle, size – pre-cut
10.	Round
11.	Brown spots, dark spots (come off), burned spots
12.	Large, brown areas, uneven color, dark color, light edges, brown/yellow edges
13.	Oil spots, greasy, oily
14.	Wavy, not flat, curved
15.	Ridges/ripples

These are typical words from individuals without any prior language development experience. In this example, the products were a set of commercially available potato snacks.

idea of assigning intensity values for the attributes. It is typical during this session for one or more subjects to ask about this while discussing the language. Assigning intensity or strength to a stimulus is a skill that requires some practice, especially when using a scale that has no numbers as depicted in Fig. 2.3. Sometimes referred to as a graphic rating scale, its effectiveness requires that subjects practice using it to familiarize themselves with the range of differences they will experience with this product set. From this time forward, subjects make all their evaluations using a scorecard with a scale associated with each attribute, an example of which is shown in Fig. 2.4. At this stage the PL requests that the subjects indicate with a show of hands which product attribute was marked further to the right. This enables each subject to see how the other subjects perceived the product attribute differences and it becomes the procedure for the remaining sessions. It serves as a kind of learning process, in which the subjects are able to see the extent to which they agree or disagree as to which attribute is stronger for that product in that session. When the scale is described, some subjects express a desire to "know where on the scale a mark should be placed," "give us an example." Experience has shown that if such information is provided, results are more variable vs when not given any information. The subjects think they can remember where on the scale they marked a particular attribute. The PL provides only that they can mark anywhere on the scale based on their impression of that product and those attributes. If the intensity is weak, they can mark toward the left end of the scale and if it is very strong they can mark toward the right. Not surprisingly, once subjects practice using the scale, the process proceeds

Table 2.4 A checklist developed from the information provided by the subjects during the first session

Red	Blue		
		Overall flavor (weak–strong)	Bland
		Potato (weak–strong)	Potato
		Starchy (weak–strong)	Starchy, uncooked starch
		Earthy (weak–strong)	Earthy, dirty, dirty potatoes
		Oily/greasy (weak–strong)	Oily, greasy, grease
		Fried (weak–strong)	Fried – Deep fat fried, fair food; french fries/fried plantains; fried potato
		Buttery (weak–strong)	Butter, buttery
		Brown (weak–strong)	Brown, cooked?
		Salty (weak–strong)	
		Sweet (weak–strong)	
		Sweet potato (weak–strong)	Sweet potato
		Nutty (weak–strong)	Nutty, nut oil; may include peanut or sesame oils
		Burned (weak–strong)	Burned, scorched, or overcooked flavors

The PL collates the information and provides this listing along with any comments from the subjects at the start of the second session with instructions to add any additional words. A discussion follows after each product, providing opportunity for more language and subject dialog.

with minimal or no difficulties. Since each test included replicates, scale location, as such, is not critical to its effectiveness. For a more detailed discussion about this scale see Stone *et al.* (2012). For these remaining sessions the subjects continue to sample products, provide votes, discuss specific attributes and add, remove, or consolidate the list of words, always based on a consensus vote of the subjects. By the end of the final session, subjects will have an agreed set of words, explanations for each word, and an evaluation protocol. An example of an agreed set of words for one modality is shown in Table 2.5. In these sessions the subjects will have evaluated at least 20 products, and have experienced a full range of the differences among the products that will be tested. Of course, if the test is only of six products there will be duplication; however, a key to the success of the method is the practice that enables

Table 2.5 Final list of attributes with their definitions

Directions: Look carefully at the chips in front of you, then evaluate	
Overall color (light–dark)	Intensity of the color of the chips, ranging from pale, yellow, and light to dark, golden brown.
Uniform color (slightly–very)	Degree to which the chips have an even or uniform color; a chip that is slightly uniform would have large brown areas, or a dark interior with lighter edges.
Skin on edges (a little–a lot)	Overall impression of the amount of potato peel or potato skin on the edges of the chips.
Browned edges (slightly–very)	Degree to which the edges of the chips look brown or toasted; includes a very brown or burned appearance.
Dark spots (few–many)	Impression of the amount of dark spots or blemishes on the chips; included dark or burned spots that come off when brushed.
Thickness (thin–thick)	Overall impression of the thickness of the chips, ranging from thin to thick; a thin chip would be transparent.
Surface bubbles (few–many)	Overall impression of the amount of air bubbles or pockets on the surface of the chip, ranging from few to many.
Size of bubbles (small–large)	Overall impression of the size of the air bubbles or pockets on the surface of the chip, ranging from small to large.
Uneven surface (smooth–bumpy)	Degree to which the surface of the chips is rough, bumpy, or uneven, ranging from smooth to rough.

The list is available to the subjects during testing. Before the next test, subjects use this list as they preview samples and make any changes. Words changed are placed in the definition. See text for an explanation.

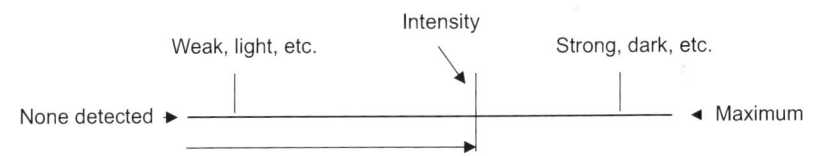

Figure 2.3 Example of the graphic rating or line scale. The subject's task is to place a vertical line across the horizontal line at that place that best represents the strength for that attribute. See text for more information as to the use of the scale.

the subjects to develop their confidence with the evaluation process, use of the scale, and the range of products that will be evaluated.

2.5.3 Data collection

Immediately after the language sessions, the actual product testing is initiated but on an individual basis in a typical booth or in an environment that reflects the specific product use, e.g. a shampoo. For most the food and beverages, a single session will take about 40–45 min with a 2 min rest between products and a maximum of eight products in a session. For frozen desserts such as ice cream, the number of samples

Directions: Look carefully at the chips in front of you, then evaluate:

Overall color	Light	————————	Dark
Uniform color	Slightly	————————	Very
Skin on edges	A little	————————	A lot
Browned edges	Slightly	————————	Very
Dark spots	Few	————————	Many
Thickness	Thin	————————	Thick
Surface bubbles	Few	————————	Many
Size of bubbles	Small	————————	Large
Uneven surface	Smooth	————————	Bumpy

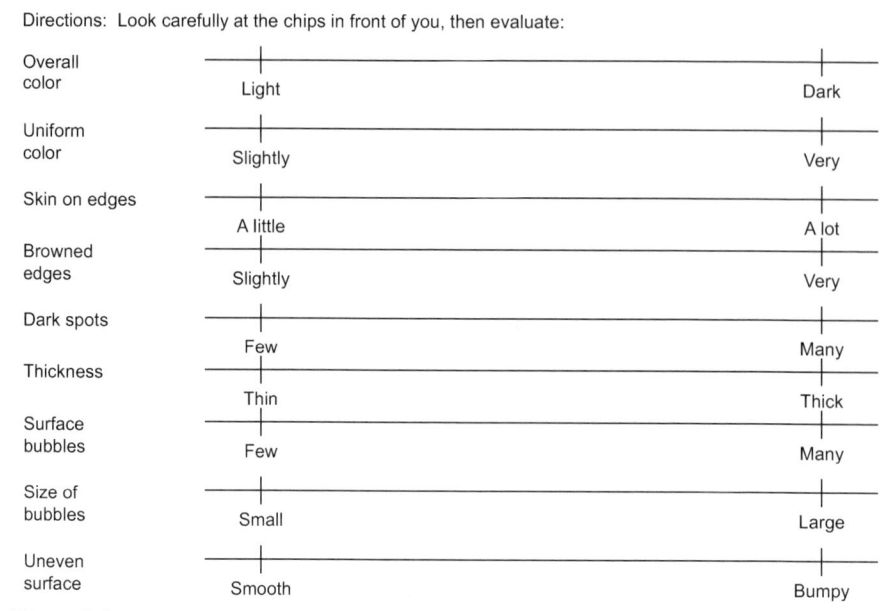

Figure 2.4 An example of the Appearance attributes for a scorecard. The subject marks each attribute until all have been scored for that product and after a timed interval, proceeds to the next product. See text for additional details.

may be reduced to six to minimize the impact of temperature on the palate, or one can extend the rest interval and use warm water for a palate rinse. The number of replicates is a PL decision based on the test objective and the priorities. For a test of nine or fewer products, each subject should evaluate each product four times; for ten or more, the number of repeated trials can be reduced to three. In either situation, the data base for each subject for each attribute and for each product enables various kinds of analyses to be applied, which is discussed in the next section. If there are time restrictions, then one could schedule two sessions in one day and a third on the morning of the next day with output is available in an hour or less, leaving time for evaluation and interpretation of the results. With an experienced panel, the time requirement is reduced and allows for a quick response. In some situations the number of replicates can be reduced from 4 to 3 or 2. This is a PL decision based on knowledge of subject skill in previous tests with a particular product category. Whether this constitutes a "rapid" method remains to be determined by those seeking a quick answer. As with any sensory test, there are alternatives, sacrificing some replication based on knowledge of the subject's past performance and the extent of product variability, but always a decision by the PL. Without any descriptive panel capability, however, the challenge of responding is difficult and the risk of inappropriate decisions is high.

2.5.4 Analysis and reporting

For any analytical test, as emphasized in several places, the design must include repeated trials and a balanced serving order. Replication is critical because without it there is no basis to consider the validity of the results and any conclusions should be viewed with caution. The fact that one obtains a significant effect does not mean that the results are valid. Achieving statistical significance is easily obtained with a large data file but the statistical power may be very low.

When subjects provide responses of the perceived intensity for a stimulus on a repeated trial basis, it allows for analyses of individual subjects, attributes, and products. A descriptive test of 10 products, 12 subjects, 3 replicates, and 40 attributes will yield 14 400 values, enabling a wide range of computations including summary statistics, the AOV, correlation of attributes, Factor Analysis, and Principle Components Analysis. In addition, the scaled values can be converted to ranks and additional analyses calculated. All these types of analyses are intended to provide different "views" of the results to verify that the product differences are consistent, that where interactions occur they have been taken into consideration. When the scaled differences and the ranks are in agreement, this adds to one's confidence with the results.

When the QDA method was being developed, several unique features were included: the first was to enable one to obtain performance measures for each subject and for each attribute; the second was to be able to test the main effect vs interaction when interaction was significant; a third was to convert the response to ranks for each subject; and the fourth was to compute an interaction value for each subject, all of which are described in Stone *et al.* (2012). These measures were focused on the internal workings of the panel, to have a way of determining which subjects were experiencing difficulties with a specific attribute, and which attributes were not useful for differentiating among products. In some instances, there was a need for a more detailed definition for an attribute, or more focus on scale direction. All this information was used by the PL during the orientation session before the next test.

The subject attribute performance measure is essential in identifying how well each subject differentiated among the products for each attribute. Since each subject scored each product attribute on multiple occasions, it was possible to use a one-way AOV to compute an F-ratio across products and its exact probability value. Since this latter value can range from 1.0 to 0.00, it represented how much that subject contributed to product differences. A value of 1.0 represented no contribution and a value of 0.00 represented a major contribution. Based on examining numerous results, a set of guidelines were developed; a value of 0.00 to 0.25 meant a large contribution, values increasing to 0.50 represented smaller contributions, and values greater than 0.50 meant that the contribution was very small. By comparing results by attribute one can identify those attributes where most subjects exhibited poor discrimination, i.e., either the differences between products were small or some subjects did not perceive the differences. The ability to examine response patterns is essential to understanding subject behavior and ultimately, the confidence with which results are described and recommendation's made. It also identifies potential topics for the next

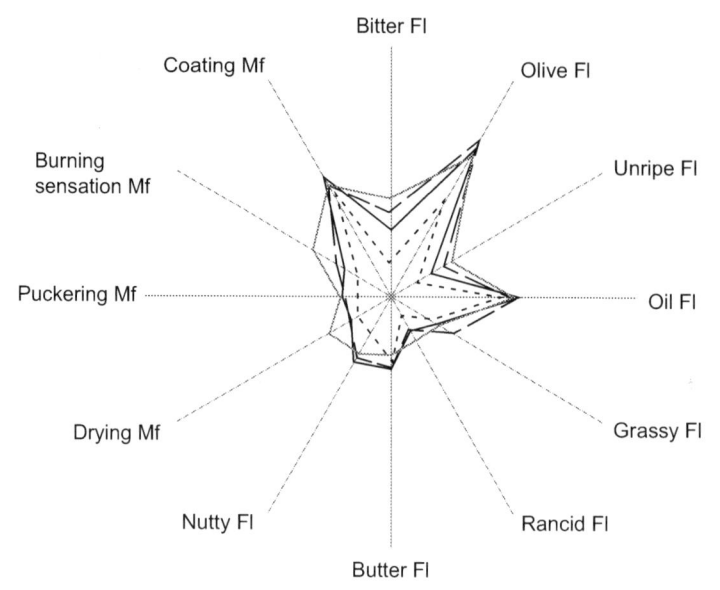

Figure 2.5 A sensory map or spider plot. The spokes are positioned equally spaced around the center point and represent the scales without anchors. Intensity values are measured from the center point (0) to that place where the line crosses that attribute for that product. Sampling product while observing a plot improves understanding of differences. Relationships among product attributes are best understood by examining the correlation values. (Fl refers to flavor; Mf refers to mouth feel.)

pre-test session. Additional discussion with examples can be found in Stone *et al.* (2012, pp. 260–4).

Reporting results is best achieved using the sensory maps with recommendations and avoiding the use of means tables unless there is a specific need. Unlike a peer reviewed article, such details can be confusing to the non-scientist looking for a specific course of action. To be effective, it is often helpful to have a project team sample the products with scorecard and definitions in hand as the results are displayed. An example of a sensory map is shown in Fig. 2.5.

2.6 Other descriptive methods

There are six or seven other descriptive analysis methods described in the sensory literature. The methods include Flavor Profile® (Cairncross and Sjöström, 1950) or its current version, Spectrum Analysis® (Meilgaard *et al.*, 2006), Texture Profile® (Brandt *et al.*, 1963), Free Choice Profiling (Williams and Arnold, 1985), and its successor Flash Descriptive Analysis (Dairou and Sieffermann, 2002). There are other methods described in the literature, but all appear to be based on methods previously described.

The primary focus of current method development, as described by Dairou and Sieffermann (2002), has been to avoid the long and costly process of using traditional methods. In the case of the Flavor and Texture Profile® methods, both of which specified 14 weeks to select and train a panel, such a time frame is too long in today's business environment. One of the reasons for developing the QDA method was this unusually long time requirement to develop a panel. The QDA method reduced that time to 8 days (or sessions) and the data collection time depended on the number of products and replications. It was also developed as a quantitative method and incorporated the repeated trials to allow for the various analyses already described. A Flavor Profile® test required a single session because no data are obtained; only a consensus judgment, a summary of the subjects' opinions reported by the PL who, by the way, also functioned as a subject.

Earlier in this chapter several key requirements for any analytical test were described and, in effect, serve as the basis for this author's approach to rapid methods. Critical to this discussion is the explanation for what is meant by "rapid." If it means testing subjects that are not qualified, then one should be concerned. If it means that there are no replicates, then one should be concerned. After all, this is no different than completing a single chemical analysis for a product. The fact that one obtains results that show differences does not mean that the results are valid. Some software will iterate until a solution is obtained and in the process will exclude some data. Flash Descriptive Analysis also relies on ranking, which is a curious choice as ranking is less sensitive than scaling and is an inefficient process. Two products are evaluated and one response is obtained vs scoring in which each product receives a score. In the paired-comparison test, the potential for sensory fatigue is high, particularly when there are many pairs. Once data collection is complete, the time requirements are the same. The qualifying and language sessions are one-time events so the speed of a method depends on the number of products being tested plus the amount of time for an orientation session. If the product category is new and a new language is needed, this usually requires three sessions.

2.7 Future trends

Sensory professionals have to establish sensory testing capabilities and be able to anticipate the kinds of informational needs required in their work. In almost all of the companies, a wide range of testing activities is usual and so the need for rapid response time means having data files that reflect subject performance, familiarity with use of different methods, and information about the range of variation typically encountered when testing one's products. Where there are gaps in capabilities, the staff need to adjust test plans to enable testing to be completed and anticipate any further requests that do not fit within the framework of typical testing procedures. As long as sensory staff has the resources, the ability to respond quickly can be more easily made with few compromises. Just as marketing research/consumer insights have developed precise specifications for the consumers they test, so too should sensory professionals be precise about the consumers they select for a sensory test.

Over the past three to four decades much progress has been achieved by sensory scientists to develop methods and continue to improve the quality of the information obtained. It is reasonable to expect that this progress will continue.

2.8 Conclusions

The idea of having available rapid methods to respond to requests for product sensory information is, as already noted, essential. However, such a resource means having subjects who have the necessary sensory skill and experience to provide information that can be relied on. This also means that one should not select subjects at random or use software for analysis that is not understood. While some legacy methods require many weeks of training, some do not. The sensory staff has to decide what is the best approach to take in developing capabilities so that best use is made of available resources. The choices are documented here along with reference to where more details can be found.

References

Bradley, R.A. and Harmon, T.J. (1964). The modified triangle test. *Biometrics*, **20**, 608–625.
Brandt, M.A., Skinner, E. and Coleman, J. (1963). Texture profile method. *J. Food Sci.*, **28**, 404–410.
Cardello, A., Lawless, H.T. and Schutz, H. (2008). Effects of extreme anchors and interior label spacing on labeled magnitude scales. *Food Qual. Prefer.*, **21**, 232–334.
Cairncross, W.E. and Sjöström, L.B. (1950). Flavor profile – a new approach to flavor problems. *Food Technol.*, **4**, 308–311.
Dairou, V. and Sieffermann, J.M. (2002). A comparison of 14 jams characterized by conventional profile and a quick original method, the flash profile. *J. Food Sci.*, **67**, 826–834.
Delarue, J. and Sieffermann, J.M. (2004). Sensory mapping using flash profile. Comparison with a conventional descriptive method for the evaluation of the flavor of fruit dairy products. *Food Qual. Prefer.*, **15**, 383–392.
Ennis, D.M. and Bi, J. (1998). The beta-binomial model: accounting for inter-trial variation in replicated difference and preference tests. *J. Sensory Stud.*, **13**, 389–412.
Ennis, J.M. and Jesionka, V. (2011). The power of sensory discrimination methods revisited. *J. Sensory Stud.*, **26** (5), 371–382.
Giradot, N.E., Peryam, D.R. and Shapiro, R. (1952). Selection of sensory testing panels. *Food Technol.*, **6**, 140–143.
Gridgeman, N.T. (1959). Pair comparisons, with and without ties. *Biometrics*, **15**, 382–388.
Groopman, J. (2009). December 17. In: *New York Review Books*, **56** (20), 22, 24.
Huitson, A. (1989). Problems with Procrustes analysis. *J. Appl. Stat.*, **16**, 39–45.
Kuesten, C.L. and Kruse, L. (2008). *Physical Requirements for Guidelines for Sensory Evaluation Laboratories*, 2nd edn., MNL60. ASTM International, West Conshohoken, PA.
Lawless, H.T. and Heymann, H. (2010). *Sensory Evaluation of Food Principles and Practices*. Springer, New York.

Mackey, A.O. and Jones, P. (1954). Selection of members of a food tasting panel: discernment of primary tastes in water solution compared with judging ability for foods. *Food Technol*, **8**, 527–530.

Meilgaard, M., Civille, G. V. and Carr, B.T. (2006). *Sensory Evaluation Techniques*, 4th edn., CRC Press, Boca Raton, FL.

Pangborn, R.M. (1981). Individuality in responses to sensory stimuli. In: *Criteria of Food Acceptance. How Man Chooses What He Eats*. J. Solms and R.L. Hall (eds), Forster Verlag, Zurich.

Peryam, D.R. and Pilgrim, F.J.(1957). Hedonic scale method of measuring food preferences. *Food Technol.*, **11**(9), 9–14.

Pfaffman, C., Schlosberg, H. and Cornsweet, J. (1954). Variables affecting difference tests. In: Peryam, D.R., Pilgrim, F.J. and Peterson, M.S. (eds.), *Food Acceptance Testing Methodology*. National Academy of Sciences/National Research Council, Washington, DC, pp. 4–17

Sawyer, F.M., Stone, H., Abplanalp, H. and Stewart, G.F. (1962). Repeatability estimates in sensory-panel selection. *J. Food Sci.*, **27**, 386–393.

Schlosberg, H., Pfaffmann, C., Cornsweet, J. and Pierrel, R. (1954). Selection and training of panels. In: Peryam, D.R., Pilgrim, J.J. and Peterson, M.S. (Eds.), *Food Acceptance Testing Methodology*. National Academy of Sciences/National Research Council, Washington, DC, pp. 45–54.

Schutz, H.G. and Cardello, A.V. (2001). A labeled affective magnitude (LAM) scale for assessing food liking/disliking. *J. Sensory Stud.*, **16**, 117–159.

Stone, H., Bleibaum, R.N. and Thomas, H.A. (2012). *Sensory Evaluation Practices*, 4th edn., Academic Press, San Diego, CA.

Stone, H. and Sidel, J.L. (1998). Quantitative descriptive analysis: developments, applications, and the future. *Food Technol.*, **52** (8), 48–52.

Stone, H. and Sidel, J.L. (2007). Sensory research and consumer-led food product development. In: MacFie, H. (Ed.), *Consumer-Led Food Product Development*. Woodhead Publishing Limited, Cambridge, UK.

Stone, H. and Sidel, J.L. (2009). Sensory science and consumer behavior. In: Barbosa-Canovas, G., Mortimer, A., Lineback, D., Spiess, W., Buckle, K. and Colonna, P. (Eds), *Global Issues in Food Science and Technology*. Elsevier, New York, pp. 67–77.

Stone, H., Sidel, J.L., Oliver, S., Woolsey, A. and Singleton, R.C. (1974). Sensory evaluation by quantitative descriptive analysis. *Food Technol.*, **28** (11), 24, 26, 28, 29, 32, 34.

Williams, A.A. and Arnold, G. (1985). A comparison of the aromas of six coffees characterized by conventional profiling, free-choice profiling and similarity scaling methods. *J. Sci. Food Agric.*, **36**, 204–214.

Williams, A.A. and Langron, S.P. (1984). The use of free-choice profiling for the evaluation of commercial ports. *J. Sci. Food Agric.*, **35**, 558–568.

Measuring sensory perception in relation to consumer behavior

J.E. Hayes

The Pennsylvania State University, University Park, PA, USA

3.1 Introduction

> *I often say that when you can measure what you are speaking about, and express it in numbers, you know something about it; but when you cannot measure it, when you cannot express it in numbers, your knowledge is of a meager and unsatisfactory kind; it may be the beginning of knowledge, but you have scarcely in your thoughts advanced to the stage of science, whatever the matter may be.*
> – Sir William Thompson, Lord Kelvin, Popular Lectures and Addresses. Volume 1, Constitution of Matter. 1889. Cambridge University Press.

The biphasic relationship between stimulus intensity and pleasure was first described by Joseph Priestly in 1775 when he noted "… a moderate degree of warmth is pleasant, and the pleasure increases with the heat to a certain degree, at which it begins to become painful; and beyond this the pain increases with the … heat, just as the pleasure had done before." A century later, Wilhelm Wundt, one of the founders of experimental psychology, included a plot of the now familiar inverted U curve in his 1874 work, "Principles of Physiological Psychology." However, Wundt also wrote "feelings, unlike sensations are not subject to exact measurement," so empirical evidence supporting Wundt's schema would have to wait until the pioneering work of Kiesow, Engel and Beebe-Center, findings which were later recapitulated by Ekman, Pangborn, Pfaffman, Moskowitz, and others.[1–5] Well into the late 1960s, such eminent psychophysicists as S.S. Stevens maintained that pleasure was too variable to be studied in a meaningful way (recounted in Reference [6]). Even today, some philosophers still debate whether it is possible to compare pleasure across individuals (e.g. Reference [7]).

In contemporary practice, much of the work sensory scientists conduct implicitly or explicitly makes four key assumptions. First, we assume that we can measure the sensations a product elicits accurately. Second, we assume we can quantify the pleasure derived from a product accurately. Third, we assume that pleasure derived from a product is a key driver of product use. Finally, we assume that we can measure consumption behavior (purchase, intake, etc.) accurately. More broadly, these assumptions can be thought of as a broader causal chain, starting with formulation and ending with use or consumption (Fig. 3.1a). Finally, those of us with an interest in health and wellness may add a fifth element, consequences of use or consumption at the far end of the chain.

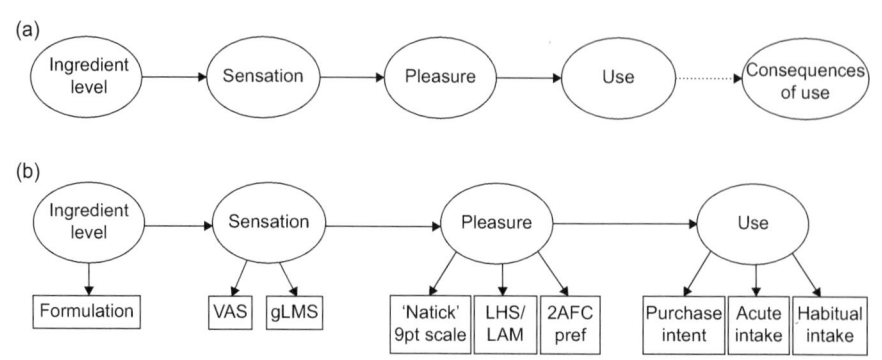

Figure 3.1 Conceptual model (a) and revised model that acknowledges the existence of measurement error (b). (VAS: Visual Analog Scale; gLMS: general Labeled Magnitude Scale; LHS and LAM: Labeled Hedonic Scale and Labeled Affective Magnitude; 2-AFC pref: two alternatives forced choice (for preference).)

Implicitly, each variable along the chain is a driver of the next variable in the chain. In daily practice, one need not measure each step of this chain, but the causal relationship remains manifest, if implicit. For example, using a designed experiment to optimize liking as a function of ingredient levels in a physicohedonic response surface model still assumes changing ingredient levels alters sensations in a systematic way (a psychophysical model), and those sensations drive liking (a psychohedonic model) (e.g. Reference [8]). Because measurement error is always present, we cannot observe the latent variables in our causal chain directly; instead, we can only observe their behavior indirectly and imperfectly from observed variables. Thus, we may refine our model, with latent variables as ovals and some examples of measured variables as rectangles (Fig. 3.1b). The remainder of this chapter will highlight the measurement of the observed variables, as well as relationships between each pair of elements in Fig. 3.1, in greater detail, giving specific attention to variation and measurement error.

3.2 Sensation

Although sensory evaluation began as an applied discipline independent from psychology, contemporary sensory science is strongly informed by psychophysics, the oldest branch of experimental psychology (see References [9–11]). Psychophysics refers to the quantitative study of the relationship between physical stimuli of known energy levels, and the sensations these stimuli produce. While these relationships were studied in their own right for many years to understand human performance, many of the same methods can be readily applied by the sensory practitioner to better understand products.

3.2.1 Accurately measuring sensations

For many years, academic psychophysicists assumed direct scaling of stimulus intensity was impossible, as sensations only occur within the mind of the observer and cannot be observed directly. This changed sometime in the late 1950s after S.S. Stevens invented the technique known as magnitude estimation. Ironically, the idea that individuals might be able to directly estimate perceptual intensity came from an argument Stevens had with a colleague,[12] who said: "You seem to maintain that each loudness has a number and that if someone sounded a tone I should be able to tell him the number," to which Stevens replied: "That is' an interesting idea. Let's try it." This approach can be contrasted with indirect scaling, where intensity distance metrics are calculated from the choice behavior of the observer. However, indirect methods are extremely slow and labor-intensive ways to generate intensity estimates via a large number of pairwise comparisons (e.g. Reference [13]). Subsequent work by Stevens and his students indicated that direct scaling via magnitude estimation was feasible (e.g. References [14–17]), and the stimulus–response relation could be described as a power function in place of the logarithmic relationship previously described by Fechner.[14] The proposed veracity of magnitude estimation ratings was further enhanced by data showing that estimates of perceived intensity across stimulus concentration are very similar to electrophysiological recordings from the chorda tympani nerve in the same individual.[18] Magnitude estimation was applied to foods by some (e.g. References [19–21]), but the applied sensory evaluation community generally remained focused on using category or line (visual analog) scales to estimate intensity. This should not be entirely surprising, as applied sensory analysts, ever the pragmatists, had already been using line scales and category scales when the academic psychophysicists were still debating direct scaling (e.g. Reference [22]).

However, magnitude estimation, visual analog scales, and category scales all suffer from the same flaw: we cannot know if your 8 is truly equal to my 8, or if your "very strong" equals my "very strong" (e.g. References [23–25]). (The interested reader is referred to References [26–30] for extensive discussion of this issue.) The proposed solution to this problem is based on cross-modality matching and magnitude matching. Specifically, J.C. Stevens and Marks demonstrated observers can consistently and accurately match intensities across modalities.[16] If we can make valid intensity comparisons across sensory modalities, it becomes possible to use a stimulus outside the domain of primary interest as a baseline. For example, if two individuals give the bitterness of grapefruit juice a 10 versus a 30 in the magnitude estimation, but both rate the remembered loudness of a car alarm near 60, we can be more confident the two individuals perceive the grapefruit differently, and the difference is not merely an artifact of how they assign numbers to sensations in the scaling task. Once these sorts of comparisons can be made, it becomes possible to study individual differences across people. For an early example, see Reference [31].

Today, generalized intensity scales (also known as "Global Sensory Intensity Scales"[32]) have largely supplanted magnitude estimation in basic chemosensory research, as they are easier to use and are believed to generate ratio-level data. The most common version currently in use is the general labeled magnitude scale (gLMS) (e.g. References [25, 30]), which was derived from the earlier LMS.[33,34] However, other generalized intensity scales, such as the general visual analog scale (gVAS), have also been proposed,[30] used (e.g. References [35, 36]), and validated.[23]

Although the gLMS is extremely common in academic usage, it has not (yet?) achieved widespread industrial usage – this may represent a traditional focus in industrial practice on the product rather than the person. The generalized context of the gLMS presumably has reduced ceiling effects, which helps to increase the validity of comparisons across individuals who live in different sensory worlds[30] and have differential experiential contexts.[37] However, this ceiling effect avoidance might potentially come at the cost of reduced product discriminability at the bottom end of the scale. That is, on the gLMS, "moderate" sensations fall in the bottom fifth of the scale: if the scale is being used properly in a global cross-modal context that includes pain, many food sensations would presumably fall into the bottom fifth of the scale. Additional research is needed to confirm whether reduced scale range (i.e. compression) negatively impacts product discrimination in practice. For further discussion of the gLMS and gVAS, their use, and ceiling and compression effects, see Reference [23]. Moreover, the gLMS has been criticized on other grounds.[38,39] Nonetheless, the gLMS has demonstrated utility when comparing individuals with clear biological differences in regard to sensation.[40–43]

3.2.2 Individual differences in sensation

Individual differences in perception have been attributed to biological (i.e. genetic) variation for over 80 years (e.g. Reference [44]), and mounting evidence from both twin studies and molecular genetics suggests the existence of numerous chemosensory phenotypes that may be highly relevant to human ingestive behavior (e.g. References [43,45–47]). Comprehensive reviews have been published recently,[48–50] so a detailed review will not be provided here. Instead, a single example will be provided for illustration.

The Arg299Cys (rs10772420) single nucleotide polymorphism (SNP) in the bitter receptor gene *TAS2R19* (HGNC:19108, formerly called *TAS2R48*) has previously been associated with the perceived bitterness of quinine,[51] and the bitterness and liking of unsweetened grapefruit juice.[40] Figure 3.2 shows bitterness intensity ratings for whole mouth (sip and spit) 0.41 mM quinine hydrochloride from 244 individuals tested in my laboratory as part of Project GIANT-CS, phase I.[43,52] The group mean for those with one or two copies of the Cys299 allele for the *TAS2R19* bitter receptor gene is 50% greater than the group mean for the Arg299 homozygotes. Whether this increase in perceived bitterness influences liking or intake of quinine containing foods such as tonic water is currently unknown.

Of course, one must always bear in mind that gene association studies are by definition quasi-experimental, as one cannot randomly assign individuals to a genotype, so significant associations may result from unmeasured third variables. This is especially important to remember for taste studies, as individual polymorphisms in taste genes often show strong linkage of disequilibrium with other polymorphisms. That is, because SNPs are not statistically independent, a statistically significant polymorphism may act as a proxy for another unmeasured causal polymorphism. Nonetheless, the point remains that individuals vary in how they perceive chemosensory stimuli,

Bitterness of quinine varies by a SNP in *TAS2R19*

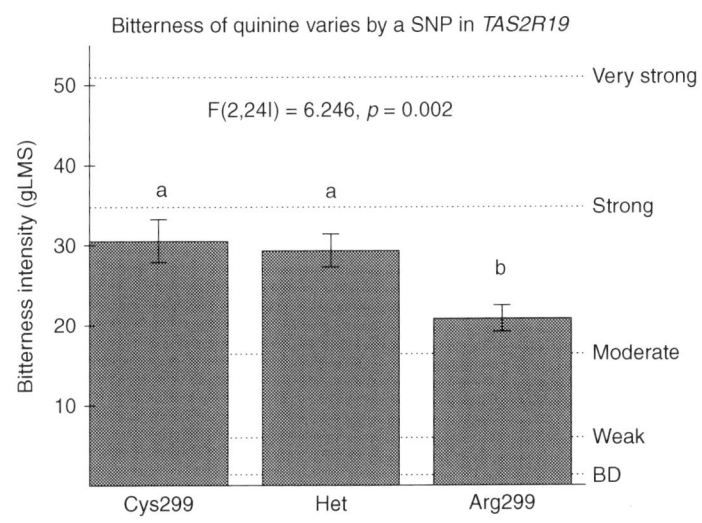

$F(2,241) = 6.246, p = 0.002$

Bar chart showing bitterness intensity (gLMS) by genotype: Cys299 ≈ 30 (a), Het ≈ 29 (a), Arg299 ≈ 21 (b). Reference levels on right axis: Very strong (~50), Strong (~35), Moderate (~17), Weak (~6), BD.

Figure 3.2 Group means by genotype: individuals with two (Cys299) or one (Het) copies of the Cys299 allele in *TAS2R19* (rs10772420) report 50% more bitterness from 0.41 mM quinine hydrochloride, on average. (The letters a and b denote that means with the same letter are not significantly different (p>0.05).)

and this variation occurs systematically as a function of biological differences, not merely prior experience or idiomatic scale usage. Such innate differences reinforce the need for generalized intensity scales when making comparisons across individuals, as a Cys299 homozygote and an Arg299 homozygote may both rate the bitterness of a quinine sample as a "6" on a classical category scale if they are each using their own personal frame of reference for bitterness.

3.2.3 Implications for classical descriptive methods

Classical panel-based profiling methods, such as Quantitative Descriptive Analysis (QDA®) and Spectrum™ Descriptive Analysis (SDA), assume a psychophysical model: as stimulus magnitude increases, the perceived intensity will increase, and it is assumed we can accurately measure these sensations. These profiling methods also assume that panelists fundamentally perceive the world in the same way, and can be taught to describe these sensations using consistent language via training. Accordingly, the issues discussed in the preceding two sections raise important questions for traditional descriptive analysis methods. For example, consider the soapy off-note some individuals report from cilantro (coriander leaf), which may be due to polymorphisms in the *OR6A2* olfactory receptor gene. In individuals who do not experience this off-note due to genetics, no amount of concept alignment and reference presentation during panel training will ever allow them to perceive or rate this quality.

This issue is not new, as Amoore described genetic variability in olfaction over 40 years ago,[53] but greater understanding of the myriad ways in which chemosensory ability varies across people (e.g. Reference [50]) suggests this may be a bigger problem than previously appreciated. Such solutions as prescreening panelists for sensitivity to the substance of interest only partially address this problem. Differences in sensation may not manifest as simple dichotomous differences (present/absence) but may instead influence the magnitude of sensations that are only observable with generalized scales. For example, some individuals report twice as much bitterness from coffee as others.[40] If these perceptual differences are real, and based on biological differences across people, it no longer makes sense to try to remove such variability via panel training, no matter how extensive. Critically, the ability to observe these differences may be blunted or absent when traditional category or line scales are used to compare individuals,[30] which may cause this problem to be underappreciated in classical profiling methods.

Moreover, individual differences in perception raise further questions about the generalizability of any specific descriptive panel to the target consumer. If I am developing a low calorie cola, do I intentionally stack my trained panel with individuals who are more responsive or less responsive to the bitterness of AceK[43]? If yes, what do I do when marketing decides to switch the product to a natural non-nutritive sweetener such as RebA, whose bitterness is unrelated to that of AceK[46]? Methods that use much larger numbers of untrained consumers in place of a low number of highly trained panelists may help address these issues, by allowing the researcher to look for different segments within the population being tested (e.g. Reference [54]). This is not to suggest large numbers of naive consumers can directly replace a highly trained panel, but at least for simple attributes, untrained participants are able to provide meaningful intensity ratings that can predict liking (e.g. References [55,56]).

3.3 Hedonics

Hedonics is a blanket term that can be used to describe the pleasure, affective valance, liking, pleasantness or preference for a stimulus. Although related conceptually, these various measures are operationalized differently in practice, which can have important implications for how resultant data are interpreted. These issues are briefly reviewed here.

3.3.1 Liking, acceptance, pleasantness, and preference

Sensory scientists have been measuring the liking and acceptability of foods with bipolar category scales since the 1950s.[57,58] Somewhat ironically, the scale first developed by Peyram and colleagues at the US Army Quartermaster Food and Container Institute to measure "food acceptance" is anchored with endpoints of "like extremely" and "dislike extremely." On its face, this suggests the scale (erroneously called the Natick 9 point scale) measures liking rather than acceptance, which would seem to be a slightly different construct. Scales for pleasantness ("very unpleasant" to "very pleasant") are also found in the literature, as "disliking" may not directly translate into other languages (see Reference [59]).

More broadly, we can define the degree of liking (the hedonic value) as the amount of pleasure one experiences when eating or using a product. In contrast, preference is typically defined as selection made within a choice paradigm. Notably, these approaches provide different information: it is possible to have a clear preference between two items that are both highly liked (cf. chocolate mousse and crème brûlée), or two items that are highly disliked (cf. Vegemite and haggis). Also, it should be noted that it is possible to prefer a product that provides less pleasure, as preferences may include non-affective factors, such as health or cost. Rozin illustrates this nicely with the example of a dieter who prefers cottage cheese to ice cream, even though the ice cream is liked more.[60] Finally, additional confusion arises from a secondary usage of "preference" to refer to rated liking for food names (as in a food preference survey (e.g. References [61,62]), in contrast to rated liking for sampled food, which is termed "acceptance" in the literature.[63] Notably, liking ratings for sampled food and food names have correlations in the range of +0.43 to +0.64.[56]

3.3.2 Liking and wanting are two related but independent constructs

Liking alone is not sufficient to explain variation in food choice, as we will discuss later. Critically, the desire to eat is distinct from the pleasure food provides.[64] Berridge has put forth a neurobiological model of reward that proposes liking and wanting are distinct constructs arising from separate neurological substrates.[65,66] In the Berridge model, liking refers to an affective/pleasantness response, while wanting is the motivation, drive or desire to consume a food. These are putatively mediated by opioid and dopamine systems, respectively, consistent with animal[67] and human data.[68]

Unlike liking, which can be determined via introspection (i.e., by simply asking how pleasant a food is), wanting is motivational and may not be a conscious process. Asking subjects to report wanting may provide a biased response if (a) they are unable to disassociate affect (pleasure) from motivation ("I like it, so I want it"), or (b) the process of answering results in "active reconstruction."[69] Thus, it appears prudent to operationalize wanting with a laboratory-based behavioral paradigm rather than relying on simple introspection (rated liking).

One approach to measuring wanting behaviorally is to determine how hard an individual will work for a food via a computerized progressive ratio task (i.e., measure its relative reinforcing value).[70–72] In an experimental paradigm developed by Epstein, participants play a game where a keypress causes three colored shapes to spin, similar to a slot machine, and, when the shapes match, a point is earned. The individual chooses to earn points for either food or an alternative reinforcer, on a variable ratio (VR) reinforcement schedule. With a VR schedule, reinforcement is delivered after a random number of responses, based upon a predetermined average. In this task, the schedule for the alternative reinforcer is held constant at VR2, meaning that on average, every second response is reinforced, while the schedule for the food increases exponentially with each trial (VR2, VR4, VR8, up to VR32). Thus, earning more food becomes increasingly difficult as the task progresses. Notably, the reinforcing value of food is independent of liking,[73] consistent with the idea that palatability and motivation arise from separate substrates. However, the Epstein paradigm is only one such approach that could be applied.

Determining the best means to measure wanting in humans (as opposed to Berridge's animal-based models) is an active, highly contentious area of research. How exactly this plays out remains to be seen, but the interested reader is referred to References [74–78] for additional discussion.

3.3.3 Using generalized scales to measure differences in pleasure across people

As in Subsection 3.2.1, substantial progress has been made on generalized hedonic scales (also known as "Global Hedonic Intensity Scales"[32]). These scales include the labeled affective magnitude (LAM) scale,[79,80] the labeled hedonic scale (LHS),[81,82] the simplified labeled affective magnitude (SLAM) scale,[83] and the hedonic gLMS.[32,84,85] All of these scales generalize the hedonic context in which ratings are made to one degree or another; the current version of the hedonic gLMS and the LHS explicitly indicate a context beyond food and oral sensation,[32,82] while the LAM has multiple versions that focus on the most pleasurable food sensation versus the most pleasurable experience of any kind.[83,86] This contrasts with the oral pleasantness unpleasantness scale (OPUS),[87] which explicitly limits its context to non-painful oral sensations.

Using generalized hedonic scales to make comparisons across products remains controversial, as all these scale variants suffer what I like to call the "best cracker in the world" problem. That is, if the participant is using a generalized hedonic scale correctly as instructed, even the world's most amazing cracker will fall in the bottom quarter of the scale when compared to the full range of life's pleasures (the birth of one's child, successfully defending one's doctoral dissertation, getting selfies from Adam Levine or Kate Upton, winning $10 000 in blackjack, etc.). Although compression of ratings on a scale is typically considered undesirable, whether compression is actually a problem for product testing also depends on whether the variance shrinks as well. That is, compression on a scale is only a problem if it obfuscates differences between products. Some reports show scale compression without loss of discrimination between products (e.g. Reference [86]) while others show compression and a small loss of discrimination between products (e.g. Reference [83]). This debate is further complicated by yet other reports that suggest generalized scales may actually increase discriminability between highly liking products by reducing ceiling effects (e.g. Reference [80]). In summary, additional work is needed to determine if generalized hedonic scales are better than traditional scales when differences between products are the primary concern. Conversely, it is becoming increasingly apparent that generalized hedonic scales are required when one is interested in making comparisons in affective experiences across individuals (see References [29,32]) who differ in terms of physiology (e.g. Reference [50]) or personality (e.g. Reference [52]).

3.4 Measuring product use and intake

Three distinct facets of product use are commonly measured in food choice research: purchase intent, acute intake, and habitual intake. The first construct, purchase intent, can be assessed in the laboratory with or without consumption or with in-home surveys, and these measures have been obtained with 5 point Likert scales, line scales with numbered tick marks, or unstructured visual analog scales (e.g. References [88–90]). However, it should be noted that purchase intent is not really a measure of use at all. Rather, it is a measure of usage intended in the future. This is an important distinction given that intention, even strong intention, may fail to predict future action. With regard to sensory practice, it should also be noted that purchase intent can be strongly influenced by external information (such as nutritional labels) that would not typically be present in blind testing (e.g. Reference [89]). The second major measure of use is acute consumption (disappearance) in the laboratory. This can be measured via plate waste in both children (e.g. References [91,92]) and adults (e.g. References [93,94]).

The third common construct, habitual intake, uses tools borrowed from dietetics and nutritional epidemiology, and may be more focused on macronutrients or patterns of eating rather than specific products. Habitual intake is typically estimated from food frequency questionnaires (FFQ) or Dietary Records. Dietary Records that include multiple 24 h food records from nonconsecutive days are the gold standard in dietary assessment. Three records are generally considered the minimum number to accurately estimate intake,[95] but with more than four consecutive records, reported intake declines because of participant fatigue.[96] It is also typical to include both weekdays and weekends to better reflect the participants' diets. However, Dietary Records also suffer from assessment reactivity, as the mere act of keeping a written record of intake can induce dietary modification (see Reference [97]). In contrast, FFQs are commonly used screening tools that ask individuals to report typical consumption of a series of foods over a specific period of time. FFQs do not provide the same level of precision as Dietary Records, due to inherent measurement error, but they are much more rapid, have a lower participant burden and may better capture habitual diet over a longer period of time. For additional information about these measures, see Reference [96].

3.5 Linking sensations, liking, and intake

As noted in the introduction, each variable in Fig. 3.1 is assumed to be a driver of the next variable in the chain. The following sections discuss data regarding this assumption.

3.5.1 Influence of sensations on liking

Given Priestly and Wundt's observations in the "Introduction", it should go without saying that altering the ingredient levels in a food to alter the sensations elicited by that food are assumed to influence liking. Indeed, countless reports over the last 60 years recapitulate this (e.g. References [8, 98–104]). As just one example, Drewnowski and Moskowitz showed that liking for a salted snack varies as both a function of salt level and level of spice.[100] Accordingly, we will not belabor this further. For another perspective and additional discussion, see Chapter 14 in this book.

What may be less apparent is that optimum liking occurs as function of sensation, not concentration. Working in multiple food matrices (vanilla pudding, yellow cake, a cherry beverage, and sucrose in water), Moskowitz and colleagues observed that maximal liking occurred at a constant sweetness level, not a constant amount of sucrose.[5] Indeed, we have repeatedly found that perceived intensity outperforms stimulus concentration as a predictor of liking.[8,55] This becomes especially relevant when we consider the genetic differences in perception described above in Section 3.2.2, as it suggests a formula that is optimal for one individual may be sub-optimal for others (e.g. Reference [8]). Moreover, the same product can evoke very different sensations across individuals and these differences in sensation predict differential liking. For example, we found that greater bitterness and lower sweetness from sampled unsweetened grapefruit juice explained 36% of the variance in liking rated with a hedonic gLMS.[85] Later, we showed in a separate cohort that differences in bitterness and liking for grapefruit juice associated with an SNP in *TAS2R19*.[40] For lager beer, variation in the endogenous sweetness and bitterness across individuals explains 27% of the variance in rated liking.[85] Likewise, similar effects are seen for kale and Brussels sprouts.[105] In contrast, the bitterness of coffee varies across individuals, but this does not predict liking.[40,85] This serves as a reminder that postingestive pharmacological consequences may, on occasion, decouple taste sensations from affective responses.

3.5.2 Does liking really predict intake?

Numerous studies support the proposition that liking is a moderate to strong predictor of intake (reviewed by Reference [59]). However, it should be noted that this relationship is heteroskedastic: as liking increases, reported intake shows more variability.[106] That is, disliking is more strongly related to disuse than liking is to use.[107] Intuitively, this makes sense: if I enjoy red wines from Château Mouton-Rothschild, I may like them very strongly, but still limit my intake due to health concerns or economic constraints. Conversely, if I dislike a single malt scotch like Lagavulin due to the peaty notes, I avoid it regardless of who is buying.

3.5.3 Exploring the causal chain within a group of individuals

Understandably, most studies do not examine the multiple levels of data shown in Fig. 3.1 within a single group of individuals, focusing instead on such narrower questions as the correlations between pairs of variables (e.g. References [92,108]). Some of the studies that do look down the chain from sensation to intake within a cohort include References [52,56,85,104,105]. A study by Lucas and colleagues on hash browns (fried potatoes) illustrates the difficulties that can arise in this approach.[104] They found that sodium chloride recognition thresholds varied across people, but these did not predict perceived saltiness or liking of hash browns (consistent with other data[109,110]). Meanwhile, varying the sodium content of hash browns influenced both perceived saltiness and liking, and saltier hash browns were liked more in a laboratory context. However, in an acute intake paradigm, they failed to observe a relationship between amount consumed and liking. Conversely, using measures of differential sensitivity across multiple chicken broth samples with varying amounts of added sodium (intensity and liking area under the curve across a concentration series), we observed a strong relationship between differential intensity and hedonic sensitivity for salty foods. Moreover, our index of hedonic sensitivity and surveyed preference for salty foods collectively explained 18% of the variance in habitual intake of 38 salty foods.[56] For other sampled foods, the strength and direction of the relationship between saltiness and liking varied across foods: potato chips were more liked as perceived saltiness increased, while soy sauce was liked less as saltiness increased.[56]

Does the absence of evidence of a liking–intake relation in the study by Lucas and colleagues imply there is not a causal relationship between liking and intake in general? Presumably not, as to assume otherwise would beg the question of why product developers and industrial sensory scientists spent countless hours and vast resources optimizing products. It may be that acute intake measurements in the laboratory suffer from sufficient contextual effects or other measurement errors that limit one's ability to observe such relationships.

Studies on habitual intake of alcohol and vegetables have demonstrated relationships between sensation and intake that are mediated via liking (e.g. References [85,105]), which potentially explain studies that directly link taste gene variation to differential intake without measuring the intermediary variables.[40,111,112] Other studies find relationships between food preferences and diet-related health outcomes, even in the absence of observed relationships with intake.[62,106] For example, liking for spicy food is strongly correlated with habitual spicy food intake,[52] and liking for spicy food associates with favorable blood pressure profiles and reduced adiposity.[106]

3.6 Summary

Either implicitly or explicitly, sensory scientists typically assume that sensations drive liking, and liking drives intake. Numerous studies support this contention, but food choice is highly multifactorial, so liking is only one driver of intake. Measurement error in each step of the chain often attenuates the ability to observe intermediary relationships, so improved measurement techniques and larger numbers of participants are needed to successfully link sensation to consumer behavior. Use of genetically informed participants has the potential to increase panel homogeneity when used as a selection criterion, or in the case of intentional stratification across a genetic variant, to increase the generalizability of a panel to the larger population.

References

1. Pfaffmann C. (1960) The pleasures of sensation. *Psychol Rev*, **67**:253–268.
2. Pfaffmann C. (1980) Wundt's schema of sensory affect in the light of research on gustatory preference. *Psychol Res*, **42**:165–174.

3. Pangborn RM. (1970) Individual variation in affective responses to taste stimuli. *Psychon Sci*, **21**:125–126.
4. Moskowitz HR. (1971) The sweetness and pleasantness of sugars. *Am J Psychol*, **84**:384–405.
5. Moskowitz HR, Kluter RA, Westerling J and Jacobs HL. (1974) Sugar sweetness and pleasantness: evidence for different psychological laws. *Science*, **184**:583–585.
6. Moskowitz H. (2004) From psychophysics to the world ... data acquired, lessons learned. *Food Qual Prefer*, **15**:633–644.
7. Klocksiem J. (2008) The problem of interpersonal comparisons of pleasure and pain. *J Value Inquiry*, **42**:23–40.
8. Hayes JE and Duffy VB. (2008) Oral sensory phenotype identifies level of sugar and fat required for maximal liking. *Physiol Behav*, **95**:77–87. Epub 2008 May 2. doi:10.1016/j.physbeh.2008.04.023.
9. Prescott J, Hayes JE, and Byrnes NK. (2014) Sensory science. In: *Encyclopedia of Agriculture and Food Systems*. Neal Van Alfen, editor-in-chief. Vol. 5, San Diego: Elsevier. 80–101.
10. Moskowitz H. (2003) The intertwining of psychophysics and sensory analysis: historical perspectives and future opportunities – a personal view. *Food Qual Prefer*, **14**:87–98.
11. Lawless HT. (2013) *Quantitative Sensory Analysis: Psychophysics, Models and Intelligent Design*: Wiley.
12. Stevens SS. (1956) The direct estimation of sensory magnitudes – loudness. *Am J Psychol*, **69**:1–25.
13. Fry JC, Yurttas N, Biermann KL, Lindley MG and Goulson MJ. (2012) The sweetness concentration-response of r,r-monatin, a naturally occurring high-potency sweetener. *J Food Sci*, **77**:S362–S364.
14. Stevens SS. (1961) To Honor Fechner and repeal his law: A power function, not a log function, describes the operating characteristic of a sensory system. *Science*, **133**:80–86.
15. Stevenson FH and Manning CW. (1962) Tuberculosis of the spine treated conservatively with chemotherapy: series of 72 patients collected 1949–1954 and followed to 1961. *Tubercle*, **43**:406–411.
16. Stevens JC and Marks LE. (1965) Cross-modality matching of brightness and loudness. *Proc Natl Acad Sci U S A*, **54**:407–411.
17. Stevens SS. (1969) Sensory scales of taste intensity. *Percept Psychophys*, **6**:302–308.
18. Borg G, Diamant H, Strom L and Zotterman Y. (1967) The relation between neural and perceptual intensity: a comparative study on the neural and psychophysical response to taste stimuli. *J Physiol*, **192**:13–20.
19. Moskowitz H. (1977) Magnitude estimation: Notes on what, how, when, and why to use it. *J Food Quality*, **1**:195–227.
20. Moskowitz HR. (1972) Subjective ideals and sensory optimization in evaluating perceptual dimensions in food. *J Appl Psychol*, **56**:60–66.
21. Giovanni ME and Pangborn RM. (1983) Measurement of taste intensity and degree of liking of beverages by graphic scales and magnitude estimation. *J Food Sci*, **48**:1175–1182.
22. Baten WD. (1946) Organoleptic tests pertaining to apples and pears. *Food Res*, **11**:84–94.
23. Hayes JE, Allen AL and Bennett SM. (2013) Direct comparison of the generalized Visual Analog Scale (gVAS) and general Labeled Magnitude Scale (gLMS). *Food Qual Prefer*, **28**:36–44.
24. Duffy VB, Hayes JE, Bartoshuk LM and Snyder DJ. (2009) Taste: vertebrate psychophysics. In: Squire LR, ed. *Encyclopedia of Neuroscience*. Oxford: Academic Press:881–886.

25. Snyder DJ, Fast K and Bartoshuk LM. (2004) Valid comparisons of suprathreshold sensations. *J Conscious Stud*, **11**:96–112.
26. Prutkin J, Duffy VB, Etter L, Fast K, Gardner E, Lucchina LA, Snyder DJ, Tie K, Weiffenbach J and Bartoshuk LM. (2000) Genetic variation and inferences about perceived taste intensity in mice and men. *Physiol Behav*, **69**:161–173.
27. Bartoshuk LM, Duffy VB, Fast K, Green BG, Prutkin J and Snyder DJ. (2003) Labeled scales (eg, category, Likert, VAS) and invalid across-group comparisons: what we have learned from genetic variation in taste. *Food Qual Pref*, **14**:125–138.
28. Bartoshuk LM, Duffy VB, Green BG, Hoffman HG, Ko C-W, Lucchina LA, Marks LE, Snyder DJ and Weiffenbach JM. (2004) Valid across-group comparisons with labeled scales: the gLMS versus magnitude matching. *Physiol Behav*, **82**:109–114.
29. Bartoshuk LM, Duffy VB, Hayes JE, Moskowitz HD and Snyder DJ. (2006) Psychophysics of sweet and fat perception in obesity: problems, solutions and new perspectives. *Philos T Roy Soc B: Biol Sci*, **361**:1137–1148.
30. Snyder DJ, Prescott J and Bartoshuk LM. (2006) Modern psychophysics and the assessment of human oral sensation. *Adv Otorhinolaryngol*, **63**:221–241.
31. Marks LE, Stevens JC, Bartoshuk LM, Gent JF, Rifkin B and Stone VK. (1988) Magnitude-matching: the measurement of taste and smell. *Chem Senses*, **13**:63–87.
32. Kalva JJ, Sims CA, Puentes LA, Snyder DJ and Bartoshuk LM. (2014) Comparison of the hedonic general labeled magnitude scale with the hedonic 9-point scale. *J Food Sci*, **79**:S238–S245.
33. Green BG, Shaffer GS and Gilmore MM. (1993) Derivation and evaluation of a semantic scale of oral sensation magnitude with apparent ratio properties. *Chem Senses*, **18**:683–702.
34. Green BG, Dalton P, Cowart B, Shaffer G, Rankin K and Higgins J. (1996) Evaluating the "Labeled Magnitude Scale" for measuring sensations of taste and smell. *Chem Senses*, **21**:323–334.
35. Pickering GJ, Moyes A, Bajec MR and Decourville N. (2010) Thermal taster status associates with oral sensations elicited by wine. *Aust J Grape Wine Res*, **16**:361–367.
36. Timpson NJ, Heron J, Day IN, Ring SM, Bartoshuk LM, Horwood J, Emmett P and Davey-Smith G. (2007) Refining associations between TAS2R38 diplotypes and the 6-n-propylthiouracil (PROP) taste test: findings from the Avon Longitudinal Study of Parents and Children. *BMC Genet*, **8**:51.
37. Dionne RA, Bartoshuk L, Mogil J and Witter J. (2005) Individual responder analyses for pain: does one pain scale fit all? *Trends Pharmacol Sci*, **26**:125–130.
38. Schifferstein HNJ. (2012) Labeled magnitude scales: A critical review. *Food Qual Prefer*, **26**:151–158.
39. Lawless HT, Horne J and Speirs W. (2000) Contrast and range effects for category, magnitude and labeled magnitude scales in judgments of sweetness intensity. *Chem Senses*, **25**:85–92.
40. Hayes JE, Wallace MR, Knopik VS, Herbstman DM, Bartoshuk LM and Duffy VB. (2011) Allelic variation in TAS2R bitter receptor genes associates with variation in sensations from and ingestive behaviors toward common bitter beverages in adults. *Chem Senses*, **36**:311–319.
41. Hayes JE and Keast RSJ. (2011) Two decades of supertasting: where do we stand? *Physiol Behav*, **104**:1072–1074.
42. Genick UK, Kutalik Z, Ledda M, Souza Destito MC, Souza MM, Cirillo CA, Godinot N, Martin N, Morya E, Sameshima K, Bergmann S and le Coutre J. (2011) Sensitivity

of genome-wide-association signals to phenotyping strategy: the PROP-TAS2R38 taste association as a benchmark. *PLoS One*, **6**:e27745.
43. Allen AL, McGeary JE, Knopik VS and Hayes JE. (2013) Bitterness of the non-nutritive sweetener acesulfame potassium varies with polymorphisms in TAS2R9 and TAS2R31. *Chem Senses*, **38**:379–389.
44. Blakeslee AF. (1932) Genetics of sensory thresholds: Taste for phenyl thio carbamide. *Proc Natl Acad Sci*, **18**:120–130.
45. Tornwall O, Silventoinen K, Kaprio J and Tuorila H. (2012) Why do some like it hot? Genetic and environmental contributions to the pleasantness of oral pungency. *Physiol Behav*, **107**:381–389.
46. Allen AL, McGeary JE and Hayes JE. (2013) Rebaudioside A and Rebaudioside D bitterness do not covary with Acesulfame K bitterness or polymorphisms in TAS2R9 and TAS2R31. *Chemosens Percept*, **6**:109–117.
47. Jaeger SR, McRae JF, Bava CM, Beresford MK, Hunter D, Jia Y, Chheang SL, Jin D, Peng M, Gamble JC, Atkinson KR, Axten LG, Paisley AG, Tooman L, Pineau B, Rouse SA and Newcomb RD. (2013) A mendelian trait for olfactory sensitivity affects odor experience and food selection. *Curr Biol*, **23**:1601–1605.
48. Knaapila A, Hwang LD, Lysenko A, Duke FF, Fesi B, Khoshnevisan A, James RS, Wysocki CJ, Rhyu M, Tordoff MG, Bachmanov AA, Mura E, Nagai H and Reed DR. (2012) Genetic analysis of chemosensory traits in human twins. *Chem Senses*, **37**:869–881.
49. Feeney E. (2011) The impact of bitter perception and genotypic variation of TAS2R38 on food choice. *Nutr Bulletin*, **36**:20–33.
50. Hayes JE, Feeney EL and Allen AL. (2013) Do polymorphisms in chemosensory genes matter for human ingestive behavior? *Food Qual Prefer*, **30**:202–216.
51. Reed DR, Zhu G, Breslin PA, Duke FF, Henders AK, Campbell MJ, Montgomery GW, Medland SE, Martin NG and Wright MJ. (2010) The perception of quinine taste intensity is associated with common genetic variants in a bitter receptor cluster on chromosome 12. *Hum Mol Genet*, **19**:4278–4285.
52. Byrnes NK and Hayes JE. (2013) Personality factors predict spicy food liking and intake. *Food Qual Prefer*, **28**:213–221.
53. Whissell-Buechy D and Amoore JE. (1973) Odour-blindness to musk: simple recessive inheritance. *Nature*, **242**:271–273.
54. Moskowitz HR and Bernstein R. (2000) Variability in hedonics: Indications of world-wide sensory and cognitive preference segmentation. *J Sens Stud*, **15**:263–284.
55. Li B, Hayes JE and Ziegler GR. (2014) Interpreting consumer preferences: physicohedonic and psychohedonic models yield different information in a coffee-flavored dairy beverage. *Food Qual Prefer*, **36**:27–32.
56. Hayes JE, Sullivan BS and Duffy VB. (2010) Explaining variability in sodium intake through oral sensory phenotype, salt sensation and liking. *Physiol Behav*, **100**:369–380.
57. Peryam DR and Pilgrim FJ. (1957) Hedonic scale method of measuring food preferences. *Food Technol*, **11**:9–14.
58. Peryam DR and Haynes JG. (1957) Prediction of soldiers' food preferences by laboratory methods. *J Appl Psychol*, **41**:2–6.
59. Tuorila H, Huotilainen A, Lahteenmaki L, Ollila S, Tuomi-Nurmi S and Urala N. (2008) Comparison of affective rating scales and their relationship to variables reflecting food consumption. *Food Qual Prefer*, **19**:51–61.
60. Rozin P and Vollmecke TA. (1986) Food likes and dislikes. *Annu Rev Nutr*, **6**:433–456.
61. Drewnowski A and Hann C. (1999) Food preferences and reported frequencies of food consumption as predictors of current diet in young women. *Am J Clin Nutr*, **70**:28–36.

62. Duffy VB, Lanier SA, Hutchins HL, Pescatello LS, Johnson MK and Bartoshuk LM. (2007) Food Preference Questionnaire as a Screening Tool for Assessing Dietary Risk of Cardiovascular Disease within Health Risk Appraisals. *J Am Diet Assoc*, **107**:237–245.
63. Cardello AV, Schutz H, Snow C and Lesher L. (2000) Predictors of food acceptance, consumption and satisfaction in specific eating situations. *Food Qual Prefer*, **11**:201–216.
64. Mela DJ. (2001) Determinants of food choice: relationships with obesity and weight control. *Obes Res*, **9** Suppl 4:249S–255S.
65. Berridge KC. (1996) Food reward: brain substrates of wanting and liking. *Neurosci Biobehav Rev*, **20**:1–25.
66. Berridge KC and Robinson TE. (2003) Parsing reward. *Trends Neurosci*, **26**:507–513.
67. Pecina S, Cagniard B, Berridge KC, Aldridge JW and Zhuang X. (2003) Hyperdopaminergic mutant mice have higher "wanting" but not "liking" for sweet rewards. *J Neurosci*, **23**:9395–402.
68. Yeomans MR and Wright P. (1991) Lower pleasantness of palatable foods in nalmefene-treated human volunteers. *Appetite*, **16**:249–259.
69. Finlayson G, King N and Blundell JE. (2007) Is it possible to dissociate "liking" and "wanting" for foods in humans? A novel experimental procedure. *Physiol Behav*, **90**:36–42.
70. Lappalainen R and Epstein LH. (1990) A behavioral economics analysis of food choice in humans. *Appetite*, **14**:81–93.
71. Saelens BE and Epstein LH. (1996) Reinforcing value of food in obese and non-obese women. *Appetite*, **27**:41–50.
72. Raynor HA and Epstein LH. (2003) The relative-reinforcing value of food under differing levels of food deprivation and restriction. *Appetite*, **40**:15–24.
73. Epstein LH, Truesdale R, Wojcik A, Paluch RA and Raynor HA. (2003) Effects of deprivation on hedonics and reinforcing value of food. *Physiol Behav*, **78**:221–227.
74. Finlayson G and Dalton M. (2012) Current progress in the assessment of "liking" vs. "wanting" food in human appetite. Comment on "You Say it's Liking, I Say it's Wanting ...." On the difficulty of disentangling food reward in man. *Appetite*, **58**:373–378.
75. Finlayson G, Arlotti A, Dalton M, King N and Blundell JE. (2011) Implicit wanting and explicit liking are markers for trait binge eating. A susceptible phenotype for overeating. *Appetite*, **57**:722–728.
76. Havermans RC, Janssen T, Giesen JCAH, Roefs A and Jansen A. (2009) Food liking, food wanting, and sensory-specific satiety. *Appetite*, **52**:222–225.
77. Havermans RC. (2011) "You Say it's Liking, I Say it's Wanting ...." On the difficulty of disentangling food reward in man. *Appetite*, **57**:286–294.
78. Havermans RC. (2012) How to tell where "liking" ends and "wanting" begins. *Appetite*, **58**:252–255.
79. Cardello AV and Schutz HG. (2004) Research note – Numerical scale-point locations for constructing the LAM (labeled affective magnitude) scale. *J Sens Stud*, **19**:341–346.
80. Schutz HG and Cardello AV. (2001) A labeled affective magnitude (LAM) scale for assessing food liking/disliking. *J Sens Stud*, **16**:117–159.
81. Lim J and Fujimaru T. (2010) Evaluation of the Labeled Hedonic Scale under different experimental conditions. *Food Qual Prefer*, **21**:521–530.
82. Lim J, Wood A and Green BG. (2009) Derivation and evaluation of a labeled hedonic scale. *Chem Senses*, **34**:739–751.
83. Lawless HT, Cardello AV, Chapman KW, Lesher LL, Given Z and Schutz HG. (2010) A Comparison of the effectiveness of hedonic scales and end-anchor compression effects. *J Sens Stud*, **25**:18–34.

84. Duffy VB, Peterson J, Dinehart M and Bartoshuk LM. (2003) Genetic and environmental variation in taste: Associations with sweet intensity, preference and intake. *Top Clin Nutr*, **18**:209–220.
85. Lanier SA, Hayes JE and Duffy VB. (2005) Sweet and bitter tastes of alcoholic beverages mediate alcohol intake in of-age undergraduates. *Physiol Behav*, **83**:821–831.
86. Cardello A, Lawless HT and Schutz HG. (2008) Effects of extreme anchors and interior label spacing on labeled affective magnitude scales. *Food Qual Prefer*, **19**: 473–480.
87. Guest S, Essick G, Patel A, Prajapati R and McGlone F. (2007) Labeled magnitude scales for oral sensations of wetness, dryness, pleasantness and unpleasantness. *Food Qual Prefer*, **18**:342–352.
88. Bower JA, Saadat MA and Whitten C. (2003) Effect of liking, information and consumer characteristics on purchase intention and willingness to pay more for a fat spread with a proven health benefit. *Food Qual Prefer*, **14**:65–74.
89. Guinard JX, SmiciklasWright H, Marty C, Sabha, RA, Taylor-Davis S and Wright C. (1996) Acceptability of fat-modified foods in a population of older adults: Contrast between sensory preference and purchase intent. *Food Qual Prefer*, **7**:21–28.
90. Mucci A, Hough G and Ziliani C. (2004) Factors that influence purchase intent and perceptions of genetically modified foods among Argentine consumers. *Food Qual Prefer*, **15**:559–567.
91. Birch LL. (1979) Preschool children's food preferences and consumption patterns. *J Nutr Educat*, **11**:189–192.
92. Caporale G, Policastro S, Tuorila H and Monteleone E. (2009) Hedonic ratings and consumption of school lunch among preschool children. *Food Qual Prefer*, **20**:482–489.
93. Vickers Z, Holton E and Wang J. (2001) Effect of ideaîâ relative sweetness on yogurt consumption. *Food Qual Prefer*, **12**:521–526.
94. Zandstra EH, De Graaf C, Mela DJ and Van Staveren WA. (2000) Short and long-term effects of changes in pleasantness on food intake. *Appetite*, **34**:253–260.
95. de Castro JM. (1994) Methodology, correlational analysis, and interpretation of diet diary records of the food and fluid intake of free-living humans. *Appetite*, **23**:179–192.
96. Thompson FE and Subar AF. (2013) Dietary assessment methodology. In: *Nutrition in the Prevention and Treatment of Disease,* Third Edition. Elsevier Inc., 5–36.
97. Macdiarmid J and Blundell J. (1998) Assessing dietary intake: Who, what and why of under-reporting. *Nutr Res Rev*, **11**:231–253.
98. Pangborn RM, Simone M and Nickerson TA. (1957) The influence of sugar in ice cream. I. Consumer preference for vanilla ice cream. *Food Technol*, **11**:679–682.
99. Moskowitz HR. (1972) Subjective ideals and sensory optimization in evaluating perceptual dimensions in food. *J Appl Psychol*, **56**:60.
100. Drewnowski A and Moskowitz HR. (1985) Sensory characteristics of foods: new evaluation techniques. *Am J Clin Nutr*, **42**:924–931.
101. Drewnowski A and Greenwood MR. (1983) Cream and sugar: human preferences for high-fat foods. *Physiol Behav*, **30**:629–633.
102. Warwick ZS and Schiffman SS. (1990) Sensory evaluations of fat-sucrose and fat-salt mixtures: relationship to age and weight status. *Physiol Behav*, **48**:633–636.
103. De Graaf C and Zandstra EH. (1999) Sweetness intensity and pleasantness in children, adolescents, and adults. *Physiol Behav*, **67**:513–520.
104. Lucas L, Riddell L, Liem G, Whitelock S and Keast R. (2011) The Influence of Sodium on Liking and Consumption of Salty Food. *J Food Sci*, **76**:S72–S76.

105. Dinehart ME, Hayes JE, Bartoshuk LM, Lanier SL and Duffy VB. (2006) Bitter taste markers explain variability in vegetable sweetness, bitterness, and intake. *Physiol Behav*, **87**:304–313.
106. Duffy VB, Hayes JE, Sullivan BS and Faghri P. (2009) Surveying food/beverage liking: a tool for epidemiological studies to connect chemosensation with health outcomes. *Ann Ny Acad Sci*, **1170**:558–568.
107. Randall E and Sanjur D. (1981) Food preferences – their conceptualization and relationship to consumption. *Ecol Food Nutr*, **11**:151–161.
108. Lucas F and Bellisle F. (1987) The measurement of food preferences in humans: do taste-and-spit tests predict consumption? *Physiol Behav*, **39**:739–743.
109. Pangborn RM and Pecore SD. (1982) Taste perception of sodium chloride in relation to dietary intake of salt. *Am J Clin Nutr*, **35**:510–520.
110. Keast RS and Roper J. (2007) A complex relationship among chemical concentration, detection threshold, and suprathreshold intensity of bitter compounds. *Chem Senses*, **32**:245–253.
111. Duffy VB, Hayes JE, Davidson AC, Kidd JK, Kidd KK and Bartoshuk LM. (2010) Vegetable intake in college-aged adults is explained by oral sensory phenotypes and TAS2R38 genotype. *Chemosens Percept*, **3**:137–148.
112. Duffy VB, Davidson AC, Kidd JR, Kidd KK, Speed WC, Pakstis AJ, Reed DR, Snyder DJ and Bartoshuk LM. (2004) Bitter receptor gene (TAS2R38), 6-n-propylthiouracil (PROP) bitterness and alcohol intake. *Alcohol Clin Exp Res*, **28**:1629–1637.

Insights into measuring emotional response in sensory and consumer research

4

M. Ng[1,2], J. Hort[2]
[1]PepsiCo Europe R & D, Leicester, UK and University of Nottingham, Nottingham, UK; [2]University of Nottingham, Nottingham, UK

4.1 Introduction

> *Emotions and feelings are not a luxury; they are a means of communicating our states of mind to others. But they are also a way of guiding our own judgments and decisions. Emotions bring the body into the loop of reason.*
>
> Professor Antonio Damasio, 1994

In understanding product performance, traditional sensory and consumer research has always tended to focus on the relationship between sensory perceptions and liking measures. In these days of extremely competitive markets, some recent studies have highlighted that using hedonic measurement alone is inadequate for measuring the consumer affective product experience (Desmet and Schifferstein, 2008a; Koster, 2009; King and Meiselman, 2010; Ng et al., 2013a). Very often, consumers rely on unconscious emotions associated with a product via sensory perceptions (Thomson et al., 2010) to make their purchase or consumption decisions (Lehrer, 2006; Walsh et al., 2011). In fact, evidence shows that, without emotions, one struggles to make decisions (Damasio, 2006). Damasio illustrated this when recounting the case of Elliot, a patient with brain damage to his ventromedial prefrontal cortex (VPFC) who suffered a consequential inability to experience emotions (Damasio, 1994). Being so "rational", Elliot had to endlessly deliberate over irrelevant details and to reason every decision he had to make – for example, whether to use a blue or black pen, and what radio station to listen to. This, Damasio said, illustrated the limitation of pure reason in decision making such that, in people with normal brains, decisions are more "weighted" by emotions, enabling them to make decisions more quickly compared to patients who suffer an inability to experience emotions. However, Damasio has also posited that emotions are not independent of rationality – they are both part of rationality and inseparably interlinked:

> "... emotions probably assist in reasoning, especially when it comes to personal and social matters involving risk and conflict. I suggest that certain levels of emotions processing probably point us to the sector of the decision making space where our reason can operate most effectively." (Damasio, 1999)

Consequently, the knowledge and capability to define, measure and understand how emotions are connected with respect to specific products and brands via sensory attributes could ultimately provide a decisive and competitive advantage in the marketplace.

This chapter begins with a general definition of emotion and discussion of the importance of measuring emotion in sensory and consumer research. General approaches that are used to measure emotions are briefly laid out. Some of the verbal self-report emotion lexicons described in the current literature, and their applications in the sensory and consumer fields, are reviewed. The authors will also discuss how consumer emotional responses can be related to the output of sensory descriptive analysis. Finally, some unresolved issues in verbal self-report emotion measurement and topics for future research are discussed.

4.2 Defining emotion

An emotion has been defined as "*a mental state of readiness that arises from cognitive appraisals of events or thoughts; has a phenomenological tone; is accompanied by physiological processes; is often expressed physically (e.g., in gestures, posture, facial features); and may result in specific actions to affirm or cope with the emotion, depending on its nature and meaning for the person having it*" (Bagozzi et al., 1999). In other definitions, emotions have been described as brief, intense and often focused on a referent (e.g. "the comment made him angry") (Clore et al., 1987; King and Meiselman, 2010). However, efforts to confirm a widely acceptable definition of emotion have proven to be unsuccessful (Panksepp, 2003). Nevertheless, emotions matter. According to Damasio (2006), they are "in the loop of reason," guiding thoughts and deeds.

Some researchers have argued that there are different types of emotions, ranging from "lower-order" through "basic" to "higher-order" on an emotional continuum (Poels and Dewitte, 2006). "Lower-order emotions," denote emotional reactions that are spontaneous and uncontrollable (LeDoux, 1996; Shiv and Fedorikhin, 1999), whereas "higher-order emotions" refer to emotional reactions that are more complex and involve cognitive processing (Frijda et al., 1989; Lazarus, 1991). Some basic emotions, e.g. fear, anger and happiness, however, are situated in between lower-order and higher-order emotions. For example, standing face to face with a lion will automatically fill an individual with lower-order "fear" but, on the other hand, it might also cause that individual to experience higher-order "fear" after conscious appraisal of the situation, i.e. fear of being attacked. Therefore, basic emotions can be experienced both automatically and after cognitive processing.

Furthermore, some researchers have proposed multidimensional circumplex models to organize human emotions (Fig. 4.1) (Russell, 1980; Watson and Tellegen, 1985; Larsen and Diener, 1992). These circumplex models are two dimensional, circular structures in which single emotions correlate highly with those emotions nearby on the circumference of the circle, but do not correlate with those emotions one-quarter away (90°). The models are used to describe the dimensionality of human emotion

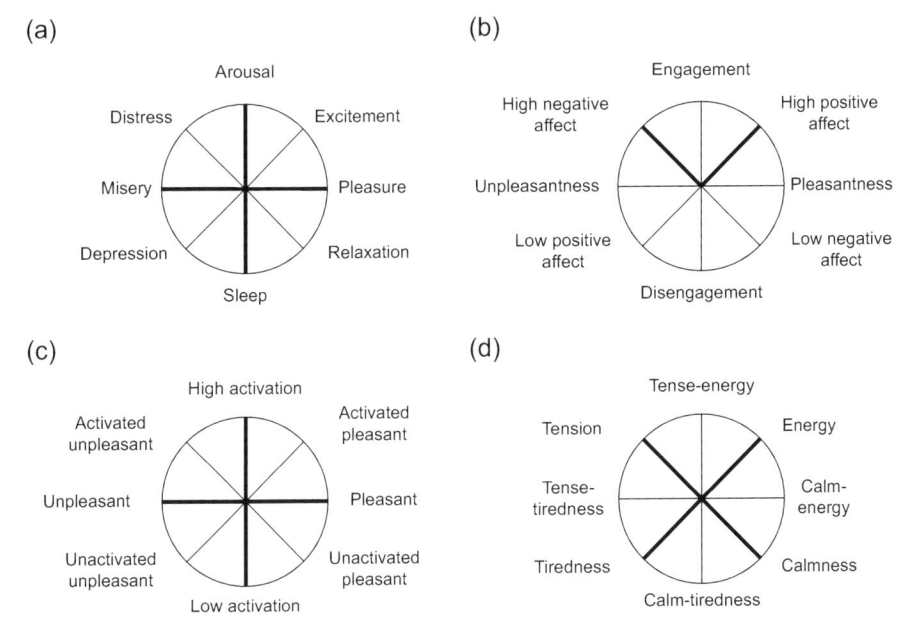

Figure 4.1 Four descriptive models of core effect: (a) Russell, (b) Watson and Tellegen, (c) Larsen and Diener, (d) Thayer.
Source: Reprinted from Russell and Barnett (1999), with permission from APA.

where the dimensions are bipolar; emotional terms represent a continuity of mood state from pleasant/positive to unpleasant/negative on one dimension and different levels of engagement/arousal on the other. Interestingly, the latter two-dimensional structure (pleasantness versus engagement/activation) has been observed within a juice squash product space (Ng et al., 2013a), as well as a beer product category (Chaya et al., *under review*).

Alternatively, some researchers have proposed appraisal theory to define and study emotional experience (see Scherer et al., 2001 for review). The main assumption of appraisal theory is *"that emotions arise, and are distinguished, on the basis of a person's subjective evaluation of an event of appraisal dimensions such as novelty, urgency, goal congruence, coping potential and norm compatibility"* (Juslin and Vastfjall, 2008). In addition, appraisal theory also claims that emotions can be elicited by physiological arousal (e.g. facial expression), or by action tendencies (e.g. hunger leading to an infant's distress) (Scherer et al., 2001).

4.3 The importance of measuring emotions in sensory and consumer research

Advances in neuroscience and psychology in recent years have not only identified some key brain regions that process emotions (i.e. the prefrontal cortex, amygdala, hypothalamus and anterior cingulate cortex), but have also evidently illustrated that emotions guide and bias our decision making (see Bechara, 2004; Dalgleish, 2004 for reviews). Without emotions, individuals make poor decisions, and in fact they struggle to make decisions at all (Damasio, 2006). In a famous study, Bechara *et al.* (1994) assessed the somatic state activation of two groups of subjects – normal subjects versus patients with VPFC damage – when making decisions during a gambling task. Somatic state activation refers to physiological reactions that have had emotion-related consequences in the past (Dalgleish, 2004). In this study, the subjects' skin conductance responses (SCR) were recorded after they picked a card and were told that they had won or lost money. The study revealed that, as the normal controls became experienced with the task, they began to generate SCRs prior to the selection of any cards, and learned to perform the task better than the patients with VPFC damage, who failed to generate any SCRs before picking a card. The study clearly demonstrated that decision making is guided by emotional signals (or somatic states), which are generated in anticipation of future events.

Not surprisingly, since the 1980s emotion research has gained renewed attention in the marketing and advertising fields as a tool to predict consumer choice behaviour measures, such as purchase intent, brand choice and actual purchase (Poels and Dewitte, 2006). Marketing researchers often sit on innovation teams together with sensory scientists: marketing researchers are responsible for consumer insights, and sensory scientists are responsible for all consumer product insights (Lundahl, 2012). Not surprisingly, sensory scientists have also started to employ emotion research in guiding food product innovation.

Food and emotions are very much linked, even from the moment a parent first offers milk to comfort and quiet a child; food has then become a way of nourishing the soul as well as the body. Given the fundamental importance of food, there are surprisingly few genetically based constraints in humans, according to Rozin (1999), *"in humans (and rats), genetic factors include: (1) biases to prefer sweet tastes and to avoid bitter taste; (2) a tendency to be interested in new potential food (neophilia), but at the same time to be cautious about trying them (neophobia); and (3) some special abilities, that allow for learning the relationship between a food and the consequences of its ingestion, which may occur hours later."* Indeed, some recent studies have also highlighted the important role of emotions in influencing our decision making concerning food. For example, Laros and Steenkamp (2005) assessed consumer emotional response ($n = 645$ Dutch) towards different food types (genetically modified food, functional food, organic food and regular food). The study revealed that different food types elicited different emotional responses and might therefore influence consumer choice behaviour, e.g. genetically modified food elicited a strong association of risk and uncertainty, leading to feelings of fear and reducing the likelihood of purchase.

Furthermore, in these days of competitive and mature markets, the emotional quality of products is becoming increasingly important for differential advantage, especially when products within the same category are now often similar with respect to quality, price (Schifferstein *et al.*, 2013) and liking. In addition, emotions evoked by products also enhance the pleasure of buying, owning and using them (Hirschman and Holbrook, 1982). Packaging should help in making the product stand out from its competitors on the shelves (Schifferstein *et al.*, 2013), because it is known to affect how the food is perceived and experienced by suggesting a certain identity for its content (Cardello, 2007; Piqueras-Fiszman and Spence, 2012).

4.4 Approaches to measuring emotional response

Different approaches are used across many disciplines, including psychology, social science, health and nutrition, and consumer research to measure emotions. These can generally be divided into three categories: autonomic measures, brain imaging techniques and self-report measures (verbal/visual) (Mauss and Robinson, 2009).

The autonomic nervous system (ANS) is a general-purpose physiological system responsible for modulating peripheral functions (Kreibig, 2010). Autonomic measures rely on bodily reactions, e.g. heart rate, skin conductance, pupil dilation and facial expression. The autonomic measures are partially beyond an individual's conscious control, and therefore should overcome the cognitive bias that is sometimes linked to self-report measures (Poels and Dewitte, 2006). However, one of the downsides of autonomic measures is that they need to be taken in a very controlled environment, as physiological and neuronal responses are affected by external or internal stimuli present during the experience (e.g. light intensity changes) (Mauss and Robinson, 2009). In addition, the accuracy of autonomic measures in depicting emotions and quantifying emotional response is questionable. For example, what emotions are being measured? Or are they just a measure of the level of anxiety? Due to the challenges in data interpretation and implementation, the use of autonomic measures is not yet fully integrated in sensory and consumer research.

On the other hand, brain imaging techniques allow scientists to visualize the regions of the brain that are activated when stimuli are presented. There are several brain imaging techniques, including functional magnetic resonance imagery (fMRI) and electroencephalography (EEG) (Mauss and Robinson, 2009). fMRI has contributed significantly to progress in cognitive neuroscience and has entered consumer research focusing on emotional aspects and decision making (Mauss and Robinson, 2009). However, this method is not only extremely expensive, in that it requires special equipment and expert knowledge; it also needs to be conducted under a very restricted and controlled environment. For example, fMRI requires the subject to adopt a supine position; hence, it is possible for the subjects to swallow small amounts of liquid, but not to consume solid food (Hort *et al.*, 2008).

Although autonomic measures and brain imaging techniques provide direct evidence of emotional engagement, they are not articulate enough to describe what or how this emotional engagement has come about. Self-report measures have the advantage of being relatively cheap and simple, as no complex instruments or programmes are required, and subjects can be fairly specific and descriptive of emotional feelings. In general, there are two types of self-report measure: visual or verbal. In

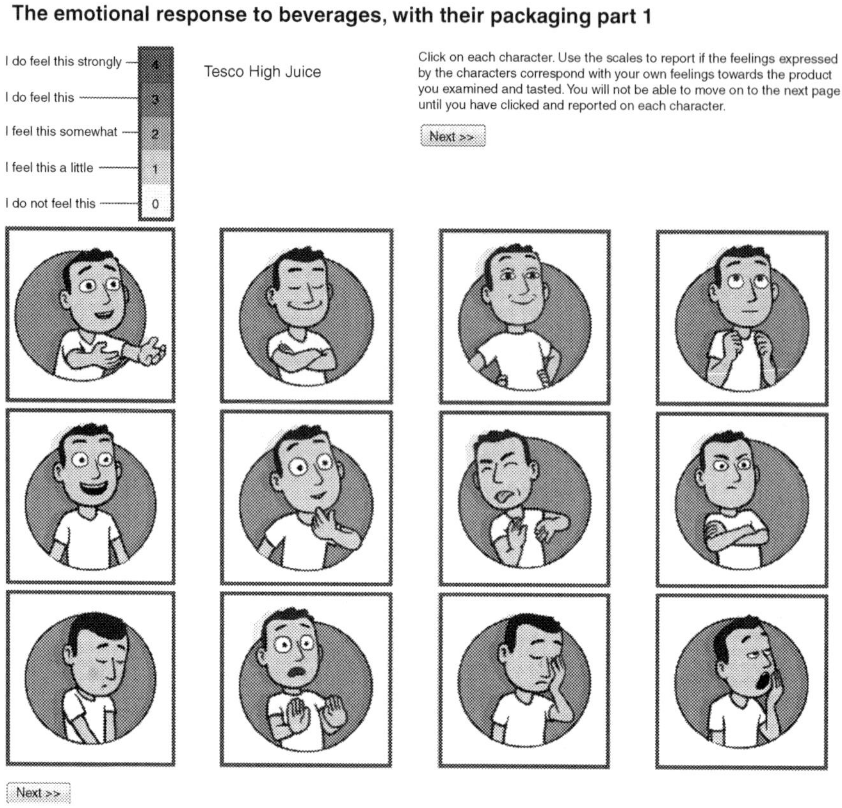

The emotional response to beverages, with their packaging part 1

I do feel this strongly	4
I do feel this	3
I feel this somewhat	2
I feel this a little	1
I do not feel this	0

Tesco High Juice

Click on each character. Use the scales to report if the feelings expressed by the characters correspond with your own feelings towards the product you examined and tasted. You will not be able to move on to the next page until you have clicked and reported on each character.

Next >>

Next >>

Figure 4.2 Using PrEmo, instead of relying on the use of words, respondents can report their emotions with the use of expressive cartoon animations (available under license via SusaGroup). More info on http://www.premotool.com.

visual self-report, subjects are asked to express their emotions by means of images or animation. Some examples of visual self-report measures include the Product Emotion Measurement Instrument (PrEmo) (Desmet *et al.*, 2000) and mood portraits (Churchill and Behan, 2010). The PrEmo programme consists of 12 different characters visually and audibly expressing six positive emotions (i.e. "desire," "satisfaction," "pride," "hope," "joy" and "fascination") and six negative emotions (i.e. "disgust," "dissatisfaction," "shame," "fear," "sadness" and "boredom") and subjects are asked to report their emotions with the use of expressive cartoon animations on a five-point scale from "I do not feel this" to "I do feel this strongly" in relation to a product or scenario (see Fig. 4.2).

In verbal self-report, subjects are asked to express their emotions verbally by means of open-ended questions, or to rate their emotions using Likert (or intensity) scales, CATA or best-worst-scaling (BWS). Unlike Likert scales, CATA questions allow subjects to simply check (or select) attributes that are relevant to them without having to be forced to rate all attributes on a scale. For BWS, subjects are presented with a set of either four or five words (quads or quins) and asked to choose the "best" as well as the "worst" words in terms of describing their emotions (Thomson *et al.*, 2010). Several researchers have also developed comprehensive emotion lexicons associated with consumption experiences (Clore *et al.*, 1987; Laros and Steenkamp, 2005; Chrea *et al.*, 2009; King and Meiselman, 2010; Thomson *et al.*, 2010) which will be discussed in more detail in the following section.

4.5 Verbal self-report emotion lexicon

Early verbal self-report emotion scales were developed for use in clinical psychiatry, e.g. the Profiles of Mood States (POMS) (McNair *et al.*, 1971). The POMS questionnaire asks subjects to rate 65 mood terms on a five-point scale measuring mood on six dimensions: tension-anxiety, depression-dejection, anger-hostility, vigour-activity, fatigue-inertia and confusion-bewilderment. Another mood questionnaire which is used extensively in clinical psychiatric settings is the Multiple Affect Adjective Check List (MAACL) (Zuckerman and Lubin, 1965), revised as the MAACL-R (Zuckerman and Lubin, 1985). It asks subjects to rate 135 mood terms using CATA methodology, and it measures mood on five dimensions: anxiety, depression, hostility, positive effect and sensation seeking.

However, as the emotion lexicons that were developed in the field of psychology do not focus on emotions experienced during product consumption, they are more applicable for clinical practice. Since the 1990s, many consumer researchers have also attempted to refine emotion terminology related to consumption experience. One key example of this is the consumption emotion set (CES), which was developed by Richins (1997) based on the work of Ortony *et al.* (1988). The CES questionnaire consists of 47 emotion terms divided into 17 categories (i.e. anger, discontent, worry, sadness, fear, shame, envy, loneliness, romantic love, love, peacefulness, contentment, optimism, joy, excitement and other items). Later in 2005, Laros and Steenkamp (2005) reviewed 173 negative emotions, 143 positive emotions and 39 basic emotions that were drawn from the literature and developed into a hierarchical model of consumer emotions. This model consists of three levels: the superordinate level with positive and negative effects; the basic level with four positive (i.e. contentment, happiness, love and pride) and four negative emotions (anger, fear, sadness and shame); and the subordinate level of specific emotions (Fig. 4.3).

They tested the structural model across different food types (i.e. genetically modified food, functional food, organic food and regular food) and revealed that, "positive and negative emotions" are the most frequently employed emotion dimension in the food consumption context. Following that, Desmet and Schifferstein (2008b) identified five main sources of positive and negative emotion related to food experience; i.e. sensory attributes, experienced consequences, anticipated consequences, personal or cultural meanings and actions of associated agents. In addition, they also showed that pleasant emotions were reported more often than unpleasant emotions in response to eating and tasting food.

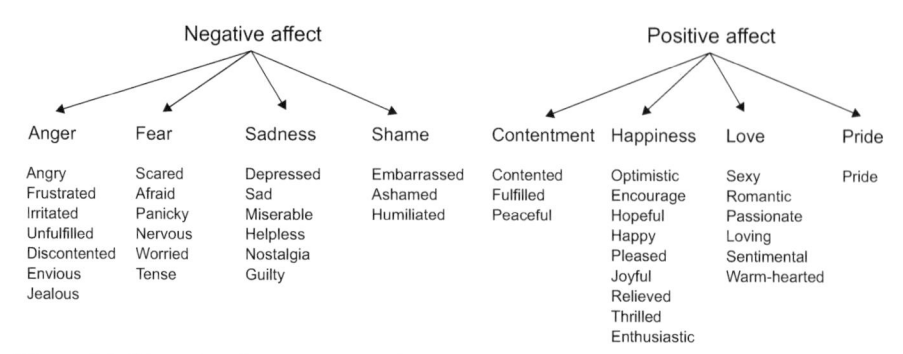

```
                Negative affect                              Positive affect

   Anger     Fear      Sadness    Shame        Contentment  Happiness   Love          Pride

   Angry     Scared    Depressed  Embarrassed  Contented    Optimistic  Sexy          Pride
   Frustrated Afraid   Sad        Ashamed      Fulfilled    Encourage   Romantic
   Irritated Panicky   Miserable  Humiliated   Peaceful     Hopeful     Passionate
   Unfulfilled Nervous Helpless                             Happy       Loving
   Discontented Worried Nostalgia                           Pleased     Sentimental
   Envious   Tense     Guilty                               Joyful      Warm-hearted
   Jealous                                                  Relieved
                                                            Thrilled
                                                            Enthusiastic
```

Figure 4.3 Hierarchy of consumer emotions.
Source: Adapted from Laros and Steenkamp (2005), Copyright © 2005, with permission from Elsevier.

4.6 Application of verbal self-report emotion techniques in the sensory and consumer field

In recent years, some verbal self-report emotion techniques have been developed in the sensory and consumer field. For example, King and Meiselman (2010) developed an emotion lexicon for EsSense Profile using adjectives from clinical psychiatry, POMS and MAACL questionnaires. Terms were validated based on a few criteria, such as frequency of use and consumer feedback, to ensure that they could be applied to a range of products. The final emotion lexicon for EsSense Profile consisted of 39 terms, which were classified as "positive," "negative" or "unclassified" (see Table 4.1). Terms were labelled as "unclassified" if more than 50% of the participants had rated them as neither positive nor negative. EsSense Profile incorporates emotion measures (five-point scales, anchored from "not at all" to "extremely") with measures of overall acceptability (nine-point scale, anchored from "dislike extremely" to "like extremely") in order to differentiate the liking and emotional responses among and within product categories. King and Meiselman (2010) have highlighted that emotion measures provide better discrimination than liking measures and can therefore provide a competitive advantage in the industry. The EsSense Profile was also validated using different food categories for its discriminating power; however, little published data are available in the literature to understand its application in a commercial context within a single product category.

However, Ng *et al*. (2013a) have argued that the list of emotional terms in the EsSense Profile is populated with mainly positive emotions and may miss emotions important to a product category, especially negative ones. Focusing mainly on positive terms can only tell us whether a person is generally having a positive experience. Work is then carried out to compare the effectiveness of the EsSense Profile to a consumer defined (CD) lexicon CATA approach in measuring consumers' emotions to commercial blackcurrant juice squash category (Ng *et al*., 2013a). Indeed, the authors have found that CD-CATA was somewhat more discriminating than the

Table 4.1 **Emotion lexicon for EsSense Profile**

Positive	Negative	Unclassified
Active	Bored	Aggressive
Adventurous	Disgusted	Daring
Affectionate	Worried	Eager
Calm		Guilty
Energetic		Mild
Enthusiastic		Polite
Free		Quiet
Friendly		Steady
Glad		Tame
Good		Understanding
Good-natured		Wild
Happy		
Interested		
Joyful		
Loving		
Merry		
Nostalgic		
Peaceful		
Pleasant		
Pleased		
Satisfied		
Secure		
Tender		
Warm		
Whole		

EsSense Profile, which is likely to be due to the better balance of positive and negative emotion terms, and the use of more focused consumer language relating to the product category. However, there are relative merits with each method. In terms of performing the experiments, the EsSense Profile was relatively easier in that it did not require the fairly labour intensive lexicon development stages, and was quicker and cheaper to perform. In addition, the quantitative results of the EsSense Profile readily lent themselves to conventional statistical analysis. However, the authors have reiterated the importance of incorporating the consumer's voice in emotion lexicon development and therefore proposed that a hybrid approach, and that an emotion lexicon combining consumer and published emotion lists would be needed for future emotional studies.

Chrea and colleagues (2009) developed the Geneva Emotion and Odour Scale (GEOS) questionnaire using adjectives from literature on emotions and on olfaction. Terms were validated based on a series of exploratory factor analyses of the data collected from consumers evaluating different odours. Terms were reduced from 480

Table 4.2 **List of emotion terms within six dimensions in the original and modified* GEOS questionnaire**

Dimension	Emotion terms
Pleasant feeling	Pleasant, wellbeing*, pleasantly surprise*, feeling awe, attracted, happiness*
Unpleasant feeling	Dirty, unpleasant, disgusted*, unpleasantly surprised*, dissatisfaction, sickening, irritated*, angry
Sensuality	Desire*, romantic*, sensual, in love*, excited, admiration, sexy
Relaxation	Relaxed*, soothed, reassured*, light, serene*
Refreshment	Revitalized, energetic*, refreshed, stimulated, invigorated*, shivering, clean*
Sensory pleasure	Nostalgic*, mouth-watering*, amusement*

*Terms that were kept for the modified GEOS questionnaire.

to 36, and were divided into six dimensions, i.e. pleasant feeling, unpleasant feeling, sensuality, relaxation, refreshment and sensory pleasure (see Table 4.2). GEOS was then modified and validated by Porcherot, Delphanque et al. (2010) whereby only the three most representative terms of each dimension (i.e. highest loadings derived from the factor analyses), and that were the most consensual (as measured with Cronbach's alpha), were selected. Instead of rating 36 terms, the modified GEOS questionnaire asks consumers to rate each of the six emotion dimensions, each dimension consisting of three terms (see terms marked with * in Table 4.2). The modified GEOS questionnaire has been applied to different perfumery and flavour products (n = 30–60) and the results have revealed that the most frequently used dimension was the "pleasant feeling," whereas the least used dimension was "unpleasant feeling" (Porcherot et al., 2010). Intriguingly, Ferdenzi et al. (2011) highlighted that emotion response to odours varies as a function of culture. The authors have developed two self-report scales, one in Liverpool (United Kingdom) and another in the city of Singapore, following the same procedure as used in the past to develop GEOS (Chrea et al., 2009). The authors named each questionnaire after the respective city: the Liverpool Emotion and Odour scale (LEOS) for Liverpool, and the Singapore Emotion and Odour Scale SEOS for Singapore. LEOS and SEOS questionnaires were found to generate a total of seven emotion dimensions, as opposed to six dimensions in GEOS. These included dimensions that were common across three cultures, i.e. "disgust," "happiness/wellbeing," "sensuality/desire" and "energy," and common to two European populations, i.e. "soothing/peacefulness." Dimensions that were culture-specific included: "sensory pleasure" for Geneva populations; "nostalgic" and "hunger thirst" for Liverpool; and "intellectual stimulation," "spirituality" and "negative feelings" for Singapore.

Whilst some lexicons are drawn from published literature, some researchers have developed emotion lexicons using consumer language. One example of this is a study conducted by Thomson, Crocker et al. (2010). Unlike previous emotion research, the authors also delved into more than just emotions. They believe that when consumers see a product, they do not just attach "emotions" to product characteristics, but also other "meanings," which they referred to as "*conceptualisations.*" These can be reduced into three broad categories: emotional (e.g. "will make me happy," "will calm me," "will annoy me"), abstract (e.g. "is sophisticated," "is trustworthy," "is feminine") and functional ("will refresh me," "will wash my clothes cleaner," "will kill germs"). Conceptual lexicons (24 words) were developed by a small group of reasonably articulate subjects who tasted and discussed the products under the guidance of a suitably qualified moderator. Subjects were then asked to rate their conceptual responses on nine sensorily differentiated UK commercial dark chocolates using BWS scales. Unlike other scales, BWS does not produce a score, so complex statistics may be needed for data analysis. For example, the most rigorous approach to analysing best-worst data is to model the probability that an individual will choose a particular best-worst pair over all other possible best-worst pairs (Thomson *et al.*, 2010). However, the BWS does provide an interesting way of visualizing the dataset by ranking emotions (Fig. 4.4). Figure 4.4 shows the basic conceptual profile of Cadbury's Bournville Deeply Dark, where conceptualizations (e.g. "sociable" and "easy-going" in particular), which scored the highest scale values (situated on the right side of the line), were the most prevalent with this chocolate. On the other hand, "arrogant" and "aggressive," scoring the lowest scale values (situated on the left side of the line), were associated least with this chocolate.

Following on the conceptual work conducted by Thomson *et al.* (2010), Ng *et al.* (2013b) explored the relative roles of sensory and packaging on consumer conceptualizations on commercial blackcurrant squashes. The results indicated that intrinsic sensory properties seem to have a stronger association with consumer liking and emotions, whereas extrinsic product characteristics seem to have a stronger association with abstract/functional conceptualization (Ng *et al.* 2013b). Figures 4.5a and 4.5b show the multi-factorial analysis (MFA) plots comparing the individual product maps obtained under the three different experimental conditions (i.e. (1) blind – consumers tasting the product blind (2) pack – consumers viewing the packaging, and finally (3) informed – consumers tasted and viewed the packaging concurrently).

Product configurations of the informed and blind conditions for liking and emotional conceptual profiles are closely aligned (Figs 4.5a and 4.5b), demonstrating that liking and emotional responses were influenced more by the sensory properties (blind condition) than the packaging cues of the products. For example, the aesthetic packaging of some products tested was found to evoke positive emotions, e.g. "interested." However, when consumed in the presence of the package, the products evoked negative emotions, e.g. "unpleasant surprise", demonstrating the power of sensory cues over expectations built through the packaging. Indeed, sensory properties of the products (blind condition) were found to evoke negative emotion, e.g. "unhappy."

On the other hand, product configurations of the informed and pack conditions for abstract/functional conceptual responses are closely aligned (Fig. 4.5c), demonstrating abstract/functional conceptual responses are more influenced by packaging cues. Indeed, during the lexicon development stage, over twice the numbers of abstract/functional conceptual terms were generated by packaging cues as compared to blind

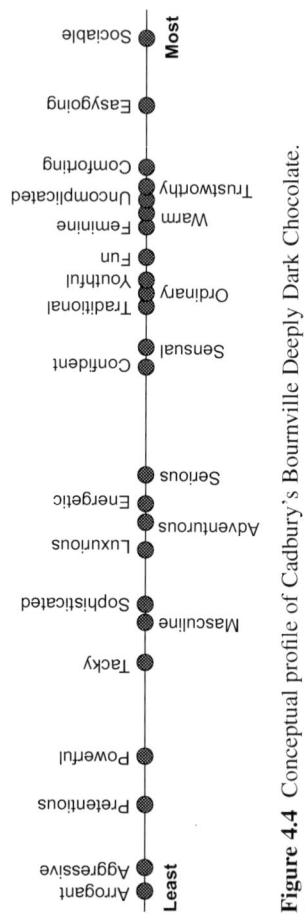

Figure 4.4 Conceptual profile of Cadbury's Bournville Deeply Dark Chocolate.
Source: Reprinted from Thomson *et al.* (2010), Copyright © 2010, with permission from Elsevier.

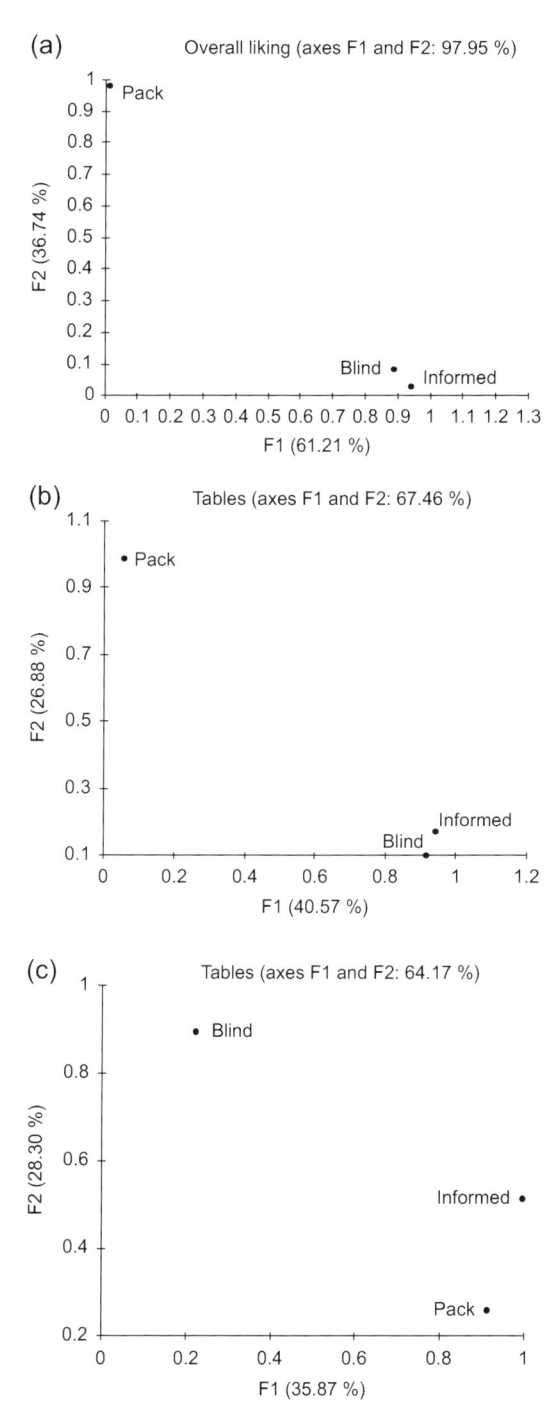

Figure 4.5 Representation of consumers' responses under three conditions considered in the first two dimension for (a) liking profiles, (b) emotional conceptual profiles, (c) abstract/functional conceptual profiles.

product assessment. Many abstract/ functional conceptual responses built from the packaging cues (e.g. "old fashioned" and "treat") were retained during the informed tasting, demonstrating that the sensory consumption experience of the products did not change many of the abstract/functional conceptual response. However, it is important to note that there were some conceptual terms (e.g. "natural") that were not retained during the informed tasting, and some of these terms seem to be influenced by sensory consumption experience of the products.

The findings of this study (Ng et al., 2013b) are fascinating, in that they could radically change the way we think about the role of sensory properties in delivering emotional impact, confirming previous findings that human senses are powerful elicitors of emotion (Gibson, 2006; Chrea, 2008; Thomson et al., 2010; Porcherot et al., 2012). In fact, touch, smell and taste are reported to be more closely connected with emotions, (Hinton and Henley, 1993), whereas vision and audition are sensory modalities that are suggested to be more closely connected to cognitive or rational thinking (Neisser, 1994). However, before generalizing these findings across all contexts, trials testing the comparative effects of intrinsic and extrinsic characteristics on conceptualizations in a more systematic manner are required, for example through conjoint studies varying sensory characteristics and aspects of packaging design.

At present, it is not clear whether one comprehensive list of emotions covers all food categories (King and Meiselman, 2010). Therefore, not surprisingly, different emotion or conceptual lexicons have been developed to measure emotion in response to the consumption experience. However, it is evident that sensory research is extending to emotion research, as emotional profiling could provide data beyond liking and, in some cases, products which were equally liked evoked different emotional profiles (e.g. King and Meiselman, 2010; Porcherot et al., 2010; Thomson et al., 2010; Ng et al., 2013a), which could lead to one product being more successful than others. However, what is key to the success of the product is being able to align the emotions projected from the product with other aspects of product, including the brand, packaging and sensory attributes. In fact, sensory attributes have been suggested to *"have the potential to communicate something of the emotionality and the functionality of the brand as well as adding distinctiveness to the brand's persona by adding a unique sensory signification"* (Thomson, 2007).

4.7 Relating sensory properties to consumers' emotional response

Lindstrom (2005) has illustrated how some product brands tune their sensory profiles to evoke emotions that best fit the brand's positioning, which could essentially help to increase consumer loyalty. Take the brands "Coke" and "Pepsi" for example: the brands differ in the way people describe their sensory profiles. Coke has been described by Coke drinkers as *"having a good blend of sweetness and sharpness,"* whereas Pepsi was described by Pepsi drinkers as being *"light sweetness, smooth, no bite or strong aftertaste."* As Lindstrom (2005) pointed out, although both sets of drinkers believe their brand to be equally distinctive, slightly more Coke drinkers agreed that they felt very positive about the taste of Coke than Pepsi drinkers did for Pepsi. Lindstrom (2005) believes that this could be due to the more challenging taste experience of Coke, which leads to a stronger emotional response in consumers. Consequently, understanding the relationship between sensory attributes and emotional responses may prove even more insightful than the traditional focus on sensory attributes and liking.

However, to the authors' knowledge, not many attempts can be found in the current sensory field to identify the relationship between sensory properties of food and consumer emotional response. Indeed, it might be commercially sensitive for companies to publish such findings. Thomson *et al.* (2010) were probably among the few in the current sensory arena who have attempted to demonstrate the relationship between sensory attributes and consumers' conceptualizations by identifying which of the sensory attributes in commercial chocolate, measured by quantitative descriptive analysis (QDA), evoked which conceptual response in consumers. For example, "cocoa" flavour (sensory attribute) was found to evoke "energetic" and "powerful" conceptualizations in consumers.

The Sensory Research Team at University of Nottingham has attempted to relate sensory properties of 11 commercial blackcurrant squash juices, measured by QDA as well as temporal dominance of sensations (TDS) (Ng *et al.*, 2012a), to consumer emotional response measured by EsSense method (Ng *et al.*, 2012b). TDS is one of the sensory techniques used to identify and quantify the intensity of dominant sensory properties over time. Not surprisingly, standard added-sugar squash juices were found to elicit positive emotions, and this was due to sensory properties "natural processed blackcurrant" and "natural sweetness" associated with the products. Indeed, the sweet taste of natural sugar was reported to elicit positive affect reaction to sensory pleasure (Steiner, 1973; Steiner *et al.*, 2001; Berridge, 2003) and therefore, not surprisingly, standard added-sugar products that were associated with positive emotions were also generally more preferred to no-added-sugar (NAS) products. However, interestingly, one of the NAS products was found to be similarly liked to other standard added-sugar products and this was due to its associated dominant sensory attribute of "minty" (as measured by TDS), as it was found to be strongly correlated with positive emotions (Ng *et al.*, 2012b). This suggests that, besides QDA, temporal information provided by TDS could be used to understand certain consumers' experiences.

Following on that, the Sensory Research Team at University of Nottingham has discovered some interesting relationships between some sensory properties in beer and consumer emotional response (Chaya *et al.*, under review). For example, increasing the carbonation level in beer, and decreasing the sweetness and alcohol content, were shown to influence a more pleasant and engaging emotional response in the batch of beer that was tested. Previous research has reported naturally sweetened solutions to be associated with positive emotions, e.g. happiness and surprise (Rousmans *et al.*, 2000; Berridge, 2003), but as bitterness is a dominant attribute in lager it is not surprising that sweetness is viewed differently in this product category.

Some of these studies have highlighted the need to relate consumer emotional response to sensory properties to gain better insight into developing and understanding the emotional signature of a product category. Therefore, emotional measurement should be used in conjunction with sensory analysis in order to understand how taste, olfactory and visual aspects of a product evoke subconscious feelings and emotions that ultimately drive hedonic measures (Ferrarini *et al.*, 2010) and choice behaviour.

4.8 Unresolved issues and topics for future research in verbal self-report emotion measurement

Emotion research is a new area of research for sensory and consumer researchers, and the impact is far reaching and there is still much to consider. Consequently, the potential for further work is considerable and general ideas for future research follow.

Many emotion lexicons published in sensory and consumer fields can be fairly lengthy, and there is a need to develop a rapid method of measuring emotion. The Sensory Research Team at University of Nottingham has done some work on comparing 12 emotion categories to a reduced emotion lexicon specific to the beer product category; and proved that even a reduced emotion lexicon could be used to highlight difference in consumers' emotional responses across cultures and between beers that are sensorily different (Eaton *et al.*, 2013). More research is needed to both develop and validate rapid methods for emotional response if they are to become common place in product development.

Very often, consumer testing of products is conducted under controlled laboratory conditions, but this setting does not represent how food and drink are consumed in reality. Food intake is usually immersed in social rituals and daily routines, and is also often related to such behaviour as preparing, consuming and sharing (Bourdieu, 1984). In addition, the whole concept of asking the question might also affect consumer responses, as it requires them to think about how they feel, instead of having them to respond at an emotional level that accurately reflects the emotional state at the time of the assessment. Interestingly, Hein *et al.* (2010) have recommended using a written scenario to evoke a consumption context in a laboratory setting. Indeed, they have reported that it was easier for subjects to indicate their product liking or disliking when using an evoked context. If this works for consumer liking, further work will be needed to gain awareness of the potential use of an evoked context in a controlled setting for eliciting consumer emotional responses. In addition, further work would also need to take account of physiological factors (e.g. hunger, satiety) that usually influence emotions and liking, and understand their impact on emotional ratings.

From a methodological perspective, an interesting question that was not investigated here is whether the positioning of the liking question at the beginning of the product evaluation influences the subsequent emotion profiles, and this warrants further investigation in a future study. In addition, one might also want to consider moving away from paper and pen to the use of more interactive and engaging data collection methods (e.g. smartphone app, tablets app, etc.) and to measure consumer emotional response in real time at point of purchase and/or consumption.

One of the issues in current emotion research is that different people have different psychological, cultural, memory and even social experiences, and therefore different emotions. Further research should focus on identifying consumer segments, considering both demographic (e.g. young versus old people, niche versus the mass market, gender) and non-demographic variables (e.g. lifestyles, occasion-based and need states). This information could yield valuable insights for exploring future target market and optimizing product positing. In addition, cross-cultural validation could be a topic of further research to explore similarities and differences with respect to emotional response of product between consumers across different countries. This is of particular interest to global companies who wish to develop a method that would work globally.

Emotions, although brief, are temporal and have an onset, duration and a short and long term end point (Lundahl, 2012), and can change. Further work needs to be considered concerning the potential use of temporal techniques to track dynamic changes in emotion over time. In addition, it is also important to consider that consumers may change their opinions and emotion of food products over longer periods of time. Researchers have shown that repeated exposure to familiar food leads to reduced liking for those foods and boredom (Porcherot and Issanchou, 1998; Rolls and Bell, 2006). Increased exposure to novel foods can lead to increased liking of the foods (Birch and Marlin, 1982). Therefore, it would be interesting to measure and monitor the dynamic changes in emotions over a number of exposures to the food.

There is still a long way to go before some of these issues are unravelled, especially as measuring a person's emotional state is one of the most vexing problems in affective science (Mauss and Robinson, 2009). However, it is clear that understanding the relationship between emotions and liking will provide industry with a much better understanding of consumer choice behaviour.

Human behaviour flows from three main sources:
desire, emotion, and knowledge.

Plato (Greek philosopher), 424/423 BC–348/347 BC

References

Bagozzi R.P., Gopinath M. and Nyer P.U. (1999) The role of emotions in marketing. *Journal of the Academy of Marketing Science* **27**:184–206.
Bechara A. (2004) The role of emotion in decision-making: Evidence from neurological patients with orbitofrontal damage. *Brain and Cognition* **55**:30–40.
Bechara A., Damasio A.R., Damasio H. and Anderson S.W. (1994) Insensitivity to future consequences following damage to human prefrontal cortex. *Cognition* **50**:7–15.
Berridge K.C. (2003) Pleasures of the brain. *Brain and Cognition* **52**:106–128.
Birch L. and Marlin D. (1982) I do not like it; I never tried it: effects of exposure on two-year old children's food preferences. *Appetite* **3**:353–360.

Bourdieu P. (1984) *Distinction: A Social Critique of the Judgment of Taste*. Cambridge: Harvard University Press.
Cardello A.V. (2007) Measuring consumer expectations to improve food product development. In: *Consumer-led Food Product Development*, pp. 223–261. H.J.H. MacFie (ed). Cambridge, UK: Woodhead Publishing Limited.
Chaya C., Pacoud J., Ng M. and Hort J. (under review) Measuring the emotional response to beer and the impact of sensory and packaging cues. *Journal of the American Society of Brewing Chemists*.
Chrea C., Grandjean D., Delplanque S., Cayeux I., Calve´ B.L., Aymard L., Velazco M.I., Sander D. and Scherer K.R. (2008) Mapping the semantic space for the subjective experience of emotional responses to odors. *Chemical Senses* **34**:49–62.
Churchill A. and Behan J. (2010) Comparison of methods used to study consumer emotions associated with fragrance. *Food Quality and Preference* **21**:1108–1113.
Clore G.L., Orthony A. and Foss M.A. (1987) Psychological foundations of the affective lexicon. *Journal of Personality and Social Psychology* **53** (4): 751–706.
Dalgleish T. (2004) The emotional brain. *Nature Reviews Neuroscience* **5**:582–585.
Damasio A. (1994) *Descartes' Error: Emotion, Reason, and the Human Brain*, Avon Books.
Damasio A. (1999) *The feeling of What Happens: Body and Emotion in the Making of Consciousness*. Orlando, Florida: Harcourt Brace.
Damasio A.R. (2006) *Descartes' Error: Emotion, Reason and the Human Brain*. London: Vintage.
Desmet P.M.A. and Schifferstein H.N.J. (2008a) Emotional influences on food choice: Sensory, physiological and psychological pathways. *Appetite* **50**:290–301.
Desmet P.M.A. and Schifferstein H.N.J. (2008b) Sources of positive and negative emotions in food experience. *Appetite* **50**:290–301.
Desmet P.M.A., Hekkert P. and Jacobs J.J. (2000) When a car makes you smile: Development and application of an instrument to measure product emotions. In: Hoch, S.J. and Meyer, R.J. (Eds.), *Advances in Consumer Research* **27**:111–117.
Eaton C., ChayaC., Hewson L., Fernandez Vazquez R., Ng M., Fernández-Ruizd V., Smart K A., Bealin-Kelly F., Fenton A. and Hort J. (2013) Happiness from hoppiness? Using consumers lexicons to compare emotional response between the UK and Spain. In *10th Pangborn Sensory Science Symposium*, Rio de Janeiro, Brazil.
Ferdenzi C., Schirmer A., Roberts S.C., Delplanque S., Porcherot C., Cayeux I., Velazco M.-I.s., Sander D., Scherer K.R. and Grandjean D. (2011) Affective dimensions of odor perception: A comparison between Swiss, British, and Singaporean populations. *Emotion* **11**:1168–1181. DOI: 10.1037/a0022853.
Ferrarini R., Carbognin C., Casarotti E.M., Nicolis E., Nencini A. and Meneghini A.M. (2010) The emotional response to wine consumption. *Food Quality and Preference* **21**:720–725.
Frijda N.H., Kuipers P. and ter Schure E. (1989) Relations among emotion, appraisal, and emotional action readiness. *Journal of Personality and Social Psychology* **57**:212–228. DOI: 10.1037/0022-3514.57.2.212.
Gibson E.L. (2006) Emotional influences on food choice: Sensory, physiological and psychological pathways. *Physiology & Behavior* **89**:53–61.
Hein K.A., Hamid N., Jaeger S.R. and Delahunty C.M. (2010) Application of a written scenario to evoke a consumption context in a laboratory setting: Effects on hedonic ratings. *Food Quality and Preference* **21**:410–416.
Hinton P.B. and Henley T.B. (1993). Cognitive and affective components of stimuli presented in three modes. *Bulletin of the Psychonomic Society*, **31**:595–598.

Hirschman E.C. and Holbrook M.B. (1982) Hedonic consumption: Emerging concepts, methods and propositions. *Journal of Marketing* **46**:92–101.

Hort J., Redureau S., Hollowood T., Marciani L., Eldeghaidy S., Head K., Busch J., Spiller R.C., Francis S., Gowland P.A. and Taylor A.J. (2008) The effect of body position on flavor release and perception: implications for fMRI studies. *Chemosensory Perception* **1**:253–257. DOI: 10.1007/s12078-008-9034-0.

Juslin P.N. and Vastfjall D. (2008) Emotional responses to music: the need to consider underlying mechanisms. *Behavioral and Brain Science* **31**:559–575.

King S.C. and Meiselman H.L. (2010) Development of a method to measure consumer emotions associated with foods. *Food Quality and Preference* **21**:168–177.

Koster E.P. (2009) Diversity in the determinants of food choice: A psychological perspective. *Food Quality and Preference* **20**:70–82.

Kreibig S.D. (2010) Autonomic nervous system activity in emotion: A review. *Biological Psychology* **84**:394–421.

Laros F.J.M. and Steenkamp J.-B.E.M. (2005) Emotions in consumer behavior: A hierarchical approach. *Journal of Business Research* **58**:1437–1445.

Larsen R.J. and Diener E. (1992) Promises and problems with the circumplex model of emotion. *Review of Personality and Social Psychology* **13**:25–59.

Lazarus R.S. (1991) *Emotion and Adaptation*. New York: Oxford University Press.

LeDoux J.E. (1996) *The Emotional Brain: The Mysterious Underpinnings of Emotional Life*. New York: Phoenix.

Lehrer J. (2006) Driven to market. *Nature* **443**:502–504.

Lindstrom M. (2005). *Brand Sense: Build Powerful Brands through Touch, Taste, Smell, Sight, and Sound*. New York: Free Press.

Lundahl D. (2012) *Breakthrough Food Product Innovation*. Elsevier.

Mauss I.B. and Robinson M.D. (2009) Measures of emotion: A review. *Cognition and Emotion* **23**:209–237.

McNair D.M., Lorr M. and Droppleman L.F. (1971) *Profile of Mood States*. San Diego: Educational and Industrial Testing Service.

Ng M., Chaya C. and Hort J. (2013a) Beyond liking: Comparing the measurement of emotional response using EsSense Profile and consumer defined check-all-that-apply methodologies. *Food Quality and Preference* **28**:193–205.

Ng M., Chaya C. and Hort J. (2013b) The influence of sensory and packaging cues on both liking and emotion, abstract and functional conceptualisations. *Food Quality and Preference* **29**:146–156.

Ng M., Lawlor J.B., Chandra S., Chaya C., Hewson L. and Hort J. (2012a) Using quantitative descriptive analysis and temporal dominance of sensations analysis as complementary methods for profiling commercial blackcurrant squashes. *Food Quality and Preference* **25**:121–134.

Ng M., Chaya C. and Hort, J. (2012b) Liking sensory attributes obtained from QDA and TDS to liking and emotional response. *Poster Presentation in 5th European Conference on Sensory and Consumer Research (Eurosense)*, Bern, Switzerland.

Ortony A., Clore G.L. and Collins A. (1988) *The Cognitive Structure of Emotions*. Cambridge, England: Cambridge University Press.

Panksepp J. (2003) Damasio's error?. *Conscious Emotion* **4**:111–134.

Piqueras-Fiszman B. and Spence C. (2012) The weight of the bottle as a possible extrinsic cue with which to estimate the price (and quality) of the wine? Observed correlations. *Food Quality and Preference* **25**:41–45.

Poels K. and Dewitte S. (2006) How to capture the heart? Reviewing 20 years of emotion measurement in advertising. *Journal of Advertising Research* **46**:18–37.

Porcherot C. and Issanchou S. (1998) Dynamics of liking for flavoured crackers: Test of predictive value of a boredom test. *Food Quality and Preference* **9**:21–29.

Porcherot C., Delplanque S., Planchais A., Gaudreau N., Accolla R. and Cayeux I. (2012) Influence of food odorant names on the verbal measurement of emotions. *Food Quality and Preference* **23**:125–133.

Porcherot C., Delplanque S., Raviot-Derrien S., Calvé B.L., Chrea C., Gaudreau N. and Cayeux I. (2010) How do you feel when you smell this? Optimization of a verbal measurement of odor-elicited emotions. *Food Quality and Preference* **21**:938–947.

Richins M.L. (1997) Measuring emotions in the consumption experience. *Journal of Consumer Research* **24**:127–146.

Rolls B. and Bell A. (2006) Effects of repeat consumption on pleasantness, preference and intake. *British Food Journal* **102**:507–521.

Rousmans, S., et al. (2000) Autonomic nervous system responses associated with primary tastes. *Chemical Senses* **25**: 709–718.

Rozin P. (1999) Food is fundamental, fun, frigtening, and far-reaching. *Social Research* **66**:9–30.

Russell J.A.(1980) A circumplex model of affect. *Journal of Personality and Social Psychology* **39**:1161–1178.

Russell J.A. and Barrett L.F.(1999) Core affect, prototypical emotional episodes, and other things called emotion: Dissecting the elephant. *Journal of Personality and Social Psychology* **76**(5): 805–819.

Scherer K.R., Schorr A. and Johnstone T. (2001) *Appraisal Processes in Emotion: Theory, Methods, Research*. New York: Oxford University Press.

Schifferstein H.N.J., Fenko A., Desmet P.M.A., Labbe D. and Martin N. (2013) Influence of package design on the dynamics of multisensory and emotional food experience. *Food Quality and Preference* **27**:18–25.

Shiv B. and Fedorikhin A. (1999) Heart and mind in conflict: The interplay of affect and cognition in consumer decision making. *Journal of Consumer Research* **26**:278–292.

Steiner J.E. (1973) The gustofacial response: Observation on normal and anencephalic newborn infants. *Symposium on Oral Sensation and Perception* **4**:254–278.

Steiner J.E., Glaser D., Hawilo M.E. and Berridge K.C. (2001) Comparative expression of hedonic impact: Affective reactions to taste by human infants and other primates. *Neuroscience and Biobehavioral Reviews* **25**:53–74.

Thomson D.M.H. (2007) SensoEmotional optimisation of food products and brands. H.J.H. Macfie (Ed.), *Consumer-Led Food Product Development*, Woodhead Publishing Limited, Cambridge, pp. 281–303.

Thomson D.M.H., Crocker C. and Marketo C.G. (2010) Linking sensory characteristics to emotions: An example using dark chocolate. *Food quality and Preference* **21**:1117–1125.

Walsh G., Shiu E., Hassan L.M., Michaelidou N. and Beatty S.E. (2011) Emotions, store-environmental cues, store-choice criteria, and marketing outcomes. *Journal of Business Research* **64**:737–744.

Watson D. and Tellegen A. (1985) Toward a consensual structure of mood. *Psychological Bulletin* **98**:219–235.

Zuckerman M. and Lubin B. (1965) *Manual for the Multiple Affect Adjective Check List*. San Diego:Educational and Industrial Testing Service.

Zuckerman M. and Lubin B. (1985) *The Multiple Affect Adjective Check List Revised*. San Diego: Educational and Industrial Testing Service.

Expedited procedures for conceptual profiling of brands, products and packaging

5

D.M.H. Thomson
MMR Research Worldwide, Wallingford, UK

5.1 Introduction

All objects, including brands, products and packaging, have perceptual (sensory) characteristics and conceptual associations. Together these determine how objects seem to us and how they impact on our feelings. Capturing and quantifying the conceptual associations that trigger the feelings that induce reward and subsequently motivate behaviour (conceptual profiling) provide a rich source of insight for guiding brand, product and packaging development.

This chapter begins by considering the role of reward in new product success or failure. It then explores the nature of concepts, the role played by conceptualisation in delivering reward and the importance of aligning the conceptual profiles of product and brand (brand–product consonance) to achieve fit-to-brand (or fit-to-concept). The relevance of fit-to-brand and its impact on reward is discussed.

An explanation as to why reward and fit-to-brand cannot (and should not) be measured directly is provided. A *derived* index of fit-to-brand, obtained by superimposing brand and product conceptual profiles, is proposed as an alternative.

The next part of the chapter describes the development of conceptual profiling as a process and explores various methodological options. Ideally, conceptual profiling would involve the development of a bespoke, category-specific conceptual lexicon and quantification via 2-way best-worst scaling (BWS). Strategies for expediting the process involving MMR's General Conceptual Lexicon® (as an alternative to a bespoke lexicon) and a quantification tool known as "Bullseye" (as an alternative to 2-way BWS), are described.

Finally, three case studies are presented that showcase aspects of these expedited procedures. The first demonstrates the use of MMR's General Conceptual Lexicon® to profile three UK "High Street" pharmacy and healthcare retailers. It also showcases the enhanced insights that these procedures can deliver. Case Study 2 demonstrates "Bullseye" in brand profiling and thereafter in obtaining a derived index of fit-to-brand in the Single Malt Scotch whisky (SMSW) category. Case Study 3 explores the relationship between our derived index of fit-to-brand (obtained via "Bullseye") versus actual and predicted sales volumes in an undisclosed savoury snack food category in the USA. This leads to the hypothesis that there may be a minimum threshold for fit-to-brand, below which commercial success in new products is unlikely.

The expedited procedures described herein provide a practical and cost effective means of conceptual profiling and consequently for obtaining a derived index of fit-to-brand, within the context of routine new product development.

5.2 Fundamentals of new product success and failure

It is widely recognised that most new products and almost all new brands are commercial failures. This is despite the fact that the form of the brand, product and pack will often have been developed and optimised using formal processes and the final launch decision will usually have been informed by research amongst target consumers.

A new product fails when the number of consumers motivated to buy it repeatedly is too small, and the frequency and volume of their consumption is too modest, to deliver profitable sales volumes. The fundamental cause of failure is therefore lack of motivation to try the new product and then buy it repeatedly.

Motivation is the force that drives us to attain our goals (Reiss, 2002; Higgins, 2009). Its influence may be conscious or non-conscious (Maslow, 1987; Ellis, 1995; Ellis and Newton, 2010). Obtaining reward is the ultimate goal that drives human behaviour (Olds, 1956; Berridge and Kringelbach, 2008; Gendolla and Brinkmann, 2009). Reward has many different manifestations. It may be positive or negative. When positive, reward motivates us to maintain or repeat the reward-triggering behaviour. Negative reward has the opposite effect.

Product purchase or other manifestations of product choice are reward-triggering behaviours. Consequently, it may be deduced that new products fail to gain initial trial when, consciously or otherwise, consumers predict that reward is unlikely to exceed a minimum threshold, when predicted reward is outweighed by the anticipated cost of acquisition, or when the new product is predicted to be less rewarding than incumbent products targeted at similar usage occasions. Post-trial, products fail to engender repeat purchasing when, consciously or otherwise, the magnitude of reward actually delivered on consumption falls short of expectation, when the new product fails to "out-reward" incumbent products targeted at similar usage occasions, or when the cost of obtaining the reward is deemed to be too great.

By the same rationale, products and brands that form an established part of consumers' repertoires must consistently deliver sufficient reward to allow them to retain this position. In addition, familiarity brings certainty of reward at a known cost. Conversely, unfamiliarity generally casts doubt over efficacy and therefore diminishes the certainty of reward and increases risk. Collectively, this means that it is fundamentally difficult for new products to displace incumbents from consumers' repertoires, even when the new product is highly regarded and much liked. This simple truth has huge implications for new product and new brand development:

- The probability of long-term financial success is intrinsically low.
- Being highly liked (as measured in research using a hedonic scale) is usually not enough to deliver success.

Conceptual profiling of brands, products and packaging 93

- Targeting new brands and products at established needs and desires brings a very low probability of success unless the ratio of benefit (reward) to cost is significantly higher than existing products.
- The greatest probability of success will come from identifying new needs and desires that are not addressed by existing products and targeting carefully optimised new brands and products at them.
- Attempting to "buy" success in a mature category by intensive advertising, by investing heavily to obtain enhanced levels of distribution, retail shelf space and in-store merchandising or by prolonged discounting, will also have a very low probability of long-term success if the new product fails to deliver enhanced levels of reward.
- For a new product to have any real chance of success, the product and the packaging must be aligned with the brand's personality and promise. This is known as consonance and it really matters! Failure to align these three key touch points creates undesirable tensions within the product (known as dissonance), which can have a really debilitating effect on new products.

From the foregoing, two "fundamentals" emerge. Firstly, delivering reward to consumers is fundamental to new product success. The more rewarding the new product is to more people, the greater the probability of success. Secondly, optimising product and pack to deliver the brand promise is also fundamental to product success. Conversely, in circumstances where new product development is product-led, branding and packaging should be developed or modified to align with the product. Either way, brands make promises to consumers that should be reinforced by the form and nature of the packaging and then delivered by the product. This is known as "fit-to-brand" (sometimes described as "fit-to-concept"). Generally, the better the "fit-to-brand," the greater the probability of new product success!

These two fundamentals are linked by virtue of the fact that brands have the potential to deliver emotional benefits to consumers, either directly or by promising desirable emotional and functional outcomes. Under certain circumstances, some consumers will find these benefits and outcomes rewarding. The more coherently and forcefully the branded product communicates its message, the greater the potential to reward and therefore the greater the motivation to try and repeatedly purchase the new product. In short, the better the fit-to-brand, the greater the potential to reward and, therefore, the greater the probability of new product success! This leads to the hypothesis that there may be a minimum threshold for fit-to-brand, below which new product success is unlikely. This threshold is considered later in the chapter. Before doing so, it is important to consider how reward and fit-to-brand might be captured and quantified.

5.3 Measurement using direct scaling

In consumer research, rapid methods often involve direct quantification using scales. The popularity of using scales is due, at least in part, to the notion that scales are apparently logical, to the observation that consumers seem to understand scales and find them easy to use, to the practical reality that scales can be incorporated into research questionnaires quickly, simply and cheaply, to the fact that scaling data can be extracted, summarised, analysed statistically and reported very easily (sometimes automatically), and to the simple truth that summarised data thus obtained are readily understood and easily communicated and disseminated by those charged with implementing research findings.

However, these apparent benefits may be misleading. The fundamental assumption underpinning the use of scales in consumer research is that the "external scale" (on-page or on-screen) corresponds directly to a quantifiable mental construction that exists in the mind of the individual concerned. Experience suggests that this assumption may hold true for liking. Degree of liking or disliking is something that comes immediately and readily to mind, for example, when we eat something. There are compelling evolutionary reasons for why this should be the case. Liking is easy to quantify internally and easy to communicate externally either verbally or on a scale. Unfortunately, this is the exception rather than the rule.

Purchase intent (or purchase interest) scales are used widely in consumer research in an attempt to gauge future purchasing, but often fail to predict longer-term purchasing behaviour unless the data are transformed and modelled. The reason for this lack of predictive validity is probably that consumers cannot bring to mind and consciously process all of the factors that determine their longer-term purchasing behaviour. Consequently, it is unlikely that a single, all-embracing mental construction could come to mind that relates directly to and provides an overview of all the influences that determine purchase intent. This means that purchase intent is likely to be fundamentally unquantifiable, and therefore consumers will inadvertently default to the nearest related phenomenon that does come to mind, probably liking!

As researchers, we need to be mindful that just because we can identify a phenomenon of interest (e.g. purchase intent), just because we can formulate a research question with an associated scale around this phenomenon, and just because consumers seem to understand the question and are able to respond using the scale, this does not necessarily mean that a corresponding mental construction should exist already or could be constructed via cognitive thought processing. This suggests that when designing questions and associated scales to incorporate into research questionnaires, we should always challenge the foregoing assumption and only use scales when the rationale for the existence of a corresponding internal construction is sufficiently strong, as it is with liking. There can, of course, never be certainties!

If this rationale is applied to the measurement of reward, there are compelling reasons for believing that consumers cannot realistically form the overview that would allow them to gauge reward as a single, quantifiable construction. This is largely because significant aspects of reward probably occur below the level of conscious awareness and therefore are likely to be inaccessible via cognitive thought processing. Moreover, there will also be circumstances when long periods of time elapse between the reward-triggering event and the ultimate delivery of positive or negative reward, making it difficult to associate cause and effect. This suggests that reward is likely to be fundamentally unquantifiable by direct questioning of consumers. Various alternatives to direct questioning have been proposed, including the coding and analysis of facial expressions (Hill, 2008) and the application of neuroscience tools to measure brain activity (Lindstrom, 2008). However, these methods have not been developed and validated to the point where they can be considered practical, proven research tools.

The case for and against measuring fit-to-brand directly using a scale is less clear cut, although it is generally recognised that unbranded products that are highly liked are invariably rated very positively for fit-to-brand when the branded concept is eventually disclosed, even when some degree of brand/product dissonance is apparent. This suggests that there is either a significant halo effect caused by the dominance of liking, or that fit-to-brand may not exist in the minds of consumers as a holistic, quantifiable construction and that respondents may simply default to liking.

Because of the likely importance of brand/product consonance in the success of new products and also because of the close phenomenological association between fit-to-brand and reward (as described above), a completely new approach for quantifying fit-to-brand has been developed within the author's organisation (MMR Research Worldwide). We describe this as a "derived index of fit-to-brand" because it avoids the problem of direct quantification and, of greater importance, it also avoids the a priori assumption that a mental construction that corresponds to fit-to-brand should exist in the mind of consumers and that this construction may be accessed via cognitive thought processing. This new approach builds on the principle that all objects have conceptual associations and therefore a conceptual profile (Thomson and Crocker, 2014), and that the degree of correlation between the conceptual profile of product (or packaging) and the corresponding brand provides a derived index of fit-to-brand.

The next sections of this chapter explore concepts, conceptualisation and conceptual profiling. This is then used as the basis for developing the idea of a derived index of fit-to-brand, which is illustrated using a number of case studies.

5.4 Concepts, conceptualisation and conceptual structure

This section describes the nature and role of concepts (Carey, 2009), sometimes referred to as implicit associations (Greenwald and Banaji, 1995; Greenwald *et al.*, 1995) or conceptual associations (Thomson, 2010), and how these might be elicited and quantified (Thomson *et al.*, 2010; Crocker and Thomson, 2014; Thomson and Crocker, 2014).

If behaviour is an output of mental activity then sensory stimulation is a triggering input. All the human senses function in essentially the same manner by taking energy in whatever form is intrinsic to the triggering stimulus and transducing this into patterns of electrical activity that are transmitted via neurons to various locations within the brain. Input analysers translate the consequent physiological activity into mental activity (Carey, 2009), much of which occurs below the level of conscious awareness (Ellis and Newton, 2010). Our first awareness of the triggering stimulus occurs when various mental representations come to mind. However, by the time this has happened, it is possible that we may already have been influenced unknowingly by associated non-conscious mental activity. The flight response that occurs in reaction to danger is an example. Kahneman (2003) referred to this type of mental activity as "System 1" thinking.

In modern psychology, the conscious aspect of object representation is typically discussed under the heading of perception. Historically, philosophers such as Duncker (1941) and Claws (1965) described three aspects of object representation: percept (what something is), concept (what it means to us) and affective reaction (how much pleasure or reward it brings). This deconstruction is helpful because it alerts us to the critical role of concepts, conceptualisation and affect in the process of creating mental representations and the effect that they may have upon us (Carey, 2009).

The distinction between percept and concept can be illustrated by considering any red object. Ignoring the form of the object, the main perceptual characteristic of the surface of the object is, by definition, redness. Closer scrutiny of the object might also reveal hints of yellowness, blueness, lightness–darkness, graininess–smoothness and glossiness–mattness. If so, these would be the perceptual or sensory characteristics of the image. Redness also has associated meaning (danger, anger, love, passion, luck, money, socialism, etc.). The nature and strength of a particular association depends on the nature of the object, the cultural background of the individual and the context. A red flag flying in the wind at the seaside would warn us of stormy seas, large waves, undercurrents and consequently, danger of drowning. This may evoke fear. In this context, danger is a concept and fear is an emotion. Experiencing fear is normally unrewarding, so we are motivated not to swim. More generally, merely thinking about an object (e.g. a red flag) can cue the associated meaning (drowning) automatically, which may trigger a particular behavioural outcome (the decision not to swim). Often this will happen unconsciously.

There are two aspects to the process of conceptualisation: the concepts (red flag and drowning), and the linkages that connect them (swimming and danger), known as intuitive theories (Carey, 2009). It is possible that other concepts may be embedded within this intuitive theory (e.g. undercurrents, large waves and the possibility of being swept out to sea). On this basis, the content of the mind may be likened to a multidimensional structure comprising a vast number of concepts interlinked via intuitive theories. We refer to this as the "conceptual architecture" of the mind. The fine definition of this mental structure is unique, although there may be commonalities based on heritable traits and common experiences across individuals. It may be modified via learning.

Brands are abstract objects comprising a cluster of related concepts interlinked via various intuitive theories. The number and nature of these concepts and the associated intuitive theories determine the brand's conceptual structure. A successful brand is likely to be rich in conceptual content that has the capacity to trigger rewarding emotional outcomes that determine behaviour.

In putting forward the proposition that all objects have conceptual content, we are tacitly suggesting that the conceptual structure of man-made creations could (and should) be a matter of design or, at the very least, that the conceptual structure of the

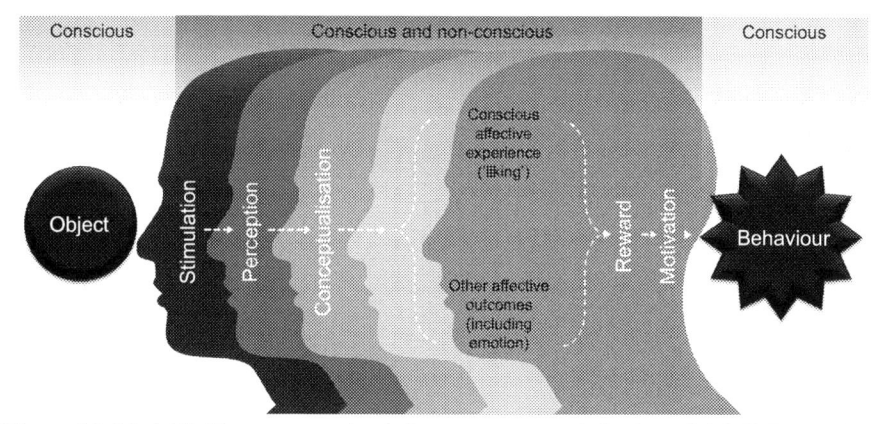

Figure 5.1 Model linking sensory stimulation to consequent behaviour. © MMR Research Worldwide.

creation should not be ignored. Whilst this is normally the case with the branding of consumer packaged goods, until recently it has rarely been the case with products and the form (shape and size) of the packaging. This is surprising because consistency of conceptual structure across these three primary touch points (which we call "conceptual consonance") should augment brand impact. Contradiction ("conceptual dissonance") would be expected to have the opposite effect. In order to create more holistic products, we require research tools for elucidating and quantifying key aspects of the conceptual structure of branding, packaging and products.

In summary, a psychological process is envisaged (Fig. 5.1) whereby physical stimulation triggers perception, conceptualisation and a conscious affective experience (liking), although not necessarily in that order. Conceptualisation triggers emotional outcomes, which may be positively or negatively rewarding. This aspect of reward may occur above or below the level of conscious awareness. It is the totality of net reward engendered via these different routes that ultimately drives behaviour.

5.5 Emotion profiling versus conceptual profiling – some theoretical considerations

In principle it should be possible to predict choice behaviour by measuring and then comparing or benchmarking the nature and extent of reward delivered by the options available to consumers in a particular choice context. However, as discussed previously, reward is fundamentally unmeasurable. As an alternative, it may be possible to learn something of consumers' likely reactions to products by elucidating and measuring the conceptualisations and emotional outcomes that engender reward (Fig. 5.1). To this end, product research tools have been developed for capturing and quantifying emotions (King and Meiselman, 2009; Cardello *et al.*, 2012; King *et al.*, 2013) and conceptualisations (Thomson *et al.*, 2010; Crocker and Thomson; Thomson and Crocker, 2014). Although these two approaches are essentially complementary, important phenomenological distinctions need to be drawn that could (and should) inform the researcher's choice of approach.

Notwithstanding the many uncertainties about what emotion is and how it should be defined (Frijda and Scherer, 2009; Thomson and Crocker, 2013), there is a consensus that an emotion is something experienced by an individual and that the emotion (or the state of action readiness that it triggers) is apparent when the event actually takes place. Emotions can be fleeting, transient and they are heavily dependent on psyche and circumstance, making it difficult to emulate the emotion of real life within research scenarios. Conceptual associations are retained in the memory so they are enduring, have some degree of permanence and pertain to the object rather than the individual. Although context will determine which particular concepts will be associated with a stimulating object, the process of conceptualisation and the existence of associated concepts in the memory of the individual are not contextually dependent. From a practical perspective the difference between emotion profiling and conceptual profiling might be likened to aiming at a fast moving and a static target, respectively.

Returning to the red flag example, danger (concept) and fear (emotion) may impact on the individual in three ways: (i) being in danger and experiencing fear; (ii) thinking about danger and experiencing fear; (iii) thinking about danger and thinking about fear without experiencing either. Whether or not the emotion is actually experienced depends on the strength of association between the emotion (fear), the concept (danger) and the stimulating object (red flag). In unbranded product research, where the strength of association between product, concept and emotion will often be weak, the probability of triggering an emotional outcome is likely to be low, yet research participants may still be able to think about associated emotions without actually experiencing them. If so, the reported emotion would be an "emotion concept" held within the conceptual architecture that forms part of that individual's memory. It follows that emotion profiling (King and Meiselman, 2009) will probably capture a mixture of emotions actually experienced and emotion concepts. Conceptual profiling focuses solely on conceptual associations but extends this beyond emotion concepts to include abstract concepts (e.g. masculine, feminine, trustworthy, traditional, prestigious) and functional concepts (e.g. healthy, energising, sanitising).

The process of defining and quantifying the conceptual content of an object, and its practical application, was first described as "conceptual profiling" by Thomson and co-workers (Thomson, 2010; Thomson *et al.*, 2010). By analogy with sensory profiling, the conceptual profile of an object may be defined as the degree of association of a series of conceptual descriptors with that object. Conceptual profiling has much in common with brand personality profiling (Aaker, 1997) in which brands are ascribed human-like personality traits. Personality traits can be considered as conceptual associations, although in many instances the "personality" terms extend well beyond what might be construed as personality traits to encompass various emotional, abstract and functional conceptualisations.

5.6 Conceptual profiling in practice

The practical considerations involved in conceptual profiling and the associated analysis tools developed by the author and his colleagues at MMR Research Worldwide are described in detail in a suite of recent, peer-reviewed scientific publications (Thomson, *et al.*, 2010; Crocker and Thomson, 2014; Thomson and Crocker, 2014).

For practical purposes, there are two aspects to conceptual profiling that need to be considered separately: concept description and quantification of degree of conceptual association.

5.6.1 Concept description – use of words and the issue of counter-intuitiveness

The object of conceptual profiling is to access, identify and quantify the conceptualisations that we associate with products, branding and packaging, including those less-than-obvious concepts that are held deep within the conceptual architecture of the mind (sometimes below the level of conscious awareness), and also those concepts that might seem somewhat counter-intuitive if we were to stop and think about them too much!

Using words to elucidate and label conceptualisations is fundamental to the process of conceptual profiling as developed by Thomson and co-workers. This is based on the principle that language evolved as a means of communicating our thoughts to each other (Chomsky, 1967) and therefore it is highly likely that as part of this process, words will have evolved to describe the concepts that are so important in describing the objects, beings, events and other creations that define life as we know it. Nonetheless, the use of words to label concepts has been criticised because of the inference that word-concept linkages are inherently rational (Penn, 2006) and therefore preclude the possibility of seemingly counter-intuitive associations.

This issue of apparent counter-intuitiveness is important and should be addressed. Consider, for example, that in many cultures milk chocolate is a comfort food. One of the key conceptualisations associated with milk chocolate is trustworthiness (Thomson *et al.*, 2010), in much the same way that an old friend would be trustworthy. Trustworthiness delivers the feeling of being comforted (emotion), which can be rewarding in certain circumstances. In spite of this clear association, it would be inappropriate to provide a research participant with a piece of milk chocolate and ask them to rate it for trustworthiness on a scale of 1–100, because the question would seem to be counter-intuitive! However, the problem does not lie with word association but with the form of questioning and the use of classical forms of direct scaling. This issue is revisited in Subsection 5.6.3.

5.6.2 Concept description – developing a conceptual lexicon

The principle underpinning development of a conceptual lexicon is broadly similar to that applied to a sensory lexicon, in that it should be comprehensive enough to fully describe and discriminate amongst the range of objects being investigated whilst avoiding replication and redundancy. Irrespective of whether conceptual profiling will be applied to fully branded products, unbranded products or packaging, lexicon development usually starts with a comprehensive, desk-based review of the key brands in the immediate and adjacent categories. Print, TV and web advertising, on-pack copy, websites and all other product and category narratives are thoroughly scrutinised to establish a draft lexicon of conceptual terms. Depending on the category, this could amount to well over 100 terms and would include emotional, abstract and functional conceptualisations. This is usually followed by several iterations of qualitative discussion where target consumers, existing category users and users of specific brands interact with unbranded products, unbranded packaging (to explore the influence of form) and fully branded products, to generate a list of conceptual terms from scratch. Towards the end of the discussion group, the draft lexicon of terms obtained via the earlier phase of desk research is introduced so that it can be integrated and reconciled with the respondents' own list. The task of the qualitative moderator is to encourage participants to remove any perceptual (sensory) and affective terms, to segregate the emotional and abstract terms from functional terms (thereby creating two lexicons), resolve any disputes, remove obvious replication and redundancy, and challenge any doubtful terms.

If conceptual profiling is to be conducted exclusively on unbranded products (which is often the case within our organisation), it is extremely important that lexicon development should focus on conceptualisations associated with between-product sensory differences (*sensory-specific effect*) as well as those that characterise the category as a whole (*category effect*). Experience suggests that this can be achieved most effectively by involving category users that have been pre-screened for sensory acuity, articulacy, communicativeness and genuine interest in the product. Specialist sensory moderators trained in conceptual profiling are normally required. In circumstances where it proves too difficult to reduce the conceptual lexicon down to a practical number of terms via qualitative discussion, quantitative data driven techniques, usually involving the check-all-that-apply (CATA) process and cluster analysis, can often be used to great effect.

Vocabulary development will typically yield lexicons of 20–30 emotional and abstract terms and 15–25 functional terms, depending on the product category. Although the actual number of terms in a lexicon is something that should be determined and not predefined, subsequent experimental design is made much easier if the number of terms is highly factorable and prime numbers are avoided. It is important that the conceptual lexicon should include relevant negative terms as brands and products are often differentiated on the basis of what they are not!

Experience obtained working in English, Mandarin, Japanese and most of the major European languages suggests that this vocabulary development procedure works equally well across all languages and cultures.

Development of a conceptual lexicon is highly category-specific and requires investment of time and resource. However, once developed, and assuming that this has been done thoroughly, the conceptual lexicon becomes an asset that can be applied repeatedly and rapidly without the need for further development.

By way of example, the conceptual lexicon developed by MMR Research Worldwide for the SMSW category is shown in Fig. 5.2 (emotion and abstract

Conceptual profiling of brands, products and packaging

Key 10		Other 15
Classy Comforting Distinctive Genuine Sophisticated Traditional Trustworthy	Arrogant Boring Cheap	Cheerful Complex Conservative Contemporary Deceptive Dominating Feminine Free-spirited Friendly Intriguing Masculine Sensual Serious Simple Youthful
Must haves for single malt Scotch whisky	Must not haves for single malt Scotch whisky	

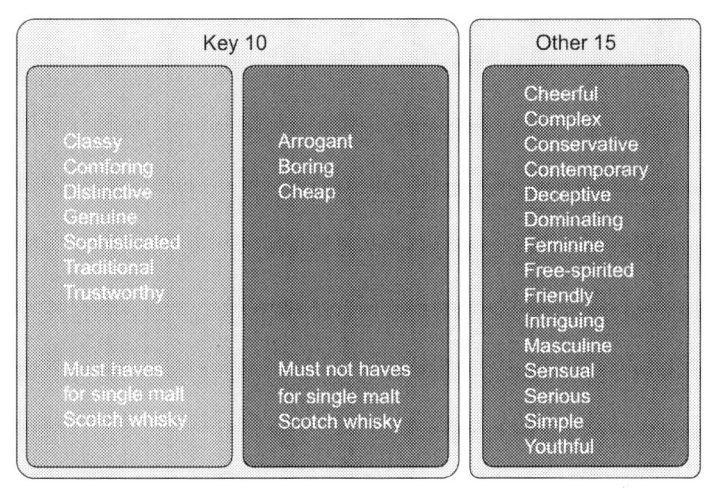

Figure 5.2 Conceptual lexicon (emotion and abstract terms only) for the single malt Scotch whisky category. © MMR Research Worldwide.

concepts only). This was developed across branding, packaging (form) and product, as described above, using a combination of moderated qualitative discussion and CATA with subsequent cluster analysis to make the final selection. The SMSW lexicon comprises 25 conceptual terms: seven terms that are fundamental "must haves" for all brands and products in the SMSW category (classy, comforting, distinctive, genuine, sophisticated, traditional and trustworthy), three terms that are fundamental "must not haves" for the category (arrogant, boring and cheap) and the remaining 15 terms that are the basis on which brands and products within the category are differentiated.

This lexicon was developed on behalf of Annandale Distillery Company Limited, owners of a historic distillery in the south of Scotland that will re-start production in the autumn of 2014 after a gap of 95 years. Over and above the practical difficulties involved in bringing a derelict distillery back into production, the company also faced the challenge of developing meaningful branding for their two expressions of SMSW that would be potent enough to differentiate them from the 100 or so other distilleries in Scotland that also produce SMSW. Although the outcome of this development work must remain confidential for the time being, simply knowing the nature of the 15 conceptualisations that they could work with, and also knowing which conceptualisations must be preserved and augmented and which must be avoided, proved enormously beneficial to the brand development team. It also meant that at various stages in the development of branding and packaging, design concepts could be tested rapidly and inexpensively using conceptual profiling. This provided valuable feedback that helped the designers and developers to stay on-brief at every stage of the process.

An application of this conceptual lexicon, as used to obtain a derived index of fit-to-brand for two branded SMSW products, is presented as a case study later in this chapter.

5.6.3 Quantification

Profiling of emotion and abstract conceptualisations is normally conducted separately from functional conceptualisations. This is especially the case with unbranded products, where the association between product and emotion/abstract conceptualisations is often much weaker (less obvious) than it is for functional conceptualisations. In practice, this means that functional conceptualisations are more readily accessible via cognitive thought processing and therefore will inevitably dominate and overwhelm the emotional/abstract conceptualisations. We have found that functional conceptualisations can be readily quantified using CATA, whereas emotion/abstract conceptualisations usually require more sophisticated methodologies. The remainder of this study focuses on emotion/abstract conceptualisations.

With conceptual profiling, the measurement processes must be able to accommodate the possibility of apparent counter-intuitiveness. As described previously (Section 5.6.1), milk chocolate is often thought of as a comfort food and trustworthiness is an important conceptualisation that is associated with comforting (Thomson *et al.*, 2010), yet it would be utterly counter-intuitive to invite respondents to taste an unbranded sample of milk chocolate and rate it for trustworthiness on an external scale. In other words, direct scaling usually invokes rational thought processing.

BWS, also known as maximum difference scaling, is an indirect method of scaling that has been found to be particularly useful for measurement of "soft" or abstract attributes that may be counter-intuitive or otherwise are not easily quantified (Flynn *et al.*, 2007; Lee *et al.*, 2008). Respondents are required to choose one item or attribute that they think is the best/largest/most in respect of some property x and one that is the worst/smallest/least of the same property from a series of sets that contain different combinations drawn from a larger master set of items. In the context of conceptual profiling, the items are conceptual descriptors (words) and the property x is the degree of association of a conceptual descriptor with the object. In a typical conceptual profiling study, respondents are presented with the object to be profiled (brand, unbranded product, pack, etc.) and a sequence of sets of four or five conceptual descriptors drawn from a lexicon of terms developed for the product or category. For each set of four or five terms they are asked to choose the terms most and least strongly associated with the object. The choice data can be analysed in various ways to yield a set of values, often called scale values or utilities, on a difference scale (Suppes and Zinnes, 1963) representing the underlying dimension of interest (Finn and Louviere, 1992; Marley and Louviere, 2005).

The application of BWS to conceptual profiling and the associated data analysis procedures are described in detail by Thomson *et al.* (2010) and by Thomson and Crocker (2014). Subsequent to this, we have developed a variant of BWS where the best-worst exercise described above is augmented by a "calibration" best-worst exercise in which the roles of the conceptual terms and objects are reversed. Referred to as 2-way best-worst scaling, this procedure represents a major methodological advancement because it overcomes the problem of relativism in BWS, thereby facilitating direct comparison of the conceptual profiles of different objects within the same study (Crocker and Thomson, 2014); something that is otherwise precluded and has hitherto restricted the usefulness of BWS.

5.6.4 Rapid methods for conceptual profiling

It is our view that 2-way BWS, using a bespoke, category-specific conceptual lexicon, is the most sensitive and precise tool available currently for conceptual profiling and, given the option, we would choose this approach in preference to all others in most circumstances.

However, we are mindful that the development of a bespoke, category-specific conceptual lexicon for each study can be quite time consuming and obviously there are associated costs. With this in mind, we have reviewed the many conceptual profiling studies conducted by MMR in the English language over the past 5 years and have identified a shortlist of 30 conceptual terms (Fig. 5.3), selected intuitively on the basis of frequency of use, general applicability and discriminatory power. We have found that this is generally adequate for initial screening, but otherwise advocate the development of bespoke, category-specific conceptual lexicons, especially for new product or brand development purposes. An application of MMR's General Conceptual Lexicon® is presented in Case Study 1.

The practicalities of 2-way BWS dictate that it is best implemented as a standalone study or as a discrete part of a larger study. Again, we are mindful that this is not always compatible with rapid and low-budget product evaluation. To this end, we have developed and validated an alternative measurement tool that we refer to as "Bullseye". This is an online implementation that was originally developed within our organisation to make continuous rating scales more engaging in online surveys. Respondents drag and drop a sequence of conceptual terms, drawn

1. Adventurous	16. Happy
2. Aggressive	17. Inspiring
3. Arrogant	18. Irritating
4. Boring	19. Masculine
5. Carefree	20. Modern
6. Cheap	21. Powerful
7. Classy	22. Pretentious
8. Comforting	23. Sensual
9. Confident	24. Serious
10. Easy-going	25. Simple
11. Energetic	26. Sophisticated
12. Feminine	27. Traditional
13. Friendly	28. Trustworthy
14. Fun	29. Unique
15. Genuine	30. Youthful

Figure 5.3 MMR's General Conceptual Lexicon®. © MMR Research Worldwide.

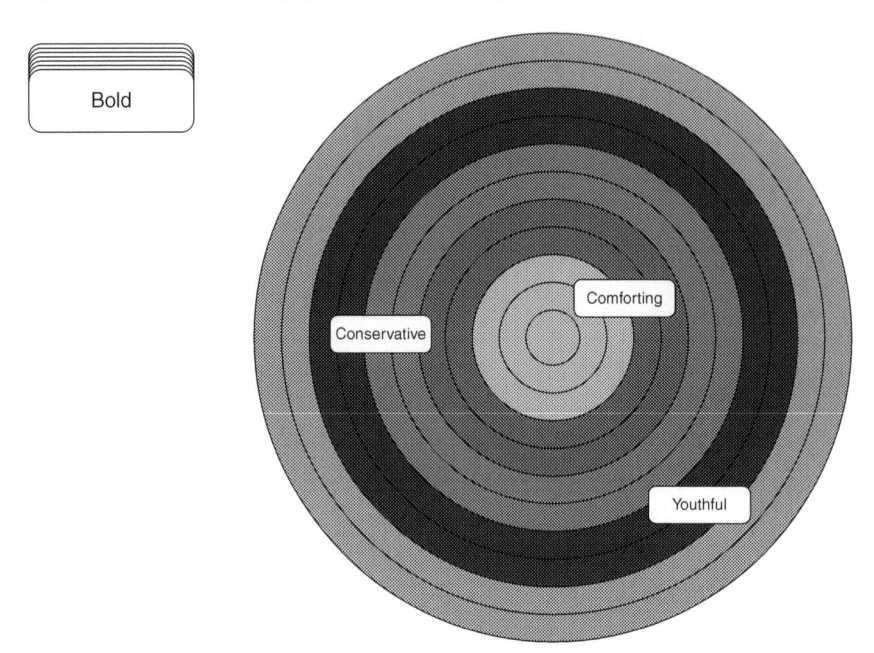

Figure 5.4 Bullseye conceptual profiling interface (coloured on computer screen). © MMR Research Worldwide.

from the conceptual lexicon and presented according to a rotated design, onto an image of concentric circles resembling an archery target (Fig. 5.4). We consider that placing a word on a target requires a fairly low level of conscious cognitive processing and therefore it goes some way to fostering an intuitive rather than a rational response. The distance from the centre of the target is taken as the strength of the association between the conceptual term and the object being profiled. There are two variants of the technique: "disappearing Bullseye," in which the target is cleared before the next term is placed; and "non-disappearing Bullseye" in which the terms are left on the target. The cards are smaller than shown in Fig. 5.4 and in the "non-disappearing" variant can be superimposed. The latter variant tends to give less "noisy" data (i.e. lower standard errors), presumably because the retained terms act as a frame of reference for the subsequent terms. This method and the associated data analysis procedures are described in detail by Thomson and Crocker (2014).

MMR has used Bullseye widely over the past few years as part of its Brandphonics® process. Whilst it is undoubtedly less discriminatory than 2-way BWS, with thoughtful application it is sufficiently discriminatory for many purposes and is highly compatible with the requirement for speed. We often combine Bullseye and BWS within a single study, where the former is used for conceptual profiling of brands and the latter for conceptual profiling of unbranded products where the conceptual content is less accessible. These data are then used to obtain a derived index of fit-to-brand. Various applications of Bullseye are presented in Case Studies 2 and 3.

5.7 Applications and case studies

Three case studies are presented in this section. The first two studies describe the application of conceptual profiling to "High Street" pharmacy retailer brands and to the development of a derived index of fit-to-brand in the Single Malt Scotch Whisky category. The third study explores the relationship between our derived index of fit-to-brand and new product sales volumes.

5.7.1 Case Study 1 – Expedited conceptual profiling of three UK "High Street" pharmacy retailers using MMR's General Conceptual Lexicon®

(Although this case study was undertaken for MMR's own "internal" purposes, the names of the retailers are not disclosed.)

This case study demonstrates MMR's General Conceptual Lexicon® (Fig. 5.3). It is used in conjunction with 2-way BWS (our preferred quantification tool) rather than "Bullseye" (our expedited quantification tool) simply because there is no real time or cost saving when less than four objects are profiled.

The research was conducted amongst 300 UK residents who regularly make purchases from High Street pharmacy/healthcare retailers. All had recent experience of two or more of the retailers in question and awareness of the third. The research was conducted online and the time taken from inception to completion of the initial analysis was approximately 1 week.

There are two parts to 2-way BWS (Crocker and Thomson, 2014). In the first phase, known as calibration, the logos of the three retailers were presented simultaneously on-screen along with one of the words, drawn from the 30-term conceptual lexicon according to a rotated design. The respondent had to indicate the retailer to whom this term seemed most applicable followed by the retailer to whom it was least applicable. In this particular case, the calibration exercise was repeated for all 30 conceptual terms per respondent although an incomplete design using a subset of about 25–30% of the terms will often suffice. In the second phase, which is known as the profiling exercise and was conducted immediately after calibration, the process is reversed so that respondent sees just one of the retailer's names along with five conceptual terms. The respondent had to indicate which of the five terms they associate most closely with this retailer and of the remaining four terms, which is least closely associated. This was repeated five more times so that all respondents saw all 30 conceptual terms in various combinations of five for this particular retailer. The order of presentation of the terms and the order of terms within the sets of five were rotated. Each respondent profiled all three retailer brands. The order of presentation of the retailers was also rotated across respondents. Most respondents were able to complete the entire task in 20–30 min. As part of our methodological development for conceptual profiling, we have determined that data quality is maintained for up to three objects but tends to deteriorate thereafter (although this depends to a large extent on the nature of the object).

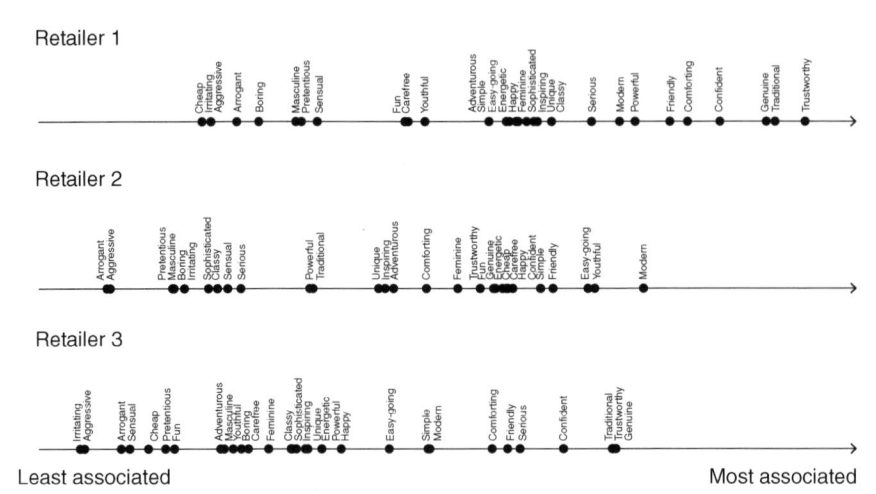

Figure 5.5 Conceptual profiles of 3 UK "High Street" pharmacy/healthcare retailers.
© MMR Research Worldwide.

The choice data for the profiling exercise were analysed using a multinomial logit (MNL), as described by Thomson *et al.* (2010). The output is a set of scale values, one for each conceptual term for each retailer. Within each retailer, the scale values for the conceptual terms were then transformed using a unique rescaling factor derived from the calibration data according to the procedure described by Crocker and Thomson (2014). The rescaled data for each of the three retailers were plotted on a common difference scale, thereby facilitating direct comparison of the scale values for the conceptual terms across retailers (Fig. 5.5).

There are three factors that need to be considered when interpreting conceptual profiles: (i) the hierarchy of the conceptual terms; (ii) the absolute magnitude of the scale values of the conceptual terms; and (iii) the spread of the conceptual terms from the most to the least associated – known as the conceptual footprint. The size of the conceptual footprint is determined by two factors; the strength of association of the conceptual terms with the object (retailer in this case), and the degree of consensus amongst respondents. In this regard, Retailer 1 has a larger conceptual footprint than the other two, suggesting greater distinctiveness.

Within Retailer 1, the conceptualisations form a clear hierarchy. The most strongly associated terms are *trustworthy, traditional* and *genuine*, followed by *confident, comforting* and *friendly* and then *powerful, modern* and *serious*. The conceptualisations least associated with Retailer 1 are *cheap, irritating* and *aggressive*. The hierarchy of most and least associated conceptualisations for Retailer 3 is broadly similar to Retailer 1, although the absolute magnitudes of the most strongly associated terms are lower for Retailer 3. This suggests that these two retailer brands are conceptualised very similarly but the strength of association and hence the strength of the brand for Retailer 3 is significantly less. Whilst this obviously represents a commanding position for Retailer 1, the lack of qualitative differentiation from Retailer 3 should perhaps be of some concern to both parties. Moreover, although *irritating, aggressive,*

Conceptual profiling of brands, products and packaging 107

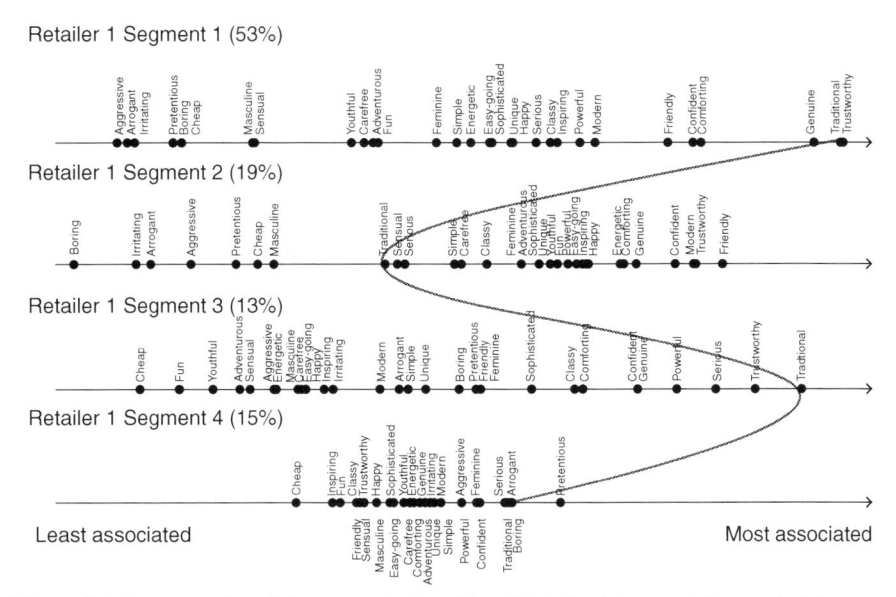

Figure 5.6 Segmentation of the conceptual profile of Retailer 1 to reveal four underlying segments. © MMR Research Worldwide.

arrogant and *pretentious* are the conceptualisations least associated with Retailer 1 (along with *cheap* and *boring*), the absolute values of these terms are greater than for Retailer 3, suggesting perhaps that there may be "issues" with Retailer 1, for some people at least. This is investigated further, below.

The hierarchy of conceptualisations associated with Retailer 2 is very different from the other two. The most strongly associated terms are *modern* followed by *youthful* and *easygoing*, then *friendly, simple, confident, happy, carefree* and *cheap*. In many respects, this is the antithesis of Retailers 1 and 3, thereby positioning Retailer 2 very differently. Although Retailers 1 and 3 currently command a very large share of this particular market, the cheap, modern and rather "happy-go-lucky" persona of Retailer 2 could prove to be a threat, especially amongst those consumers who may be looking for something different.

Just as people can have very different views about what they like or dislike, consumers may also conceptualise objects differently. However, with branding, the object is to have a very single-minded and distinctive conceptual profile that differentiates it from competitors and motivates consumers to buy (preferably at a premium). This requires single-mindedness in positioning, marketing and advertising the brand. As a consequence, a well-executed brand strategy should bring a considerable degree of homogeneity in how a particular brand is conceptualised amongst target consumers. The same rationale obviously applies to retailer brands.

By way of example, a form of intrinsic segmentation (using latent class analysis) was applied to the conceptual profile of Retailer 1 in order to explore conceptual homogeneity. This process yielded four underlying segments (Fig. 5.6); one very large segment (Segment 1), comprising 53% of the respondents, and three smaller segments. Unsurprisingly, in view of its dominance, the conceptual profile of Retailer 1 within Segment 1 resembles that of the "total sample" (Fig. 5.5), except that the conceptual footprint is much larger, reflecting greater homogeneity of opinion due to extraction of the other three, contradictory segments.

The conceptual profile of Retailer 1 in Segment 3 (13%) differs from Segment 1 in that there is greater emphasis on *serious* and *powerful* and less on *comforting* and *friendly*. However, of greater significance, Segment 3 associates negative conceptualisations such as *pretentious, boring, arrogant, aggressive* and *irritating* quite strongly with Retailer 1. This goes some way to explain the earlier observation across all respondents (Fig. 5.5) that the absolute scale values for *pretentious, aggressive* and *arrogant* are higher for Retailer 1 than they are for Retailer 3, in particular.

Respondents in Segment 2 (19%) have a very different opinion of Retailer 1, conceptualising it as *friendly* and *modern*. Segment 4 respondents conceptualise Retailer 1 very negatively and otherwise seem very disinterested and disengaged with the category.

Returning to the issue of conceptual homogeneity for the "Retailer 1 brand," it is fairly clear that respondents in Segments 1 and 3, amounting to 66% of this population, share the view that Retailer 1 is *traditional* and *trustworthy*. This combined with the absolute size of Segment 1 (53%) suggest a considerable degree of brand homogeneity. However, it should be of some concern that consumers in Segments 3 and 4 (28%) seem to associate various negative conceptualisations quite strongly with this particular retailer brand. These issues need to be addressed by the retailer. Conversely, the fact that Segment 2 respondents (19%) seem to conceptualise Retailer 1 in a very positive manner, combining *trustworthy* and *genuine* with *friendly, modern, confident, comforting* and *energetic*, should be encouraging, assuming of course that this is part of this retailer's brand strategy. It would certainly differentiate the brand from Retailer 3 (its closest competitor) and Retailer 2. That said, 80% of these customers still need to be transitioned to this new positioning.

This case study demonstrates the level and depth of insight that can be obtained when MMR's General Conceptual Lexicon® is used in the conceptual profiling of brands. In spite of the fact that the lexicon is not category specific, it has provided fine differentiation across "objects" (retailers in this case) and is easily sensitive enough to facilitate segmentation of consumers on the basis of their differences in opinion. Notwithstanding the complexities of the analysis procedures, using MMR's General Conceptual Lexicon® to expedite the research process meant that this study was designed and implemented in 1 week.

5.7.2 Case Study 2 – Developing a derived index of fit-to-brand using conceptual profiling with expedited quantification

This case study describes the application of conceptual profiling to obtain a derived index of fit-to-brand in the SMSW category. In this case, "Bullseye", our expedited quantification tool, is used as an alternative to 2-way BWS.

There are two primary factors that determine the sensory characteristics of SMSW; i.e. whether or not the damp barley (malt) is dried using hot air or peat smoke post-malting and whether the new-make spirit is matured in ex-bourbon or ex-sherry oak barrels. The process of drying damp malt with peat smoke creates SMSWs that have very distinct sensory characteristics variously described as *smoky, medicinal, phenolic* and *peaty*. Ex-sherry casks produce whiskies that have a fairly *dark, brown fruit, sweet, sherry* and *winey* character. Ex-bourbon casks impart a strong *vanilla, biscuit, dessert* and *confectionery* characteristics. Consequently, the range of different flavours produced within the SMSW category is quite extreme.

On the basis of prior knowledge, we were also aware that the conceptual profiles of the many individual brands and sub-brands of SMSW (of which there are several hundred) are not well differentiated. If so, and if it also transpires that the conceptual profiles of smoky/peaty and non-smoky/peaty SMSWs are very different from each other, some degree of brand/product dissonance (i.e. the opposite of consonance) should be anticipated.

To explore this, a large sample of UK SMSW users were recruited online. All respondents were aware of multiple SMSW brands, had consumed SMSW in the past 3 months, had purchased a bottle of SMSW in the past 12 months and were able to articulate some degree of understanding of the differences between SMSW and blended Scotch whisky.

The brand-set comprised standard distillery expressions of 23 SMSWs and three premium blended Scotch whisky brands. The blended Scotch whiskies were included for benchmarking purposes. Respondents were given the opportunity to browse through the brand stimuli prior to the commencement of conceptual profiling. For each respondent, six brands were then extracted from the set according to a rotated design and presented in a sequence of three brands, followed by a short break and then another three brands. Conceptual profiling was conducted using MMR's "Bullseye" expedited quantification tool (see Section 5.6.4 and Fig. 5.4), in conjunction with the bespoke, category-specific conceptual lexicon developed previously for SMSW (see Fig. 5.2). The project was left in-field until such time that approximately 100 data sets per brand had been collected. (MMR has established that stable conceptual profiles for brands are usually obtained with approximately 50 sets of data.)

A subset of 15 from 23 of the SMSWs previously included in the branded study was taken forward for unbranded conceptual profiling. All of the whiskies were diluted to 20% alcohol by volume; an industry-wide standard procedure amongst Scotch whisky master blenders. The diluted whiskies were served in 25 mL "shots" in Glencairn whisky nosing glasses (www.glencairn.co.uk) at approximately 15°C.

Conceptual profiling of the unbranded Scotch whiskies was conducted face-to-face in a central location in Reading, UK. Initial respondent selection criteria were similar to the brand study (above) but were more stringent. In particular, respondents were selected who were able to spontaneously describe from memory, various aspects of the major sensory differences across the SMSW category. Prospective respondents were further screened for competence in terms of their intrinsic capabilities in conceptual profiling. This involved informal conceptual profiling of international cities presented merely as names (e.g. London, Paris, New York, Shanghai) and monochromatic line drawings of different types of flowers (e.g. roses versus lilies) and animals (e.g. cats versus dogs). Those who demonstrated competence in these tasks then profiled coloured images of various different Scottish tartans (plaids) using the SMSW conceptual lexicon (Fig. 5.2) and "Bullseye". This second task served to introduce the quantification process and the SMSW conceptual lexicon and also as a final screener. The final selection of respondents returned on multiple occasions over an extended period of time to profile the unbranded whiskies in sets of three according to a rotated design. About 30 data sets were collected per whisky.

Conceptual profiling of unbranded products, as described above, can be expensive and difficult to expedite. MMR has established that it is much more efficient to work with relatively small datasets obtained from appropriate and competent respondents (as above) than larger datasets obtained from less apt respondents who are selected less stringently. Although the data are usually similar, the signal-to-noise ratio is much greater when select respondents are used.

The data for the brands and the unbranded liquids were analysed as described by Thomson and Crocker (2014).

Only the data for The Glenlivet® (www.theglenlivet.com) and Lagavulin® (www.lagavulin.com) are presented here. The Glenlivet® is produced using non-smoky malt and is matured largely in ex-bourbon casks (see above). Lagavulin® is produced using very smoky malt (amongst the highest levels in the Scotch whisky industry) and is matured using a relatively large proportion of ex-sherry casks. Consequently, in production terms and in sensory terms, The Glenlivet® and Lagavulin® are the antithesis of each other.

As anticipated, the conceptual profiles of the two unbranded SMSWs are very different from each other (Fig. 5.7). The Glenlivet® is strongly associated with *friendly, comforting* and *cheerful* and is definitely not associated with *distinctive* or *masculine*. Conversely, Lagavulin® is strongly associated with *traditional, masculine, distinctive, genuine* and *serious*.

In sharp contrast, the conceptual profiles of The Glenlivet® and Lagavulin® brands are very similar (Fig. 5.8). Indeed, the only minor difference between the two is in terms of the relative emphasis of *comforting* (The Glenlivet®) and *dominating* (Lagavulin®). This is in spite of the fact that these are two of the best-known brands in the SMSW category and most of the respondents would have been aware of the difference in smokiness. We have subsequently repeated part of this branded study amongst frequent users of SMSW who know the category very well and who are very aware of the sensory differences between The Glenlivet® and Lagavulin®. In spite of this heightened awareness, the differences in the conceptual profiles of the unbranded products greatly exceeded the differences in the branded conceptual profiles. Indeed, it seems that branding actually has a dampening rather than a differentiating effect, perhaps because the branding triggers an all-pervading recollection of the category profile that swamps the other information. This is most unusual!

In Figs 5.9 and 5.10 the conceptual profiles of the unbranded products are plotted versus their corresponding brands. The correlation coefficient is used as a derived index of fit-to-brand. For Lagavulin® there is a high degree of consonance across all 25 conceptual terms ($r = 0.80$). This is even greater when the ten key conceptualisations

Different sensory characteristics – distinctly different conceptual profiles

Product (left, The Glenlivet):
- Friendly
- Comforting
- Cheerful
- Traditional
- Conservative
- Simple
- Genuine
- Youthful
- Trustworthy
- Boring
- Cheap
- Contemporary
- Feminine
- Free-spirited
- Serious
- Sensual
- Complex
- Intriguing
- Distinctive
- Sophisticated
- Deceptive
- Masculine
- Classy
- Arrogant
- Dominating

○ Must have
● Must not have

Product (right, Lagavulin):
- Traditional
- Masculine
- Distinctive
- Genuine
- Serious
- Conservative
- Complex
- Comforting
- Dominating
- Trustworthy
- Sophisticated
- Friendly
- Intriguing
- Classy
- Cheerful
- Free-spirited
- Sensual
- Arrogant
- Youthful
- Simple
- Deceptive
- Contemporary
- Boring
- Feminine
- Cheap

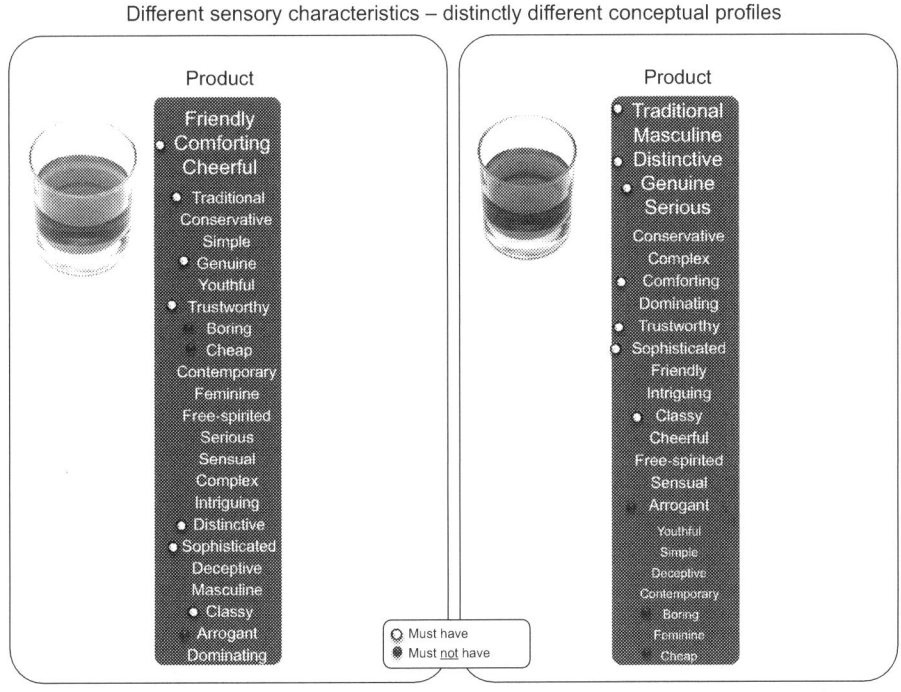

Figure 5.7 Summaries of the unbranded product conceptual profiles of The Glenlivet® (left) and Lagavulin® (right). © MMR Research Worldwide.

are considered ($r = 0.92$). This implies that whatever Lagavulin's® branding may be suggesting, it is delivered to a considerable extent by the whisky itself and therefore there is a high degree of brand–product consonance (and hence fit-to-brand) with Lagavulin®. For The Glenlivet® there is no significant correlation ($r = 0.27$) across all 25 terms or the subset of ten key conceptual terms ($r = 0.31$). Clearly the sensory characteristics that define unbranded The Glenlivet® do not confer "fit-to-brand." Indeed, we observed the same effect for most of the non-smoky SMSWs included in this study (e.g. Glenfiddich®, Glenmorangie®, Cardhu®). This is most unusual, especially since these are category-leading brands. One possible explanation might be that consumers are attracted to the category rather than to specific brands and if they enjoy a particular non-smoky whisky, they use the branding merely as a tag to aid future selection. If so, these non-smoky SMSW brands must surely be vulnerable to attack from category entrants or repositioned brands that have brand–product consonance.

Notwithstanding the practical difficulties experienced in obtaining the conceptual profiles for (unbranded) SMSW liquids, this case study illustrates the speed and efficiency of MMR's "Bullseye" quantification procedure and the simplicity of using correlation between product and brand conceptual profiles as a derived index of fit-to-brand.

Brands – the conceptual profiles are very similar

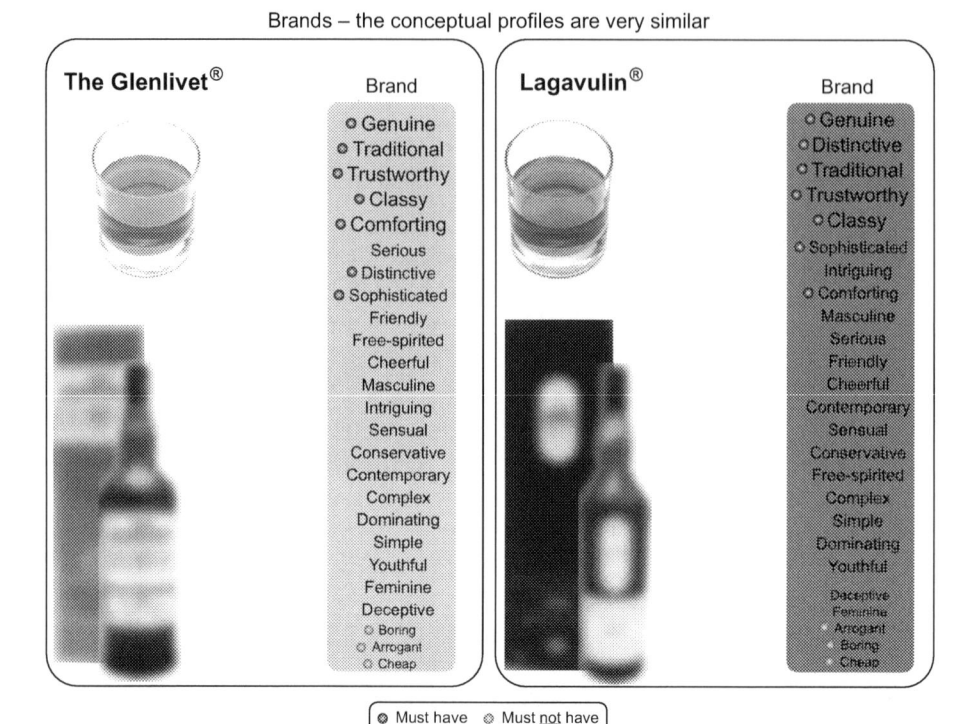

The Glenlivet®   Brand
- ○ Genuine
- ○ Traditional
- ○ Trustworthy
- ○ Classy
- ○ Comforting
- Serious
- ○ Distinctive
- ○ Sophisticated
- Friendly
- Free-spirited
- Cheerful
- Masculine
- Intriguing
- Sensual
- Conservative
- Contemporary
- Complex
- Dominating
- Simple
- Youthful
- Feminine
- Deceptive
- ○ Boring
- ○ Arrogant
- ○ Cheap

Lagavulin®   Brand
- ○ Genuine
- ○ Distinctive
- ○ Traditional
- ○ Trustworthy
- ○ Classy
- ○ Sophisticated
- Intriguing
- ○ Comforting
- Masculine
- Serious
- Friendly
- Cheerful
- Contemporary
- Sensual
- Conservative
- Free-spirited
- Complex
- Simple
- Dominating
- Youthful
- Deceptive
- Feminine
- ● Arrogant
- ● Boring
- ● Cheap

● Must have   ○ Must *not* have

Figure 5.8 Summaries of the conceptual profiles of The Glenlivet® (left) and Lagavulin® (right) brands. © MMR Research Worldwide.

5.7.3 Case Study 3 – exploring the relationship between the derived index of fit-to-brand and new product sales volumes

Case Study 2 demonstrated a simple procedure for obtaining a derived index of fit-to-brand. It also demonstrated (for the SMSW category at least) that fit-to-brand is not necessarily a prerequisite for success in long-established brands, but we speculate that brand–product dissonance (lack of fit-to-brand) will make them vulnerable. However, as discussed previously, new products need all the help they can get, so minimising brand–product dissonance (thereby maximising fit-to-brand) is highly desirable.

Several years ago, we had the opportunity to put this idea to the test in a particular savoury snack foods category (which we are unable to disclose) in the USA. This was exploratory research, tagged onto the end of a pre-launch product and concept test involving 15 new products. All of the products were launched under the same brand name. Expedited conceptual profiling of the unbranded products was conducted using MMR's "Bullseye" procedure in conjunction with a heavily truncated conceptual lexicon selected intuitively by the researchers. Liking and purchase intent were measured after the branded concept was revealed. Conceptual profiling of the (single)

Conceptual profiling of brands, products and packaging 113

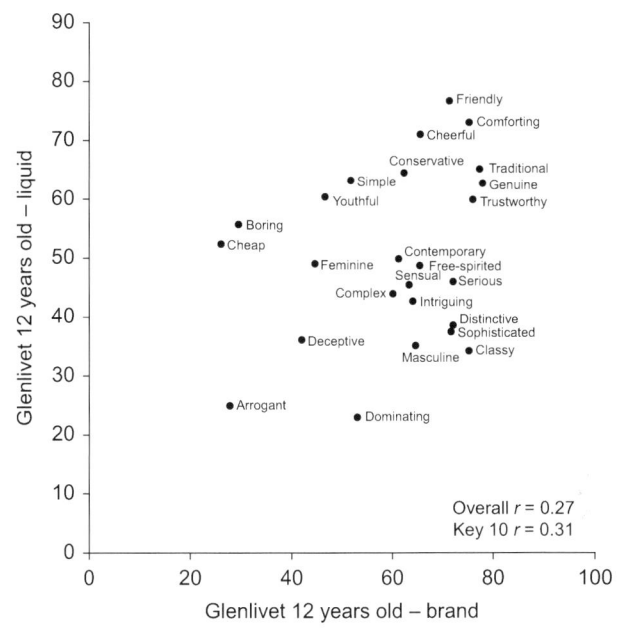

Figure 5.9 The Glenlivet® – conceptual profiles of product versus brand. © MMR Research Worldwide.

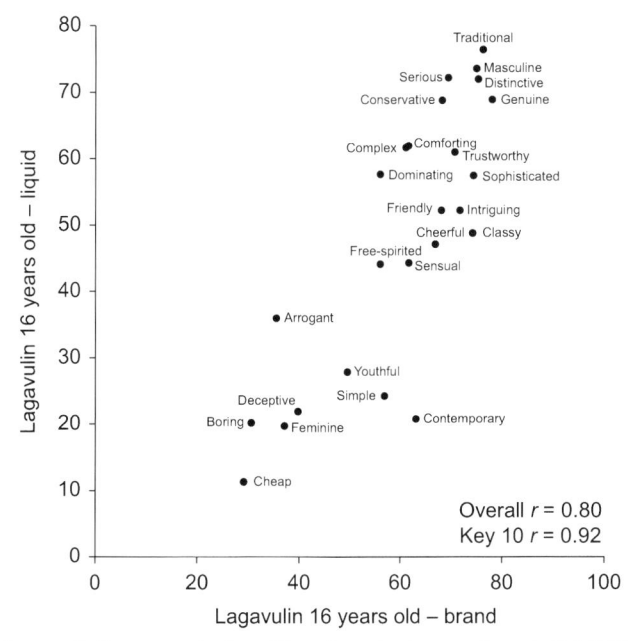

Figure 5.10 Lagavulin® – conceptual profiles of product versus brand. © MMR Research Worldwide.

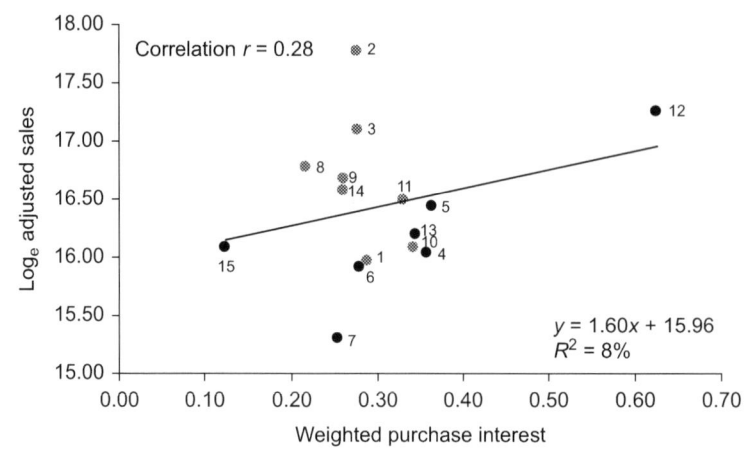

Figure 5.11 Savoury snack food category – adjusted 52-week sales versus weighted purchase interest. © MMR Research Worldwide.

brand was conducted separately. To explore fit-to-brand the conceptual profile of each unbranded product was plotted against the conceptual profile of the brand and the correlation coefficient used as a derived index of fit-to-brand.

Although this research design was undoubtedly suboptimal, it was the best that could be achieved under very constrained commercial circumstances. However, 12 months later we were given access to the actual 52-week sales data achieved in-market. This gave us the opportunity to explore some of the ideas expounded during the course of this chapter.

Figures 5.11 and 5.12 plot the exponential logarithm ($\log_e$) of 52-week sales (adjusted for differences in distribution and merchandising) versus weighed purchase intent (a transformation of the proportion of "definitely would buy" and "probably would buy") and liking (the proportion of respondents rating the product in the top three boxes on a 9-point hedonic scale). These metrics are fairly standard in volumetric research.

For purchase intent (Fig. 5.11), the coefficient of determination (R^2) is 0.28 (nowhere near significant) with only 8% of variation in the data explained by the model fitted. With liking (Fig. 5.12), R^2 is 0.38 (not significant) with only 15% of variation explained. If these findings are generalisable, they would indicate that using threshold values of purchase intent or liking as action standards in new product evaluation would be suboptimal.

Figure 5.13 plots the relationship between adjusted sales versus the derived index of fit-to-brand for the 15 products. The R^2 value at 0.46 is significant (albeit at >90% level of confidence) and the variance explained is 21%. Whilst this is hardly a superb model, the relationship is on the cusp of significance and the variance explained is greater than for liking and purchase intent. We believe that a much better fit could have been obtained had the execution of the conceptual profiling not been so heavily compromised.

Conceptual profiling of brands, products and packaging 115

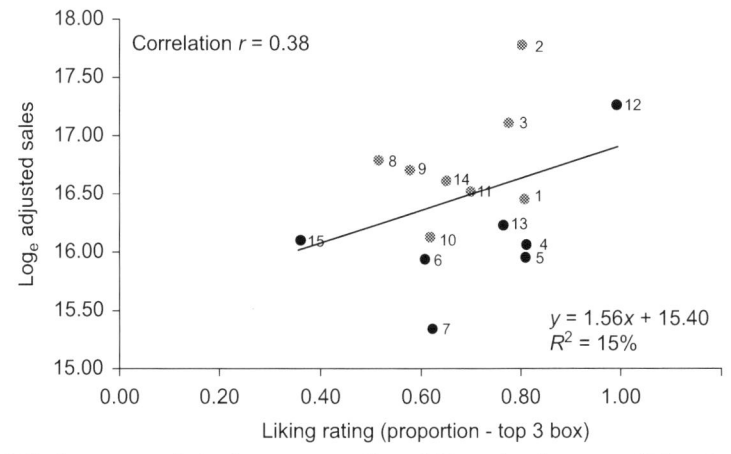

Figure 5.12 Savoury snack food category – adjusted 52-week sales versus liking. © MMR Research Worldwide.

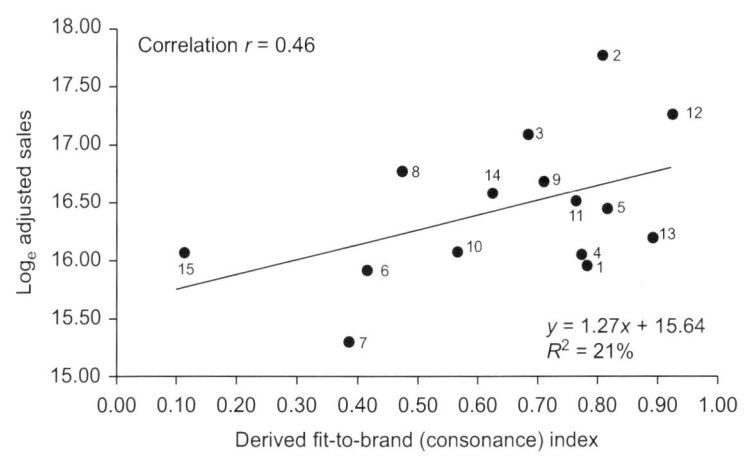

Figure 5.13 Savoury snack food category – adjusted 52-week sales versus derived index of fit-to-brand. © MMR Research Worldwide.

In a typical new product development scenario, we self-evidently would not have the luxury of actual sales data prior to launch. To complete the picture we used an Eskin-type trial and repeat volumetric model to predict sales volumes for the same 15 products, as would typically be done at the end of the new product development process. The sales volume predictions were then modelled against the corresponding values for our derived index of fit-to-brand. The exponential function shown in Fig. 5.14 fits very well.

On the basis of this model, we hypothesise that a fit-to-brand index of about 0.8 or above is indicative of the level of brand–product consonance required to give a

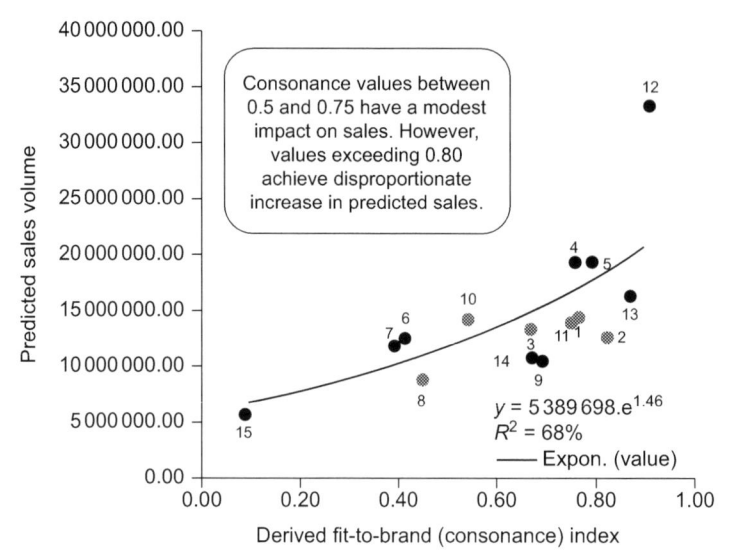

Figure 5.14 Savoury snack food category – predicted sales versus derived index of fit-to-brand. © MMR Research Worldwide.

Plot annotation: Consonance values between 0.5 and 0.75 have a modest impact on sales. However, values exceeding 0.80 achieve disproportionate increase in predicted sales.

Fitted curve: $y = 5\,389\,698 \cdot e^{1.46}$, $R^2 = 68\%$ — Expon. (value)

new product a real chance of commercial success. In practice, achieving this level of fit-to-brand is no mean feat and many of the newly launched products that we have monitored over the past few years have fallen well short of this threshold. Perhaps this explains why so many new products fail!

Whether or not our derived index of fit-to-brand could be used as a rapid alternative to volumetric prediction in new products is something that would need to be validated with further research (and submitted to peer-review in a credible scientific journal). However, it is an interesting possibility that could save a great deal of time and money.

5.8 Conclusion

This chapter attempts to provide a theoretical basis for the role of conceptualisation in influencing human behaviour. On this basis, it also makes a case for the inclusion of conceptual profiling as a fundamental part (possibly the most important part) of brand, product and packaging evaluation during product development. It has also been shown that although the creation of a bespoke, category-specific conceptual lexicon is highly desirable and although 2-way BWS is undoubtedly the most sensitive quantification tool, the process of conceptual profiling can be expedited with minimal compromise in some circumstances by using MMR's General Conceptual Lexicon® and MMR's "Bullseye" quantification tool.

New products need all the help they can get, so overcoming dissonance between the conceptual profiles of brand and product, thereby optimising fit-to-brand, is a prerequisite for new product success. This chapter also demonstrates how conceptual profiling is used to obtain a derived index of fit-to-brand, by comparing the conceptual profiles of brand versus unbranded product. The resulting correlation coefficient is used as the index of fit-to-brand. We believe that achieving an index of ≥ 0.8 is necessary to give a new product the best possible chance of success. Conversely, we would be very pessimistic about a new product that did not achieve this level of brand–product consonance. To this end, conceptual profiling could and should be used at various key stages of brand, product and packaging development, to direct the whole process towards the required degree of conceptual convergence. Without this, the new product has little or no chance of commercial success. The procedures described herein for expediting conceptual profiling mean that a derived index of fit-to-brand can be obtained relatively quickly and cheaply!

Acknowledgements

I am grateful to colleagues at MMR Research Worldwide, particularly Dr Chris Crocker, Toby Coates, Phiala Mehring and Dr Steve Ferris, for being trusty and valued companions on the "conceptualisation journey". Thanks are also due to Annandale Distillery Company Ltd for allowing us to use their data in Case Study 2.

References

Aaker, J.L. (1997). Dimensions of brand personality. *Journal of Marketing Research*, **34**, 347–356.
Berridge, K.C. and Kringelbach, M.L. (2008). Affective neuroscience of pleasure: Reward in humans and animals. *Psychopharmacology*, **199**, 457–480.
Cardello, A.V., Meiselman, H.L., Schutz, H.G., Craig, C., Given, Z., Lesher, L.L. and Eicher, S. (2012). Measuring emotional responses to foods and food names using questionnaires. *Food Quality and Preference*, **24**, 243–250.
Carey, S. (2009). *The Origin of Concepts*. New York: Oxford University Press.
Chomsky, N. (1967). A review of B.F. Skinner's Verbal Behavior. In L.A. Jakobovits and M.S. Miron (Eds.) *Readings in the Psychology of Language* (pp. 142–143). New York: Prentice Hall.
Claws, P. (1965). *The Philosophy of Science: A Systematic Account*. New York: Van Nostrand.
Crocker, C. and Thomson, D.M.H. (2014). Anchored scaling in best worst experiments: A process for facilitating comparison of conceptual profiles. *Food Quality and Preference*, **33**, 37–53.
Duncker, K. (1941). On pleasure, emotion and striving. *Philosophical and Phenomenological Research*, **1**, 391–430.
Ellis, R.D. (1995). *Questioning Consciousness: The Interplay of Imagery, Cognition and Emotion in the Human Brain*. Amsterdam: John Benjamins.
Ellis, R.D. and Newton, N. (2010). *How the Mind Uses the Brain (to Move the Body and Image the Universe)*. Chicago: Open Court.
Finn, A. and Louviere, J.J. (1992). Determining the appropriate response to evidence of public concern: The case for food safety. *Journal of Public Policy and Marketing*, **11**(1), 12–25.

Flynn, T.N., Louviere, J.J., Peters, T. J. and Coast, J. (2007). Best–worst scaling: What it can do for health care research and how to do it. *Journal of Health Economics*, **26**, 171–189.
Frijda, N.H. and Scherer, K.R. (2009). Emotion theories and concepts (philosophical perspectives). In D. Sander and K.R. Scherer (Eds.), *Emotion and the Affective Sciences* (pp. 142–144). Oxford: Oxford University Press.
Gendolla, G.H.E. and Brinkmann, K. (2009). Reward. In D. Sander and K.R. Scherer (Eds.), *Emotion and the Affective Sciences* (pp. 344–347). Oxford: Oxford University Press.
Greenwald, A.G. and Banaji, M.R. (1995). Implicit social cognition: Attitudes, self-esteem, and stereotypes. *Psychological Review*, **102**, 4–27.
Greenwald, A.G., Klinger, M.R. and Schuh, E.S. (1995). Activation by marginally perceptible ("subliminal") stimuli: Dissociation of unconscious from conscious cognition. *Journal of Experimental Psychology: General*, **124**, 22–42.
Hill, D. (2008). *Emotionomics: Winning Hearts and Minds*. New York: Adams Business and Professional.
Higgins, E.T. (2009). Motivation. In D. Sander and K.R. Scherer (Eds.), *Emotion and the Affective Sciences* (pp. 365–367). Oxford: Oxford University Press.
Kahneman D. (2003). A perspective on judgement and choice. *American Psychologist*, **58**, 697–720.
King, S.C. and Meiselman, H.L. (2009). Development of a method to measure consumer emotions associated with foods. *Food Quality and Preference*, **21**, 168–177.
King, S.C., Meiselman, H.L. and Carr, B.T. (2013). Measuring emotions associated with foods: Important elements of questionnaire and test design. *Food Quality and Preference*, **28**, 8–16.
Lee, J.A., Soutar, G. and Louviere, J. (2008). The best–worst scaling approach: An alternative to Schwartz's values survey. *Journal of Personality Assessment*, **90**(4), 335–347.
Lindstrom, M. (2008). *Buyology: How Everything We Believe About Why We Buy is Wrong*. New York: Broadway Books.
Marley, A.A.J. and Louviere, J.J. (2005). Some probabilistic models of best, worst, and best–worst choices. *Journal of Mathematical Psychology*, **49**, 464–480.
Maslow, A.H. (1987). *Motivation and Personality*, pp.3–14. New York: Addison-Wesley.
Olds, J. (1956). Pleasure centers in the brain. Scientific American, October 1956. Reprinted in Coopersmith, S. (ed.), *Frontiers of Psychological Research* (1966), pp. 54–59. San Francisco: W.H. Freeman and Company.
Penn, D. (2006). Looking for the emotional unconscious in advertising. *International Journal of Market Research*, **45**, 515–524.
Reiss, S. (2002). *Who Am I?: The 16 Basic Desires that Motivate Our Actions and Define Our Personalities*. New York: Berkley.
Suppes, P. and Zinnes, J.L. (1963). Basic measurement theory. In R. D. Luce, R. R. Bush and E. Glanter (Eds.), *Handbook of Mathematical Psychology*, I, pp. 1–76. New York: Wiley.
Thomson, D.M.H. (2010). Reaching out beyond liking to make new products that people want. In H.J.H. MacFie and S.R. Jaeger (Eds.), *Consumer Driven Innovation in Food and Personal Care Products*. Cambridge: Woodhead Publishing Limited.
Thomson, D.M.H. and Crocker, C. (2013). A data–driven classification of feelings. *Food Quality and Preference*, **27**, 137–152.
Thomson, D.M.H. and Crocker, C. (2014). Development and evaluation of measurement tools for conceptual profiling of unbranded products. *Food Quality and Preference*, **33**, 1–13.
Thomson, D.M.H., Crocker, C. and Marketo, C.G. (2010). Linking sensory characteristics to emotions: An example using dark chocolate. *Food Quality and Preference*, **21**, 1117–1125.

Part Two

Rapid methods for sensory profiling

Flash Profile, its evolution and uses in sensory and consumer science

6

J. Delarue

UMR1145 Ingénierie Procédés Aliments, AgroParisTech, INRA, Cnam, Massy, France

6.1 The method and its origins

Among rapid descriptive methods, Flash Profile (FP) is certainly the method most closely related to the conventional profile, as it is based on the quantitative evaluation of products by means of sensory attributes. It was imagined and developed by Sieffermann (1995, 2000), at a time when such methods as free choice profiling (FCP) (Williams and Langron, 1984) – from which it derives – and Repertory Grid (Kelly, 1955) had gained interest among sensory scientists. These two methods, introduced several years earlier, opened possibilities of using individual attributes elicited by the panellists themselves without seeking a semantic consensus. The development of multidimensional techniques of data analysis, and especially generalized Procrustes analysis (GPA) (Gower, 1975), made these methods more readily available in the practice of sensory labs. These methods had further useful spin-off in facilitating the use of untrained subjects, and hence opening the way to conducting sensory profiling with consumers (Thomson and McEwan, 1988; Jack and Piggott, 1992; Piggott and Watson, 1992). However, as they were still relatively time-consuming, their use remained rather limited and, apart from research studies, very few companies actually used them.

FP was the first descriptive method to emphasize rapidity. Actually, the very name "Flash Profiling" reflects the idea of taking a snapshot picture of a product set as it is perceived by human subjects. FP was thus very innovative because it was the first sensory profiling method to be thought as a one-shot measurement with emphasis on the relative sensory positioning of the products being evaluated. In that sense, there is a clear convergence with holistic methods such as free sorting and projective mapping. But contrary to these non-verbal methods, FP does primarily rely on quantitative description. From this perspective, attributes become only a means to collect sensory data. Their definition and precise meaning is no longer central. As will be shown in this chapter, the downside is that the interpretation of sensory dimensions is less straightforward, since one cannot rely on precise definitions of consensual attributes.

This chapter first presents the principles of FP, before the main methodological points are illustrated through a simple example of a description of dark chocolates. Further methodological issues will then be discussed and additional guidelines are provided, together with a presentation of advantages and limitations of FP. The last part of the chapter presents the evolution in the use of FP and examples of potential application in R&D and market research are given.

6.1.1 Principle

The aim of FP is to provide quick access to the relative positioning of a set of products. Its principle is very simple. It consists of the combination of a free choice of attributes, as in free choice profiling (FCP), and of a comparative evaluation of the samples for each chosen attribute (attribute-by-attribute protocol) and quantification by the means of ranks. This usually goes with a simultaneous assessment of the whole product set and direct focus on inter-product differences.

FP can thus be seen as an extension of FCP. However, its originality lies in the emphasis on rapidity and on relative sensory positioning. A number of methodological features arise from these two points. First is the absence of training in the traditional sense. In order to make the FP faster than regular FCP, it was indeed proposed not to train panellists specifically for the evaluation of the product set or product category under consideration but to use experienced subjects instead (what is meant by "experienced subjects" will be discussed next).

Aside from the recruitment of experienced subjects and the free choice of attributes, the need for training is further reduced by the comparative evaluation mode. The fact that assessors have simultaneous access to the whole sample set indeed allows direct comparative evaluation and forces the assessors to focus on the differences they perceive. This leads them to generate discriminant attributes only. This also means that, in comparison to many conventional profiling methods, more autonomy is left to the assessors.

Always with a view to increase rapidity of the acquisition of sensory data, FP is designed with a single evaluation session. The duration of the FP session may of course vary depending on the number and the type of products, but it would typically last between 40 min and 2 h. Note that this is probably longer than usual evaluation sessions in conventional profiling and may seem sometimes discouraging. However, in practice, instead of multiple tasting sessions that usually come along with constraining schedules in conventional descriptive analysis, the fact that the study consists of a single individual session proves to be better suited to standard professional work rhythms. This opens ways for a flexible organization based on individual appointments, and makes FP particularly well adapted for working with professional experts and internal panellists.

6.1.2 Subjects

6.1.2.1 Selection criteria

In the original FP procedure, sensory evaluation experts – not necessarily product experts – are selected to participate in the panel. Such sensory evaluation experts are meant to be experienced subjects who are able to understand experimenter's instructions and to generate discriminant and non-hedonic attributes (Delarue and Sieffermann, 2004a). Experienced subjects could be subjects who have previously participated in several descriptive evaluation tasks. They could also be professionals who are used to sensorily evaluating products in their daily work. It was indeed anticipated that such subjects' ability to focus on their own perceptions and to communicate them quantitatively would allow them to complete the descriptive task more efficiently. In practice, we have often observed that this point is indeed of crucial importance in the quality of the results. Note that these subjects do not need to be trained on a specific product set.

6.1.2.2 Number of panellists

Although there is no absolute rule regarding the number of panellists that should participate in an FP, we generally consider four or five to be the minimum. This allows yielding relatively stable and complementary sensory information about the products. Naturally, recruiting more panellists is worthwhile, but not strictly necessary.

It must be noted that in the case of FP, the logic behind recruiting several panellists is slightly different from that of conventional profiling. Because of inter-individual differences in sensitivity and performances, it is necessary in conventional profiling to stabilize the panel's outcome by the use of several panellists. In FP, however, the panellists' responses are seen as complementary measures instead of repeated measures. Actually, the use of an FCP procedure accounts for the fact that subjects' perceptions may differ in nature, not only in terms of sensitivity. Consequently, the selection criteria are perhaps more important to the quality of the description than the number of recruited subjects.

6.1.2.3 Consumers

Conducting descriptive analysis with consumers is extremely appealing as it is a potential way to assess the diversity of consumer perception of their products (Jack and Piggott, 1992; Faye *et al.*, 2006; Thamke *et al.*, 2009). However, most sensory scientists have long been reluctant to use "untrained consumers" to perform conventional descriptive analysis. The use of a common vocabulary and subsequent training were indeed seen as a limitation and contradictory to consumer testing. Like other rapid techniques, FP could overcome these limitations, although it was initially designed to be conducted with sensory evaluation experts, hence not with consumers. Sensory scientists and market researchers in the industry have thus started conducting FP with consumers as a way to capture consumers' perceptions with perhaps a greater focus on subjects than on products. These attempts to conduct FP studies with consumers have proved to be feasible and have yielded insightful results. This is consistent with the observations of Husson *et al.* (2001) and Worch *et al.* (2010), who have shown that descriptive analysis with consumers could provide reliable data.

In this perspective, it is important to recruit targeted and representative consumers. From the literature, the most frequent panel size is of about 40–50 participants. Yet, depending on the objectives of the study, FPs with consumers have been conducted with panels ranging from 24 (Moussaoui and Varela, 2010) to 200 participants (Ballay *et al.*, 2006). The sampling of consumers in this case could also be related to sampling principles in qualitative surveys (Glaser and Strauss, 1967; Marshall, 1996; Gibbs *et al.*, 2007). Even though qualitative methods do not necessarily imply a large number of participants, one way to increase the validity is to diversify sampling and to attempt to reach theoretical saturation or informational redundancy (Sandelowski, 1995). Further developments in the use of FP with consumers are presented at the end of this chapter.

6.2 Flash Profile (FP) methodology through an example: evaluation of dark chocolates

The FP methodology is probably best understood through a case study. Below is thus presented a simple study of dark chocolates. Although this example conducted with students is relatively straightforward, it will provide a good grasp of the key methodological points to take into account when implementing and analysing a Flash Profile.

6.2.1 Products

In this example, 11 dark chocolates from the French market were analysed (Table 6.1). As can be seen, this selection includes chocolates made with various cocoa contents. Some products are from major brands, as well as from private labels, and two products are fair trade and organic chocolates.

6.2.2 Panel

The panel was composed of seven students (five men, two women). All of them were trainee flavourists aged between 22 and 24. As part of their education, these

Table 6.1 **The dark chocolate samples evaluated by FP**

Product name	Cocoa content (%)	Short name
Poulain noir extra	47	PoulainExtra-47
1848 76% noir degustation	76	Poulain1848-76
1848 86% noir degustation	86	Poulain1848-86
Lindt Excellence 85% "extra dark"	85	Lindt85
Lindt Excellence 90% "supreme dark"	90	Lindt90
Nestlé Grand chocolat 70%	70	Nesté70
Côte d'Or degustation 70% noir intense	70	CotOr70
Monoprix Gourmet Saint Domingue	72	Monop-StDom72
Monoprix chocolat noir	50	Monop50
Alter Eco noir intense Pérou*	58	AltEco-Peru58
Ethiquable 68% Pérou*	68	Ethiq-Peru68

*Fair trade and organic products.

students were trained to use their senses (especially their sense of smell) in their work of evaluating and creating flavours. Although they are not trained panellists in the usual sense (i.e. they had never participated in a descriptive sensory panel), they were experienced in sensory evaluation. Note that they had not received any specific training for the evaluation of dark chocolates, and were not particularly familiar with this set of products before the study. The evaluation was individual and took place in standard sensory booths.

6.2.3 Sample preparation

The samples, one square of each chocolate per subject, were presented in white plastic cups coded with letters randomly assigned from A to K. In this way, the subjects could evaluate the samples and physically rank them in front of them before reporting the ranks on paper forms. If needed, additional samples were provided upon request.

Note that using letters is usually preferred to three-digit random numbers because such a simpler coding lessens the risk of mistakes when reporting the ranks on the paper forms. In addition, the simultaneous evaluation makes the use of random numbers less relevant than with a monadic sequential design. This is of course no longer a problem when data are acquired using sensory software. However, even with a screen interface, the subjects should have the possibility to physically rank the samples in front of them before reporting the ranking positions on the screen. In our experience, spatially organizing the samples indeed helps the subjects in their comparative evaluation.

6.2.4 Instructions and course of the session

The evaluation consisted of a single session. The subjects were asked to manipulate and taste the products in order to identify any non-hedonic attributes by which the samples could be discriminated and then to rank the samples on each of these attributes, with ties allowed. The number of attributes was left open. Subjects were told to proceed at their own pace. Some subjects preferred to first identify all discriminant attributes before ranking the samples, while others completed the task in a more integrated manner. It was also noted that one subject had to use scores on a draft as a reminder to facilitate the comparative evaluations, although most subjects directly ranked the products on each attribute.

The ranks were to be reported on a form, one for each attribute. The form typically consists of a sheet of paper with a printed arrow where the sample codes are to be reported in ranked order, from the weakest intensity on the left-hand side to the strongest intensity on the right-hand side (Fig. 6.1). The subjects might also decide to specify anchors they would find more appropriate (e.g. *from milk chocolate* to *dark chocolate*). The panellists were instructed that the relative distance on this "scale" is not relevant and that only the ranking scores would be used for data analysis.

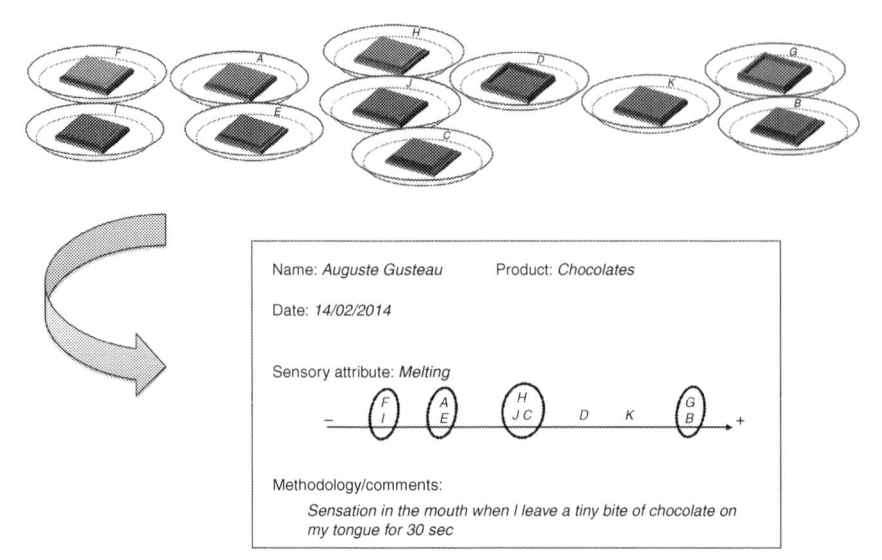

126 Rapid Sensory Profiling Techniques and Related Methods

Figure 6.1 Example of the transcription of the ranking of the samples (dark chocolates) for one attribute. The panellists were asked to circle ties in order to avoid any confusion. Note that letter codes are used to identify the samples.

The subjects were advised to rinse their mouths between samples, and to take breaks for few minutes between attributes, which most of them did. Overall, they took between 30 and 50 min to complete the session.

6.2.5 Results

6.2.5.1 Generated attributes

The subjects used between 4 and 12 attributes (mean 6.4) for a total at the panel level of 45 attributes, 27 of which were semantically different: 1 for odour, 1 for aspect, 19 for aroma, 2 for taste and 4 for texture. Obviously, the dominance of flavour attributes reflected these subjects' culture in olfaction and hence their inclination to describe flavour. Subjects from different fields of expertise might for instance yield more texture attributes. For this reason, in most of our studies, we try to recruit panellists with complementary expertise.

Data can be first analysed at the individual level using principal component analyses (PCA). The resulting sensory mapping might be of some interest, but more important are the resulting scree plots that can be used to assess the dimensionality of each assessor's description. This can be seen as a way of evaluating subjects' descriptive efficiency (number of significant eigenvalues vs number of attributes). As shown on Fig. 6.2, the descriptions provided by subjects S1 and S6, who respectively used 7 and 12 attributes, was mainly expressed on the first principal component, which accounts respectively for 80% and 60% of total

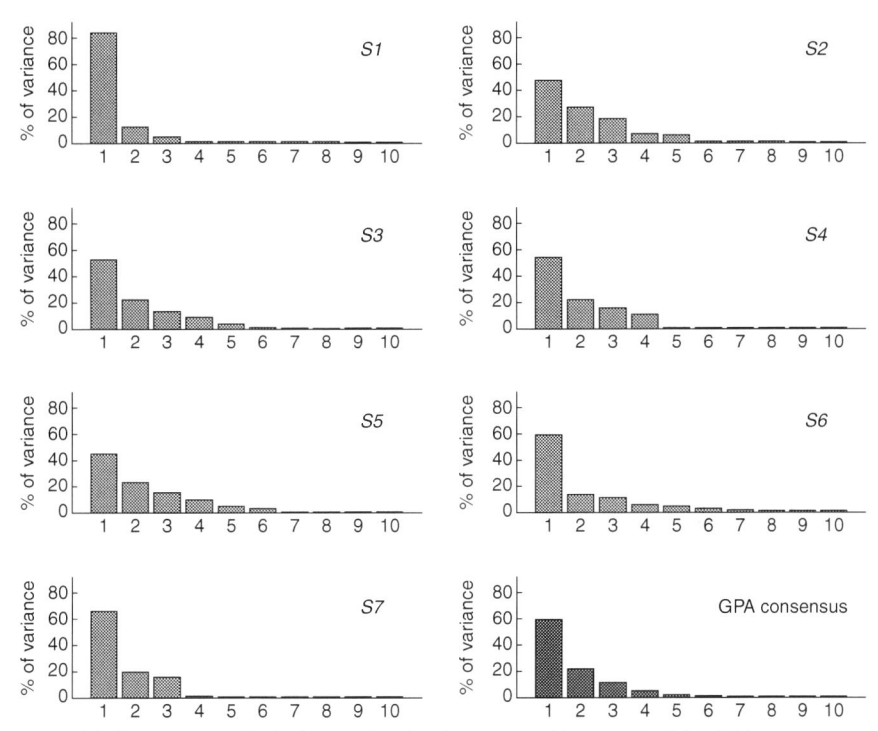

Figure 6.2 Scree plots of individual PCA for the seven subjects and of the GPA consensus for the whole FP panel.

variance. From a strict mathematical point of view, these two subjects are thus relatively inefficient. On the other hand subjects S2 and S5, who both used six attributes, yielded a description that was more evenly distributed over principal components.

6.2.5.2 Sensory map and additional interpretation aids

The main result provided by FP is a sensory map. To this end, the individual data tables can be compiled and analysed by the means of a multi-block data analysis technique, such as generalized Procrustes analysis (GPA) (Gower, 1975; Dijksterhuis, 1996), multiple factor analysis (MFA) (Escofier and Pagès, 2008) or STATIS (Lavit, 1988; Lavit et al., 1994). Figure 6.3 shows the map of the first two principal components obtained by GPA.

As can be seen from this map, which accounts for 78% of the total variance, the products are well separated. The three chocolates with the highest cocoa content (Lindt85, Lindt90 and Poulain1848–86) are clustered together and separated from other products on the first axis (Poulain 1848–76 may seem to be positioned with them but is actually well separated on the third axis). Interestingly, one cluster gathers the two fair trade chocolates from Peru, together with the CotOr70. Eventually, the

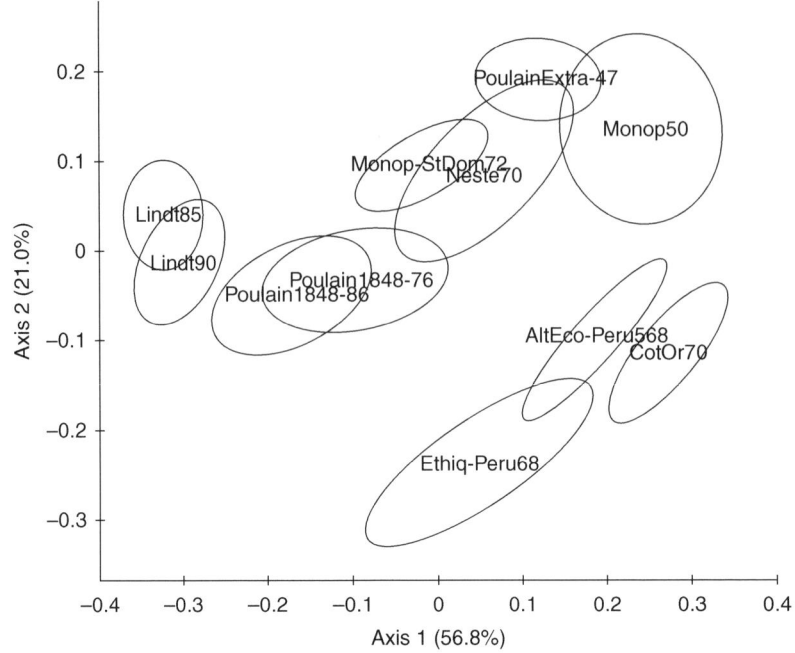

128   Rapid Sensory Profiling Techniques and Related Methods

Figure 6.3 FP sensory map (i.e. score plot from the GPA consensus) with 95% confidence ellipses for each product. Two-digit numbers in product names represent the cocoa content.

two products with the lowest cocoa content (PoulainExtra-47 and Monop50) which may also be considered as lower-range products are very closely positioned on this map. In addition to sensory map, hierarchical cluster analysis may also be performed following GPA (Fig. 6.4). This is a convenient way to look at clusters of products, taking into account all sensory information. The resulting dendrogram is very consistent with the score plot from the GPA. In some cases, such charts are found to be easier to communicate to stakeholders.

In order to test whether the products are discriminated, it is possible to run a multivariate analysis on variance (MANOVA) or a discriminant analysis on the rotated data after GPA (i.e. the *(product*subject)* × *GPA axes* table), using *product* as a dependent variable. The general discrimination level can then be estimated using multivariate F-ratio (Wilks' λ). Note that pair-wise differences can also be tested in a multivariate way, which may be useful for decision making. Alternatively, confidence ellipses can be calculated and drawn on the sensory map (Husson *et al.*, 2005), as in Fig. 6.3.

6.2.5.3 Attribute loadings and semantic interpretation

Although FP was not designed for semantic interpretation, the sensory positioning of the products being considered as the main output, the analysis of the loading of attributes after GPA can still be informative. However, the loading plots from

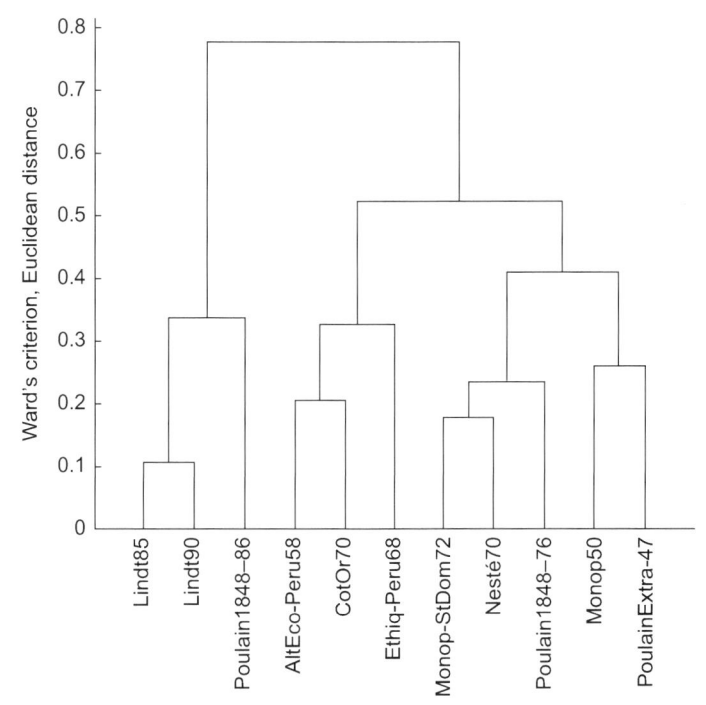

Figure 6.4 Dendrogram from the hierarchical cluster analysis of the product set (Euclidean distance, Ward's criterion) performed after GPA on FP data.

FP usually bristle with attributes, and it may be difficult to find one's way through this massive amount of information. This example is no exception, even though the number of attributes is relatively moderate (Fig. 6.5).

Nevertheless, careful attention shows that attributes that are semantically similar are clustered together and interpretation can be attempted. It comes as no surprise that the first axis seems to be driven by a cocoa flavour/bitter dimension, which makes sense given the positioning of the high cocoa content products negatively on this first axis. Besides, the two chocolates from Peru and CotOr70 can be described as more fruity, and the chocolates with lowest cocoa content (and therefore higher sugar content) are obviously sweeter.

This example is actually quite straightforward because the level of consensus in the use of those semantically similar attributes is relatively good. This is however not always the case. If we extract groups of attributes with apparently similar meanings, we can observe various cases that are instructive (Fig. 6.6). For instance, all subjects used the attribute *fruity* and, from what can be seen in Fig. 6.6, these attributes are well correlated on the GPA plot. This means that all the subjects used *fruity* in a very similar way, except maybe for one subject (S4). The same goes for *sweet*, that is used by three subjects and as can be noted, these attributes *sweet* are strongly anti-correlated to *bitter*. Other attributes can also be grouped thanks to their

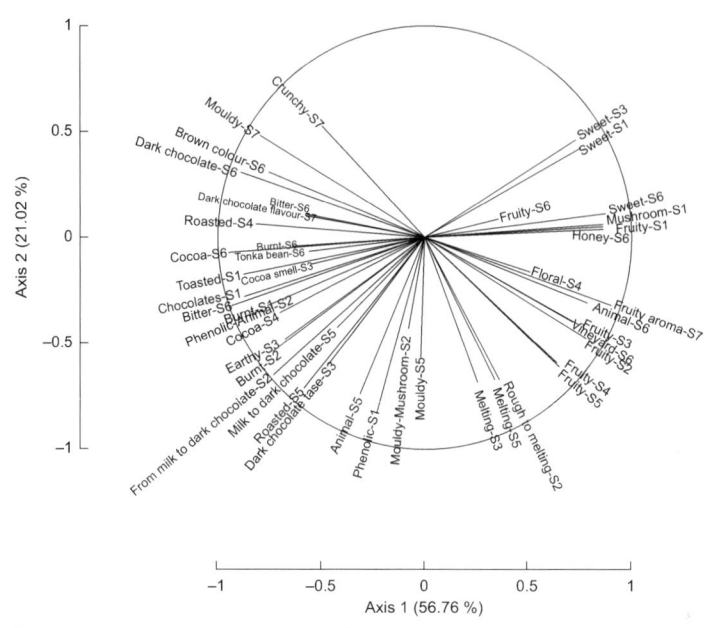

Figure 6.5 Loadings of individual attributes from the FP of dark chocolates on the first two principal components after GPA.

semantic resemblance. Although they are not strictly identical, they seem to convey similar meaning, as can be deduced from their pointing directions: (*roasted, toasted* and *burnt*) or (*cocoa, chocolate, dark chocolate, from milk to dark chocolate*). In the present case, subjects' background in flavour evaluation has certainly facilitated this consensus. Beside these obvious results, however, one may observe that the subjects also referred to notes such as *mouldy, mushroom* and *earthy*, as well as *phenolic* and *animal*. In this case, interpretation is much less clear. *Mouldy* was used twice, but the two corresponding vectors are almost orthogonal on the loading plot, which means that these two subjects may not describe the same thing even though they use the same word. The same goes for *animal*. Alternatively, it is very likely that some subjects have used totally different words (for example, *animal* (S5) and *phenolic* (S1)) to express the same sensory notion. In my experience, these two situations occur very frequently, if not all the time. As a result, the interpretation is much more challenging, but it can also be seen as an advantage when one wants to investigate subjects' diversity in their perception and sensory description.

6.2.6 Conclusions from the chocolate study

The most striking result from this study is perhaps the fact that it took less than 3h to complete the whole process, from sample preparation to data analysis. Results are consistent and compared fairly well to a conventional sensory profile (quantitative descriptive analysis (QDA®)-like) conducted with another group of students (data not

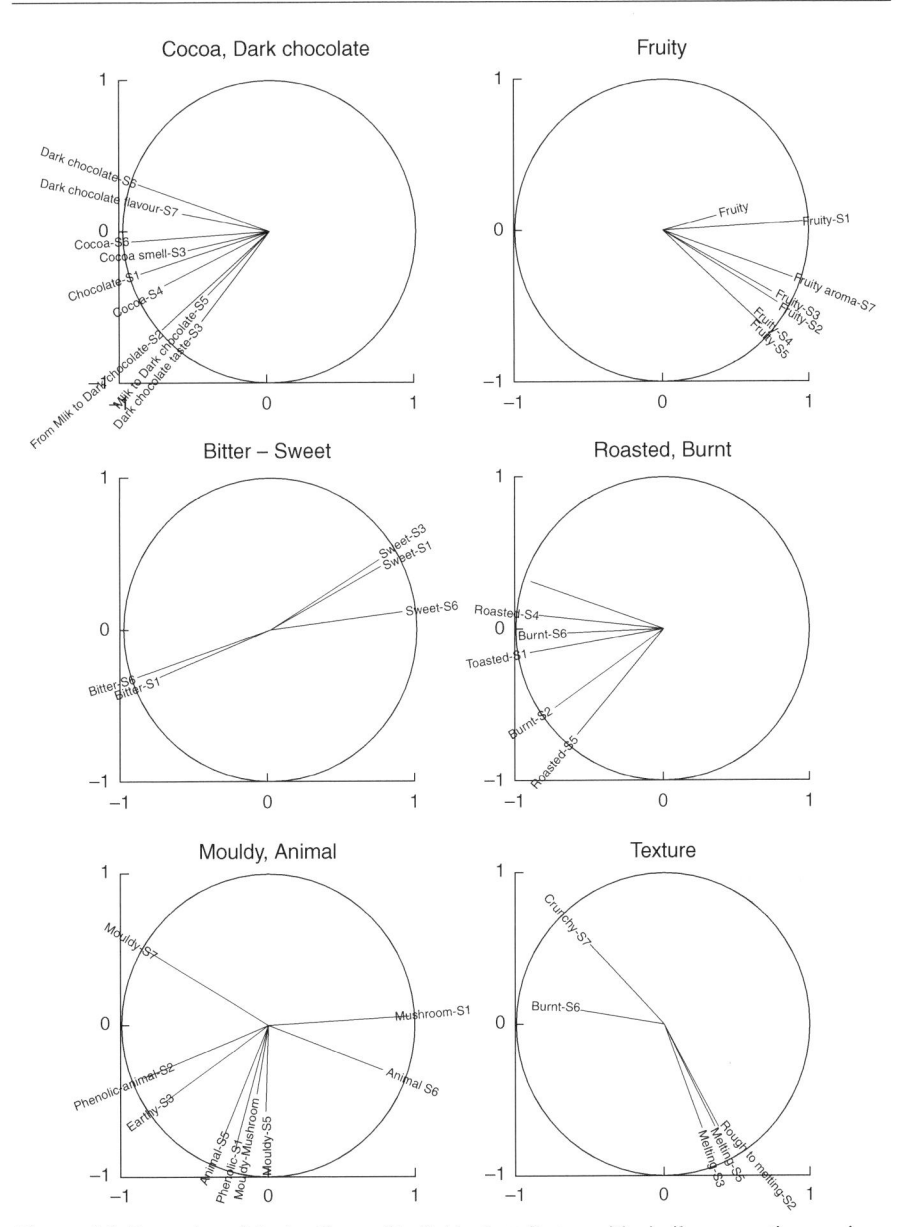

Flash Profile: evolution and uses in sensory and consumer science   131

Figure 6.6 Examples of the loadings of individual attributes with similar semantic meaning. Groups of attributes are represented on separate plots for better legibility although they all result in the same GPA. Each attribute appears with the corresponding subject's number.

shown) which necessitated about 20 h. Naturally, one may always consider that the descriptive evaluation was facilitated by the fact that these FP subjects were already experienced in sensory evaluation. Well, that is exactly the point of FP! Product-specific training is replaced by more general sensory expertise.

Of course, results from FP are not always as accurate as those of conventional profiling. As was discussed above, the absence of definition and specific training could lead to uncertain semantic interpretation. However, the trade-off between accuracy and rapidity should be considered by the sensory practitioner depending on his/her objectives and constraints.

6.3 Further methodological considerations

In addition to the example presented above, the experimenter may be willing to adapt part of the methodology. Below are three additional considerations regarding the setting up of an FP.

6.3.1 Attribute elicitation

In our experience with FP, attribute elicitation is usually not a problem when working with experienced subjects or with product experts. Such subjects are indeed used to describing their sensations using descriptive terms. The only pitfall we met is when conducting an FP with panellists from a conventional profile panel heavily trained to describe the same category of products. Those subjects indeed tended to stick to the list of attributes they had learned before, and did not really adapt to the product set under study.

The attribute elicitation issue may be considered differently when working with consumers. For instance, should one be interested in consumers' spontaneous description, it would seem more pertinent to let the participants use their own terms without any suggestions. This may result in the elicitation of rather obvious attributes, although in past studies the vocabulary used by consumers was found to be rich and diverse (Veinand et al., 2011). On the contrary, providing a list may facilitate description (as in the Check-all-that-apply technique), but it may also bias consumers' responses. However, these two options have not been formally compared yet, and ultimately the choice really has to be made with respect to the goals of the study.

6.3.2 Preliminary session

In order to facilitate elicitation of attributes, the initial FP procedure (Dairou and Sieffermann, 2002; Delarue and Sieffermann, 2004a) included a preliminary session during which the panellists were asked to individually generate descriptive terms based on a first assessment of the whole product set. In the same objective, it was also possible to provide help in the form of pre-existing lists of attributes from the literature or from previous studies. These lists might indeed help the subjects to put words on their sensations.

Following this preliminary session, the experimenter collected all the terms generated by the panel and gathered them into a common list. Terms on this list were organized according to their corresponding sensory modality and sorted alphabetically. The compiled list was then provided to the panellists during the main evaluation session, together with their own initial list in order to help them in adjusting their final choice of attributes, and possibly finding attributes they might have either neglected or named with difficulty. In our experience, however, this happens very rarely and subjects tend to stay with their initial list. As a result, we now mostly skip this step, considering that its added value is relatively limited. Besides, applying two sessions considerably affects the quickness and responsiveness of the methods.

The preliminary session might nevertheless be useful when the differences to be described are subtle, or when practical constraints make a first evaluation round necessary in order to get a global picture of between-product differences. The extreme scenario would be the evaluation of products like cars, when one has to drive few kilometres to draw some sensations from this experience and then has to switch to another car and drive again, and so on.

Should the experimenter decide to run the FP with a preliminary session, it must be made very clear to the panellists that the objective at this stage is to elicit discriminant attributes only. It would indeed be useless to generate an exhaustive list of attributes. In very few cases, it has indeed been observed that subjects who are used to attribute elicitation in QDA may misunderstand the task and generate a list of all potentially applicable attributes, which would be quite counterproductive.

6.3.3 Training a panel dedicated to FP

Like other rapid methods, FP is ideally suited for one-time use, according to the needs of the study. However, when it comes to regular use of FP the question may arise whether it is worth dedicating a panel to FP. It would indeed be convenient to have panellists at hand for running FP studies. Informally, it is nearly the way we work in our lab with subjects (i.e. colleagues) who we regularly recruit. They have complementary fields of expertise in food science (e.g. flavour chemistry, rheology, food processing, packaging), which has been found to be of great added value and they now have extensive experience of sensory testing. Similar situations may be encountered both in academia and in the industry. For example, the food ingredient company Puratos has implemented such a panel for regular use of FP. Testimony of this experience is given by Petit and Vanzeveren in Chapter 15. It must be noted that maintaining such a panel only makes sense when there is an actual need for evaluating finite sets of products in a very short time, and if the product sets are different. On the contrary, if the panel always has to evaluate the same category of products, it would be much wiser to train a panel for conventional profiling. As soon as it comes to routine evaluation, the decision of whether to run FPs or to set up a conventional profile (or another technique) has to be made on a broader perspective.

A directly related question is whether it is possible to train such a panel dedicated to FP. To date, this question has not been investigated, and there is also no standard method for such training. Should one want to do so, with all the precautions that have just been made, it would be advisable to train the subjects in the view of developing their attention, introspective sensory focus, and general quantification abilities, rather than training them on specific attributes, or specific product sets. Should this latter option come into consideration, it would probably mean that conventional profile is a better option.

6.4 Metrological properties of Flash Profile

Although FP data may be relatively unusual to the user of conventional descriptive profile, a series of metrological properties (accuracy, repeatability, sensitivity, robustness and level of consensus) may provide a good indication of the quality of the results. These properties, how to estimate them and their pertinence in the case of FP are discussed in this section.

6.4.1 Accuracy

The accuracy or validity of FP results can first be estimated by comparison with a reference technique, i.e. a conventional profile such as QDA. Note that when conducting an FP with consumers from a market research perspective, this comparison criterion may not hold and following discussion on validity takes on a different meaning.

Studies comparing FP with conventional profiling have shown that very similar configurations were obtained with both methods, and that in addition to this very good match, FP could reach a similar or even better discrimination level (Dairou and Sieffermann, 2002; Delarue and Sieffermann, 2004a). Comparison is, however, essentially limited to configurational analysis, i.e. comparison of sensory maps. Although a semantic comparison may also be attempted (Dehlholm *et al.*, 2012b), it remains highly interpretative.

In addition to this, one may want to assess the quality of FP measurements in terms of panel performance. In this regard, it should first be mentioned that panel performance cannot be understood as in conventional consensus-based methods, since panellists do not use the same attributes, and even if they do use the same words these words may not be assumed to convey the exact same sensory notion. Also, one should bear in mind that FP results are dependent on subjects' descriptive skills, levels and fields of expertise. As already mentioned, should one wish to get the most comprehensive description of a product set, it is advisable to recruit experienced subjects with complementary fields of expertise.

As a result, most panel monitoring techniques that have given rise to a significant body of literature are not really applicable here. Nevertheless, a number of criteria can be reviewed and discussed briefly.

6.4.2 Repeatability

Repeatability in FP could be assessed by means of a full repetition of the evaluation task, or by the use of duplicates. A full repetition would imply repeated evaluation sessions. In the past for instance, we ran FP evaluation sessions in triplicate and found a good repeatability of the panel's results over the three repetitions (Delarue and Sieffermann, 2004a). Besides, in an unpublished earlier study ran in triplicate we found a slight drift from evaluation session 1 to the evaluation session 3, which we explained by an increased familiarity of the subjects with the product set along with repetitions. However, an evolution of the products in that time frame (10 days from first session to the last) could not be excluded. Eventually, in current practice, running multiple sessions is rare, since it is quite contradictory to the objective of rapidity sought with the use of FP.

The second option is to use duplicates. Duplicates are indeed often used as way to probe repeatability in rapid descriptive methods. The analysis of the position of the duplicate on the sensory map gives an indication of how repeatable is the sensory measurement with the view that the closer are the duplicates on the map, the more repeatable are the subjects. Although this is a simple and efficient way to test repeatability, this solution cannot be seen as completely satisfactory because of the comparative nature of FP data. As such, the interpretation of the position of the duplicates on a sensory map may be misleading. In effect, when two duplicated samples are distant on the sensory map, the most obvious interpretation would be that the panel is not repeatable. However, two replicated samples that are positioned apart on a sensory map could mean that all products in the product set are confusable and that inter-product distances are not larger than the difference between two samples of the same product. The experimenter should thus consider this ambiguity and question the interpretation of the results in each case and in the light of prior knowledge of the existing differences in the product set. This also relates to the issue of the sensitivity of the measurement, and that will be discussed next.

Besides, it must be noted that in any case duplicated samples would only provide an index and not an evaluation of the full repeatability. Of course, this could be improved by duplicating more than one product. However, it is wise not to multiply the number of duplicates up to the point where the product set will be distorted and where the subjects will implicitly be asked to rank too many identical stimuli. Second, the panellists must not, and at any cost, suspect that duplicates are present. They will otherwise pay extra attention in "finding" the duplicates and their responses will otherwise be automatically biased.

6.4.3 Sensitivity and robustness

Sensitivity of FP measurement essentially reflects in its discriminant power. As mentioned above, comparison studies have shown that FP leads to a level of product discrimination at least as good as that of conventional profile. Of course, these are results based on few examples only, and cannot be taken for granted. Some methodological points may, however, explain why FP may be more discriminant than conventional profiling.

First, Ishii *et al.* (2007) have provided evidence that attribute-by-attribute evaluation is easier for untrained judges and that, using this protocol, they reached criterion performance more rapidly than when using a serial monadic protocol. In addition to this, the ranking procedure, combined with this comparative evaluation, certainly helps in that it is easier to complete and allows re-tasting the samples (Kim and O'Mahony, 1998; Rodrigue et al., 2000).

Second, in the FP procedure, the subjects are told to focus on the differences between the products. This implies that the subjects directly prioritize on major differences between the products and choose attributes accordingly. Dairou and Sieffermann (2002) have argued that this point certainly makes the attribute elicitation step more efficient.

Third, it can also be hypothesized that subjects better discriminate products on individual attributes they have personally chosen than on consensual attributes that they had to learn, or for which they had to align with other subjects, and hence for the definition of which they might have had to compromise.

6.4.4 Level of consensus among subjects

As already discussed, the level of consensus cannot really be a quality criterion in the case of FP since in its very principle it is assumed that different subjects may not perceive and hence describe the product set in the same way.

It is nevertheless informative to assess the level of consensus among the subjects. as it gives an idea of the diversity of perceptions. It can be estimated in several ways, depending on the data analysis technique that is initially used. Most of the time, the level of agreement is primarily examined in terms of data configuration.

When using GPA, for instance, the relative distance of each subject's configuration to the consensus is given by the Procrustean residual. Accordingly, several authors have proposed tests of the level of consensus from GPA (King and Arents, 1991; Wakeling et al., 1992). See also Dijksterhuis and Heiser (1995) and Xiong et al. (2008) for a discussion of permutation tests for GPA. On the score plot, one may also look at how the product positions in each subjects' configuration connect to the product positions in the consensus configuration (Dijksterhuis, 1996). Alternatively, and as presented in the example above, some authors have also proposed to represent confidence ellipses on the score plot from the GPA (Hunter and Muir, 1995). When using MFA instead of GPA, the level of consensus can also be tested, and inter-subjects similarities can be evaluated using RV coefficients (Robert and Escoufier, 1976; Escofier and Pagès, 2008). Dehlholm et al. (2012a) and Cadoret and Husson (2013) have also proposed ways to calculate confidence ellipses.

In addition to this, one may want to detect possible subgroups and outliers, as suggested by Gains and Thomson (1990) in the case of FCP. To this end, cross-subject RVs or Procrustes residuals can be used. An example of such analysis is given in Section 6.3.

Eventually, Blumenthal et al. (2000) have argued that when experts differ in their description of the product set, it is wiser to look at their sensory maps separately. In such cases, they showed that individual sensory maps could be used selectively in order to improve the goodness of fit in preference mapping studies.

6.5 Limitations of Flash Profile

The comparative nature of FP is an advantage but it also bears limitations. Some of them relate to practical aspects (product availability, control of evaluation conditions) while others are more fundamentally sensory pertaining to the difficulties of the evaluation task. Eventually, the relative nature of the data acquired may limit the use of FP for some applications.

6.5.1 Product availability

Because FP relies on the simultaneous presentation of the product set, all the products must be available at the same time. This may be a major limitation to the use of FP if some products are not available together with the rest of the product set. It means they cannot be included in the test. When planning an FP, the experimenter must thus pay attention to production constraints (if samples have to be made), availability on the market, etc.

6.5.2 Control of evaluation conditions

As in any sensory evaluation, one should ensure that all the products are presented in the same conditions of age (for short shelf life products), quantity, temperature, etc. In FP, presenting all the samples simultaneously at the same temperature can be quite challenging for "unstable" products such as hot beverages and ice-creams. In such cases, it is especially important to make sure that the samples remain stable while the subjects perform the evaluation task. An option is to renew the product set with "fresh" samples when needed in the course of the session. In regard to control of the conditions of evaluation, one could also argue that the fact that all products are evaluated in the same session ensures that they are evaluated in the same contextual conditions (Albert *et al.*, 2011).

6.5.3 Difficulty of the task and number of products

Of course, presentation order effects (carry over, halo, etc.) cannot be controlled as in a monadic sequential protocol and can thus be a difficulty. Subjects' experience plays a role here and partly circumvents these issues, as subjects can manage their own evaluation by adapting their tasting, mouth-rinsing and comparison protocols. Nevertheless, one should pay attention to sensory adaptation, sensory fatigue, etc. Besides, one should keep in mind that the way a session is conducted in FP is flexible, and can be designed so that the evaluation task stays feasible. For example, it is possible to plan long breaks between attributes or, if necessary, to set up separate sessions of two or three attributes only.

This raises the question of the number of products that can be evaluated with FP. The answer, of course, depends on the nature of the products to be evaluated. The comparative evaluation task is believed to limit the number of samples that can be evaluated in a single session. Considering the theoretical number of pair-comparisons that would be necessary to accurately rank all samples, the task would indeed be expected to be extremely tiring and time-consuming (Ishii *et al.*, 2008). But in practice subjects adapt their cognitive strategies and use both direct comparative evaluation and their memory of previously tasted samples, which make them much more efficient. For this reason, it is important that the subjects can physically position the samples in front of them so that they can proceed to ranking in an adaptive manner.

In our experience, it seems that this number is often underestimated by sensory practitioners. For instance, Tarea, Cuvelier and Sieffermann (2007) have shown it was possible to get relevant sensory information from the FP of as many as 49 different apple and pear purées.

6.5.4 Cross-evaluation comparison and build-up of database

The comparative nature of FP also makes it poorly adapted to quality control and to building up a sensory database. The fact that there is no "absolute" quantification with the use of calibrated scales indeed prevents data comparison from one study to another. Solutions to this limitation may, however, be sought by the use of reference samples that would be similar to separate studies. This would be similar to using "poles," as in polarized sensory positioning (PSP) (Teillet *et al.*, 2010). From this perspective, Teillet *et al.* (2013) have proposed a Flash-PSP approach.

6.5.5 Additional considerations

FP stands midway between the conventional attribute-based profiling methods and holistic non-verbal methods. This is certainly a very good trade-off in terms of rapidity, quantification and richness of descriptive information. However, when compared separately to each type of method, it could be noted that FP has neither the accuracy of conventional profiling, nor the advantage of holistic methods that allow access to subjects' perceptual space without interference of attributes.

The main assumption behind sensory profiling is indeed that perception can be split up into separate attributes, which may sometimes be challenging (Lawless, 1999). In addition to this, relying on attributes implies an analytical approach that may be conducive to results that are somewhat distinct from the global sensory image that forms in our brains when normally consuming or using a product. It could also be argued that verbal-based methods rely on the ability to translate sensations into words and hence on subjects' linguistic aptitudes. As noted by Valentin *et al.* (2012), it is likely that product aspects difficult to verbalize will be overlooked by these methods. Using an FCP protocol and relying on subjects' complementarity may partly compensate for this potential weakness.

6.6 Evolution in the use of Flash Profile

Thanks to its emphasis on rapidity, FP has been an early success. Some sensory scientists have first seen it as a quicker and cheaper alternative to conventional profiling

(Valentin et al., 2012), and sensory practitioners have started to use it in situations where sensory positioning of the products is central and more relevant than attribute per attribute accurate ratings. In addition to this, practitioners have started to realize that rapidity comes with flexibility which is perhaps an even more interesting feature. This has opened ways for the use of descriptive analysis in situations where it was previously not adapted or even not possible to apply. It must be stressed that in these cases FP is no longer a mere cheaper alternative to conventional profiling, but actually a complementary measurement tool.

FP would indeed be particularly well suited to situations where one has to deal with a finite set of products at a given point in time. This would often be the case in the frame of research or R&D projects when there is a need for sensory information in the absence of a trained panel for the product category under investigation. Besides, FP can also be seen as a consumer profiling method. The idea of using FCP with consumers dates back to Jack and Piggott (1992), and several studies in this direction have been conducted since. In order to gain insights into consumers' perceptions, sensory and market researchers in the industry have thus started applying fast sensory descriptive methods to relatively large consumer panels.

Below are listed the cases where FP would be most useful (Delarue et al., 2004):

- In any one-shot analyses, linked to one project with no pre-existing descriptive tool and no need for a trained panel after the project is completed.
- In research projects, especially in first stages of the project in order to gain faster sensory knowledge and make informed decisions in project management. This may apply to new product development (NPD) projects in R&D as well as to sensory research projects.
- When the products to be analysed have very short shelf lives.
- In market research, when one wants to get rapid sensory insights into a market for a given product category.
- In sensory consumer research, in order to understand the diversity of consumers' perceptions.

Examples of such situations are presented in this section with emphasis on three major directions: FP with design of experiment (DoE) in early stages of a research project, FP with professionals and FP with consumers.

6.6.1 Use of FP with a design of experiment (DoE)

A typical situation where FP can be of great added value is when conducting research or development projects on a product set that is defined following a DoE. DoEs are used for a number of applications in formulation, in optimization, in psychophysics, etc. In such cases, the product set under study essentially consists of prototypes that are, by definition, new in some way. Thus, even if pre-existing sensory profiling resources (i.e. a trained panel on the same category of products) are available, they may not be directly applicable, and further adaptation of the attributes with corresponding training may be needed. If such resources are not available, setting up a conventional profile dedicated to the evaluation of that specific product set may prove to be difficult, or even impossible, depending on product fabrication and budget constraints. What follows is an example of such a use of a DoE for the study of flavour perception in yogurts (Delarue, 2002).

Table 6.2 **Factorial design used for the formulation of strawberry yogurts to be evaluated by FP**

Product number	Factor levels			Concentrations		
	Sugar	Lactic acid	Citric acid	Sugar (%)	Lactic acid (°D)	Citric acid (g kg^{-1})
1	−1	−1	−1	7.50	9.0	0.5
2	−1	−1	1	7.50	9.0	2.5
3	−1	1	−1	7.50	13.0	0.5
4	−1	1	1	7.50	13.0	2.5
5	0	0	0	8.25	11.0	1.5
6	1	−1	1	9.00	9.0	2.5
7	1	−1	1	9.00	9.0	2.5
8	1	1	−1	9.00	13.0	0.5
9	1	1	−1	9.00	13.0	0.5

Yogurts are living and short shelf life products and they cannot be preserved stably over the time that would be necessary for training a panel. Therefore, unless pilot products are made ad hoc, conducting a conventional profile is impossible and alternatives must be sought should one still wish to gain sensory insights.

In this study we investigated the effect of sugar and various acids on the flavour perception of strawberry yogurts. A set of nine products was thus made according to a 2^3 factorial design based on three factors: *total sugar content*, *titratable acidity* (i.e. concentration of lactic acid in the yogurt base expressed in degrees Dornic (° D)), and *concentration of citric acid* in the fruits (Table 6.2). The levels of each factor were defined according to literature and prior knowledge, but before the study was conducted the effect of these factors was not well-known and one crucial question was whether the range of variation for these factors would be appropriate with respect to the study of flavour perception of this type of product.

These nine products were thus evaluated by means of an FP with seven experienced subjects recruited from among graduate students and researchers at AgroParisTech. These subjects used between four and eight attributes, for a total of 27 semantically different attributes. The data were pooled and subsequently analysed by GPA. The sensory map derived from the first two axes of the GPA is shown on Fig. 6.7, with symbols representing the factor levels (citric acid is not represented). As can be seen from this plot, the yogurts are well separated according to their titratable acidity and sugar content. It can even be noticed that these two factors seem to have independent effects. The effect of titratable acidity is mainly loaded on the first axis, which accounts for 76% of variance, while the effect of sugar content reflects on the second axis, which accounts for 10% of variance. This gives us an idea of the relative weight of these factors on sensory perception. The effects of experimental factors can further be tested by principal component regression (Table 6.3). This analysis confirms visual interpretation and indicates that citric acid also has an impact on the sensory positioning of the products.

Flash Profile: evolution and uses in sensory and consumer science 141

Table 6.3 Effect of experimental factors on the sensory perception of strawberry yogurts

	Standardized coefficients	F	p-value
Axis 1 (75.6%): $R^2 = 0.85$; $F_{model} = 91.5$ ($p < 0.0001$)			
Lactic acid	−0.94	214.7	<0.0001
Sugar	0.30	29.2	<0.0001
Citric acid	−0.16	6.4	0.014
Axis 2 (9.7%): $R^2 = 0.43$; $F_{model} = 12.5$ ($p < 0.0001$)			
Lactic acid	0.35	7.9	0.007
Sugar	0.56	27.7	<0.0001
Citric acid	0.32	6.9	0.012

These effects are assessed by a multiple linear regression on each of the first two principal components of FP data after GPA.

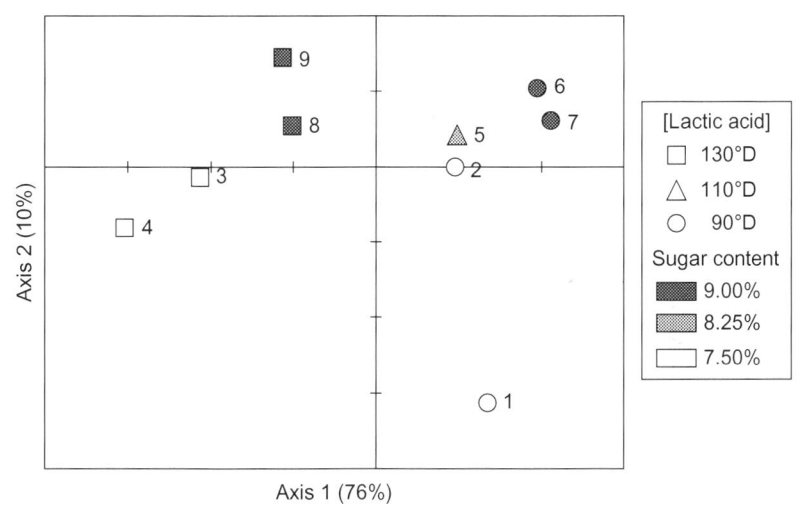

Figure 6.7 Sensory map (GPA consensus configuration) of the FP of strawberry yogurts made with varying levels of lactic acid and sugar.

In this example, analysing the relative sensory positioning of the products in relation to the DoE provides insightful information regarding the role of experimental factors on perception and on the pertinence of the DoE. At this stage of the study, this information gathered within a week may be sufficient. Of course, the analysis of the loading plot may also give some insights into what sensory characteristics are influenced by these factors but this information would come next. In this case, the most striking result is that the first axis could be interpreted as a *sour to sweet* dimension (Fig. 6.8). Interestingly, this dimension is mostly driven by titratable acidity and also correlates with other flavour and texture attributes.

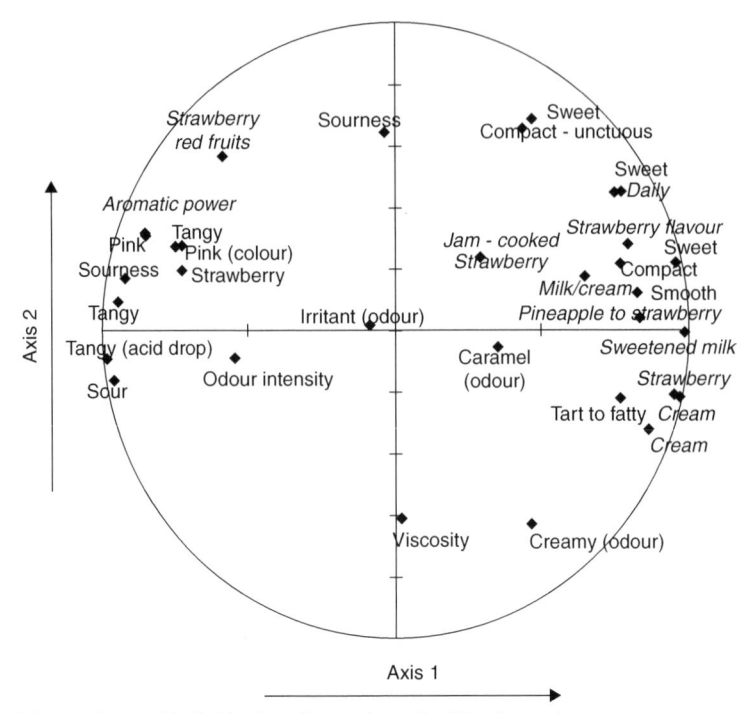

Figure 6.8 Loadings of individual attributes from the FP of strawberry yogurts.

For about 10 years, FP was used in our team in combination with DoE in a number of studies (see for example: Dairou *et al.*, 2003; Taréa *et al.*, 2003; Delarue and Sieffermann, 2004b; Coustel, 2005; Blancher *et al.*, 2007). By extension, such a way of working can be adopted in early stages of research or development projects where rapid sensory feedback is needed.

6.6.2 FP with professionals

As already mentioned FP is in line with the professional work rhythms and the evaluation could very well be conducted in the frame of a working meeting. Its flexibility and the fact that evaluation consists of only one session make it particularly well adapted for working with professional experts and internal panellists.

Professional experts are of at least two types. First are those experts, such as oenologists, flavourists, perfumers, chefs, acousticians, who have developed acute sensory skills. Together with their recognized sensory expertise, these professionals have the ability to relate their sensations to technical specifications of the products (e.g. quantity of a given ingredient in a formula, tuning of a technical parameter in a car), to the quality of raw materials (e.g. vineyard and grape variety; quality and origin of cocoa beans) or even to the way the products were made (e.g. cooking process, fermentation management, ageing). These experts, who usually work individually, are often reluctant to participate in conventional descriptive analysis. The main reason for this, besides time constraints, is the restrictive nature of consensus-based descriptive techniques that limit the use of attributes to a lowest common denominator. A number of chapters in this book illustrate how FP has been successfully used with such experts.

The second type refers to all those experts like product developers who have developed a specific product-wise expertise. Although they are usually not so trained as the first type of experts, whose sensory perceptions are central in their work, they are used to evaluate products in their daily work and in many cases they use their senses to make decisions and to orientate their work. They also need to communicate their sensory perceptions to colleagues (i.e. other project team members) and to stakeholders. The problem often met with these professionals is that they lack a methodological framework that would help them in structuring and quantifying their own descriptions in an efficient way. Naturally, for those professionals as well, time is a major constraint. Techniques like FP can prove to be very useful to them. In a study on the relationship between instrumental and sensory characterization of bread, we have included bakers and other professionals from the baking industry in the sensory panel (Lassoued *et al.*, 2008). These professionals quickly generated a great number of sensory attributes. Their good knowledge of this kind of product made this step easy. However, they needed more time to complete their evaluation session than other sensorily experienced panellists.

6.6.3 FP as a consumer-orientated method

6.6.3.1 Evaluation task with consumers

Although FP was initially designed to be conducted with experienced panellists, it can also be seen as a consumer-orientated method. FP is potentially better adapted than other attribute-based profiling methods because ranking is a simpler task than rating (Ishii *et al.*, 2007), and because only one evaluation session is needed. According to Ballay *et al.* (2004) and to Moussaoui and Varela (2010), it is possible for totally untrained consumers to perform the descriptive task without any difficulty. However, in our experience specific guidance must be provided with very didactic examples. It would also require extra attention on the part of the experimenter, as in face-to-face interviews, even though several participants could be managed simultaneously (Veinand *et al.*, 2011).

6.6.3.2 Subject segmentation

When conducting an FP with a relatively large number of consumers, it is doubtful whether they will all have a similar view of the product set. It may thus be pointless to analyse their data in the same analysis as is usually done with experts. In an FCP study conducted with consumers, Gains and Thomson (1990) have proposed to use subject-wise Procrustes residuals to detect possible subgroups and outliers. In the same line of thinking, we have suggested applying cluster analysis in between subjects RVs in order to identify possible segmentation in our sample of consumers (Delarue *et al.*, 2011). In that study, 42 consumers evaluated a set of ten paper tissues using FP. As is usually the case in studies like this, the loading plot from the GPA gives a massive amount of information but is just not legible. It was thus decided to analyse whether subgroups of subjects could be identified and could hopefully lead to simpler, more pertinent and more actionable interpretation (Fig. 6.9).

In order to identify a possible segmentation, we first calculated cross-subject RV coefficients which were pooled into a square matrix of RVs. This matrix is then turned into a dissimilarity matrix of [1-RV] and submitted to hierarchical cluster analysis (Fig. 6.10).

Three groups of subjects can be clearly distinguished on the dendrogram. It thus makes sense to analyse these three groups of subjects separately. Accordingly, we ran three separate GPA, and three different sensory maps, one for each group, are obtained. The configurations obtained on these three maps indeed differ widely, but what is more, the semantic interpretation for each group becomes more evident.

The interest of this subgroup analysis can be illustrated with the example of the attribute *thickness*, which is the most frequently used (26 consumers used it). At first sight, one might think that the thickness of paper tissues is an obvious notion and that it would not lead to much disagreement among subjects. However, these data show quite the opposite. Indeed, as can be seen on the left-hand side of Fig. 6.11, individual *thickness* attributes point in every direction on the loading plot from the pooled GPA. When looking at the GPA output for the subgroups (on the right-hand side of Fig. 6.11) it can be seen that *thickness* is more consensual (i.e. attributes point in a narrower angle) for two of these groups (Class 1 and Class 3). With respect to product positioning, one can tell that for these two groups *thickness* indeed conveys different notions. For Class 1, *thickness* opposes product A (being the least thick) to products F and I (being the thickest), whereas for Class 3 *thickness* seems to be mostly driven by products G, F and H. The same goes with other attributes that are used in different ways by many consumers, but that are consensual within a subgroup (data not shown).

This type of analysis of consumer FP data may thus be of great added value should a typology of configurations exist. It is of course not always the case, and there are some situations where no subgroup can be distinguished and where the attempts at semantic interpretation remain fuzzy. In such cases, one should thus stick to data interpretation in terms of sensory positioning of the products at the group level.

6.6.3.3 Beyond sensory description

Once FP has been successfully conducted with consumers, researchers have attempted to go beyond sensory description in the strict sense. First, it is observed that when consumers participate in descriptive analysis studies, they tend to use some hedonically orientated terms, as well as composite notions that may refer to benefits, which are consequences of consuming the product (e.g. thirst quenching, filling up) (Veinand et al., 2011). They may also use attributes that are more integrative than those used by experts. The French adjective "onctueux" (creamy and smooth) would typically fall in this category. Such attributes may be referred to as meta-descriptors (Frøst and Janhøj, 2007) or "second-order sensory attributes."

Market researchers have thus envisaged using FP to evaluate how consumers perceive products not only in terms of sensory attributes, but also in terms of evocations,

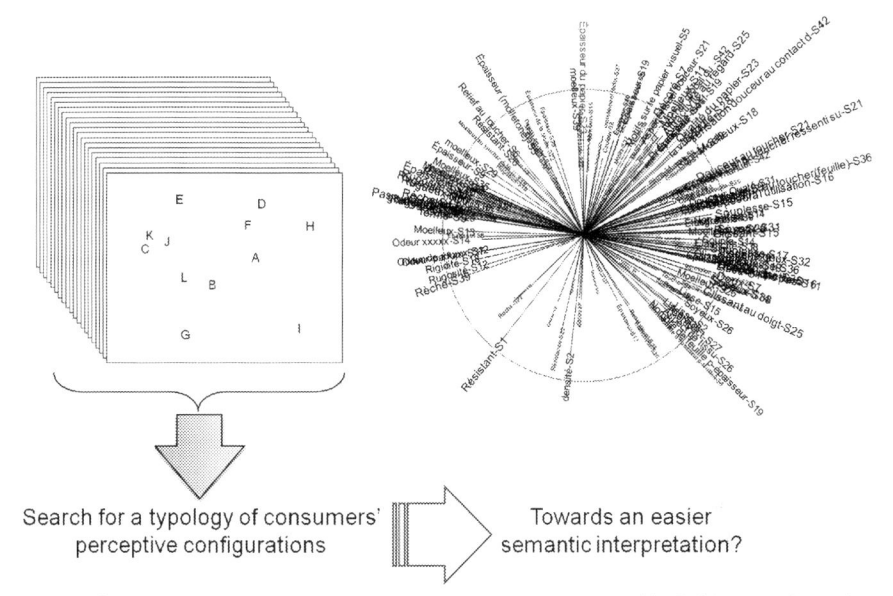

Flash Profile: evolution and uses in sensory and consumer science   145

Figure 6.9 When an FP is conducted with consumers a typology of individual configurations may be sought for easier and more pertinent interpretation.

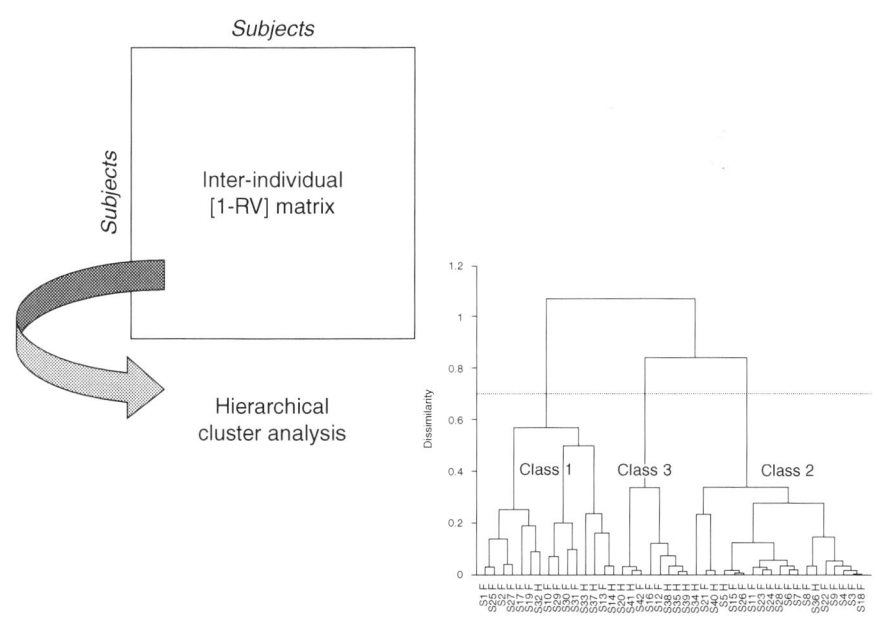

Figure 6.10 Principle of the cluster analysis of FP subjects based on cross-subject RV coefficients. Each class obtained can be submitted to a separate GPA.

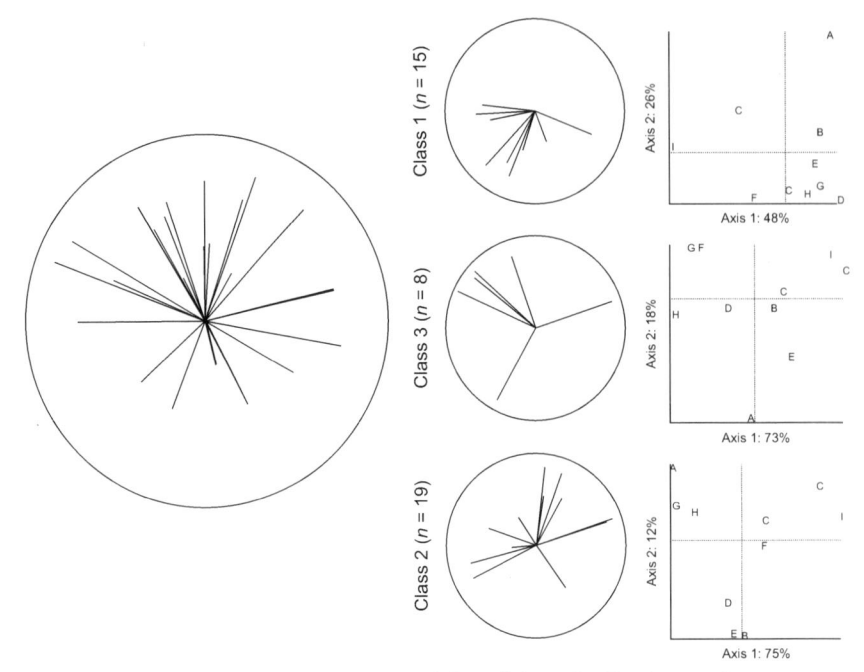

Figure 6.11 Loadings plots for the attribute "thickness" from the FP of a set of ten paper tissues. The plot on the left-hand side represents the loadings from the pooled GPA for the whole panel. On the right-hand side are the results (loadings and score plots) from three separate GPAs performed respectively on three classes of subjects defined from cluster analysis.

lifestyles, etc. Ballay *et al.* present a very nice example of how FP can be adapted to investigate these aspects in Chapter 19 of this book.

Accordingly, it is worth mentioning that FP can be adapted and combined with other rapid methods depending on the goal of the study. For example, in a study conducted on ten makeup remover wipes (Delarue *et al.*, 2008), we adapted the FP procedure to evaluate whether sensory claims found on such products' packaging related well to sensory perception of the products. The study was conducted in two stages.

In the first stage, all the claims, terms and expressions, found on the packaging were collected and presented to 34 consumers on small printed cards. Using a free sorting protocol, these subjects had to group the terms that had similar meaning, according to their own semantic perception. They were allowed to make as many groups as they wanted. Then they had to give a name to each group, with the possibility of using one of the terms from the group under consideration. Each subject thus built her own structured list of terms.

In the second stage, the groups defined individually were to be used as attributes in order to sensorily evaluate and rank the samples, as in FP, except that the list of attributes was constrained by the initial terms. Sufficient time was given to the participants so that they could evaluate the samples with limited fatigue and irritation.

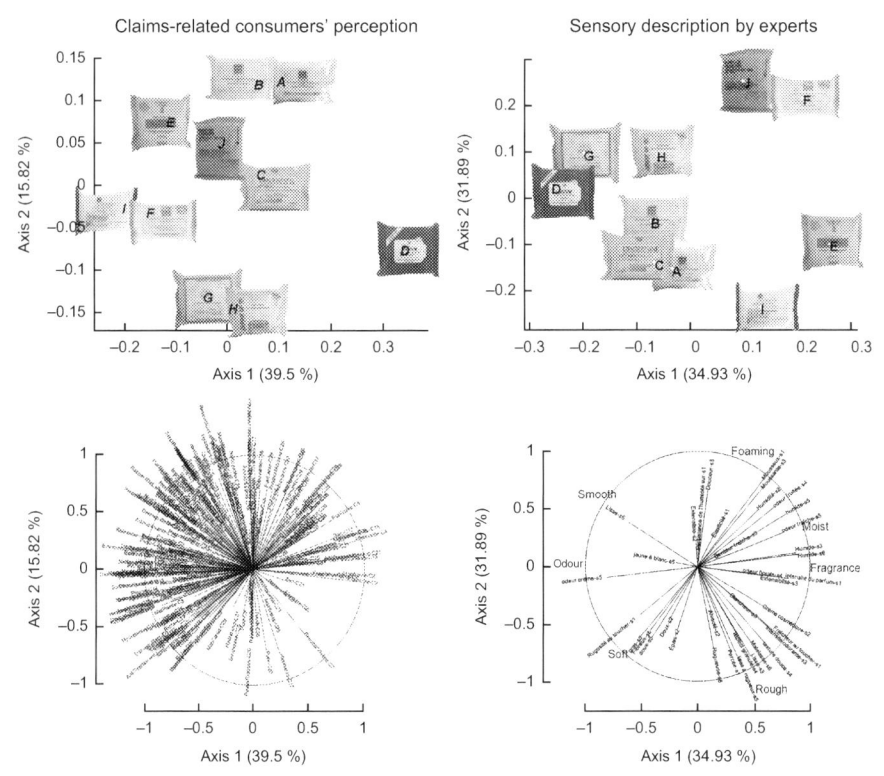

Flash Profile: evolution and uses in sensory and consumer science   147

Figure 6.12 Use of an adapted FP procedure to map consumers' perceptions of makeup remover wipes in relation to packaging sensory claims (left), and comparison with pure sensory description as obtained from a "conventional" FP with experts (right).
Source: Adapted from Delarue *et al.* (2008).

Results were collected and analysed using GPA. The resulting score plot was compared to that obtained with a "conventional" FP conducted with six experienced panellists on the same products (Fig. 6.12).

These data show an important discrepancy between experts' sensory descriptions and consumers' perceptions according to benefits and claims praised on packaging. A key finding was that wipes' fragrance seemed to play a major role in product perception, although it is very poorly communicated to the consumers on the packaging. In addition, it must be noted that the outcome of the consumer data is very consistent internally. In conclusion, the pertinence of the claims that are printed on such products may be seriously qualified when it comes to sensory perception. As a result, consumers may be confused by claims and benefits that do not match their product sensory properties. Developers and marketers could probably take advantage of such insights into consumers' perception of their products. This example shows that the initial FP procedure can be adapted to better address issues pertaining to consumer perceptions.

6.7 Conclusions and future trends

As has been shown through various examples, FP can be used for a number of applications. In current practice, FP is primarily used for its rapidity and responsiveness. As regards the issue of rapitodity, attention must be paid to the actual time spent in a broad perspective. In many cases, FP would indeed allow accessing to sensory information faster, providing that panellists can be easily recruited and that data analysis tools are available and mastered. When there is no pre-existing trained panel, there is no doubt that FP would be faster than any conventional profiling technique and trade-off would only have to be made between rapidity on the one hand and accuracy and ease of interpretation on the other hand. On the other hand, when one has to make repeated descriptive analyses on the same category of products, and/or when there is a trained panel available (for the category of products under consideration), a conventional profiling approach would be both more accurate and more rapid on the long run.

Besides, it should be noted that FP flexibility is also a key advantage that makes it a very adaptable tool in sensory science. Accordingly, the protocol could be modified so that measurement is better adapted to the aims and needs of the study. Combinations with other methods may also be considered, either to address current limitations of FP (for example by the development of Flash-PSP) or to better adapt to one's constraints and objectives. In future, it is most likely that FP will be increasingly used in the first stages of sensory studies or, alternatively, in complement to conventional profile when for example one wants to have a more consumer-orientated description of a product set.

References

Albert, A., Varela, P., Salvador, A., Hough, G. and Fiszman, S. (2011). Overcoming the issues in the sensory description of hot served food with a complex texture. Application of QDA®, flash profiling and projective mapping using panels with different degrees of training. *Food Quality and Preference,* **22**, 463–473.

Ballay, S., Sieffermann, J. M., Gazano, G. and Mahe, C. (2004). Flash profile with consumers: a new method to understand specific Japanese moisturizing expectations on cosmetic products. *23rd Congress of the International Federation of Societies of Cosmetic Chemists.* Orlando, FL.

Ballay, S., Warrenburg, S., Sieffermann, J.-M., Glazman, L. and Gazano, G. (2006). A new fragrance language: intercultural knowledge and emotions. *24th IFSCC Congress – Integration of Cosmetic Sciences,* 16–19 October 2006 Osaka, Japan.

Blancher, G., Chollet, S., Kesteloot, R., Hoang, D. N., Cuvelier, G. and Sieffermann, J. M. (2007). French and Vietnamese: How do they describe texture characteristics of the same food? A case study with jellies. *Food Quality and Preference,* **18**, 560–575.

Blumenthal, D., Dairou, V., Sieffermann, J. M. and Danzart, M. (2000). How to improve sensory information provided by free choice profiling in preference mapping using individual maps? *The 5th Sensometrics Meeting.* Columbia, Missouri.

Cadoret, M. and Husson, F. (2013). Construction and evaluation of confidence ellipses applied at sensory data. *Food Quality and Preference,* **28**, 106–115.

Coustel, J. (2005). *Mise au point du contre typage d'échantillons de rhum par voie organoleptique et analyses physico-chimiques.* PhD Thesis, ENSIA.

Dairou, V., Priez, A., Sieffermann, J.-M. and Danzart, M. (2003). An original method to predict brake feel: a combination of design of experiments and sensory science. *Society of Automotive Engineers World Congress.* Detroit, MI.

Dairou, V. and Sieffermann, J.-M. (2002). A comparison of 14 jams characterized by conventional profile and a quick original method, the Flash Profile. *Journal of Food Science,* **67**, 826–834.

Dehlholm, C., Brockhoff, P. B. and Bredie, W. L. P. (2012a). Confidence ellipses: A variation based on parametric bootstrapping applicable on multiple factor analysis results for rapid graphical evaluation. *Food Quality and Preference,* **26**, 278–280.

Dehlholm, C., Brockhoff, P. B., Meinert, L., Aaslyng, M. D. and Bredie, W. L. P. (2012b). Rapid descriptive sensory methods – Comparison of free multiple sorting, partial napping, napping, flash profiling and conventional profiling. *Food Quality and Preference,* **26**, 267–277.

Delarue, J. (2002). *Influence de l'acide et du sucré sur la perception de la flaveur. Application au cas du yaourt à la fraise.* PhD Thesis, ENSIA.

Delarue, J., Beurier, F. and Sieffermann, J. M. (2008). Mapping claims according to consumers' sensory perception of cosmetics. *25th IFSCC congress.* Barcelona, Spain.

Delarue, J., Blumenthal, D., Danzart, M. and Sieffermann, J.-M. (2011). Finding one's way through consumers' sensory descriptions. A simple clustering-based methodology. *9th Pangborn Sensory Science Symposium.* Toronto.

Delarue, J., Danzart, M. and Sieffermann, J.-M. (2004). Flash profile gives insights into human sensory perception. *5th International Multisensory Research Forum.* Barcelona, Spain.

Delarue, J. and Sieffermann, J.-M. (2004a). Sensory mapping using Flash profile. Comparison with a conventional descriptive method for the evaluation of the flavour of fruit dairy products. *Food Quality and Preference,* **15**, 383–392.

Delarue, J. and Sieffermann, J.-M. (2004b). Use of 2(-4-methoxyphenoxy)propanoic acid (Na-PMP) to investigate flavour interactions in real food products. *14th International Symposium on Olfaction and Taste (ISOT/JASTS).* Kyoto, Japan.

Dijksterhuis, G. (1996). Procrustes analysis in sensory research. *In:* Naes, T. and Risvik, E. (eds.) *Mulitvariate Analysis of Data in Sensory Science.* Amsterdam: Elsevier Science, pp 185–219.

Dijksterhuis, G. B. and Heiser, W. J. (1995). The role of permutation tests in exploratory multivariate data analysis. *Food Quality and Preference,* **6**, 263–270.

Escofier, B. and PAGÈs, J. (2008). *Analyses factorielles simples et Multiples, objectifs méthodes et interprétation,* Paris, Dunod.

Faye, P., Bremaud, D., Teillet, E., Courcoux, P., Giboreau, A. and Nicod, H. (2006). An alternative to external preference mapping based on consumer perceptive mapping. *Food Quality and Preference,* **17**, 604–614.

Frøst, M. B. and Janhøj, T. (2007). Understanding creaminess. *International Dairy Journal,* **17**, 1298–1311.

Gains, N. and Thomson, D. M. H. (1990). Sensory profiling of canned lager beers using consumers in their own homes. *Food Quality and Preference,* **2**, 39–47.

Gibbs, L., Kealy, M., Willis, K., Green, J., Welch, N. and Daly, J. (2007). What have sampling and data collection got to do with good qualitative research? *Australian and New Zealand Journal of Public Health,* **31**, 540–544.

Glaser, B. G. and Strauss, A. L. (1967). *The Discovery of Grounded Theory: Strategies for Qualitative Research,* Chicago, Aldine Publishing.

Gower, J. C. (1975). Generalised procrustes analysis. *Psychometrika,* **40**, 33–51.

Hunter, E. A. and Muir, D. D. (1995). A comparison of two multivariate methods for the analysis of sensory profile data. *Journal of Sensory Studies,* **10**, 89–104.

Husson, F., Le Dien, S. and Pagès, J. (2001). Which value can be granted to sensory profiles given by consumers? Methodology and results. *Food Quality and Preference,* **12**, 291–296.

Husson, F., Lê, S. and Pagès, J. (2005). Confidence ellipse for the sensory profiles obtained by principal component analysis. *Food Quality and Preference,* **16**, 245–250.

Ishii, R., Chang, H.-K. and O'Mahony, M. (2007). A comparison of serial monadic and attribute-by-attribute protocols for simple descriptive analysis with untrained judges. *Food Quality and Preference,* **18**, 440–449.

Ishii, R., Stampanoni, C. and O'Mahony, M. (2008). A comparison of serial monadic and attribute-by-attribute descriptive analysis protocols for trained judges. *Food Quality and Preference,* **19**, 277–285.

Jack, F. R. and Piggott, J. R. (1992). Free choice profiling in consumer research. *Food Quality and Preference,* **3**, 129–134.

Kelly, G. A. (1955). *The Psychology of Personal Constructs,* New York, Norton.

Kim, K. and O'Mahony, M. (1998). A new approach to category scales of intensity I: traditional versus rank-rating. *Journal of Sensory Studies,* **13**, 241–249.

King, B. M. and Arents, P. (1991). A statistical test of consensus obtained from generalized procrustes analysis of sensory data. *Journal of Sensory Studies,* **6**, 37–48.

Lassoued, N., Delarue, J., Launay, B. and Michon, C. (2008). Baked product texture: Correlations between instrumental and sensory characterization using Flash Profile. *Journal of Cereal Science,* **48**, 133–143.

Lavit, C. (1988). *Analyse conjointe de tableaux Quantitatifs,* Paris, Masson.

Lavit, C., Escoufier, Y., Sabatier, R. and Traissac, P. (1994). The ACT (STATIS method). *Computational Statistics and Data Analysis,* **18**, 97–119.

Lawless, H. T. (1999). Descriptive analysis of complex odors: Reality, model or illusion? *Food Quality and Preference,* **10**, 325–332.

Marshall, M. N. (1996). Sampling for qualitative research. *Family Practice,* **13**, 522–525.

Moussaoui, K. A. and Varela, P. (2010). Exploring consumer product profiling techniques and their linkage to a quantitative descriptive analysis. *Food Quality and Preference,* **21**, 1088–1099.

Piggott, J. R. and Watson, M. P. (1992). A comparison of free-choice profiling and the repertory grid method in the flavor profiling of cider. *Journal of Sensory Studies,* **7**, 133–145.

Robert, P. and Escoufier, Y. (1976). A unifying tool for linear multivariate statistical methods: The RV-coefficient. *Journal of the Royal Statistical Society. Series C (Applied Statistics),* **25**, 257–265.

Rodrigue, N., Guillet, M., Fortin, J. and Martin, J.-F. (2000). Comparing information obtained from ranking and descriptive tests of four sweet corn products. *Food Quality and Preference,* **11**, 47–54.

Sandelowski, M. (1995). Sample size in qualitative research. *Research in Nursing and Health,* **18**, 179–83.

Sieffermann, J.-M. (1995). *Etude comparative de méthodes descriptives en analyse sensorielle – Application à l'évaluation de l'efficacité du profil sensoriel libre.* Thèse de Doctorat, ENSIA.

Sieffermann, J.-M. Le profil flash – Un outil rapide et innovant d'évaluation sensorielle descriptive. AGORAL (2000)- XIIèmes rencontres "L'innovation: de l'idée au succès," 2000 Montpellier, France. 335–340.

Taréa, S., Danzart, M., Sieffermann, J. M. and Cuvelier, G. (2003). A simple way to analyze sensory profiling data in order to select relevant experimental factors in food formulation. *The 5th Pangborn Sensory Science Symposium.* Boston, MA.

Teillet, E., Petit, C. and Delarue, J. (2013). Combining PSP and Flash Profiling... Why? How does it work? *10th Pangborn Sensory Science Symposium.* Rio de Janeiro.

Teillet, E., Schlich, P., Urbano, C., Cordelle, S. and Guichard, E. (2010). Sensory methodologies and the taste of water. *Food Quality and Preference,* **21,** 967–976.

Thamke, I., Dürrschmid, K. and Rohm, H. (2009). Sensory description of dark chocolates by consumers. *LWT – Food Science and Technology,* **42,** 534–539.

Thomson, D. M. H. and McEwan, J. A. (1988). An application of the repertory grid method to investigate consumer perceptions of foods. *Appetite,* **10,** 181–193.

Valentin, D., Chollet, S., Lelièvre, M. and Abdi, H. (2012). Quick and dirty but still pretty good: a review of new descriptive methods in food science. *International Journal of Food Science and Technology,* **47,** 1563–1578.

Veinand, B., Godefroy, C., Adam, C. and Delarue, J. (2011). Highlight of important product characteristics for consumers. Comparison of three sensory descriptive methods performed by consumers. *Food Quality and Preference,* **22,** 474–485.

Wakeling, I. N., Raats, M. M. and Macfie, H. J. H. (1992). A new significance test for consensus in generalized procrustes analysis. *Journal of Sensory Studies,* **7,** 91–96.

Williams, A. A. and Langron, S. P. (1984). The use of free-choice profiling for the evaluation of commercial ports. *Journal of the Science of Food and Agriculture,* **35,** 558–568.

Worch, T., Lê, S. and Punter, P. (2010). How reliable are the consumers? Comparison of sensory profiles from consumers and experts. *Food Quality and Preference,* **21,** 309–318.

Xiong, R., Blot, K., Meullenet, J. F. and Dessirier, J. M. (2008). Permutation tests for generalized procrustes analysis. *Food Quality and Preference,* **19,** 146–155.

Free sorting as a sensory profiling technique for product development

P. Courcoux,[1] E.M. Qannari,[1] P. Faye[2]

[1]L'UNAM University, ONIRIS and INRA, Nantes, France; [2]L'UNAM University, ONIRIS and PSA Peugeot Citroen, Nantes, France

7.1 Introduction

Categorization tasks based on sorting objects into groups are routinely used in the cognitive and social sciences. One of the earliest references is Hulin and Katz (1935) that describes a study on judgment of facial expressions by sorting a set of 72 pictures. In the domain of psychology, sorting tasks are usually performed for investigating the cognitive process of categorization, a fundamental and natural process of classification of objects. The items presented to the subjects are usually exemplars (words or pictures) relative to a concept and the subjects are asked to sort them according to their similarities and/or dissimilarities. The practitioner is interested in how the subjects categorize the stimuli and, more importantly, in the properties of the stimuli that the subjects use to assign them to categories.

Sorting tasks were introduced in the field of sensory sciences in the late 1980s. The main objective is generally the assessment of the perception of a set of products by a panel of subjects. In this context, the categorization focuses more on the sensory characterization of the stimuli (product-oriented studies) than on the analysis of the strategies that the subjects use to sort the products or the differences between subjects (subject-oriented studies). For this reason, the sorting task is frequently presented as an alternative to usual descriptive profiling techniques. The groups formed by the subjects are considered as a means of assessing the dissimilarities between stimuli, and the main outcome of the experiment is the representation of these dissimilarities on a map, interpreted as the representation of the perception of the products.

The sorting procedures may also offer more possibilities than the sensory characterization of a set of products. Indeed, the simplicity of the procedure and the non-verbal nature of the task make it possible to perform such experiments with consumers (i.e. non-trained subjects), with subjects of different cultures or languages or even with children. In this context, the objective may be to describe both perceptual and cognitive representation of the stimuli. This leads to gaining a better insight into the perceptions of consumers or shedding light into differences in subjects' perceptions or behaviors.

7.2 The free sorting task

7.2.1 The free sorting procedure

7.2.1.1 Perceptive task

Among the various categorization procedures, the free sorting task is the most frequently used in sensory sciences. Subjects are provided with a set of stimuli and are instructed to sort them in groups, with the understanding that the samples assigned to the same group are supposed to be similar and that different groups should contain samples that they perceive as different. Subjects are generally free to form as many groups as they deem necessary. Sometimes they are instructed to make at least two groups and not to set each product in a separate group. Generally, no instruction is given concerning the strategy the subjects should use to sort the stimuli into groups.

Depending on the objective of the study, free sorting may be performed considering one or several sensory modalities and the criterion for grouping the objects may be specified (texture, odor, etc.). In contrast, if the global similarity between stimuli is of interest, the subjects are advised that they are free to use any characteristic of the products to form coherent and homogeneous groups.

The initial order of presentation of the samples may have an effect on the partitioning given by the subjects. In order to minimize this bias, samples should be presented to each subject in a randomized order or, preferably, using a design balanced for order effect (as a Latin Square design).

7.2.1.2 Verbalization task

Once the sorting task has been completed, the subjects may be instructed to describe the different groups they have formed. In this case, they are asked to give a description of each group using their own vocabulary. For each group, the subjects provide a list of terms (or attributes) that characterize the products in the group under consideration. This verbalization task is a crucial step for interpreting the perceptual dimensions underlying the categorization process.

The verbalization task is generally not easy to perform, particularly with untrained subjects who are not familiar with sensory evaluation. Alternatively, a list of terms can be provided to the subjects who are asked to check all the terms that apply to each group they have formed. Lelièvre et al. (2008) compared the two techniques of verbalization task (free verbalization vs a predefined list of attributes) after a sorting task of beers and showed differences in the results obtained under the two conditions. Particularly for untrained assessors, the number of attributes was larger with the list than for the free condition without, however, enhancing the accuracy of the description of the products. These results confirm the findings of Hughson and Boakes (2002), who showed that the length and the composition of the pre-established list have an effect on the attributes generated by the subjects. The use of a list of terms is less time consuming in terms of data collection and analysis than free verbalization. However, the interpretation of the terms in a predefined list may be difficult since the choice of such terms can be dictated by an association of ideas, words or a specific context, and thus may have little bearing to the stimuli involved in the experiment. Furthermore, the list may be incomplete or contain technical or specific terms, unsuitable for the subjects performing the task. By contrast, free verbalization yields a spontaneous description of the stimuli. Thanks to the diversity in consumer perception, the richness of the products' descriptions generally increases with the number of subjects involved in the experiment.

7.2.1.3 Size of the panel

A large number of studies reported different panel sizes. When working with trained assessors or professional experts of the products, the number of subjects is relatively small (between 10 and 20). For studies involving untrained panelists, the number of subjects ranges between 9 and 389 (Chollet *et al.*, 2011; Teillet *et al.*, 2010; Varela and Ares, 2012). Using resampling techniques, Faye *et al.* (2006) showed that the product configuration becomes stable when the size of the panel is more than 25 consumers. Blancher *et al.* (2012) proposed a method for investigating the stability of sorting maps in connection with the number of subjects in free sorting task, and recommended working with an initial number of 30 untrained assessors. It is worth noting that the minimum number of untrained subjects in a free sorting experiment highly depends on the homogeneity of the panel perception (Faye *et al.*, 2013) and the diversity of the product space.

7.2.2 Overview of the use of free sorting

The objective of the sorting task in product perception is two-fold: the assessment of the similarities between products, and the understanding of how the subjects form categories. Clearly, most of the applications involving sorting tasks are both product- and subject-oriented. In an early work on sorting fragrances, Lawless was interested not only in the generation of a perceptual mapping of odors, but also in investigating the structure of odor categories and the categorization process involved during the procedure (Lawless, 1989).

Sorting tasks have been applied to a large variety of products. Concerning food products, several types of products have been studied: cheese (Lawless *et al.*, 1995), yogurts (Saint-Eve *et al.*, 2004), jellies (Blancher *et al.*, 2007; Tang and Heymann, 1999), breakfast cereals (Cartier *et al.*, 2006), soaps (Sinesio *et al.*, 2010), snacks (King *et al.*, 1998), beers (Chollet *et al.*, 2011; Lelièvre *et al.*, 2008), drinking waters (Falahee and MacRae, 1997; Teillet *et al.*, 2010) and wine (Ballester *et al.*, 2008). Recently, applications have been reported on other materials as different as clothing fabrics (Soufflet *et al.*, 2004), automotive fabrics (Giboreau *et al.*, 2001), plastics (Faye *et al.*, 2004), aromas and perfumes (Cadoret *et al.*, 2009; Courcoux *et al.*, 2014), bottles of olive oil (Santosa *et al.*, 2010), wine glasses (Faye *et al.*, 2013) and sound of a car door closing (Parizet and Koehl, 2012).

Concerning the sensory modalities, most of the studies aim at assessing the overall perceptual differences among a set of products. However, some studies were focused on specific sensory modalities, such as vision, taste, texture, odor and sound.

The sorting task is often presented as an alternative to quantitative descriptive analysis. Considering some of the advantages of this procedure (easiness of the task, less time consuming, no need for training the subjects), it has been advocated as a quick procedure to set up a perceptual mapping of products (Cartier et al., 2006; Faye et al., 2004; Saint-Eve et al., 2004). Several studies have compared the maps derived from the two procedures (sorting and sensory profiling), and authors have generally observed that the groups of products and their description by the subjects bear high similarities with both techniques (Cartier et al., 2006; Deegan et al., 2010; Faye et al., 2004; Saint-Eve et al., 2004; Sinesio et al., 2010). Therefore, it was concluded that sorting task is a reliable procedure for revealing the main sensory dimensions of the product space. Other studies have highlighted some differences in the description of the perception between the two methods. Following Saint-Eve et al. (2004) and Blancher et al. (2007), these differences can be explained by the nature of the two tasks: conventional profiling is an analytical procedure, whereas the sorting task is a holistic approach based on the evaluation of global similarities between the products. Complex interactions between different sensory modalities (visual/texture for instance) may be difficult to describe by decomposing perception into separate independent attributes. In this case, free sorting task appears to be a relevant alternative to analytical procedures to better describe a sensory perception.

The sorting task can also be applied to achieve a selection of products prior to a descriptive analysis. Piombino et al. (2004) developed a two-step strategy for the analysis of wine flavor. The sorting of a large number of samples by untrained subjects led to the selection of a representative subset of products that were evaluated, in a second step, by descriptive techniques.

7.2.3 Variants and extensions of the sorting task

Several variants of the free sorting task have been developed for specific objectives. Since they differ in terms of the instructions given to the subjects and the type of collected data, they can be considered as separate procedures.

7.2.3.1 The Q-sort methodology

The Q-sort methodology (Coxon, 1999; Rosch, 1976) originates from psychology and consists in sorting the stimuli (conceptual or perceptive) into categories defined a priori. This task is generally used in order to test the existence of concepts shared by the subjects and to identify the properties linked to these concepts.

Jaffré et al. (2009) investigated the concept of "vin de garde" through the perception of aging potential of red wines. Forty-one wine professionals were asked to taste 16 commercial Burgundy red wines and assign each of them to one of the two categories: "with an aging potential" or "without an aging potential." The authors noted a consensual understanding of the aging potential of wines by wine professionals.

The relationships between sensory modalities can also be analyzed by Q-sort. Ballester et al. (2009) presented 18 wines in dark glasses to 49 subjects (23 experts and 26 novices) and asked them to sort the wines into three categories: "red wine," "rosé wine" and "white wine." They showed that odor representations of red and white wines exist even without a visual activation.

7.2.3.2 The constrained sorting (or fixed-sorting)

In some instances, the number of categories is fixed, without any labeling of these categories (Coxon, 1999). Chollet *et al.* (2011) compared the results of the free sorting task and the constrained sorting of beers into five categories (hoppy beers, alcoholic beers, Stout beers, Gueuze beers and alcohol-free beers). They concluded that there was no impact of the instructions (free number of groups vs sorting into five groups) on the classification of the products. Contrariwise, Lawless (1989) showed that the configuration of products varied in the sense that a more detailed structure of the products was obtained by means of the unconstrained procedure.

7.2.3.3 The free multiple sorting procedure

In the free multiple sorting procedure, the subject is allowed several trials to sort the stimuli. As long as it makes sense for him, each subject is encouraged to perform a new sorting of the same set of products, using different criteria from those for the previous sorts. Rosenberg and Kim (1975) showed that multiple sorting performed better than single sorting in an experiment on kinship terms categorization. They advocated using multiple sorting as a routine procedure instead of (simple) sorting, on the grounds that in the latter procedure, subjects are likely to overlook the most obvious attributes.

Recently, Dehlholm *et al.* (2012) used free multiple sorting in the field of food science. They described a study of nine samples of liver pâté and reported a number of different sortings ranging between 3 and 14 per assessor. They also report that using multiple sorting provides more precise results compared to single sorting.

The methodology of multiple free sorting is detailed in Chapter 8 of this book.

7.2.3.4 The hierarchical sorting procedure

The usual sorting procedure presupposes an organization of the stimuli according to an inclusive–exclusive relationship. Each subject gives a partition of the set of stimuli into exclusive groups, that is, a stimulus belongs to one and only one group. This means that similarities between objects (and between groups of objects) are binary (0/1). A hierarchical structure would be a more flexible way to depict the relationships between objects, allowing the distances between them to be more precisely quantified. Various techniques for eliciting a hierarchical organization of the stimuli have been developed with two variants: agglomerative and divisive hierarchical procedures.

The agglomerative hierarchical procedure consists in presenting the whole set of objects to each subject. He or she is firstly asked to group the two objects that are considered the most similar and, step by step, merge two objects or groups that are deemed to be close until all the objects are clustered in the same group (Coxon, 1999). Another way to construct a hierarchical structure is to start from the partition given by the subject in a free sorting task. In a second step, the subject is asked to lump together the two groups that he or she perceives as the closest. This step is repeated iteratively until only two groups remain. This approach has been proposed by Courcoux *et al.* (2012) and named "taxonomic free sorting." A case study of 14 brands of milk chocolates showed that the configuration of products was more stable and more discriminant. This finding can be explained by the fact that this technique of evaluation gives more insight into the dissimilarities between the stimuli than the usual free sorting task.

By contrast to the taxonomic sorting, a hierarchical structure of stimuli may be obtained from subjects by using a divisive variant of the free sorting task. After the first grouping of the objects, the subjects are asked to divide the groups they have formed into further subgroups. Divisive hierarchical sorting has been applied to the tasting of 16 sauces by 11 trained subjects (Blancher *et al.*, 2008), to visual assessment of 25 olive oil bottles by 31 consumers (Santosa *et al.*, 2010) and to sorting of 16 cards by 89 children (Cadoret *et al.*, 2011). In these three situations, the authors consider the second stage of the sorting task as a refinement giving more information on the structure and the description of the products.

7.2.4 Interests and limitations

7.2.4.1 Interests

One of the main characteristics of the sorting task is the non-verbal nature of the procedure. In the course of the categorization step, there is no need for the subjects to rate any attribute, or even describe or elicit the criteria they are using to construct their grouping structure. For this reason, this task can easily be performed by untrained subjects, and may be considered as an efficient technique for understanding consumers' perceptions. Moreover, it can be readily used for comparing groups of assessors or consumers with different languages or cultures, without the bias that might result from misunderstanding of the attributes. Chrea *et al.* (2004) compared perception of odors by three groups of subjects of different cultures (French, Vietnamese and American). Free sorting task showed agreement in how the subjects perceive odors, and highlighted differences related to fruit and flower categories. It also made it possible to investigate the relationships between culture and odor categorization. Another noteworthy benefit of free sorting is that it can be performed by subjects with different levels of expression, or with very young children (Ricciuti *et al.*, 2006).

Several authors have emphasized the simplicity of the sorting task. It is generally reported that the subjects easily understand the instructions, probably because the sorting process corresponds to a natural mental activity. Coxon (1999) points out that the subjects usually enjoy performing this task. However, this does not mean that the task is always easy for them (Ares *et al.*, 2011). During some experiments with saturating products, subjects may express fatigue at the end of the task. This may be due to difficulties in memorizing all the different stimuli. Obviously, this is likely to occur when the experiment involves a large number of products.

The holistic nature of free sorting confers on this task several advantages over more analytical procedures. Indeed, there is no need to generate an exhaustive list of attributes or a shared understanding of descriptive terms to perform a sorting task. For this reason, this task provides a subtle approach for assessing consumers' perceptions of a set of products. Faye *et al.* (2006) proposed using perceptive maps generated by consumers for modeling their preferences, as an alternative to usual external preference mapping, based on descriptive profiling by trained assessors.

The fact that the categorization process is separated (at least partly) from the verbalization process is one of the main interests of the sorting task (Dubois, 2009). Very often, this free verbalization yields a spontaneous vocabulary with a large number of terms referring to descriptive, hedonic and usage properties of the stimuli.

7.2.4.2 Limitations

The interpretation of product configuration using the terms provided by the subjects is usually insightful. However, the encoding of the terms collected during a free verbalization task is generally time consuming and laborious (identification of synonymous terms, discarding sparse terms, etc.). In order to counteract this impediment, a constrained version of the verbalization task consists in providing a predefined list of terms to the subjects. However, such a strategy may result in the introduction of a bias in the assessment of the products, since the predefined terms are likely to narrow the range of the characteristics that the subjects should investigate (Hoc and Leplat, 1983).

Free sorting task is a comparative experimental procedure that entails the stimuli being simultaneously presented to the subjects in a single session. Therefore, the set of stimuli in the study is constrained by the number of products that a subject can evaluate during a session. However, if it is assumed that the number and the complexity of the products may cause fatigue to the subjects, there is the possibility of recourse to an incomplete design by presenting only a subset of products to each subject. In this case, the use of a balanced incomplete block design is recommended, in order to have the same number of comparisons for all the pairs of products.

One of the main limitations of the free sorting task lies in the binary response of each subject. At the individual level, the distances between products are coded in a dichotomic way: they either belong to the same category, or they do not. The free sorting task does not give access to the individual decision criterion, or to a threshold used by the subject for deciding whether two products belong to the same group or should be set in different categories.

In order to differentiate the products within a group, some authors use another task from the free sorting procedure. At the end of the verbalization task, the subject may be asked to indicate, for each group they formed, the stimulus that they perceive as the most representative of this group (Coxon, 1999). The objective is to determine the prototypical member of each group (Tardieu *et al.*, 2008). An alternative is to assess the degree of typicality of all stimuli relative to a given group (Rosch and Mervis, 1975). This method, widely used in experimental psychology, has been applied more recently in sensory evaluation (Chrea *et al.*, 2009).

The multiple sorting task and the taxonomic sorting task are variants that allow the assessment of a more refined evaluation of the dissimilarities between stimuli at the subject level than the (single) free sorting task.

7.3 Statistical treatment of free sorting data

7.3.1 General overview

As mentioned above, a free sorting test yields a family of partitions of the same set of products, each partition being associated with a particular subject. For the statistical treatment of free sorting data, several strategies of analysis have been proposed. We refer to the book by Coxon (1999) for a comprehensive review of these strategies. We also refer to a French paper by Faye *et al.* (2011) for a comparison of several methods of analysis on the basis of a case study.

Broadly speaking, the statistical strategies of analysis can be classified into two families of methods, namely: (i) factor analytical methods including, in particular, multidimensional scaling (MDS) and multiple correspondence analysis (MCA); and (ii) methods pertaining to cluster analysis and additive trees. As is usually the case, the choice of one method over another depends on several factors: (i) the domain of application (i.e. traditionally, some methods are more popular than others in each particular domain of application); (ii) the individual preferences and background of each practitioner; and (iii) the availability of appropriate (and user-friendly) software.

As an illustration, let us mention that the use of additive trees for analyzing free sorting data is relatively popular in psychology (Dubois, 1991), whereas in psychoacoustics, MDS methods are more often used (Gygi *et al.*, 2007; MacAdams *et al.*, 1995). In the field of sensory analysis that particularly interests us here, the mainstream is to use MDS methods (Faye *et al.*, 2004; King *et al.*, 1998; Lawless *et al.*, 1995; Parr *et al.*, 2007). However, alternative methods of analysis, such as multiple correspondences and allied methods, have also been proposed in this framework (Cadoret *et al.*, 2009; Qannari *et al.*, 2009; Takane, 1981, 1982).

In order to gain more insight into these methods, let us discuss how the data from a free sorting task could be encoded. Let us suppose that J subjects take part in a free sorting task, in the course of which they were instructed to sort n products into groups. Let us assume that subject j ($j = 1, ..., J$) has sorted the stimuli into K_j clusters or groups. As mentioned above, the data can be seen as a set of J partitions of the products, each partition being associated with a particular subject. Alternatively, the data for each subject are often expressed in a symmetric (products by products) matrix containing ones and zeroes. More precisely, an entry of this matrix contains one if the products associated with the row and the column of this entry are sorted into the same group, otherwise the entry contains a zero (Fig. 7.1b). This matrix is usually interpreted as a similarity matrix between products and can be easily transformed into a dissimilarity matrix by changing the zeroes into ones and vice versa (Fig. 7.1c). At the panel level, the individual dissimilarity matrices can be averaged across subjects, yielding a global dissimilarity matrix between products. As discussed below, these strategies of encoding the data entail using methods of analysis pertaining to MDS. Advocates of MCA consider yet another encoding of the data. For subject j ($j = 1, ..., J$) who has sorted the products into K_j groups, we associate a matrix X_j (n rows by K_j columns) called the matrix of indicator variables (or dummy variables) that indicate for each product (row) the group (column) to which it belongs (Fig. 7.1d).

(a)

Products	Clusters
P1	C1
P2	C1
P3	C1
P4	C2
P5	C2

Raw data of a fictitious subject who sorted n = 5 products in K = 2 clusters

(b)

Products	P1	P2	P1	P4	P5
P1	1	1	1	0	0
P2	1	1	1	0	0
P3	1	1	1	0	0
P4	0	0	0	1	1
P5	0	0	0	1	1

Similarity matrix between products

(c)

Products	P1	P2	P1	P4	P5
P1	0	0	0	1	1
P2	0	0	0	1	1
P3	0	0	0	1	1
P4	1	1	1	0	0
P5	1	1	1	0	0

Dissimilarity matrix between products

(d)

	Clusters	
Products	C1	C2
P1	1	0
P2	1	0
P3	1	0
P4	0	1
P5	0	1

Matrix of indicator variables

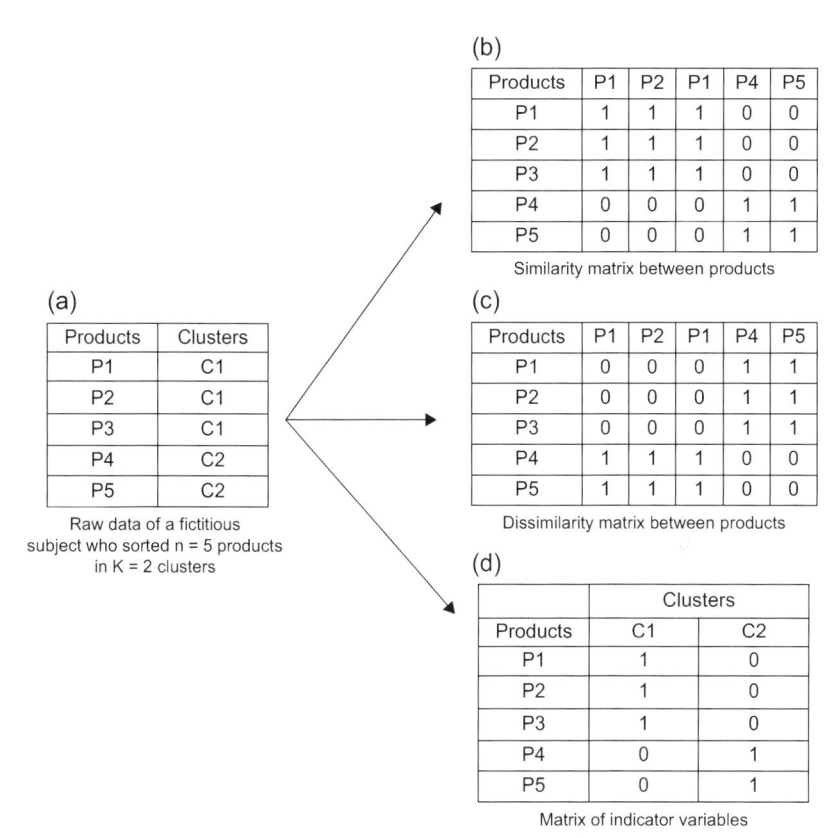

Figure 7.1 Transformation of the raw data (a) given by a fictitious subject into a similarity matrix (b), dissimilarity matrix (c), and a matrix of indicator variables (d).

With no claims to exhaustiveness, we present in the Table 7.1 some of the most important strategies of analysis that have been proposed to analyze free sorting data.

7.3.2 Factor analytical techniques

7.3.2.1 General overview

The general aim of factor analytical method is to depict the relationships among the products on the basis of a low dimensional space. This is achieved by seeking few axes (typically, two or three) that recover as much variation in the data as possible. In other words, the representation of the products on the basis of these few axes should be as consistent as possible with the representation in the original space (i.e. when all the original data are used). From a conceptual point of view, the axes (called sometimes factors) can be interpreted as latent (or hidden) variables that structure the perceptive space. These latent variables may be considered as the underlying criteria that govern the categorization process. From a practical point of

Table 7.1 General overview of the statistical methods of analyzing free sorting data

Family of methods	Subfamily	Input data	Methods	Bibliographic references
Factor analytical methods	Methods related to MDS	Overall (across panelists) dissimilarities among products	Metric MDS	Kruskal, 1964; Schiffman et al., 1981; Gygi et al., 2007
			Non-metric MDS	Lawless, 1989; Falahee et al., 1997; Faye et al., 2004
		Individual dissimilarities among products	Metric INDSCAL	Carroll and Chang, 1970
			DISTATIS (metric)	Abdi et al., 2007
			Non-metric INDSCAL	Takane et al., 1977
	Methods related to MCA	Matrices of indicator (or dummy) variables	MCA, MDSORT	Takane, 1981; Van der Kloot and Van Herk, 1991; Cadoret et al., 2009
			IDSORT, CCSort	Takane, 1982; Qannari et al., 2009
Classification methods		Global dissimilarities	Hierarchical classification	Lawless, 1989; Coxon, 1999; Giboreau et al., 2001; Lebart et al., 2006
			Additive trees	Sattah and Tversky, 1977; Bartholomew in Dubois, 1991
		Individual partitions	Central partitions	Marcotorchino and Michaud, 1982; Guenoche, 2011; Courcoux et al., 2014
			Latent partitions	Wiley, 1967

view, the interpretation of these axes is usually achieved by using physico-chemical variables related to the products or the terms used by the subjects in the course of the verbalization task.

7.3.2.2 Multidimensional scaling (MDS) on the overall dissimilarity matrix

The input of the MDS technique (Kruskal, 1964; Torgerson, 1958) is a dissimilarity matrix between the products. The aim is to seek a low dimensional space to investigate the relationships among the products, with the understanding that the distances between products computed on the basis of this low dimensional space (usually visualized on a graphical display) are highly consistent with the dissimilarities in the original dissimilarity matrix.

In free sorting task, let us consider the matrix $\Delta = (\delta_{ij})$ of global dissimilarities between products. As stated above, the dissimilarity δ_{ij} between *products i and j* (*i, j* = 1, ..., *n*) is equal to the number of subjects who did not set these two products in the same group. Let us denote by X the representation space of dimension k (say). We suppose for the time being that the parameter k is fixed. We denote by $d_{ij}(X)$ the Euclidean distance between products *i* and *j* obtained from the coordinates of products *i* and *j* in the representation space, X. The optimal representation space is obtained by minimizing a criterion called "Stress" (Kruskal, 1964):

$$\text{Stress} = \left[\sum_i \sum_j (\delta_{ij} - d_{ij}(X))^2 / \sum_i \sum_j d_{ij}(X)^2 \right]^{1/2}$$

The rationale behind this criterion is clear enough since it shows that we are seeking a representation space, X, such that the Euclidean distances, $d_{ij}(X)$, between the products in this space are as close as possible to the original dissimilarities in matrix Δ.

This minimization problem is solved using an iterative algorithm. A commonly used algorithm is the so-called *SMACOF* (De Leeuw, 1988; Heiser, 1990). It goes without saying that the closer the Stress is to zero, the better is the solution. Kruskal (1964) claimed that a value of Stress criterion less than 0.10 indicates a good quality of the representation space (good fit between the original dissimilarities and the Euclidean distances on the representation space). Contrariwise, values of the Stress criterion larger than 0.20 indicate a poor fit. These remarks can be used to advantage to select the appropriate number of axes (i.e. k, dimension of the representation space X). To do so, it is recommended to display the curve showing the decrease of the Stress criterion as k increases. At a certain point, this curve ceases to significantly decrease and starts to form a plateau. This indicates that there is no significant gain in increasing k (Fig. 7.2). Therefore, the value corresponding to the starting point of the plateau should be retained, provided that the corresponding Stress criterion is small enough (e.g. smaller than 0.10).

It is important to note that several techniques of MDS can be used, depending on the nature of the dissimilarities at hand. The MDS strategy that we have just sketched pertains to the so-called *metric* MDS since the original dissimilarities have been directly used in the Stress criterion. By contrast, *in non-metric* MDS, we assume that the meaningful information is not the original dissimilarity matrix per se but rather the rank order of its entries. Therefore, we seek a representation space of low dimension that preserves as much as possible the rank order of the dissimilarities in Δ. More formally, in *non-metric* MDS (Borg and Groenen, 2005), we seek a representation space X in such a way as to minimize the following Stress criterion:

$$\text{Stress} = \left[\sum_i \sum_j (f(\delta_{ij}) - d_{ij}(X))^2 / \sum_i \sum_j d_{ij}(X)^2 \right]^{1/2}$$

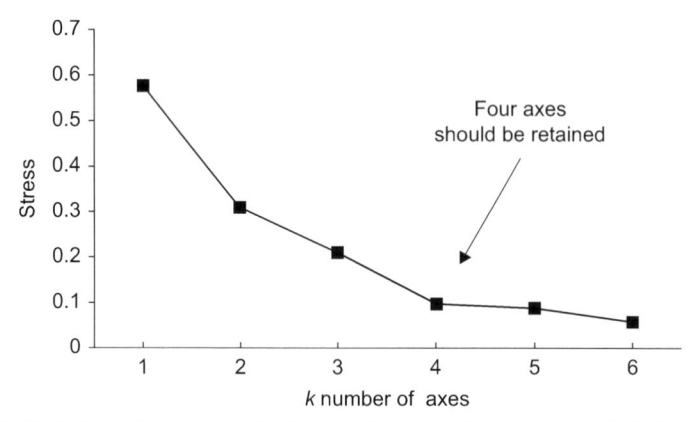

Figure 7.2 Evolution of the stress criterion as a function of the number of MDS axes.

where f is a function (to be determined), which is supposed to be monotonic nondecreasing. The introduction of this function entails that the original dissimilarities can be transformed at will provided we preserve their rank order.

We claim that *non-metric* MDS is better suited to analyzing free sorting data than *metric* MDS (Faye *et al.*, 2004; Lawless *et al.*, 1995). Our claim is backed up by the fact that the dissimilarities in matrix Δ are computed by counting the number of subjects who did not set the products in the same groups and, therefore, do not stem from interval or ratio scale data.

7.3.2.3 MDS on the individual dissimilarity matrices

The strategy of analysis based on submitting the overall dissimilarity matrix to MDS does not take account of the individual differences among the subjects (Lawless *et al.*, 1995). This may be a drawback, particularly when there is a high disagreement among these subjects. The MDS method on individual dissimilarities based on *INDSCAL* model (Carroll and Chang, 1970) allows us to overcome this impediment. Similarly to the general approach in MDS, *INDSCAL* seeks a small number of underlying axes to represent the products. In addition, *INDSCAL* model assumes that each axis may be differentially weighted by the various subjects. More precisely, the weights assigned to each axis reflect the importance that the subjects attach to this axis. This means that, although the products' representation space is common to all the subjects, each axis is stretched or shrunk for each subject depending on whether the subject under consideration assigns a large or a small weight to this axis (Fig. 7.3).

It is clear that the *INDSCAL* model is very flexible and yields very useful information regarding the relationships among products and the differences between the subjects. However, it is worth noting that its metric nature is ill-suited to the treatment of free sorting data. This is the reason why we recommend using a non-metric version of *INDSCAL* (Takane *et al.*, 1977). To the best of our knowledge, this technique of analysis has never been performed on free sorting data.

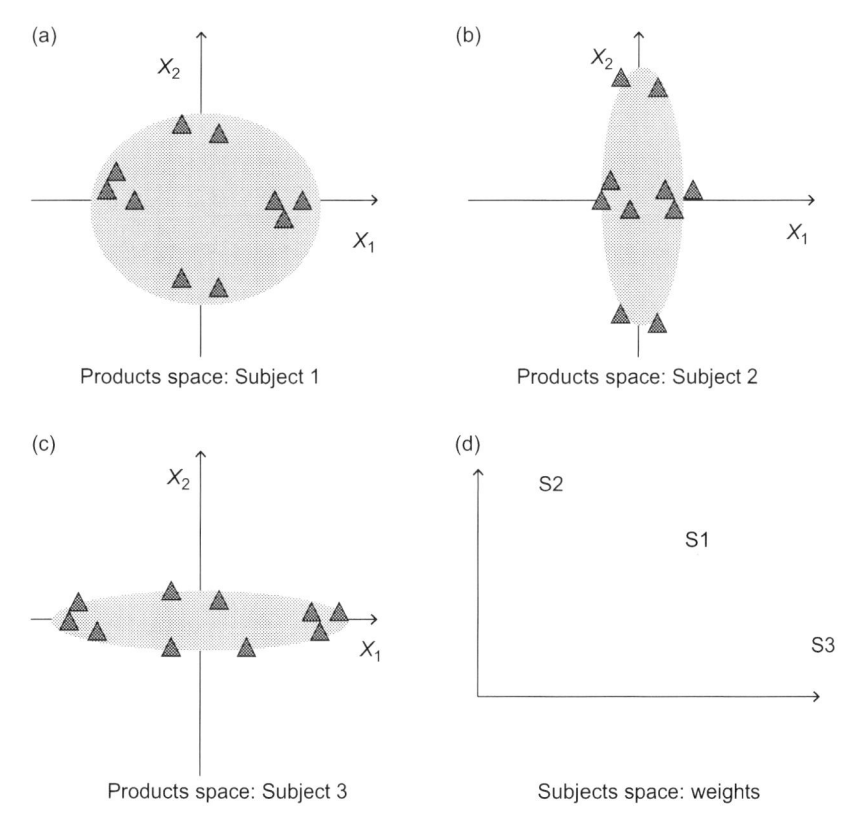

Figure 7.3 Products spaces associated with three subjects S1, S2 and S3 (Figs 7.3a–7.3c) showing the configuration of ten products on *INDSCAL* axes X_1 and X_2. (d) Subjects space showing the representation of the three subjects on the basis of the weights they assign to axes X_1 and X_2.

In this section related to the treatment of the individual dissimilarities by an approach akin to MDS, it is worth including the *DISTATIS* approach, proposed by Abdi (1990). This method of analysis was designed as a generalization of the MDS method to the case of dissimilarities organized in three ways (products × products × subjects). *DISTATIS* can also be seen as an extension of the *STATIS* method (Lavit, 1988; Schlich, 1996) to the analysis of several dissimilarity matrices. This involves calculating the RV coefficients between the individual configurations. These coefficients reflect the mutual agreement of the subjects' configurations. Thereafter, weights assigned to the subjects reflecting their agreement with the panel's general point of view are computed. Subsequently, these weights are used to derive a (weighted) group average configuration, which is submitted to principal components analysis in order to investigate the relationships among the products.

It is clear that, similarly to the *INDSCAL* model, *DISTATIS* makes it possible to shed light on the differences among subjects. However, *INDSCAL* appears to be more flexible than *DISTATIS* since it investigates the agreement of the subjects with respect to each underlying dimension, whereas *DISTATIS* investigates the agreement at a global level.

7.3.2.4 Multiple correspondence analysis

MCA is a multivariate factor analytical method that was designed as an extension of (simple) correspondence analysis (Lebart *et al.*, 2006). Its use in the framework of free sorting task dates back to the early 1990s (Van der Kloot and Van Herk, 1991). MCA can be regarded as a principal components analysis on categorical data (Greenacre, 1993). With free sorting data, we consider that there are as many categorical variables as subjects where the categories for each variable (subject) are the groups formed by the subject under consideration. Thereafter, the data from each subject are expressed as a matrix of indicator variables (see Fig. 7.1d). These individual matrices are horizontally merged to form a super-matrix formed of dummy variables. As stated above, column j (say) is associated with a category, that is, a group of products associated with a given subject. The average p_j of this column reflects the proportion of products contained in the group under consideration. Subsequently, column j is standardized by dividing it by $\sqrt{p_j}$. This kind of standardization is very typical with correspondence analysis, and is backed up by several considerations, the discussion of which is beyond the scope of this paper. As a final step, principal components analysis is performed on the standardized super-matrix of dummy variables.

It is worth noting that Takane (1981) proposed a method of analysis called *MDSORT* for analyzing sorting data. The rationale behind this method is very appealing. Suppose that we seek an axis to depict the products. Naturally, we expect that, for a given subject, products that are in the same group should be very close to each other, whereas products in different groups should be far removed from each other. In other words, the representation axis should discriminate as much as possible the groups given by the subject under consideration. At the panel level, we may seek a representation axis that, on average, discriminates as much as possible the groups given by the subjects. Subsequent axes may be sought following the same strategy of analysis after imposing orthogonality constraints between the successive axes. From the derivation of the solution to this discrimination problem, it turns out that, as a matter of fact, we are led to the same solution as MCA. This remark was also stressed by Van der Kloot and Van Herk (1991), who stated that their program for running MCA gave outcomes that are identical to those of *MDSORT*. Takane (1982) designed yet another procedure, called *IDSORT*, for analyzing sorting data. *IDSORT* can be seen as a refinement over *MDSORT* since it makes it possible to take account of individual differences among the subjects. This is done by adopting a strategy of analysis that combines MCA and *INDSCAL* methods. The bottom line is that *IDSORT*, similarly to *INDSCAL*, yields a representation space for the products and a set of weights associated with the subjects that reflect the importance they attach to the various dimensions of this representation space. In the same vein, Qannari *et al.* (2009) proposed a method of analysis called *SORT-CC*.

7.3.3 Methods pertaining to cluster analysis

7.3.3.1 General overview

The aim of cluster analysis is to depict the dissimilarity among a set of products as a hierarchical tree (also called dendrogram) or achieve, on the basis of a dissimilarity matrix, a partitioning of the products into a given number of groups (Duda *et al.*, 2000).

The psychometric rationale behind cluster analysis is to assume that there is a latent partition of the products that each subject is trying to retrieve in the course of the sorting task. Thus, it follows that an appealing feature of such a strategy of analysis is that the outcomes (i.e. a partition or a hierarchically organized set of partitions) are of the same nature as the inputs (i.e. partitions of products associated with the various subjects).

Strategies of cluster analysis are divided in mainly two families of methods: (i) partitioning algorithms; and (ii) hierarchical algorithms. In the first family of methods, the aim is to seek a partition formed of homogeneous groups of products. This entails that products in the same group should be closer to each other than products belonging to different groups. A typical partitioning algorithm is the so-called k-means algorithm (Duda *et al.*, 2000). This algorithm starts with a given set of k centroids, k being the number of clusters in the partition to be determined. Then, the products are assigned to the closest centroid, thus forming k classes. Thereafter, the centroids are updated by considering the centroids of the new classes, and the process is iterated until the classes are stabilized. The second family of methods aims at setting up a hierarchically organized set of partitions. For instance, the aggregative strategy of analysis, which is more used than the divisive strategy of analysis, consists in starting from the situation where each product forms a group by itself and moving on, step by step, until all the products are in a single group. At each step, we merge those two products or groups of products that are deemed to be the closest according to a given aggregation criterion. Once the hierarchical tree is set up, it only remains to cut it at an appropriate level to obtain a partition of the products. Indeed, the structure of the dendrogram usually gives a hint as to which levels it is more appropriate to cut it.

It is worth noting that the two families of methods are not exclusive since one can take advantage of both of them to derive an optimal partition of the products. This is usually done by: (i) performing a hierarchical clustering algorithm; (ii) from the structure of the hierarchical tree, choosing an appropriate number of classes for the final partition; (iii) performing a partitioning algorithm where the starting points are set as the centroids of the classes obtained by cutting the tree at the appropriate level.

The inputs of both the hierarchical and the partitioning algorithms are the overall dissimilarities obtained at the panel level. We will also outline strategies of analysis where the input data are the individual (i.e. subjects) dissimilarity matrices. In this strategy of analysis, the aim is to obtain what is usually referred to as a consensus partition. That is, a partition that agrees as much as possible with the partitions given by the subjects.

We will also outline the representation of the products by means of an additive tree. Indeed, this is an appealing concept, which has been used in several studies related to the free sorting task.

7.3.3.2 Ascending hierarchical analysis

The input data are the overall dissimilarities $\Delta = (\delta_{ij})$. The ascending hierarchical analysis is an aggregative strategy, which starts by considering that each product forms a group by itself. Then, step by step, groups of products are formed by merging, at each step, two products or groups of products that are deemed to be the closest according to a predefined criterion. The most commonly used agglomeration criterion is certainly the so-called Ward criterion. The rationale behind this criterion is that, at each step, the merging of two groups of products inevitably results in an increase of the within groups total variance, which reflects the degree of heterogeneity within the groups. The Ward criterion corresponds to the variation of the within groups total variance between two successive steps. Naturally, it is advocated to merge those two groups of products that result in the smallest variation of this criterion (i.e. smallest increase of the heterogeneity within the groups). When the whole hierarchical tree is set up, we can trace back the Ward criteria at the various steps. If, at a given step, there is significant jump of this criterion, this means that we are merging two heterogeneous groups and that we better cut the tree at this level.

7.3.3.3 Consensus partition

As stated above, the aim herein is to derive a partition of the products that is as close as possible to the individual partitions given by the various subjects. This is termed the consensus partition, but this terminology is inappropriate since the partition does not represent a consensus of views of the panel. Rather, it represents a group average partition that may conceal a wide range of individual differences. This is the reason why Courcoux *et al.* (2014) advocated investigating the agreement among the subjects particularly by means of cluster analysis of the subjects. Notwithstanding, the computation of a consensus partition entails the assessment of the agreement between two partitions. This problem has been widely studied and numerous criteria have been introduced for evaluating the proximity between two partitions of the same set of objects. A comprehensive review of different measures of agreement between partitions is given by Youness and Saporta (2004). Among these indices, we single out the so-called Rand index (Rand, 1971).

Let us consider two partitions of the same n products. For instance, these could be the partitions of the same stimuli operated by two subjects in the course of a free sorting task. The Rand index (Rand, 1971) is defined by:

$$\text{Rand} = \frac{a+d}{T}$$

where a is the number of pairs of products that are set in the same group in both partitions, d is the number of pairs of products that are not set in the same group in both partitions, and T is the total number of pairs of products (i.e. $T = (n(n-1)/2)$). Obviously, the Rand index lies between 0 and 1. In particular, it is equal to 1 in case of perfect agreement (i.e. identity of the two partitions).

Notwithstanding its intuitive appeal, the Rand index suffers from several pitfalls. In particular, it highly depends on the number of groups in the two partitions and the number of stimuli at hand. So much so that, in some situations, it can take high values even for two random partitions (Courcoux et al., 2014; Santos and Embrechts, 2009; Youness and Saporta, 2004). In order to cope with this problem, an adjusted Rand index (ARI) was proposed as a form of the Rand index that is corrected for the grouping of the stimuli by chance (Albatineh and Mihalko, 2006; Hubert and Arabie, 1985).

The problem of finding a consensus partition that reflects the general point of view of a set of partitions has been studied by several authors. One of the first works is due to Régnier (1983). A bibliographical review of this problem was undertaken by Leclerc and Cucumel (1987).

Let us denote by U_1, U_2, ..., U_J the partitions of the n products respectively associated with the J subjects. We denote by U the consensus partition of the products that we seek to obtain. Choosing the ARI as a measure of agreement between partitions, the problem of determining the consensus partition leads us to the following criterion to be maximized:

$$C(U) = \sum_{j=1}^{J} ARI(U_j, U)$$

For solving this optimization problem, Courcoux et al. (2014) proposed an iterative algorithm that consists in starting with an initial partition with k classes (k being fixed beforehand). Thereafter, the products are allowed to move from one class to another if this transfer results in an improvement of criterion C. The algorithm stops when there is no further improvement. This technique is an adaptation of the procedures proposed by Krieger and Green (1999) and Guénoche (2011).

Since the number, k, of classes of the consensus partition is not known a priori, Courcoux et al. (2014) suggest performing the previous algorithm with increasing values of k and retaining the value k that corresponds to the maximum of criterion C. An illustration of this strategy of analysis is outlined when discussing the case study.

7.3.3.4 Additive trees

The reason for dwelling on additive trees in this chapter is that, as stated above, they have been quite often used for the analysis of free sorting data (Blancher et al., 2007; Chollet et al., 2011; Chrea et al., 2004). Furthermore, they yield a very appealing graphical display that highlights the proximities among products.

The starting point is the overall dissimilarity matrix among products Δ. The outcome is a tree (or more precisely, a graph) that connects the various products (Fig. 7.4). The shortest path from one product to another reflects the dissimilarity between these two products.

```
              P2
              /
        P1 —<
              \
               \— P4
               /
              P3
```

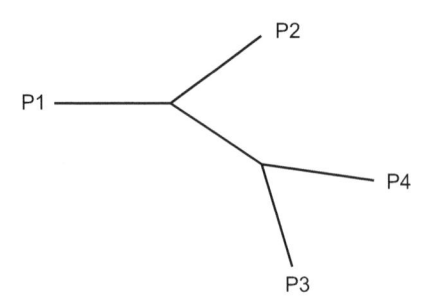

Figure 7.4 Illustration of an additive tree highlighting the similarities between the products P1 to P4.

The determination of an additive tree is based on the four-points condition, which states that a distance, d (say), is a tree distance (i.e. can faithfully be represented by a tree) if and only if for any four products i, j, k and l, we have:

$$d(i,j) + d(k,l) \leq \max\ (d(i,k) + d(j,l), d(i,l) + d(j,k))$$

Very likely, the original dissimilarity matrix Δ is not a tree distance. Thereupon, we undertake to approximate it by a tree distance. Several approximation methods are discussed in the literature. For a comprehensive review, we refer to Abdi (1990).

The strategy of analysis is illustrated on the basis of the data from the case study outlined in a subsequent section.

7.4 A case study in the automotive industry: understanding the consumer perception of car body style

7.4.1 Context and issues

In the car industry, the supply is constantly increasing at an international level and, quite naturally, so is the competition among car manufacturers. To face this aggressive and dynamic competitive environment, car companies develop different strategies, such as the launching of new brands and the diversification of car body styles. To better understand consumers' expectations, their perceptions and their judgments of new car designs, clinical tests are conducted early in the development process of new projects. Unfortunately, a large number of proposals cannot be compared in these tests, since they are carried out on prototypes, which are of the same size as actual cars. Moreover, the development time of new projects is getting shorter and shorter, and the designers need early consumer data for defining the architectural parameters of new car concepts.

In this context and to complete the clinical test procedure, we have explored the possibility of studying several car design proposals using 2D-representation of cars in a reduced scale. Another advantage is to collect consumer insights early enough in the car development process before designing the first prototypes. This constraint is shared by numerous industrial domains, but may be particularly important in the car industry due to a long development process.

The aim of this study is to better understand the consumers' representations and perceptions of car body styles, and assess the effect of the physical parameters of cars on the consumers' perceptions. For this purpose, we have developed a methodological approach using experimental designs and free sorting task procedure.

7.4.2 Material and methods

7.4.2.1 Samples

In order to control the physical parameters, samples were generated according to an experimental design. The choice of this approach was suggested by the results of a previous study focused on the perception of existing and current car models. The heterogeneity of vehicles on the market, the irregular variations of physical parameters, and the fact that consumers may recognize the existing cars, limited the scope of the investigation. Therefore, the use of an experimental design is adapted to set up a homogeneous product space and to control the variations of the physical parameters. Car designers selected eight parameters of the cars' side face to define the body style of vehicles (Table 7.2). Figure 7.5 depicts the definition of these parameters.

Thirty-four car body styles (Fig. 7.6) were generated according to a D-optimal experimental design. Six of these 34 car body styles are considered as typical body style categories by the car designers: hatch, estate, SUV ("Sport Utility Vehicle") (short and long) and MPV ("Multi-Purpose Vehicle") (squared rear and rounded rear).

Table 7.2 Description of the eight physical parameters and their modalities

Code	Morphological parameters	Modalities	Coding of modalities
P1	Front overhang	Long/short	P1_L/P1_S
P2	Wheelbase	Long/short	P2_L/P2_S
P3	Rear overhang	Long/middle/short	P3_L/P3_M/P3_S
P4	Windshield position *vs* front rear axis	Advanced/middle/reclined	P4_A/P4_M/P4_R
P5	Rear window tilt	Rear straight/rear reclined	P5_S/P5_R
P6	Body height	High/low	P6_H/P6_L
P7	Window height over body height ratio	R1 – R2 – R3	P7_R1/P7_R2/P7_R3
P8	Size of the rims	S1 – S2 – S3 – S4	P8_S1/P8_S2/P8_S3/P8_S4

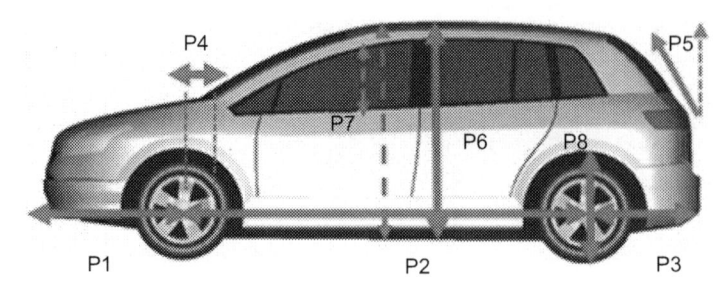

Figure 7.5 Definition of morphological parameters.

Car body styles are presented to the subjects on color print paper, glued on cardboard in order to strengthen the samples. As the experiment focuses on the physical and structural characteristics of the cars, pictures do not include any element of style, and present several similarities in their representation: gray body color, same trims and head lights design.

Consumers received 30 samples, each coded with a three-digit number, in a single session. For each subject, the 30 samples were selected from the 34 samples according to a balanced incomplete block design (BIBD). The order of presentation of the samples was randomized for each subject.

7.4.2.2 Subjects

Ninety-two car owners were recruited to perform this experiment. They owned a middle sized car purchased in the previous 5 years. They were recruited according to their gender, age and the type of car they owned (Table 7.3).

7.4.2.3 Sorting procedures

The sorting procedures consist of three successive steps.

- *Free sorting task*: subjects were asked to sort the samples into mutually exclusive groups based on perceived similarities and differences.
- *Verbalization task*: subjects were asked to freely describe with their own words and expressions the groups they formed.
- *Exemplar task*: subjects were asked to select the most representative sample of each of their groups.

7.4.3 Results

7.4.3.1 Number of groups

Figure 7.7 shows the distribution of the number of groups that the consumers set up in the course of the free sorting task. It can be seen that this number ranges between 2 and 17, with an average number of seven groups with a relatively high variability (StD = 2.9).

Free sorting as a sensory profiling technique   173

Figure 7.6 Car body styles generated by experimental design.

Table 7.3 **Frequency table of owned car body type, models, age and gender of the panel ($N = 92$)**

Body type	Hatch (25%)	Estate (25%)	MPV (25%)	SUV (25%)
Models	Peugeot 307 (17%), Volkswagen Golf (15%), Renault Megane (11%), Audi A3 (11%), Citroën Xsara (17%), Opel Astra (6%), Ford Focus (6%), Other models (17%) (Seat Leon, Alfa 147, Toyota Corolla, Nissan Almera, Fiat Stilo)	Peugeot 307SW (47%), Ford Focus (24%), Renault Megane (14%), Citroën Xsara (10%), Opel Astra (5%)	Citroën Picasso (38%), Renault Scenic (25%), Opel Zafira (16%), Volkswagen Touran (16%), Toyota Corolla (5%)	Nissan Xtrail (44%), Toyota RAV 4 (33%), Hyundai Santa Fe (11%), Subaru Forester (6%), Honda CRV (6%)
Age	Less than 35 years (12%) 35–49 years (29%) More than 50 years (59%)	Less than 35 years (29%) 35–49 years (33%) More than 50 years (38%)	Less than 35 years (21%) 35–49 years (33%) More than 50 years (46%)	Less than 35 years (0%) 35–49 years (39%) More than 50 years (61%)
Gender	Male (82%) Female (18%)	Male (90%) Female (10%)	Male (80%) Female (20%)	Male (83%) Female (17%)

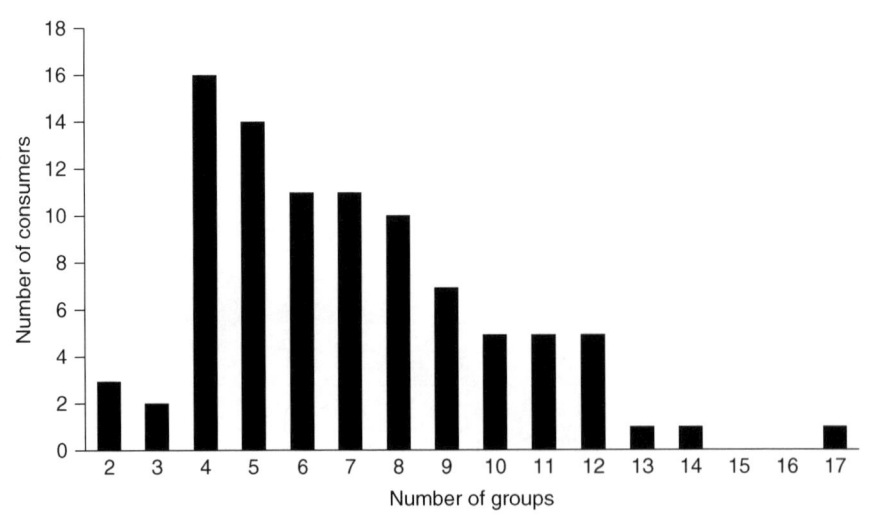

Figure 7.7 Number of groups formed by the consumers.

7.4.3.2 Consensus partition

As pointed out in the section devoted to the statistical treatment of free sorting data, the determination of a consensus partition is based on the maximization of a criterion that reflects the extent to which the subjects agree with this consensus partition. Figure 7.8 shows the evolution of the optimization criterion according to the number of groups of the consensus partition. The maximum value is obtained with seven groups of car body styles.

Table 7.4 describes the seven groups of the consensus partition, illustrated by the most representative exemplar of each group (the body style that was the most frequently cited as representative of its group). This table also gives the five most cited terms generated by the consumers. The groups of cars are differently described according to their shape (e.g. rounded rear, large window, long, small), overall judgment (e.g. Hatch, MPV, compact, sporty, beautiful design) or usage (e.g. family use).

Five of these groups are established around the typical body styles defined by the car designers:

- Long MPV cars: straight rear, described as long (group A)
- Estate cars: long and straight rear car with large trunk and for family use (group C)
- Hatch: small car, with rounded rear and small hood (group E)
- SUV: short car with straight rear for family use (group F)
- Short MPV: compact car with rounded rear (group G)

The two typical SUVs (long and short) are grouped in the same category, showing that the consumers do not perceive these two cars as belonging to different categories of car body styles.

Categories B and D can be considered as new hybrid types of cars, as they are described with different body styles terms. Group B is described as an estate car with

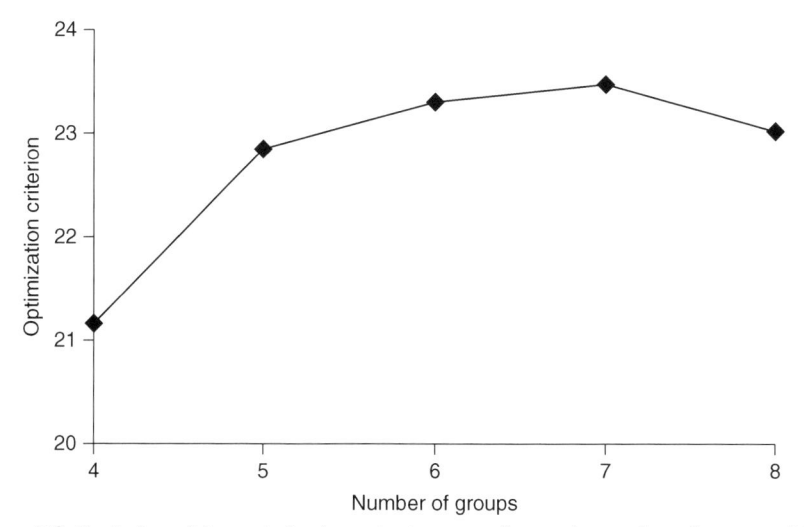

Figure 7.8 Evolution of the optimization criterion according to the number of groups of the consensus partition.

Table 7.4 Description and illustration of the seven groups of the consensus partition

Category	Code	Illustration	Five most cited terms
A	MPV Squared Rear, 9, 14, 17, 19, 20, 34	14	Straight rear (146), estate (110), long (89), important glass area (65), MPV (62)
B	18, 21, 23, 24	24	Rounded rear (87), long (34), estate (30), long hood (23), compact (21)
C	Estate, 16	Estate	Long (49), estate (38), straight rear (33), large trunk (17), family oriented (13)
D	15, 22	15	Rounded rear (53), long (44), estate (31), MPV (19), small hood (15)
E	Hatch, 7, 11, 25, 29, 30	25	Rounded rear (200), long (83), small hood (75), compact (39), small (39)
F	SUV Short, SUV Long, 13, 26, 31, 33	SUV short	Straight rear (158), MPV (46), short (44), family oriented (39), SUV (34)
G	MPV Rounded Rear, 8, 10, 12, 27, 28, 32	10	Rounded rear (240), compact (72), small hood (32), small (60), hatch (56)

Codes refer to those given in Fig. 7.6.

rounded rear and large windows, while group D lumps together long cars considered as hybrid between estates and MPVs.

7.4.3.3 Proximities between the groups of car body style

Figure 7.9 shows the additive tree resulting from the analysis of dissimilarities between car body styles. The seven groups of the consensus partition are superimposed on the tree.

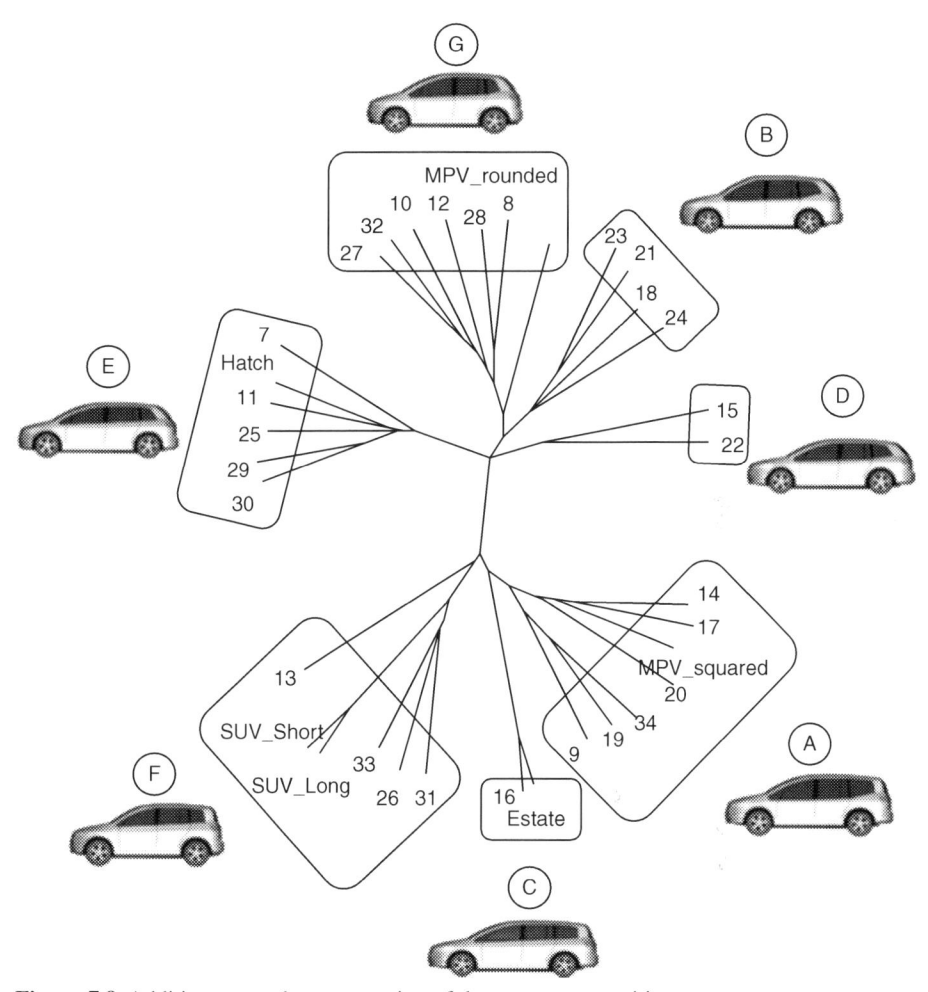

Free sorting as a sensory profiling technique 177

Figure 7.9 Additive tree and representation of the consensus partition groups.

The structure of the additive tree is highly consistent with the groups of the consensus partition. On the top of the tree, the four groups with rounded rear car body styles are clearly perceived as different types of cars: hatch (group E), rounded MPV (group G) and the two hybrid groups B and D. The bottom of the tree includes the squared rear car body styles: squared MPV (group A), Estate (group C) and SUV (group F). One can notice that the outer edges of the group C are shorter than the outer edges of the other groups, indicating that the consumers are more consensual in the grouping of samples "16" and "estate" in the same category, than in the grouping of the other car body styles.

7.4.3.4 Spatial representation of the cars

The sorting data from each consumer were expressed as a dissimilarity matrix among products. These individual dissimilarity matrices were averaged over consumers.

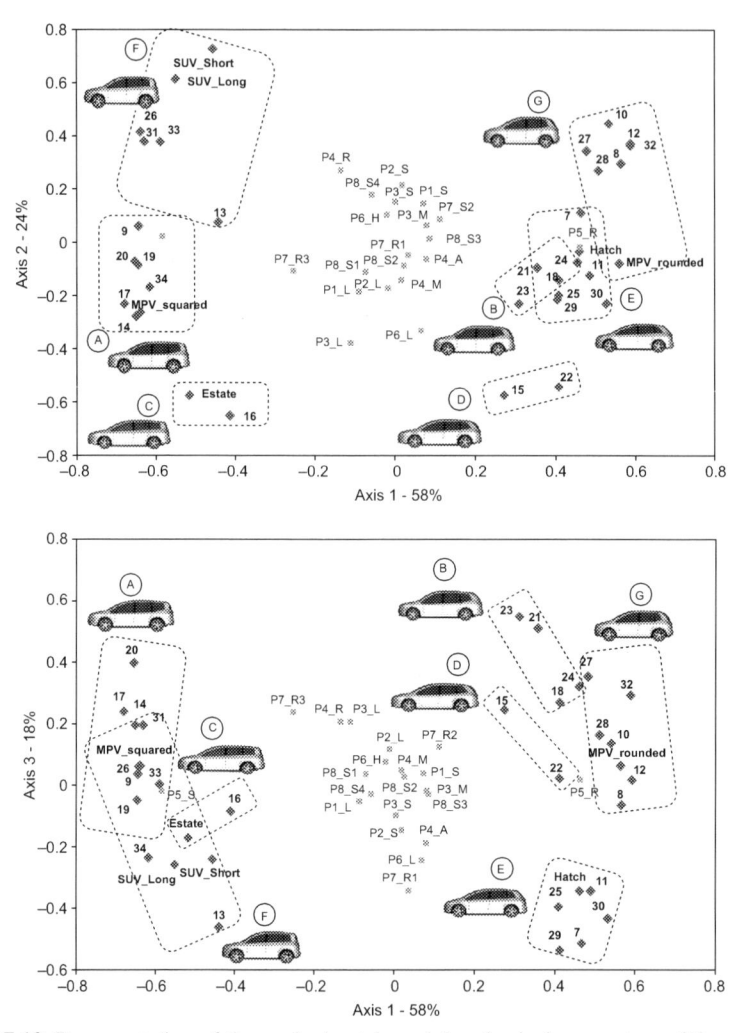

Figure 7.10 Representation of the car body style and the physical parameters of the experimental design on the basis of the first three MDS dimensions. The seven groups of the consensus partition are superimposed using short-dashed lines.

Thereafter, the average dissimilarity matrix was submitted to a non-metric MDS. With a Stress value equal to 0.096, a three-dimensional MDS configuration is retained. The three axes of the configuration recover respectively 58%, 24% and 18% of the total variance.

The configuration of car body styles resulting from the MDS solution is given in Fig. 7.10. The modalities of physical parameters defining the different cars are superimposed by using barycentric representation (Cadoret *et al.*, 2009). Figure 7.11 shows the correlations of the terms generated by means of the verbalization task with the MDS dimensions. The groups are clearly differentiated on the two factorial maps, corresponding to (axis 1 vs axis 2) and (axis 1 vs axis 3), respectively.

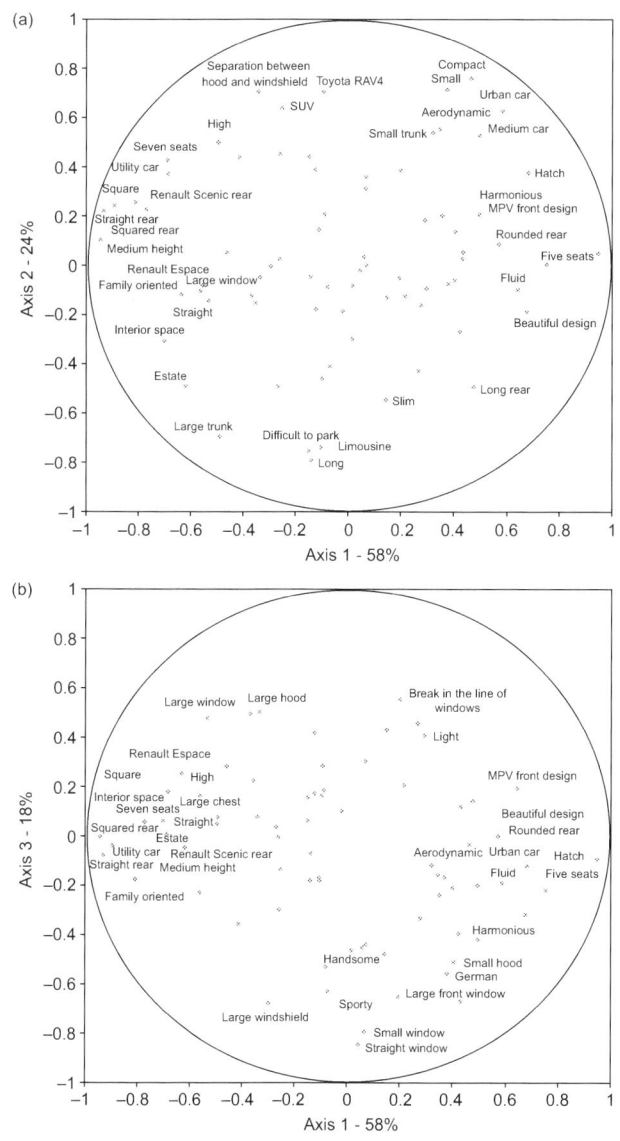

Free sorting as a sensory profiling technique 179

Figure 7.11 Correlations of the terms ($r > 0.5$) with the three dimensions of MDS configuration: (a) axis 1 vs axis 2 and (b) axis 1 vs axis 3.

The first dimension reveals a large perceptive difference between cars with squared rear (P5_S) and small windows (P7_R3) and cars with rounded rear (P5_R) and larger windows (P7_R1 and R2). The former category corresponds to vehicles belonging to groups A, C and F. These cars are described as square, having straight rear, medium height vehicles and they are qualified as "family oriented", "seven seats" and "utility cars". On the opposite side, cars belonging to groups B, D, E and G are described as having rounded rear, five seats, hatch and MPV vehicles, and qualified as having a fluid line and a beautiful design.

The second axis enables us to distinguish the short cars (short front overhang, short wheel base, short and medium rear overhang) with high body height (groups F and G described as compact, small trunk, SUV, urban car) from long and low cars (groups C and D described as long, slim, limousine, difficult to park, large trunk). At the top left of this map, group F (SUV group) appears to be typical of this opposition: these vehicles with reclined windshield (P4_R) and large rims (P8_S4) are qualified as utility cars.

Finally, the third dimension splits up the groups B (described as having a break in the line of windows), E (described as small, with small windows, small hood, handsome) and G (described as urban car, hatch, five seats). This axis separates the different modalities of the parameters P4 and P7 (position of the windshield and window height).

7.4.4 Discussion

Consumer perception of car body styles is organized around the five typical and classical car categories, "Hatch," "Estate," "MPV long," "MPV short," "SUV" and also around two new hybrid categories: "rounded rear estate car" and "hybrid cars between estate and MPV". The size of the rims and the rear overhang length are less important in the perception of the car body style than the other physical parameters. Although this experiment was conducted on reduced scale samples, it made it possible to identify the physical characteristics of the car body style that are important in the consumer perception, and also to better understand how the consumers perceive and conceptualize the car universe.

Designing the set of stimuli by experimental design prior to a free sorting task is an efficient strategy to investigate the relationships between the characteristics of samples and their perception by consumers. This investigation is of paramount importance for product development, since it highlights the respective importance of the design parameters.

7.5 Conclusion

The free sorting task gives insight into the perceptual organization of a set of products. This holistic perceptive task investigates the sensory characterization of the products, highlighting those categories of products that are meaningful for them. The cognitive processes implied by the task allow us to integrate not only the products' characteristics but also the subjects' experience and involvement toward the products in the perceptual responses of the subjects. Some properties used by the subjects are descriptive and directly related to the physico-chemical characteristics of the products. Other properties are inferred on the basis of descriptive properties (usage, hedonic, or qualifying properties). As such, they cannot be pinpointed by analytical approaches such as the sensory profiling techniques. Thus, the free sorting task may be considered as an interesting approach to highlight the consumers' preference drivers. This is of paramount importance in product development and process optimization.

In the first step of a product development process, this task can be used for defining a product space that is meaningful to the consumers. The task can be performed on a set of products selected after a benchmark of the market, in order to focus in a subsequent development step on a more homogeneous product space, based on consumers' perceptions.

Free sorting task can also be used to assess the respective importance of design parameters in an optimization process of product development. In this context, the designer can focus on parameters that define the different categories of the products inferred from the consumers' perceptions.

References

Abdi, H. (1990). Additive-tree representations (with an application to face processing). *Lecture Notes in Biomathematics*, **84**, 43–59.

Abdi, H., Valentin, D., Chollet, S. and Chrea, C. (2007). Analyzing assessors and products in sorting tasks: Distatis, theory and applications. *Food Quality and Preference*, **18**, 627–640.

Albatineh, A. N. and Mihalko, D. (2006). On similarity indices and correction for chance agreement. *Journal of Classification*, **23**, 301–313.

Ares, G., Varela, P., Rado, G. and Giménez, A. (2011). Are consumer profiling techniques equivalent for some product categories? The case of orange-flavoured powdered drinks. *International Journal of Food Science and Technology*, **46**, 1600–1608.

Ballester, J., Abdi, H., Langlois, J., Peyron, D. and Valentin, D. (2009). The odor of colors: Can wine experts and novices distinguish the odors of White, Red, and rosé wines? *Chemosensory Perception*, **2**, 203–213.

Ballester, J., Symoneaux, R. and Valentin, D. (2008). Conceptual vs perceptual wine spaces: Does expertise matter? *Food Quality and Preference*, **19**, 267–276.

Blancher, G., Chollet, S., Kesteloot, R., Nguyen Hoang, D., Cuvelier, G. and Siefferman, J. M. (2007). French and Vietnamese: How do they describe texture characteristics of the same food? A case study with jellies. *Food Quality and Preference*, **18**, 560–575.

Blancher, G., Clavier, B., Egoroff, C., Duineveld, K. and Parcon, J. (2012). A method to investigate the stability of a sorting map. *Food Quality and Preference*, **23**, 36–43.

Blancher, G., Mattei, B., Oelhafen, N. and Adam, C. (2008). A comparison of free sorting and hierarchical sorting tasks. A case study with sauces. *In:* Sfds, ed. 10th European Symposium on Statistical Methods for the Food Industry, Louvain-la-Neuve (Belgium).

Borg, I. and Groenen, P. (2005). *Modern Multidimensional scaling. Theory and Applications*, New York: Springer.

Cadoret, M., Lé, S. and Pagès, J. (2009). A factorial approach for sorting task data (FAST). *Food Quality and Preference*, **20**, 410–417.

Cadoret, M., Lé, S. and Pagès, J. (2011). Statistical analysis of hierarchical sorting data. *Journal of Sensory Studies*, **26**, 96–105.

Carroll, J. D. and Chang, J. J. (1970). Analysis of individual differences in multidimensional scaling via an N-way generalisation of ̄Eckart-Young decomposition. *Psychometrika*, **35**, 283–319.

Cartier, R., Rytz, A., Lecomte, A., Poblete, F., Krystlik, J., Belin, E. and Martin, N. (2006). Sorting procedure as an alternative to quantitative descriptive analysis to obtain a product sensory map. *Food Quality and Preference,* **17**, 562–571.

Chollet, S., Lelièvre, M., Abdi, H. and Valentin, D. (2011). Sort and beer: Everything you wanted to know about the sorting task but did not dare to ask. *Food Quality and Preference,* **22**, 507–520.

Chrea, C., Valentin, D. and Abdi, H. (2009). Graded structure in odour categories: A cross cultural case study. *Perception,* **38**, 292–309.

Chrea, C., Valentin, D., Sulmon-Rossé, C., Ly May, H., Nguyen, D. H. and Abdi, H. (2004). Culture and odor categorization: Agreement between cultures depends upon the odors. *Food Quality and Preference,* **15**, 669–679.

Courcoux, P., Faye, P. and Qannari, E. M. (2014). Determination of the consensus partition and cluster analysis of subjects in a free sorting task experiment. *Food Quality and Preference,* **32**, 107–112.

Courcoux, P., Qannari, E. M., Taylor, Y., Buck, D. and Greenhoff, K. (2012). Taxonomic free sorting. *Food Quality and Preference,* **23**, 30–35.

Coxon, A. M. (1999). *Sorting Data: Collection and Analysis,* California: Sage.

De Leeuw, J. (1988). Convergence of the majorization method for multidimensional scaling. *Journal of Classification,* **5**, 163–180.

Deegan, K. C., Koivisto, L., Näkkilä, J., Hyvönen, L. and Tuorila, H. (2010). Application of a sorting procedure to greenhouse-grown cucumbers and tomatoes. *Food Science and Technology,* **43**, 393–400.

Dehlholm, C., Brockhoff, P. B., Meinert, L., Aaslyng, M. D. and Bredie, W. L. P. (2012). Rapid descriptive sensory methods. Comparison of Free Multiple Sorting, Partial Napping, Napping, Flash Profiling and conventional profiling. *Food Quality and Preference,* **26**, 267–277.

Dubois, D. (1991). *Sémantique et cognition. Catégories, Prototypes, typicalité,* Paris 7 Editions du CNRS.

Dubois, D. (2009). *Le Sentir et le Dire. Concepts et méthodes en psychologie et linguistique Cognitives,* Paris 7 Editions L'Harmattan.

Duda, R. O., Hart, P. E. and Stork, D. G. (2000). *Pattern Classification,* 2nd edition, New York: Wiley.

Falahee, M. and Macrae, A. W. (1997). Perceptual variation among drinking waters: The reliability of sorting and ranking data for multidimensional scaling. *Food Quality and Preference,* **8**, 389–394.

Faye, P., Brémaud, D., Durand-Daubin, M., Courcoux, P., Giboreau, A. and Nicod, H. (2004). Perceptive free sorting and verbalization tasks with naive subjects: An alternative to descriptive mappings. *Food Quality and Preference,* **15**, 781–791.

Faye, P., Brémaud, D., Teillet, E., Courcoux, P., Giboreau, A. and Nicod, H. (2006). An alternative to external preference mapping based on consumer perceptive mapping. *Food Quality and Preference,* **17**, 604–614.

Faye, P., Courcoux, P., Giboreau, A. and Qannari, E. M. (2013). Assessing and taking into account the subjects' experience and knowledge in consumer studies. Application to the free sorting task of wine glasses. *Food Quality and Preference,* **28**, 317–327.

Faye, P., Courcoux, P., Qannari, E. M. and Giboreau, A. (2011). Méthodes de traitement statistique des données issues d'une épreuve de tri libre. *La revue de Modulad,* **43**.

Giboreau, A., Navarro, S., Faye, P. and Dumortier, J. (2001). Sensory evaluation of automotive fabrics: The contribution of categorization tasks and non verbal information to set-up a descriptive method of tactile properties. *Food Quality and Preference,* **12**, 311–322.

Greenacre, M. J. (1993). *Correspondence Analysis in Practice,* London: Academic Press.
Guénoche, A. (2011). Consensus of partitions: A constructive approach. *Advanced Data Analysis and Classification,* **5**, 215–229.
Gygi, B., Kidd, G. R. and Watson, C. S. (2007). Similarity and categorization of environmental sounds. *Perception and Psychophysics,* **69**, 839–855.
Heiser, W. J. (1990). A generalized majorization method for least squares multi-dimensional scaling of pseudodistances that may be negative. *Psychometrika,* **56**, 7–27.
Hoc, J. M. and Leplat, J. (1983). Evaluation of different modalities of verbalization in a sorting task. *International Journal of Man-Machine Studies,* **18**, 283–306.
Hubert, L. J. and Arabie, P. (1985). Comparing partitions. *Journal of Classification,* **2**, 193–218.
Hughson, A. and Boakes, R. A. (2002). The knowing noise: The role of knowledge in wine expertise. *Food Quality and Preference,* **13**, 463–472.
Hulin, W. S. and Katz, D. (1935). The Frois-Wittmann pictures of facial expression. *Journal of Experimental Psychology,* **18**, 482–498.
Jaffré, J., Valentin, D., Dacremont, C. and Peyron, D. (2009). Burgundy red wines: Representation of potential for aging. *Food Quality and Preference,* **20**, 505–513.
King, M. C., Cliff, M. A. and Wall, J. W. (1998). Comparison of projective mapping and sorting data collection and multivariate methodologies for identification of similarity-of-use of snack bars. *Journal of Sensory Studies,* **13**, 347–358.
Krieger, A. M. and Green, P. E. (1999). A generalized Rand-Index method for consensus clustering of separate partitions of the same data base. *Journal of Classification,* **16**, 63–89.
Kruskal, J. B. (1964). Nonmetric multidimensionnal scaling: A numerical method. *Psychometrika,* **29**, 115–129.
Lavit, C. (1988). *Analyse conjointe de tableaux Quantitatifs,* Paris. Edition Masson.
Lawless, H., Sheng, N. and Knoops Stan, S. C. P. (1995). Multidimensional scaling of sorting data applied to cheese perception. *Food Quality and Preference,* **6**, 91–98.
Lawless, H. T. (1989). Exploration of fragrance categories and ambiguous odors using multidimensionnal scaling and cluster analysis. *Chemical Senses,* **14**, 349–360.
Lebart, L., Morineau, A. and Piron, M. (2006). *Statistiques Exploratoires Multidimensionnelles,* Dunod 4ème édition.
Leclerc, B. and Cucumel, G. (1987). Consensus en classification: Une revue bibliographique. *Mathématiques et Sciences Humaines,* **100**, 109–128.
Lelièvre, M., Chollet, S., Abdi, H. and Valentin, D. (2008). What is the validity of the sorting task for describing beers? A study using trained and untrained assessors. *Food Quality and Preference,* **19**, 697–703.
Macadams, S., Winsberg, S., Donnadieu, S., De Soete, G. and Krimphoff, J. (1995). Perceptual scaling of synthesized musical timbres: Common Dimensions, specificities and latent subject classes. *Psychological Research,* **58**, 177–192.
Marcotorchino, J. F. and Michaud, P. (1982). Agrégation de similarité en classification automatique. *Revue de Statistique Appliquée,* **30**, 21–44.
Parizet, E. and Koehl, V. (2012). Application of free sorting tasks to sound quality experiment. *Applied Acoustics,* **73**, 61–65.
Parr, W. V., Green, J. A., White, K. G. and Sherlock, R. R. (2007). The distinctive flavour of New Zeland Sauvignon blanc: Sensory characterisation by wine professionals. *Food Quality and Preference,* **18**, 849–861.
Piombino, P., Nicklaus, S., Le Fur, Y., Moio, L. and Le Quéré, J.-L. (2004). Selection of products presenting given flavor characteristics: An applicaton to wine. *American Journal of Oenology and Viticulture,* **55**, 27–34.

Qannari, E. M., Cariou, V., Teillet, E. and Schlich, P. (2009). SORT-CC: A procedure for the statistical treatment of free sorting. *Food Quality and Preference,* **21**, 302–308.

Rand, W. (1971). Objective criteria for the evaluation of clustering method. *Journal of the American Statistical Association,* **66**, 846–850.

Régnier, S. (1983). Sur quelques aspects mathématiques des problèmes de classification automatique. *Mathématiques et Sciences Humaines,* **82**, 13–29.

Ricciuti, H. N., Marney, T. and Ricciuti, A. E. (2006). Availability and spontaneous use of verbal labels in sorting categorization by 16–23 months olds. *Early Childhood Research Quaterly,* **21**, 360–373.

Rosch, E. (1976). Classification d'objets du monde réel: Origines et représentations dans la cognition. *Bulletin de Psychologie,* 242–250.

Rosch, E. and Mervis, C. B. (1975). Family resemblances: Studies in the Internal structure of categories. *Cognitive Psychology,* **7**, 573–605.

Rosenberg, S. and Kim, M. P. (1975). The method of sorting as a data-gathering procedure in multivariate research. *Multivariate Behavioral Research,* **10**, 489–502.

Saint-Eve, A., Paçi Kora, E. and Martin, N. (2004). Impact of the olfactory quality and chemical complexity of the flavouring agent on the texture of low fat stirred yogurts assessed by three different sensory methodologies. *Food Quality and Preference,* **15**, 655–668.

Santos, J. M. and Embrechts, M. (2009). On the use of the adjusted Rand index as a metric for evaluating supervised classification. *Lecture Notes in Computer Science,* **5769**, 175–184.

Santosa, M., Abdi, H. and Guinard, J.-X. (2010). A modified sorting task to investigate consumer perceptions of extra virgin olive oils. *Food Quality and Preference,* **21**, 881–892.

Sattah, S. and Tversky, A. (1977). Additive similarity trees. *Psychometrika,* **42**, 319–345.

Schiffman, S. S., Reynolds, M. L. and Young, F. W. (1981). *Introduction to multidimensional scaling. Theory, methods and Applications,* Orlando. Academic Press.

Schlich, P. (1996). Defining and validating assessor compromises about product distances and attribute correlations. *In:* Naes, T. and Risvik, E. (eds.) *Multivariate Analysis of Data in Sensory Science.* Amsterdam: Elsevier.

Sinesio, F., Peparaio, M., Moneta, E. and Comendador, F. J. (2010). Perceptive maps of dishs varying in glutamate content with professional and naive subjects. *Food Quality and Preference,* **21**, 1034–1041.

Soufflet, Y., Calomnier, M. and Dacremont, C. (2004). A comparison between industrial experts' and novices' haptic perceptual organization: A tool to identify descriptors of the handle of fabrics. *Food Quality and Preference,* **15**, 689–699.

Takane, Y. (1981). MDSORT: A special purpose multidimensional scaling program for sorting data. *Behavior Research Methods and Instrumentation,* **13**, 698.

Takane, Y. (1982). IDSORT: An individual differences multidimensional scaling for sorting data. *Behavior Research Methods and Instrumentation,* **14**, 546.

Takane, Y., Young, F. W. and De Leeuw, J. (1977). Nonmetric Individual differences multi-dimensional scaling: An alternating least squares method with optimal scaling features. *Psychometrika,* **42**, 8–67.

Tang, C. and Heymann, H. (1999). Multidimensional Sorting, similarity scaling and free choice profiling of grapes jellies. *Journal of Sensory Studies,* **17**, 493–509.

Tardieu, J., Susini, P., Poisson, F., Lazareff, P. and Macadams, S. (2008). Perceptual study of soundscapes in train stations. *Applied Acoustics,* **69**, 1224–1239.

Teillet, E., Schlich, P., Urbano, C., Cordelle, S. and Guichard, E. (2010). Sensory methodologies and the taste of water. *Food Quality and Preference,* **21**, 967–976.

Torgerson, W. S. (1958). *Theory and Methods of Scaling,* New York: Wiley.

Van der Kloot, W. A. and Van Herk, H. (1991). Multidimensional scaling of sorting data: A comparison of three procedures. *Multivariate Behavorial Research,* **26**, 563–581.

Varela, P. and Ares, G. (2012). Sensory Profiling, the blurred line betwee sensory and consumer science. A review of novel methods for product characterization. *Food Research International,* **48**, 893–908.

Wiley, D. E. (1967). Latent partition analysis. *Psychometrika,* **32**, 183–193.

Youness, G. and Saporta, G. (2004). Some measures of agreement between close partitions. *J. Student,* **5**, 1–12.

Free multiple sorting as a sensory profiling technique

C. Dehlholm
Danish Technological Institute, Aarhus, Denmark

8.1 Introduction

This chapter focusses on the application and background of the method of "free multiple sorting" (FMS). Sorting, as such, is the easily understood task of grouping products according to their "similarity." However, the methodology of the sorting exists as various processes. Traditional "free sorting" does not impose any constraints on the assessor, who simply divides the products at hand into a number of groups based on free-of-choice criteria. Each assessor is typically instructed to sort the sample set once. This results in a number of sorting categories per assessor. A number of sorting variations exist that differ from "free sorting." For instance, each assessor may be instructed to sub-sort the sample set (e.g. ascending/descending/hierarchical/taxonomic sorting). This results in a number of weighted categories per assessor. Another sorting variation is FMS, which consists of multiple free sortings where the number of sortings is up to the individual assessor. Hence, each assessor re-sorts the same sample set a number of times with freedom of choice, resulting in one multidimensional sample space per assessor.

This chapter begins with a brief overview of the methodology (Section 8.2), highlighting with bullet points the most important aspects of FMS, from aim to application. The theoretical framework of the method is then outlined (Section 8.3), to give the reader a more profound basis for applying the approach.

The following sections of the chapter deal with practical work with FMS (Sections 8.4–8.7). These sections describe how to prepare a study, how to execute it and how to collect data, and suggest an approach for analysing the data. The chapter concludes by exemplifying possible applications of FMS, as well as its advantages and disadvantages.

Before the References section, the chapter is completed with a comment on future trends (Section 8.8).

8.2 Overview of free multiple sorting (FMS)

Method overview:

- The aim of FMS is to achieve a free-of-choice number of groupings of samples and their possible naming from the assessor. Typically, multiple categorizations of samples are

derived from each assessor, resulting in a whole sample space per assessor rather than a single categorization as in classical "free sorting."
- FMS is a sensory evaluation technique, which, combined with the naming of the sorted groups of samples, provides holistic descriptive sample information.
- Evaluation instructions are: "A number of samples are placed in front of you. Interact with the samples and arrange them in groups according to their perceived differences. Note down which samples are grouped together and, if possible, note why. Then, pool all samples and re-group them, if there are further meaningful ways in which they could be grouped. You decide according to your own discrimination criteria, and you are allowed to make additional groupings as long as it makes sense to you." The rationale is that the assessor is not limited to a single forced categorization of the samples, as in classical "free sorting."
- The resulting data are categorical and can, with advantage, be treated with multivariate multi-block analysis.
- FMS is regarded as being a rapid, as well as a low-resource, methodological approach. Studies show that FMS does not yield sample information as detailed as the more laborious conventional profiling. However, the information derived has proven multidimensional, and the approach is applicable to various situations.
- FMS is applicable to both food and non-food products.

8.3 Theoretical framework

The principle of sorting originates from the field of psychology, where it was applied for measuring personality traits (Hulin and Katz, 1935; Steinberg, 1967). These measurements of (dis)similarity were applied to multidimensional scaling in 1981 (Schiffman et al., 1981) and later applied to examining odours (Lawless, 1989). As the variation "free sorting" (Rosenberg et al., 1968), the methodology was introduced to the field of food sensory science (Lawless et al., 1995). As mentioned, the assessor only sorts samples into groups once in a single free sorting. The original sorting variation from Steinberg (1967) was "multiple sorting," built on the principle that assessors performed additional, hence multiple, sortings on the same sample set, as long as it made sense to the individual assessor. This variation was introduced to foods by Dehlholm et al. (2012).

In general, sorting is an easily applicable approach, opening up a range of variations. Within the field of food science, sorting has been applied in variants including "restricted sorting" (Lawless, 1989), "descendant hierarchical sorting" (where the assessor is asked to continuously sub-sort each sorted group of products (Egoroff, 2005)), the reverse "ascendant hierarchical/taxonomic free sorting" (where assessors continuously group together their initial groups of products (Qannari et al., 2010)), "directed sorting" (Ballester et al., 2009), "sorted Napping" (Pagès et al., 2010) and "labelled sorting" (Bécue-Bertaut and Lê, 2011). Sorting procedures in general are rapid and holistic approaches, and applicable to large product sets. By "holistic" it means that the product elicitation is initially based on an overall product assessment rather than that of e.g. the more analytical conventional descriptive profiling.

The sorting methodology as such is not descriptive unless a descriptive step is added as part of the proceedings. This is straight forward, as the assessor is asked to name the motive for grouping the products. The assessor might have grouped products based on a concept that is hard to express. In other, and most, cases, the assessor is able to verbalize why the groupings were constructed in a particular way, and group labels are recorded as valuable descriptive data. Data for one assessor consist of a data table with one column for each sorting, where the grouped products have exactly the same name. According to the statistical program in use, it may be possible to apply either numbers or the actual name as the group name.

The following chapters contain a walkthrough of the application of FMS and, as such, can be used as a guide for practical implementation.

8.4 Practical framework and design of experiments

FMS is a relative simple methodological approach to plan and carry out, and it is relatively simple for a sensory professional who is familiar with conventional profiling. Since only a few studies have implemented FMS in the sensory evaluation of food, there has been little discussion and variation of its application. Before a FMS is carried out, a few factors have to be considered. These factors and their implications are discussed in the following.

The assessors produce the raw data that hold all information for drawing the product profiles. Hence, the sensory professional must choose the type of assessor most suited to answer the research question. Naïve assessors, e.g. consumers, are more likely to organize products based on meta-attributes. As FMS is described as a holistic method, based on its product elucidation, it can be relevant to recruit naïve assessors for a FMS task. Nevertheless, the assessors must always be chosen according to the aim of the analysis. If the aim is to obtain more functional product information, product users might be a better starting point, while a previously trained sensory panel will tend to give attribute-related information. In general, the assessor will name groupings according to earlier trained or learned constructs. The number of assessors to involve will vary according to the type of product in hand and to the type of assessor chosen.

The sensory professional could decide to restrict the assessors in their product groupings. In other holistic methods, e.g. the projective mapping technique, it is an option to instruct assessors only to report on one type of answer, e.g. textural. This is not recommended for FMS, unless its focus is on specific non-functional terms, e.g. emotional, as the assessors will independently choose the personally most relevant discriminators.

It is not reasonable to work with repetitions of the whole FMS session. The method is designed in such a way that the assessor highlights all relevant groupings with use of the same product set. Therefore, it is not relevant to present another set of the same recoded samples. If assessors were asked to perform repetitions, there is a chance that some would focus more on memorizing and recalling earlier responses and product groupings rather than on responding to their spontaneous perceptions. Hence, if the goal of the evaluation is to retrieve measures of spontaneous perception, it is more rational to perform one evaluation session of one sample set per assessor only.

The number of samples to include in an evaluation must be selected according to the nature of the samples. The smallest number of samples to include in an FMS task is three in theory, as this is the smallest amount from which it is possible to form a separate group. Practically, it is more rational to include a higher number of samples. The upper number of samples must be determined with sensory fatigue in mind. As FMS is a holistic procedure, it is open for a generally higher number of samples than more laborious procedures, such as conventional profiling. If the product under investigation is a non-food, and tasting or smelling is not a part of the assessment, there might be opportunity for a higher number of samples.

The FMS is a rapid methodology with a highly (dis)similarity-based product elicitation. The samples are pooled, after the first free sorting is performed, to form the sample base for a new sorting. New recoded sample sets are not provided, although additional sample material should be available. As the assessor is allowed to re-taste (re-evaluate) and compare the samples in any order, a balanced and randomized evaluation order is not highly beneficial. The samples are presented simultaneously as a set; hence, variation in the arrangement of the presented set is applicable. If so, the assessors must be instructed to have the first taste of samples in a specific order, or samples could be presented one-by-one. Nevertheless, FMS is an approach that primarily elicits significant differences between samples. For that reason, the author would suggest the presentation design as optional. In addition, this adds to the concept of FMS as an easy method to administer. Instead, duplicates can be integrated as part of the sample set and function as validation samples.

Samples for the evaluation will, in most food product cases, have to be served blinded in such a way that the assessor cannot recognize the original product brand or origin. If, on the other hand, overall product concepts are to be evaluated, the brand, product fact sheets, labels or the actual product packaging can be applied. Samples are best identified by three-digit random numbers.

As a holistic approach, FMS is a good tool for eliciting holistic constructs and contrasts present in the sample set. As more holistic constructs are better elicited by naïve assessors as consumers, training sessions of assessors are not a part of the method process. If focus is on highlighting attributes from a predefined vocabulary, a previously trained panel can be applied with benefit.

8.5 Implementation and data collection

FMS is based on the original idea of multiple sorting by Steinberg (1967), whereby the assessors are asked to sort the samples in front of them into groups in a way that makes sense to the individual assessor only. The samples are placed in front of the assessor, who is instructed on how to evaluate the set. The instructions are: "A number of samples are placed in front of you. Interact with the samples and arrange them in groups according to their perceived differences. Note down which samples are grouped together and, if possible, note why. Then, pool all samples and re-group them as long as there are other meaningful ways in which they can be grouped. You decide on your own discrimination criteria and are allowed to make additional groupings

Free multiple sorting as a sensory profiling technique 191

	Sample	A1-S1	A1-S2	...	A1-S_y	A2-S1	A2-S2	...	A2-S_y	...	Ax-S1	Ax-S2	...	Ax-S_y
Samples n	P_1	S1-a	S2-b	...	S_y-b	S1-a	S2-a	...	S_y-c	...	S1-a	S2-b	...	S_y-a
	P_2	S1-a	S2-a	...	S_y-a	S1-b	S2-a	...	S_y-b	...	S1-b	S2-a	...	S_y-a
	P_3	S1-b	S2-b	...	S_y-c	S1-b	S2-a	...	S_y-a	...	S1-c	S2-b	...	S_y-a
	P_4	S1-b	S2-c	...	S_y-a	S1-b	S2-b	...	S_y-a	...	S1-c	S2-a	...	S_y-b
	...	...	...	...	...	...	...	...	...	...	...	...	...	...
	P_n	S1-n	S2-n	...	$S_{y\text{-}n}$	S1-n	S2-n	...	$S_{y\text{-}n}$	...	S1-n	S2-n	...	$S_{y\text{-}n}$

Columns grouped as: Assessor A1's groupings S, Assessor A2's groupings S, Assessor Ax's groupings S.

Figure 8.1 Data table structure of FMS data prepared for analysis with multiple factor analysis in R.

as long as it makes sense to you." In each sorting, they are allowed to generate two or more groups containing one to "all except one" sample. They are instructed to continue until they feel that all meaningful discrimination possibilities have been covered. As the groupings are comparative judgements, is it allowable to go back and forth re-evaluating the samples? Assessors do not need to be able to explain why they sort as they do, provided it makes sense to them as individuals. If they have a clear idea of why they are sorting as they do, they are encouraged to write down the reason. Except for method instructions, there is no preliminary training, and the FMS should be carried out in less than an hour.

For each sorting task, data collected for a sample consist of the description of the group that the sample belongs to. Data can be structured as illustrated in Fig. 8.1. The data table is constructed so that one row corresponds to one sample. The first row corresponds to assessor identification, while the first column consists of the sample names. The following columns of the data table consist of all assessors' groupings, with all groupings from one assessor placed in adjacent columns. In one column, samples belonging to the same groups need to bear exactly the same name or identification. Depending on the applied statistical software, some programs are able to handle semantics as cell values, while others need a number as the category identifier.

8.6 Data analysis

As FMS is a rapid approach, it tends to highlight main dimensionality and differences among products. For that reason, statistical methods that are overall comparable will tend to generate overall comparable results. Since the sets of variables cannot be assumed to be of a similar nature, a data analysis taking account of such varying variable sets is desirable.

In recent years, the free statistical software R (Ihaka and Gentleman, 1996; R Development Core Team, 2010) and specially developed sensory function packages have gained in popularity. The described data table structure is suited for data analysis with Multiple Factor Analysis as implemented in the FactoMineR package (Lê et al., 2008; Husson et al., 2010) in R. The FactoMineR package contains relevant algorithms for professionals who work with multivariate exploratory data analysis.

Multiple Factor Analysis is a two-step analysis wherein the first step is, in the case of categorical data, a Multiple Correspondence Analysis (Escofier and Pagès, 1994). This step is performed to normalize each set of variables. The second step is a Principle Component Analysis on the concatenated results, which ends up with the global configuration (Abdi et al., 2013). The dataset as described consists of one variable for each sorting performed. Each variable is categorical, and consists of the name of the group to which each sample belongs. There is a varying number of sortings per subjects; a recent study reports on individual assessors applying between 3 and 14 sortings to the same set of samples (Dehlholm et al., 2012). The advantage of grouping sortings from each assessor is that all groups will be equally weighted and independent of the number of variables. In this way, the assessor with 14 sortings will not dominate the product space compared with the assessor with three sortings. In addition, the multidimensionality of individual product spaces will be preserved when the spaces are joined in the global Principal Component Analysis.

The Multiple Factor Analysis results in a number of standard pictorials, of which four are shown in Fig. 8.2. In this example, nine products were assessed by 14 assessors using FMS. The global individual factor map (Fig. 8.2a) shows, as the most ordinary plot (the global Principal Component Analysis), the placement of the samples in two selected dimensions. The figure shows groupings of products that have most often been categorized together; the products P1, P3, P4 and P9 are more similar, opposed to the products P5, P6 and P8, while P2 and P7 show distinct characteristics. It is optional to display the individual assessors for each product. In Fig. 8.2c, dashed lines represent each individual assessor's judgement of the respective product. It can be used to visually examine the variation among assessors, which in this example is larger for product P2 than the others. The partial axes plot (Fig. 8.2d) shows the correlations of the individual assessor's dimensions one and two. It can be used to identify similarities and differences among the individual product spaces. As here, when more assessors are involved in the study the plot gets crowded and harder to interpret. It is possible to display the category variables in a plot. Nevertheless, as the assessments often result in a large number of sortings, the display of all in a single plot results in an overcrowded and hardly readable plot. For this reason, it is valuable to support the multivariate results with some way of looking into the descriptive information. Significant discriminators can be found with frequency analysis. FactoMineR includes a function for the purpose (*catdes*) which will show the semantics that are applied significantly more for any given sample than for residual samples. Before such an analysis, lemmatization is an important preparation step. Take into account that such a step may rely on subjective decisions of the analyst and, for that reason, must only be performed in obvious cases.

There are various ways to examine uncertainty. One is the graphical application of confidence ellipses around sample means, superimposed onto the individual factor

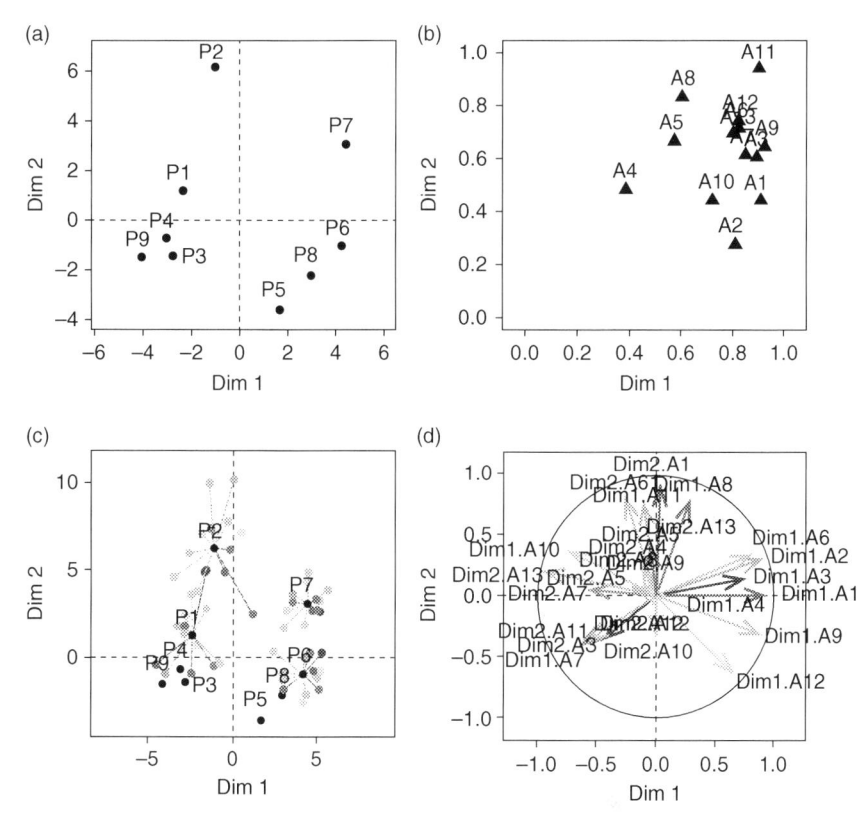

Figure 8.2 Classical multiple factor analysis output based on an FMS study. (a) Individual factor map of samples/products. (b) Groups (assessors) weight representation. (c) Individual factor map with assessor distributions shown for samples P1, P2, P7 and P8. (d) Partial axes for each group (assessor).

map. A script for easy implementation in R has been developed (Dehlholm et al., 2012), which makes it possible to construct confidence ellipses for one-repetition methodologies such as FMS. Based on an R Multiple Factor Analysis result object, the procedure bootstraps the standardized groups of variables generated by the initial part of the Multiple Factor Analysis and generates 95% confidence ellipses as well.

The non-graphical RV coefficient (Robert and Escoufier, 1976) can be used for individual assessor validation as well as for comparison of overall sample configurations. The coefficient is a multivariate correlation coefficient that lies between 0 and 1, where 1 is a complete match, and it is implemented in FactoMineR as the function *coeffRV*. The function also provides significance testing of the coefficient's difference from zero according to Josse et al. (2008). For example, it may be applied to express the distance between principal components of two individual assessors, between an individual assessor and the average configuration, or between average configurations from various assessor groups.

8.7 Advantages, disadvantages and applications

Categorized as a rapid sensory profiling technique, FMS is often compared to such methods as projective mapping (Napping) and pick-any (check-all-that-apply) variants. This categorization is solely time-wise, as rapid approaches in general differ as much from each other as from the "slower" conventional approaches. The term "rapid" relates mainly to the relatively short time it takes for the assessor to perform the test. Different rapid techniques use different schemes for sample preparation and data treatment. The FMS is rapid as well as easy to plan and perform; it is a simple and understandable task for the assessor, and the responses are fast to collect for the sensory professional. If the responses are given on paper sheets, it will take a while to transfer data into a spreadsheet. Nevertheless, the most time consuming part of working with FMS is the data treatment. Vocabularies are often excessive and diverse in nature. Despite lemmatization, the vocabularies are potentially idiosyncratic and ideal for the individual assessor only.

With FMS (as with projective mapping), assessors are allowed to separate samples on the bases of hunches or feelings not easily expressed. Focus is kept on the sample rather than a predefined set of attributes, and assessors are not asked to describe their actions until after the sorting. Compared to conventional sensory profiling where a vocabulary needs to be established before sample evaluation, the FMS approach is said to be more holistic and individual. This is an advantage when applied to conceptual tasks (Fig. 8.3.), as distinct from the more constructual tasks that focus on specific sample attributes.

As stated, FMS suits holistic and conceptual tasks open for the use of intermodal and complex semantics. Such data are often noisier and show larger confidence ellipses around the samples in an individual factor map than data from a conventional profile. As shown in Fig. 8.3, more conceptual task would often involve the use of naïve assessors, would need less concept alignment (assessor training), and would hence require a larger number of assessors.

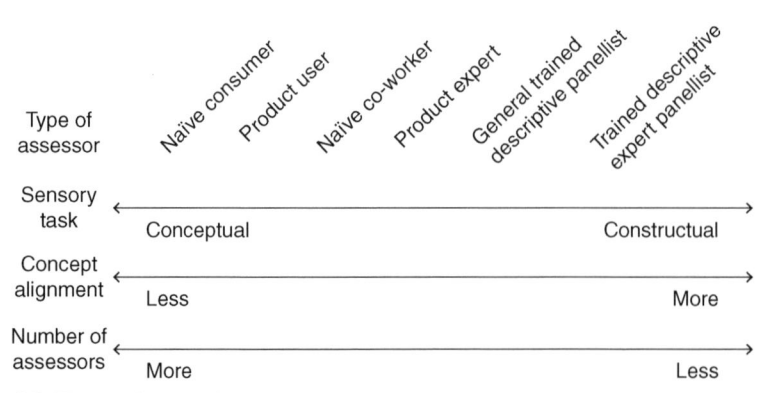

Figure 8.3 The continuum of sensory assessors.
Source: Reprinted from Dehlholm (2012), Copyright 2012.

The described set-up is ideal for tasks where a low amount of available time with assessors needs to result in a quantitative mapping of a number of samples. This situation occurs within companies that work with product development, where FMS can be applied with internal or external assessors. It especially applies to comparisons within or between product categories, in the exploration and idea generation phase of product development and to small-scale market tests of pilot products.

As FMS consumes little time and can be performed in less than an hour, it is also a low-resource approach. It therefore applies to smaller companies with smaller R&D budgets. Most important is that the method is well understood before its application rather than it being applied because it is "easy."

8.8 Future trends and further information

A recent study compares FMS with other rapid profiling techniques as well as conventional profiling (Dehlholm *et al.*, 2012). The study examines product spaces, semantic output and practical performance in detail based on experiments where the same sample sets are evaluated with several assessor panels. This is the only present study that applies FMS with food products. The straightforward execution of FMS, both for the assessor and the sensory professional, would advocate its further use. It is especially suitable for low budget product development teams, e.g. in SMEs. Implemented in such environments, FMS will, like other sorting methods, probably not be the subject of concern for sensory research groups. Instead, FMS will adapt through use with regard to each special case. In this way, most rapid sensory profiling methods have evolved.

References

Abdi, H., Williams, L. J. and Valentin, D. (2013). Multiple factor analysis: principal component analysis for multitable and multiblock data sets. *Wiley Interdisciplinary Reviews: Computational Statistics,* **5**, 149–179.

Ballester, J., Abdi, H., Langlois, J., Peyron, D. and Valentin, D. (2009). The odor of colors: can wine experts and novices distinguish the odors of white, red, and rosé, wines? *Chemosensory Perception,* **2**, 203–213.

Bécue-Bertaut, M. and Lê, S. (2011). Analysis of multilingual labeled sorting tasks: application to a cross-cultural study in wine industry. *Journal of Sensory Studies,* **26**, 299–310.

Dehlholm, C. (2012). *Descriptive Sensory Evaluations: Comparison and Applicability of Novel Rapid Methodologies.* Copenhagen: SL grafik.

Dehlholm, C., Brockhoff, P. B. and Bredie, W. L. P. (2012). Confidence ellipses: a variation based on parametric bootstrapping applicable on multiple factor analysis results for rapid graphical evaluation. *Food Quality and Preference,* **26**, 278–280.

Dehlholm, C., Brockhoff, P. B., Meinert, L., Aaslyng, M. D. and Bredie, W. L. P. (2012). Rapid descriptive sensory methods – comparison of free multiple sorting, partial napping, napping, flash profiling and conventional profiling. *Food Quality and Preference,* **26**, 267–277.

Egoroff, C. (2005). How to measure perception: a case study on automotive fabrics. *ESN Seminar,* 25–26 May 2005, Madrid, Spain.

Escofier, B. and Pagès, J. (1994). Multiple factor-analysis (Afmult Package). *Computational Statistics and Data Analysis,* **18**, 121–140.

Hulin, W. S. and Katz, D. (1935). The Frois-Wittmann pictures of facial expression. *Journal of Experimental Psychology,* **18**, 482–498.

Husson, F., Josse, J., Lê, S. and Mazet, J. (2010). FactoMineR: multivariate exploratory data analysis and data mining with R. *R package version 1.14.http://CRAN.R-project.org/package=FactoMineR.*

Ihaka, R. and Gentleman, R. (1996). R: a language for data analysis and graphic. *Journal of Computational and Graphical Statistics,* **5**, 299–314.

Josse, J., Pagès, J. and Husson, F. (2008). Testing the significance of the RV coefficient. *Computational Statistics and Data Analysis,* **53**, 82–91.

Lawless, H. T. (1989). Exploration of fragrance categories and ambiguous odors using multidimensional-scaling and cluster-analysis. *Chemical Senses,* **14**, 349–360.

Lawless, H. T., Sheng, N. and Knoops, S. S. C. P. (1995). Multidimensional-scaling of sorting data applied to cheese perception. *Food Quality and Preference,* **6**, 91–98.

Lê, S., Josse, J. and Husson, F. (2008). FactoMineR: an R package for multivariate analysis. *Journal of Statistical Software,* **25**, 1–18.

Pagès, J., Cadoret, M. and Lê, S. (2010). The sorted napping: a new holistic approach in sensory evaluation. *Journal of Sensory Studies,* **25**, 637–658.

Qannari, E. M., Courcoux, P., Taylor, Y., Buck, D. and Greenhoff, K. (2010). Statistical issues relating to hierarchical free sorting task. *10th Sensometric Meeting: Past, Present, Future,* 25–28 July 2010, Rotterdam, The Netherlands.

R Development Core Team (2010). R: a language and environment for statistical computing. *R Foundation for Statistical Computing, Vienna, Austria. ISBN 3-900051-07-0, URL http://www.R-project.org/.*

Robert, P. and Escoufier, Y. (1976). Unifying tool for linear multivariate statistical-methods – Rv-coefficient. *Journal of the Royal Statistical Society Series C-Applied Statistics,* **25**, 257–265.

Rosenberg, S., Nelson, C. and Vivekana, P. S. (1968). A multidimensional approach to structure of personality impressions. *Journal of Personality and Social Psychology,* **9**, 283–294.

Schiffman, S. S., Reynolds, M. L. and Young, F. W. (1981). *Introduction to Multidimensional Scaling: Theory, Methods, and Applications.* New York: Academic Press.

Steinberg, D. D. (1967). The word sort: an instrument for semantic analysis. *Psychonomic Science,* **8**, 541–542.

Napping and sorted Napping as a sensory profiling technique

9

S. Lê[1], T.M. Lê[1], M. Cadoret[2]
[1]Agrocampus Ouest, Rennes, France; [2]Kuzulia, Plabennec, France

9.1 Introduction

Can (sorted) Napping be considered a rapid method? Can it be considered as an alternative to descriptive analysis? One of the main objectives of this chapter is to shed new light on Napping and sorted Napping in order to provide some clues for answering these two questions. In particular, we will stress the intrinsic nature of the data collected when using Napping and the link between Napping and sorted Napping.

This chapter is divided into two main parts. The first is dedicated to Napping, while the second is dedicated to sorted Napping. We will see finally how close Napping and sorted Napping are, and how they can easily be confused in a large number of situations. Each part is illustrated by an example.

The first example used to illustrate Napping was chosen for two main reasons. First, it will allow the reader to understand the intrinsic nature of the data collected when performing Napping, and the natural evolution from Napping to sorted Napping. Second, although the example is not directly related to sensory profiling, it is typical of the sensations sensory researchers are confronted with when they develop a new product, as it deals with emotions.

The second example is directly focused on the measurement of sensory perceptions of food products. It is a pedagogical example that illustrates perfectly well how sorted Napping can be used as a sensory profiling technique. In this example, eight smoothies were chosen according to two main factors: the type of manufacturer (two categories) and the flavour (four categories).

9.2 From projective tests to Napping

9.2.1 An introduction to projective tests

In psychology, notably in clinical psychology, *projective tests* were conceived to assess the personality of a patient. Such tests allow subjects to indirectly reveal their personalities by experiencing different, vague, ambiguous stimuli, and by projecting themselves through their responses to the stimuli, hence the adjective *projective*. One of the most famous and controversial projective tests is certainly the Rorschach inkblot test.

This family of tests has been adapted for, and then frequently used in, marketing research. We can cite for instance the word association task in which "the subjects are asked to read a list of words and to indicate the first word that comes to mind." For a short and easily available review of projective tests in consumer research, refer to Donoghue (2000).

In the sensory field, in 1994, Risvik *et al.* published a paper entitled "Projective Mapping: a tool for sensory analysis and consumer research": the idea of asking subjects to reveal themselves through the way they are positioning stimuli (products) on a sheet of paper based on their perceived similarities arose in sensory analysis. However, from our point of view, this brilliant idea was not exploited for its proper use. While many different angles could have been approached, the authors focused on a representation of the stimuli obtained by using generalized Procrustes analysis (GPA), and its comparison with other representations issued from sensory profiling, and dissimilarity scaling. This is, somehow, not very surprising, as the reasoning of Risvik *et al.* (1994) seems to be mainly based on the stimuli and their differences, not on the subjects and the way they perceive differences between stimuli.

From our point view, the idea that "when an assessor evaluates the products as a whole he/she will concentrate on the most obvious differences when making a judgement" requires further discussion. Beyond the differences that may exist between stimuli, it seems that one crucial dimension has been overlooked: the individual variability amongst the subjects. In other words, what may be obvious for one subject may not matter for another. Considering these differences of perception between subjects is particularly important in our context, as the perception of the stimuli is *holistic*, on the one hand, while the way this perception is expressed is *projective*, on the other hand. As explained in the next section, this is the starting point of Napping.

9.2.2 The intrinsic nature of Napping data

Napping arrived in 2005, in "Collection and analysis of perceived product inter-distances using multiple factor analysis (MFA): Application to the study of ten white wines from the Loire Valley" (Pagès, 2005). In its basic version, Napping consists in asking subjects to position stimuli on a "nappe" (i.e., French word for tablecloth; originally, the "nappe" was simply a sheet of paper with specific dimensions) according to the way they perceive their similarities. In practice, each subject uses his own criteria to position stimuli. Two stimuli should be closer to the degree they were perceived to be similar, and more distant as they were perceived to be different. There are no good or bad answers.

With respect to the way data are collected, the two methods, Projective Mapping and Napping, look very much alike. However, the context in which each method was conceived, and their respective motivations, are very much different. Napping was developed in a period where consumers began to be used to providing sensory data. At that time, the paradigm that would strictly separate sensory data provided exclusively by trained panellists, and hedonic data provided by consumers, was evolving and consumers were more and more solicited for providing sensory information. Not any kind of sensory information, as you do not expect consumers to be as accurate as trained panellists on a list of sensory descriptors, but a more qualitative one that would give insights about what consumers can sensorily perceive.

As mentioned by Pagès in his article, when using a conventional sensory profile based on a given list of sensory descriptors, "one criticism of this methodology is that the weights given to the descriptors do not necessarily correspond to their real importance for the subjects". As evoked previously, this was the main starting point of Napping: to find a way that would reveal what is most important (or obvious) for subjects, in particular consumers, in terms of perceived differences amongst stimuli. How to get this information without asking? The answer lies in the dimensions of the "nappe".

The dimensions of the "nappe," its width and its height, are of utmost importance. The "nappe" has a rectangular shape, with a width of 60 cm and a length of 40 cm. Consequently, as subjects are naturally influenced by the X-axis, which is by construction 1.5 visually more important than the Y-axis of the "nappe," subjects are prioritizing the reasons why they perceive the stimuli differently. More importantly, this hierarchy is obtained unconsciously and spontaneously. In other words, the rectangle in which stimuli are positioned leads subjects to privilege one dimension (i.e., the X-axis) and to separate stimuli with respect to that dimension, then to use the second dimension (i.e., the Y-axis), in a hierarchical order. The idea of Napping is, first, to collect, through the positioning of the stimuli, the first and the second *reason* why stimuli are perceived as different, for each subject; then, to summarize all these reasons over all the subjects. To do so, Napping data need to be analysed specifically, with the proper method. In that sense, Napping can be seen as the combination of the way data are collected and analysed, as will be explained in the next section.

9.2.3 Analysing Napping data

As a result of Napping, each subject provides two vectors of coordinates of dimension $I \times 1$ each (one for the X-axis, one for the Y-axis), where I denotes the number of stimuli to be positioned on the rectangle. Hence, the final data set to be analysed, denoted X, is obtained by merging the N pairs of vectors of coordinates, where N denotes the number of subjects. In other words, X can be seen as a data set structured into N groups of two variables each. Typically, the statistical analysis of such data set X should take into account the "natural" partition on the variables.

MFA (Escofier and Pagès, 1994; Pagès, 2004, 2013) was precisely conceived in order to balance the part of each group of variables within a global analysis. When variables are quantitative (which is our case), MFA can be defined as an extension of principal component analysis (PCA), which takes into account a group structure on the variables, and which balances the part of each group: in that sense, MFA can be seen as a weighted PCA. Technically, each group j of variables is weighted by the inverse of λ_1^j, where λ_1^j denotes the highest eigenvalue issued from the PCA performed on the variables of group j.

The rationale behind this weighting scheme is to exhibit dimensions that are: (1) common to as many groups of variables as possible; and (2) specific to some groups of variables. From this perspective, MFA can be defined as a variant of Generalized Canonical Analysis (Carroll, 1968).

When analysing Napping data, in order to respect the intrinsic nature of the data, λ_1^j is obtained by running a PCA on the covariance matrix, not on the correlation matrix. In other words, the separate analyses of each group of coordinates are not performed on standardized variables: the X-coordinates and the Y-coordinates of each group are not scaled to unit variance. Otherwise, the information about the relative importance of the dimensions of variability amongst the stimuli would be removed: this information is precisely what the experimenter is looking for when she/he is collecting data using Napping.

9.2.4 Example: revealing oneself with Napping, through emotions as stimuli

The data presented in this section are part of a cross-cultural study of which the main purpose is to compare two very supposedly different countries, Vietnam and France, through their respective emotional spaces (i.e., the way emotions are related to each other). For this comparison, we asked Vietnamese and French subjects to perform a Napping on a set of 53 emotions. The emotions were chosen in order to obtain a corpus of words that would be as diverse and comprehensible as possible. Pre-tests were conducted to make sure that the corpus was easily understandable. This study can be divided into two parts, corresponding to two main objectives, namely (1) to understand how emotions were perceived and structured within each country, and (2) to compare Vietnam and France in terms of emotions. To illustrate how Napping can be used to reveal oneself, through emotions as stimuli, we present here results issued from the first part of the study, restricted to the French data (100 French subjects).

Figures 9.1 and 9.2 show two completely different individual configurations. The configuration of Subject 31 (cf. Fig. 9.1) is very structured: emotions seem to have been divided into two groups, along the X-axis, the positive emotions (on the right-hand side of the plane) and the negative ones (on the left-hand side of the plane). The configuration of Subject 96 (cf. Fig. 9.2) seems to be structured as well, but in a different way. Subject 96 provided small clusters of emotions, but the positioning of the clusters seems to be hardly interpretable. Subject 96 may not have understood the Napping task, and hence comes the question of getting a representation of the stimuli based on all subjects.

To answer that question, an MFA was performed on the whole of the data. In detail, MFA is performed on groups of two variables each (i.e., the X- and Y-coordinates of the stimuli on each subject's rectangle). Let us remind ourselves that the separate analysis of each group is based on unstandardized variables.

Figure 9.3 corresponds to the representation of the 53 emotions (considered as statistical units) obtained by using MFA. This representation is the consensus over the subjects on the so-called first factorial plane, constituted by the first two main dimensions of variability. The first dimension (i.e., the horizontal one), denoted Dim

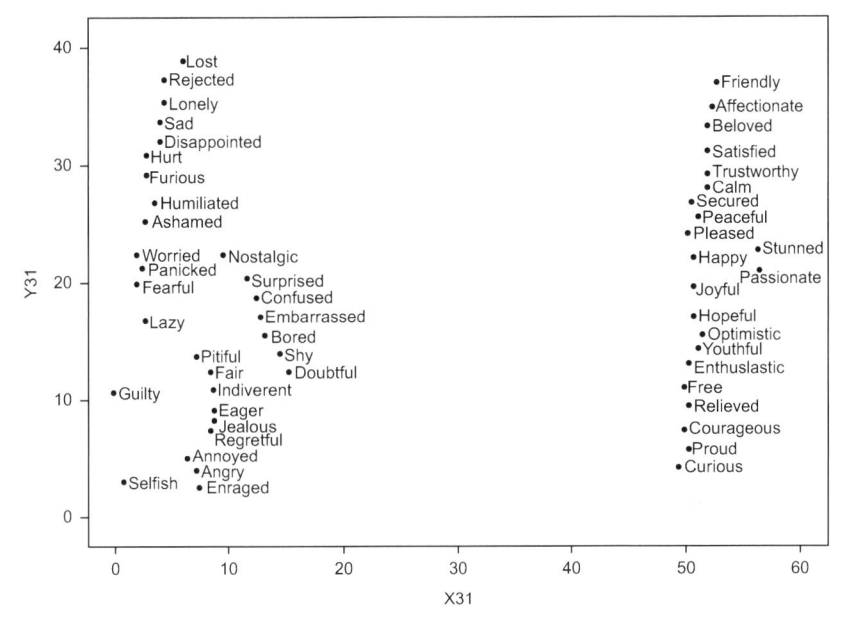

Napping and sorted Napping as a sensory profiling technique

Figure 9.1 Representation of the emotions provided by Subject 31.

1 in Fig. 9.3, explains 37.36% of the total variance. This relatively high value (considering the number of individuals and variables in the analysis) indicates that there is a consensus amongst subjects.

This first dimension clearly opposes positive emotions such as "happy" and "joyful" (on the right-hand side of the plane) to rather negative emotions such as "lost" and "sad" (on the left-hand side of the plane).

The second dimension, denoted Dim 2 in Fig. 9.3, explains 6.69% of the total variance. Although this value seems apparently low, this dimension will be kept in the analysis since it has been tested as statistically significant by using bootstrap technique (not presented in this chapter), and is obviously interesting in terms of interpretation.

Indeed, amongst the so-called negative emotions, this second dimension opposes emotions such as "angry" and "furious," which could be described as activated, to emotions such as "worried" and "lost," which could be described as deactivated.

Let us notice the singular position of the emotion "surprised" located almost on the centre of gravity of the plane. It seems that it has been isolated by the subjects, and that it has been perceived as neither positive nor negative: one can be "surprised" in positive and negative ways; "surprised" could easily be qualified as "neutral."

In terms of dispersion, we can notice the difference between the group of so-called positive emotions and the group of so-called negative emotions: positive emotions

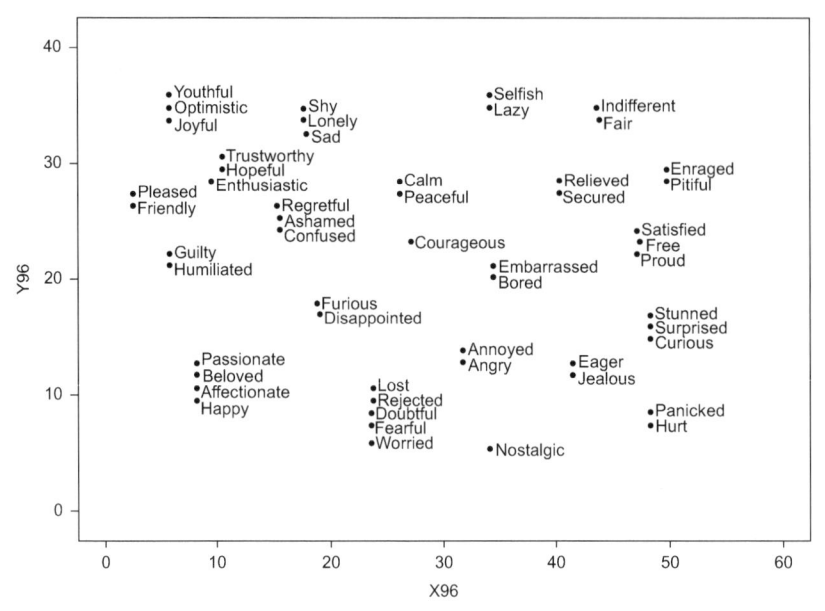

Figure 9.2 Representation of the emotions provided by Subject 96.

seem to be more homogeneous than negative emotions, which are more heterogeneous. Positive emotions are first perceived as a whole, which may be due to their strong common character (i.e., positiveness) perceived by the subjects; whereas negative emotions are perceived more as part of some kind of continuum. Actually, the third dimension (not shown here) also represents the positive emotions more as a continuum.

To supplement the previous findings, Table 9.1 shows that the first eigenvalue associated with Dim 1 equals 49.43, for a maximum value of 100 if all hundred subjects had expressed exactly the same first dimension (due to the weighting, the maximum value of the highest eigenvalue in MFA equals the number of active groups considered in the analysis). This value is indicative of the fact that subjects share a common important dimension that opposes positive emotions to negative ones.

The consensual representation in Fig. 9.3 can also be understood regarding the way each subject has positioned the emotions during the experiment. To do so, the results of the separate analyses of each group of coordinates (one group corresponding to one subject) can be projected as supplementary information on the axes provided by MFA.

In the case of our two Subjects 31 and 96 (cf. Figs 9.1 and 9.2), Figure 9.4 shows that the main axis of variability of Subject 31 (denoted Dim 1.S31) is almost perfectly correlated to the main axis of variability of MFA based on all subjects. It is also true, to a lesser degree, for the second dimension of Subject 31 (denoted Dim

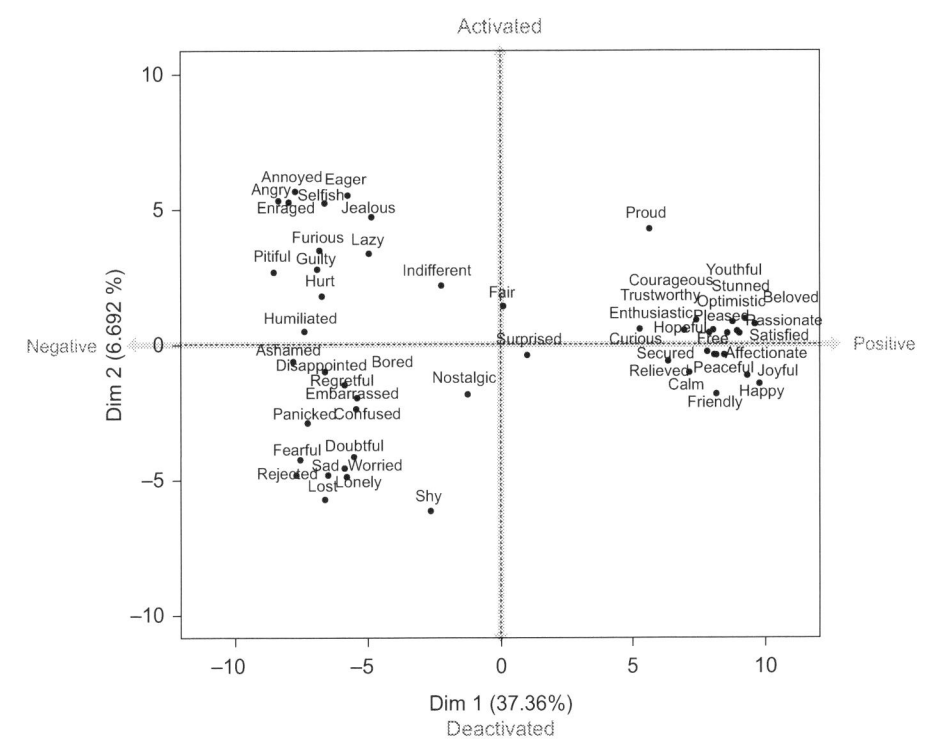

Napping and sorted Napping as a sensory profiling technique 203

Figure 9.3 Representation of the emotions issued from MFA.

Table 9.1 **Decomposition of the total variance on the first ten dimensions**

	Eigenvalue	Percentage of variance	Cumulative percentage of variance
Dim 1	49.43	37.36	37.36
Dim 2	8.85	6.69	44.05
Dim 3	5.58	4.22	48.27
Dim 4	4.85	3.67	51.94
Dim 5	4.03	3.05	54.98
Dim 6	3.79	2.86	57.85
Dim 7	3.66	2.76	60.61
Dim 8	3.36	2.54	63.15
Dim 9	3.23	2.44	65.59
Dim 10	2.71	2.05	67.64

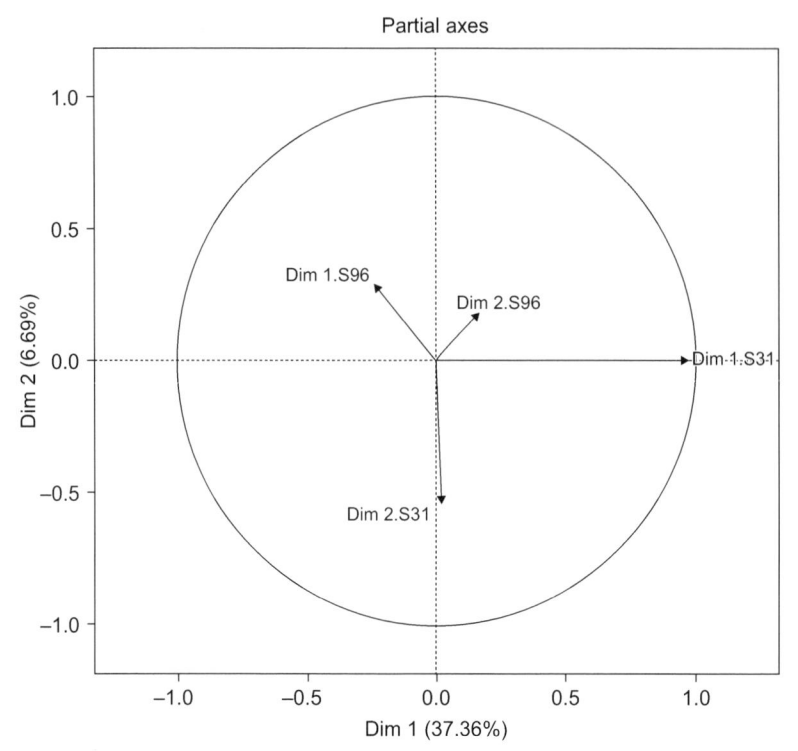

Figure 9.4 Representation of the axes issued from the separate analyses of the groups associated with Subjects 31 and 96.

2.S31), whereas it is completely false for the dimensions of Subject 96 (Dim 1.S96 and Dim 2.S96).

Complementary to that output, MFA provides a representation of the groups of variables, the subjects in a Napping context (cf. Fig. 9.5). This representation is based on the axes provided by MFA.

Beyond the optimal properties of this representation that will not be detailed here, coordinates of a subject on an axis lie between 0 and 1, and can be interpreted as the importance of the axis given by the subject with respect to his own stimuli configuration. When this value equals 0, as is almost the case for Subject 96 on the first axis, it means that the structure induced by the axis provided by MFA has nothing to do with either of the two coordinates provided by the subject. On the other hand, when this value equals 1, as is almost the case for Subject 31 on the first axis, it means that the structure induced by the axis provided by MFA can be assimilated to the principal dimension of variability of the subject. The opposition between positive and negative emotions on the first axis of MFA corresponds to the main axis of variability for Subject 31, while it does not mean anything for Subject 96. According to Fig. 9.5, it seems that, for a majority of consumers, the

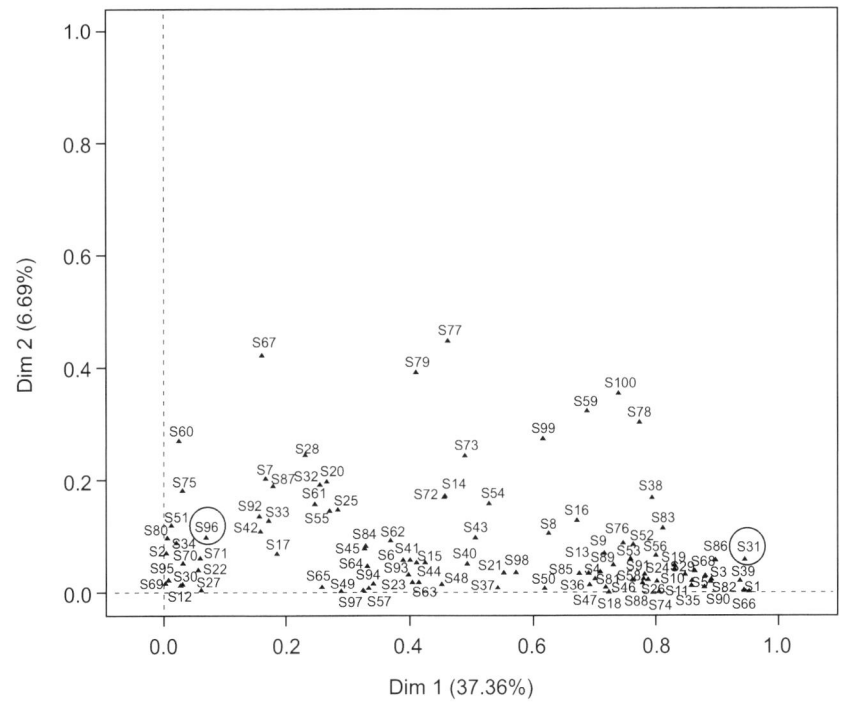

Figure 9.5 Representation of the groups of coordinates in MFA (i.e., subjects in a Napping context).

opposition between positive emotions and negatives ones represents an important dimension of variability.

This example is remarkable in the sense that dimensions are directly interpretable in terms of stimuli. This is mainly due to the nature of the stimuli, which have a clear definition and can be interpreted at will. By analogy, a representation of a product space is hardly interpretable without expert knowledge about the products; in the case of a sensory profile, without the representation of the sensory descriptors.

Supplementary information appears to be essential for interpreting the relative positioning of the stimuli (usually the products), as this sole information is generally not self-contained. Hence came the idea to supplement the original information from Napping (i.e., the only coordinates of the stimuli, as in the example), by asking subjects to add comments on the stimuli, once the placement of the stimuli had been done: this is what has been called verbalization, literally to express thoughts, feelings and emotions in words.

Originally, this information was aggregated over the subjects and coded into a contingency matrix (stimuli × words), where rows correspond to stimuli, columns correspond to words, and at the intersection of one row and one column, the number of times that the word was used to qualify the stimulus. Then, this matrix was usually considered in MFA as a supplementary group. In other words, this matrix was added to the matrix of coordinates, then treated as a supplementary group, while the groups of coordinates were treated as active groups. The information related to the contingency table was represented on a correlation circle, as if each column was assimilated to a continuous variable.

The representations of the stimuli and of the verbalization data provided by the subjects are actually very similar (including in terms of usage) to the representations issued from classical sensory profile data analysed by PCA. From this perspective, Napping combined with verbalization can be seen as a sensory profiling technique (as, indeed, Napping provides a sensory profile).

Still, intrinsically, the information provided by Napping is unique: this information is based on a multitude of sources of differences amongst the stimuli, these sources being important for the subjects. Without wanting to sound polemical, Napping is neither a rapid method, nor an alternative to descriptive analysis. More precisely, at a subject level, Napping is intrinsically a rapid method, as each subject is considering the stimuli globally, and as we expect from each subject a spontaneous answer corresponding to her/his most striking differences amongst the stimuli. But to fully exploit this unique information, the number of subjects usually involved is rather high, which extends the length of the experiment. The uniqueness of the information provided by the subjects makes us believe that Napping is not *sensu stricto* an alternative to descriptive analysis: we do not expect the representations of the stimuli from a classical descriptive analysis and from a Napping to be similar; the first is based on a fixed list of descriptors, while the second is based on a multitude of sources of differences amongst stimuli, unknown *a priori*.

Now that we have understood that verbalization is of utmost importance, and that it should be systematically asked of subjects to interpret the results, we will see in the next section where sorted Napping originated.

9.3 From Napping to sorted Napping

9.3.1 *Some behavioural considerations*

The idea of sorted Napping arose from the observation of subjects during the verbalization phase (Cadoret *et al.*, 2009; Pagès *et al.*, 2010). In effect, it appeared that most of the subjects were not verbalizing stimulus by stimulus. Instead of an individual description of the stimuli, we observed that subjects tended to provide a description of the stimuli at a more global level, with respect to clusters of stimuli. In practice, when they were asked to verbalize, subjects were naturally making clusters of stimuli by circling them with their pen, and were describing the stimuli at a cluster level. We named this natural response pattern "sorted Napping" (derived from the sorting task).

This behavioural pattern may have several explanations. One of them could be that the intrinsic nature of Napping is holistic, and that is precisely what is expected. In a sensory profile context, Napping will naturally reveal the main sensory dimensions of variability amongst the stimuli (the ones perceived by the subjects). But when it comes to describing the stimuli, subjects have to switch to a more analytical way of thinking: putting words on something that was latent and that has just been revealed may be difficult for the subjects. Hence, the recurrent behavioural pattern that has been observed, and that has shown that subjects were more likely to verbalize at a cluster of stimuli level.

Ultimately, sorted Napping can be assimilated to Napping followed by a verbalization phase of elicited clusters.

9.3.2 Analysing sorted Napping data

One of the main drawbacks when using a contingency matrix (stimuli × words) to analyse verbalization data is that data are aggregated over subjects. We thus lose a lot of information, as the individual variability amongst the subjects is lost. As a consequence, the other main disadvantage is that the information given by a subject when she/he clusters stimuli in order to describe them is not reported properly. Basically, this information is absolutely meaningful, as it reinforces the notion of distance between stimuli based on Napping only. In other words, the analyst has at his disposal two types of information, a continuous one based on the distance provided by Napping, and a categorical one based on the groups provided when circling or sorting the stimuli when verbalizing. In that sense, each subject provides a couple of vectors of coordinates for the stimuli, and eventually a categorical variable that reports the way the subject has gathered the stimuli and has described them.

The example used to illustrate sorted Napping is based on an experiment in which eight smoothies were chosen according to two main factors, the type of manufacturer (two categories: generic products from retailers such as *Casino* and *Carrefour*, and products from national brand name manufacturers such as Innocent and Immedia) and the flavour (four categories: Strawberry-Raspberry-Blueberry denoted "SRB," Pineapple-Banana-Coco denoted "PBC," Strawberry-Banana denoted "SB" and Mango-Passion fruits denoted "MP"). Twenty-four subjects were involved in the experiment. The experiment was conducted by a group of Master students from Agrocampus Ouest, France, who collected the data and who exploited them to win the Syntec Trophy in 2009 (Trophée Syntec des études Marketing & Opinion), one of the most competitive marketing contests in France.

Figure 9.6 represents the rectangle provided by Subject 5, who positioned the eight smoothies first according to the way he perceived their similarities, and then sorted them by adding a comment to each group of smoothies. Table 9.2 illustrates the way the information collected in Fig. 9.6 is formatted into a data set.

In terms of analysis, it seems rather natural to first try to find a distance amongst the stimuli that combines both continuous and categorical information. Then, once this distance is found, in order to get a global representation, it seems also natural to balance the part of each subject within a global analysis. This is precisely what

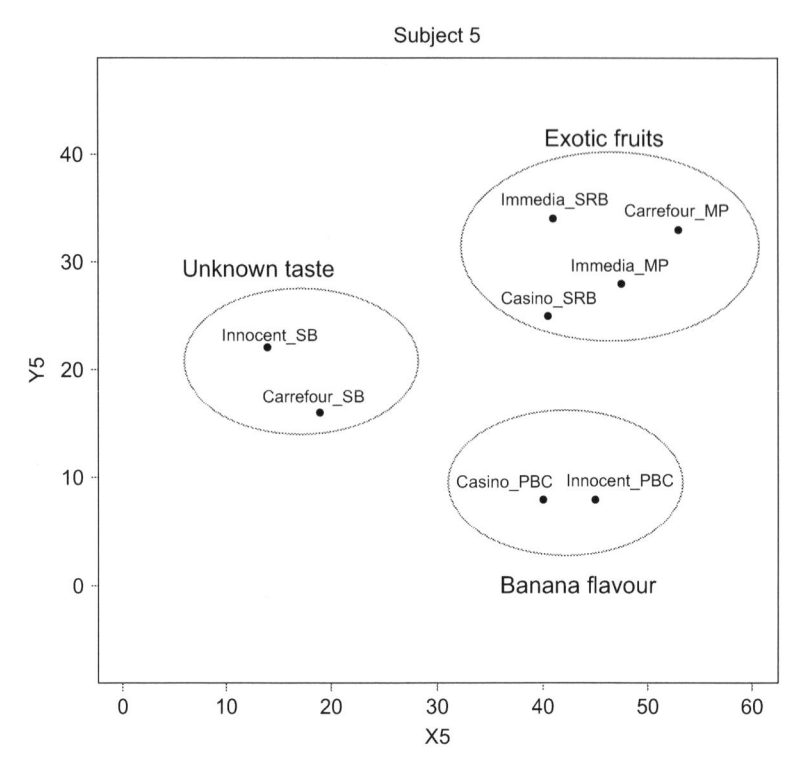

Figure 9.6 An example of data collected during sorted Napping.

Table 9.2 Data provided by Subject 5, the coordinates and the verbalization data as groups of stimuli

	X5	Y5	C5
Immedia_MP	47.5	28	Exotic fruits
Carrefour_MP	53.0	33	Exotic fruits
Immedia_SRB	41.0	34	Exotic fruits
Casino_SRB	40.5	25	Exotic fruits
Innocent_PBC	45.0	8	Banana flavour
Casino_PBC	40.0	8	Banana flavour
Innocent_SB	14.0	22	Unknown taste
Carrefour_SB	19.0	16	Unknown taste

hierarchical multiple factor analysis (HMFA, Le Dien and Pagès, 2003) does, when considering two levels of partition on the variables:

1. a first one, with respect to the subjects, in which each subject is a group of three variables;
2. a second one (at the subject level), with respect to Napping on the one hand, and the sorting on the other hand.

9.3.3 Example: (sorted) Napping as a sensory profiling technique

In this part, we will focus exclusively on the representation of the individuals provided by HMFA, but the graphical outputs and other numerical indicators provided by MFA are provided as well by HMFA, as HMFA is an extension of MFA.

As mentioned previously, the global representation of the individuals (i.e., the smoothies) provided by HMFA is hard to interpret without any supplementary information (cf. Fig. 9.7). In this example, the supplementary information the analyst has at his disposal concerns the two factors based on which the smoothies were selected (type of manufacturer and flavour).

In Fig. 9.7, we can see that the first factorial plane issued from HMFA differentiates three groups of products. SRB smoothies and Mango-Passion fruit smoothies are opposed to PBC smoothies along the first dimension. Strawberry-Banana smoothies are opposed to the rest of the products along the second dimension. In that sense, we can say that the main structure of variability amongst the smoothies is based on their flavour, not on the type of manufacturer.

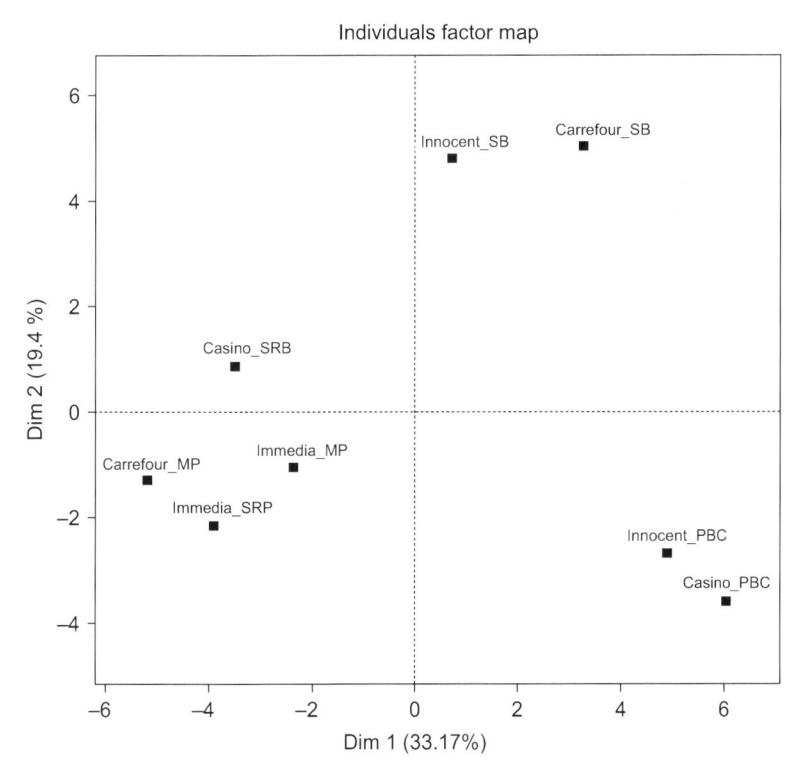

Figure 9.7 Representation of the smoothies based on sorted Napping data.

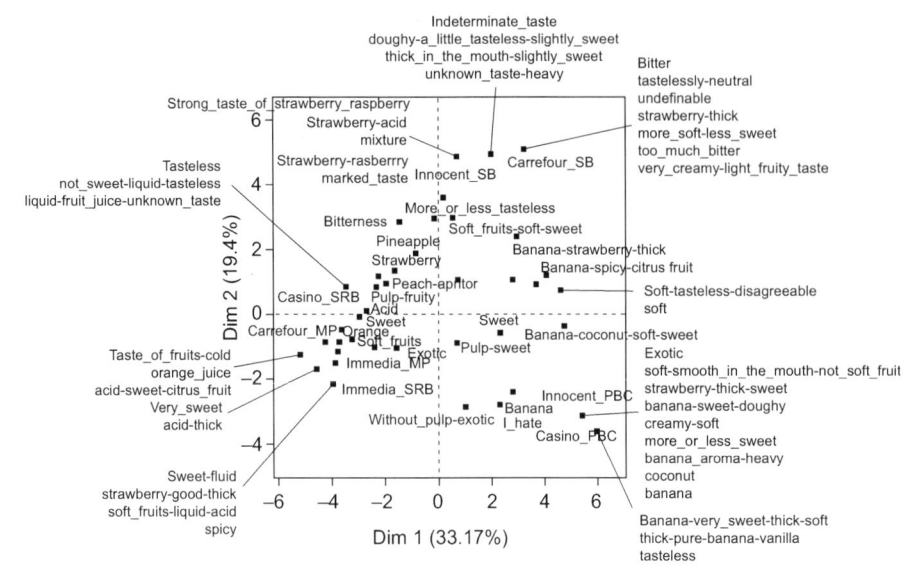

Figure 9.8 Superimposed representation of the smoothies and the comments provided by the subjects.

Beyond the information based on the factors, the information provided by the subjects on the groups appears to be essential to interpret the factorial plane. This information is represented as categories are represented in Multiple Correspondence Analysis (MCA, Husson *et al.*, 2010). In other words, as shown in Fig. 9.8, smoothies and description of the groups provided by subjects are represented on the same factorial plane, which is an important and very convenient feature.

Figure 9.8 speaks for itself and demonstrates the capacity of sorted Napping (or Napping combined with verbalization) to be used as a sensory profiling technique. The comments provided by subjects are rich enough to interpret the dimensions of variability amongst smoothies.

The positive side of the first dimension is characterized by comments such as *doughy, sweet, sickly, I don't like, banana, coco, soft, very soft, exotic fruits, creamy texture*.

The negative side of the first dimension is characterized by comments such as *acid, strong, litchi taste, blueberry, raspberry, with small grains, like guava*.

As we can see, smoothies are not all about flavour. An important aspect of this beverage has a lot to do with texture (*creamy, soft, thick, liquid, doughy*), which depends essentially on the composition of the smoothies.

9.4 Analysing Napping and sorted Napping data using the R statistical software

The analysis was performed using the statistical software R, the FactoMineR package, an R package dedicated to exploratory multivariate analysis, and the SensoMineR package, an R package dedicated to sensory data analysis.

For Napping data, the data set comprises 53 rows (statistical individuals) and 200 columns (variables). In other words, 53 emotions and 100 pairs of coordinates, one pair of coordinates corresponding to one subject. In the analysis, each subject is considered as a group of unstandardized variables. Data were analysed using the *MFA()* function of the FactoMineR package. Practically, the code that corresponds to the analysis is the following:

> MFA(emotions, group = rep(2,100), type = rep("c",100))

The first argument corresponds to the name of the data set, the second corresponds to the way variables are structured into groups and the third corresponds to the way each group is considered in the analysis (as standardized continuous data, as unstandardized continuous data, as categorical data, as a contingency table). The data set emotions on which the analysis has been performed is constituted of a sequence of 100 groups of two variables each (2 has been repeated 100 times), where each group has been considered as a group of unstandardized continuous variables ("c", standing for unstandardized continuous, has been repeated 100 times).

Sorted Napping data were analysed using the *fasnt()* function of the SensoMineR package. Practically, the code that corresponds to the analysis is the following:

> fasnt(smoothies, first = "nappe," sep.word = ";")

The first argument corresponds to the name of the data set. The second corresponds to the way variables are structured into groups: here Napping data come first, and sorting data come second. The third argument corresponds to the way comments are coded: ";" between each different comment. This function is really powerful, as it provides a complete description of the products, globally, and individually through the comments (cf. Fig. 9.9 and Table 9.3).

The core of the function is based on HMFA. To understand what the function *fasnt()* does, we recommend running an HMFA on the same data. To do so, it is essential to specify the hierarchical structure on the variables, as well as the type of data considered (standardized continuous variables, unstandardized continuous variables and categorical variables). In the case of sorted Napping, the analyst has to specify the fact that subjects provide information based on two unstandardized continuous variables and one categorical variable. When the number of consumers equals 24, for instance, the hierarchical structure is defined in the following way:

> list(rep(c(2,1),24), rep(2,24))

The first level of the hierarchy is constituted of a sequence of 2 and 1 repeated 24 times (one group of two variables, one group of one variable; repeated 24 times). The second level of the hierarchy is constituted of a sequence of two repeated 24 times (one subject corresponds to the aggregate of two consecutive groups).

Then, the analysis is specified the following way:

> HMFA(smoothies, H = list(rep(c(2,1),24), rep(2,24)), type = rep(("c","n"), 24))

The first argument corresponds to the name of the data set, the second corresponds to the hierarchical structure on the variables and the third corresponds to the type of

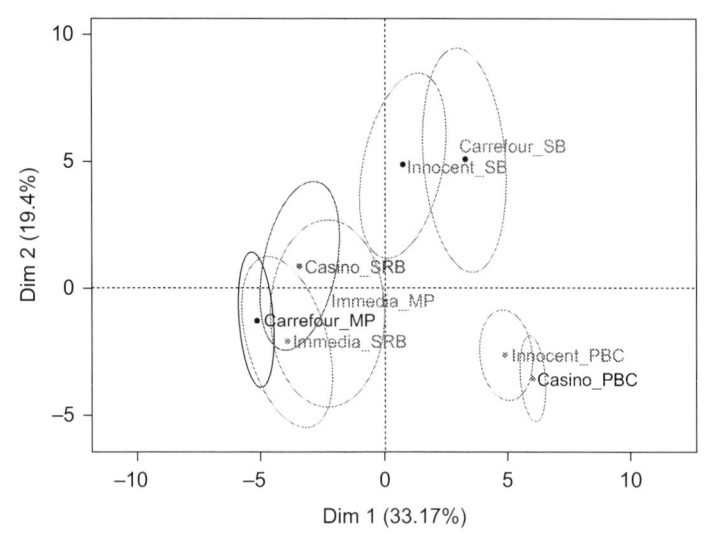

Figure 9.9 An example of graphical representation provided by the *fasnt()* function, representation of the products supplemented by confidence ellipses.

Table 9.3 **The *fasnt()* function provides an automatic description of the products in terms of verbalization data**

Product	Word	Intern %	Glob %	Intern freq	Glob freq	*p*-value
Immedia_SRB	Acid	34.78	15.66	8	26	0.02
Innocent_PBC	Acid	0	15.66	0	26	0.01
Innocent_SB	Strawberry	22.22	6.62	4	11	0.03

Immedia_SRB has been described as acid, in contrast to Innocent_PBC, the two smoothies being opposite according to the first dimension. Innocent_SB has been described by the word *strawberry*.

data considered in our case a sequence of one group of two unstandardized continuous variables ("c") and one group of one categorical variable ("n" stands for nominal).

9.5 Conclusion

If used as a sensory profiling technique, Napping has to be supplemented by a verbalization phase. Otherwise the latent information provided by the subjects through the positions of the stimuli may never be revealed. Observing subjects doing Napping combined with verbalization led us naturally to sorted Napping and its statistical analysis. In that sense, sorted Napping can be seen as a natural extension of Napping combined with verbalization. In terms of statistical analysis, HMFA can perfectly deal with a mix of "pure" Napping data, "pure" sorting data, and sorted Napping data. Ultimately, we would recommend that subjects use the task they feel most comfortable with, followed by a verbalization phase.

References

Cadoret, M., Lê, S. and Pagès, J. (2009, August 8th) "Combining the best of two worlds, the sorted Napping," *Presentation at the SPISE2009 Food Consumer Insights in ASIA: Current Issues & Future*, Ho Chi Minh City, Vietnam.

Carroll, J. D. (1968) "Generalization of canonical correlation analysis to three or more sets of variables," *Proceedings of the 76th Convention of the American Psychological Association*, **3**, 227–228.

Donoghue, S. (2000) "Projective techniques in consumer research," *Journal of Family Ecology and Consumer Sciences*, **28**, 47–53.

Escofier, B. and Pagès, J. (1994) "Multiple factor analysis (AFMULT package)," *Computational Statistics & Data Analysis*, **18**, 121–140. doi: 10.1016/0167-9473(94)90135-X.

Husson, F., Lê, S. and Pagès, J. (2010) *Exploratory Multivariate Analysis by Example Using R*, London, Chapman and Hall.

Le Dien, S. and Pagès, J. (2003) "Hierarchical multiple factor analysis: application to the comparison of sensory profiles," *Food Quality and Preference*, **18**(6), 453–464. doi: 10.1016/S0950-3293(03)00027-2.

Pagès, J. (2004) "Multiple factor analysis: main features and application to sensory data," *Revista Colombiana de Estadística*, **27**, 1–26.

Pagès, J. (2005) "Collection and analysis of perceived product inter-distances using multiple factor analysis: Application to the study of 10 white wines from the Loire Valley," *Food Quality and Preference*, **16**, 642–649. doi:10.1016/j.foodqual.2005.01.006.

Pagès, J., Cadoret, M. and Lê, S. (2010) "The sorted Napping: a new holistic approach in sensory evaluation," *Journal of Sensory Studies,* **25**(5), 637–658. doi: 10.1111/j.1745-459X.2010.00292.x.

Pagès, J. (2013) *Analyse Factorielle Multiple Avec R*. Paris, EDP Sciences.

Risvik, E., McEwan, J. A., Colwill, J. S., Rogers, R. and Lyon, D. H. (1994) "Projective mapping: A tool for sensory analysis and consumer research," *Food Quality and Preference*, **5**(4), 263–269. doi: 10.1016/0950-3293(94)90051-5.

Polarized sensory positioning (PSP) as a sensory profiling technique

10

E. Teillet
SensoStat, Dijon, France

10.1 Introduction

This chapter presents polarized sensory positioning (PSP) as a way to describe the sensory characteristics of a product set. This method relies on a very natural way to describe things which is to compare them with standards or well-known references. PSP is thus based on the comparison with a constant set of references defined as "poles". In this first section, we will see how the method was first imagined in application to the description of mineral water.

10.1.1 The birth of polarized sensory positioning (PSP)

The PSP methodology was developed by Eric Teillet (SensoStat, Dijon, France) and Pascal Schlich (INRA, Dijon, France) during a PhD, one of the purposes of which was to determine the taste of water (partnership CNRS/Lyonnaise des Eaux; Teillet, 2009, in French language). The aim for Lyonnaise des Eaux (French supplier of tap water) was to routinely determine the taste of one or more water samples. This "historic" example of the study of the taste of water will be used in this chapter in order to present the PSP methodology. Nevertheless, adaptation of PSP to other product spaces is possible and will be discussed.

Water, if not contaminated by pollutants, is only composed of H_2O molecules and minerals (K^+, Na^+, Cl^-, Ca^{2+}, SO_4^{2-}, Mg^{2+}, etc.). At first glance, water is thus an easy product for sensory analysis (no odour, no texture, no pungency, almost no persistence, etc.). Nevertheless, stimuli induced by water are generally so weak that water represents a real challenge for sensory analysis, in terms of attention and sensitivity. At the time of that study, the best results (in terms of discrimination between samples and sensory positioning) were obtained by comparative methods such as the free sorting task (see Teillet *et al.*, 2010a for these results). Nevertheless, free sorting does not enable data aggregation and comparison with additional samples once the evaluation has been done. In order to analyse a new sample, the entire set of products should be presented again. For more details on the results that were obtained with different sensory methodologies (quantitative analysis, Temporal Dominance of Sensations, free sorting task) see also Teillet *et al.* (2010b).

It was therefore decided to develop a comparative methodology which would circumvent the limitations of other rapid comparative methods and which would enable data aggregation from several studies. Ideally, this method should not require the presentation of an excessive number of samples in each new analysis, which would amount to running the evaluation on the full product set again. Besides, this methodology should be possibly performed by subjects without special training, as in free sorting.

10.1.2 The nature of PSP

PSP is not strictly a sensory methodology for which protocols would be well defined and standardized. It simply relies on a philosophy of using well-known standards or reference products defined as "poles" in order to position samples relatively to these poles in a sensory point of view. Naturally, these "poles" should be stable over time.

PSP methodology is thus based on the comparison of samples to sensory "poles." The principle is actually to determine similarities (or dissimilarities) between samples to be studied and the poles that are properly chosen in the initial product set. The poles should be different prototypes from the whole product set, representing the main differences that could be encountered in the product category of interest.

In the case of the taste of water, these poles have been chosen among French natural mineral waters in order to best represent the three main tastes of water, according to previous knowledge: the metallic and bitter taste for low mineral content waters (e.g. Volvic), the neutral taste and the sensation of freshness for medium mineral content waters (e.g. Evian) and, finally, the more salty and astringent taste for highest mineral content waters (e.g. Vittel). An interesting feature of natural mineral waters is that their mineral composition and organoleptic qualities are legally required to be stable over time. The data aggregation from several PSP studies is thus facilitated.

10.2 Polarized sensory positioning (PSP) methodologies

We imagined several methods enabling collecting the similarities (or dissimilarities) between samples and poles. The first one directly collects dissimilarities to poles, as rated on a continuous scale, and could be called "rating-PSP." The second proposes deducting similarities and dissimilarities to poles from co-occurrences. This could be seen as a triad test (MacRae *et al.*, 1990) conducted in a polarized way, or say a "polarized triad" test. A third, called "napping PSP," consisting in locating samples in a map where poles are already located, has also been tried. Nevertheless, since it was quite difficult to understand by subjects, it will not be developed here. Note that similarly, Ares *et al.* (2013) applied the PSP approach to projective mapping (Polarized Projective Mapping (PSM)). Although they did not report such difficulties in the understanding of the task, they also found that PSM was more difficult to apply than triad-PSP and rating-PSP.

Polarized sensory positioning (PSP) as a sensory profiling technique   217

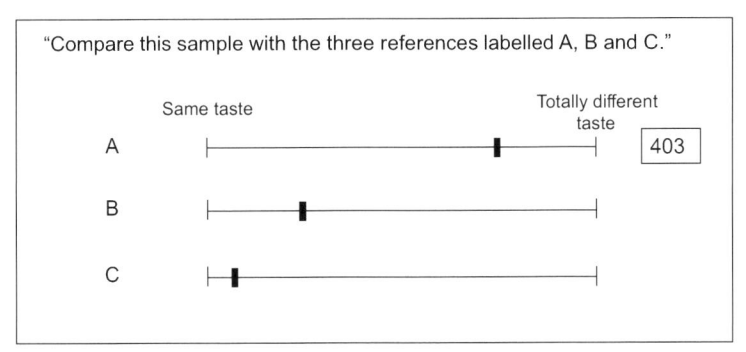

Figure 10.1 Example of continuous scale used for PSP.

These methodologies have been the first to be applied, but the reader should consider them as only some of the possibilities for using the "polarized" approach. Following these propositions, experimenters are encouraged to adapt this approach to their own protocol. The principle will remain the same, providing that similarities or dissimilarities between product and poles are collected.

10.2.1 Continuous scale

The first method is based on rating scores which reflect the distance between samples to be tested and the poles. For each sample, the assessors are asked to rate the similarity to each predefined pole using an unstructured linear scale anchored from "same taste" to "totally different taste" (Fig. 10.1).

In the example given in Fig. 10.1, a product coded 403 would have been judged very different from A, and closer to C than to B.

10.2.2 "Triad PSP"

In the case of "triad" PSP (Fig. 10.2), a panellist is simply asked which pole a sample resembles most, and which pole it resembles least. In the example given in Fig. 10.2, the water sample coded 403 most resembles C and least resembles A.

10.3 Data analyses

The comparison with a series of constant references yields relatively unusual sensory data that require dedicated data analysis approaches. Below we suggest data analysis techniques for rating-PSP and for triad-PSP.

10.3.1 Continuous scale

In a first step, data from the continuous scales can be coded from 0 for "same taste", to 10 for "totally different taste". By analogy with the free sorting task (see the

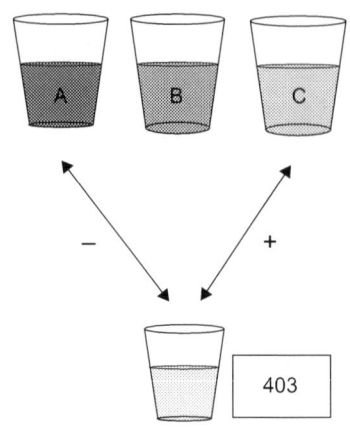

Figure 10.2 Principle of "triad PSP".

corresponding chapter), PSP data can be seen as dissimilarity matrices sample*poles (see Fig. 10.3). PSP data can thus be analysed by multidimensional scaling (MDS) techniques. But dissimilarity matrices are here rectangular and cannot be processed with classical MDS. "MDS unfolding" algorithms must be used instead (Green et al., 1989). These methods are an extension of MDS to matrices where products in rows can be different from products in columns, but the purpose still remains to locate products, of which 2-by-2 distances are known, in a space. As with classical algorithms of MDS, unfolding algorithms can incorporate non-metric transformations.

Until recently, unfolding algorithms presented convergence problems to trivial solutions (equidistant products). But this problem seems to have been solved (Busing et al., 2005). Still rarely used in sensory analysis, unfolding appears as the most natural method to process PSP data from a continuous scale.

Nevertheless, data can be seen another way. If we consider the poles as "meta-descriptors" – for the taste of water, Volvic could be a "metallic and bitter" descriptor, Evian "tasteless and cool" and Vittel "salty and astringent" – data can be encoded this way from 0 for "totally different taste" to 10 for "same taste". We now consider the intensities of each sample on several descriptors, and "classical" factorial analyses such as Principal Component Analysis (PCA), Multiple Factorial Analysis (MFA), Statis or Generalized Procrustes Analysis (GPA) can be processed (Fig. 10.3).

It should be noted that the use of the scale is certainly very different from one subject to another. What is a "totally different taste" is indeed subjective, and may vary depending on each person's individual criterion. For that reason, three-way analyses are highly recommended (Statis, GPA, MFA).

PSP on continuous scales also enables hypothesis testing. Classical ANOVAs (Analysis of Variances) applied to descriptive analysis can be used to infer significant differences between products, which is very convenient. Ongoing studies are also looking for the application of the "Mixed Assessor Model" (MAM; Brockhoff et al., 2012) in order to take into account the expected high scaling effect between subjects (especially untrained subjects).

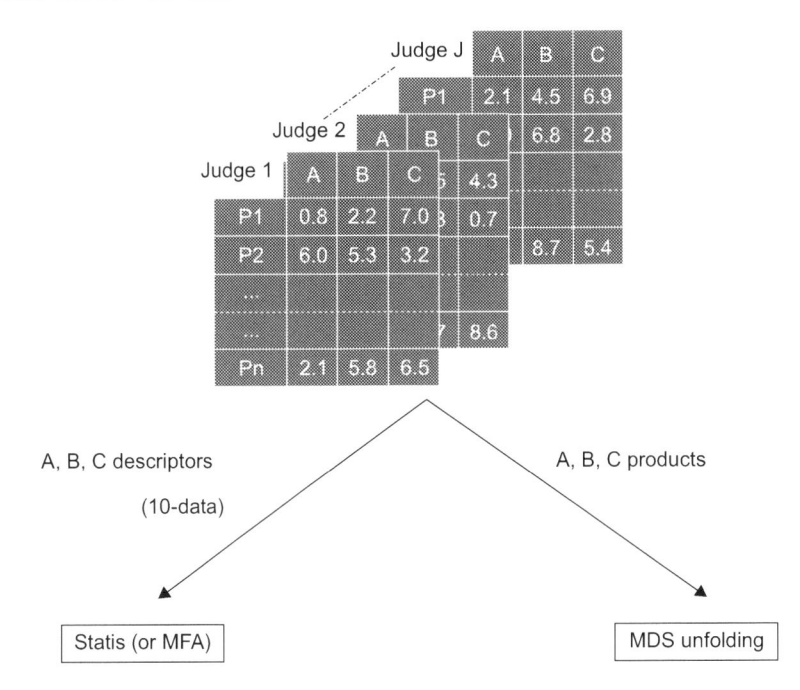

Polarized sensory positioning (PSP) as a sensory profiling technique 219

Figure 10.3 Two ways to process PSP data.

10.3.2 "Triad PSP"

In the case of triad-PSP, the information "the product 403 most resembles C" and "the product 403 least resembles A" can be considered as an occurrence between the product 403 with qualitative variables that we could name "C+" and "A−". Using such a coding system, it is possible to build a global co-occurrence matrix product × variables (A+, B+, C+, A−, B−, C−) where occurrences are summed up over subjects. This matrix can therefore be analysed by correspondence analysis (CA).

10.4 PSP and the taste of water

PSP was initially developed to describe the taste of water. In this section, we describe our very first experiment with PSP, before we present a comparison study between rating-PSP and triad-PSP. Finally, we will show how PSP can be used to aggregate data from an incomplete block design.

10.4.1 A first example

A first study, dedicated to validate the application of PSP to the taste of water, has been conducted among 32 panellists experienced in scoring attributes on scales, and ten samples of water.

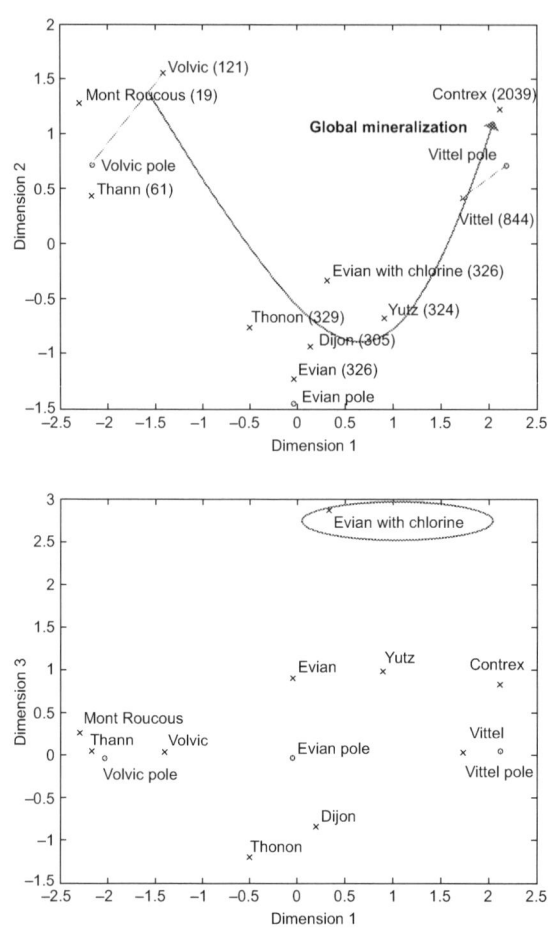

Figure 10.4 Product space, axes 1–2 and 1–3. Sensory maps obtained from the "continuous scale" approach are presented (axes 1–2 and 1–3). The global mineralization in mg/L is presented in parentheses on axes 1–2.

Waters were chosen in order to span the range of total mineralization found in French mineral waters. Volvic, Evian and Vittel have been chosen as poles and have been respectively marked "A", "B" and "C" during the trials. The poles Volvic, Evian and Vittel were also blind tested in order to verify the consistency of PSP data. A product with off-flavour (Evian with chlorine) has been added to the product space.

The unfolding analysis determined a three dimensional space. The first two axes are driven by the gradient of mineralization (Fig. 10.4). Poles are located close to their respective blind tested sample. PSP data thus seem to be consistent.

Note that on this first plan, the off-flavour "Evian with chlorine" seems to be closer to Evian than other poles, but this product is in fact isolated on dimension 3. The Statis approach has also been processed on these data, and conclusions were equivalent (Teillet *et al.*, 2010b).

Even though the data obtained were consistent, validation of these PSP results is based only on prior knowledge of the products (mineral content measured in prior studies). Other performance criteria should be defined in future research. The results in terms of interpretation are nevertheless equivalent to those obtained with the different classical sensory methods we tested. In particular, PSP better discriminates the tastes of low and high mineralized water than does sensory profiling.

10.4.2 Comparison between "rating PSP" and "triad-PSP"

A second study on eight water samples was conducted with 88 "naïve" consumers, in order to compare the two PSP approaches developed ("rating-PSP" versus "triad-PSP").

The same consumers completed both tasks. One half of them began with the continuous scale, and the other half began with the triads. It was thus possible to compare the results obtained for each methodology. Bootstrap sampling (Efron, 1979) has been performed in order to determine the variability of these results. These bootstrap methods (random sampling with replacement) allow estimation of the sampling distribution of almost any statistic, and are often useful for sensory methodologies where no inference can be done (e.g. the location of products in a sensory map, for free sorting task, Napping or PSP). We will not develop these techniques here, but they lead to confidence ellipses in the sensory map (90% confidence ellipses were chosen in the present case).

Thousand bootstrap samples have been simulated for 88, 80, 70,…,20, 10 consumers and finally enabled to compare the variability of continuous scale and triad methods. A minimum number of consumers have also been suggested for both methods.

Figure 10.5 presents the results obtained. As can be seen, the maps obtained by rating-PSP and triad-PSP are very similar, and in both cases the positioning is driven by the global mineralization of waters, as in the first study. What is more, as can be deduced from the size of ellipses, is that the variability is equivalent for both methods, even if triad-PSP seems to better discriminate some waters (less overlapping of ellipses).

In addition to comparing both PSP approaches, the bootstrap allows evaluating the sensitivity of the method to sample size (in terms of the number of panellists). Our results show that variability remains very much the same from 88 to 40 consumers, but increases dramatically with fewer consumers. In the case of water, a minimum of 40 untrained panellists would thus seem to be required for a PSP study.

It would be very interesting to study the effect of training on this variability, as one might expect training to reduce this, and improve the discrimination between products.

To conclude with this comparison, the two methodologies seem to provide equivalent results. However, participants declared that the triad-PSP was easier to perform. On the interpretation side, it should be remembered that ANOVAs cannot be performed with the triad-PSP.

10.4.3 An example of data aggregation

As PSP was initially imagined as a way to aggregate data from separate evaluations, we conducted a third study in order to assess the feasibility of PSP data aggregation.

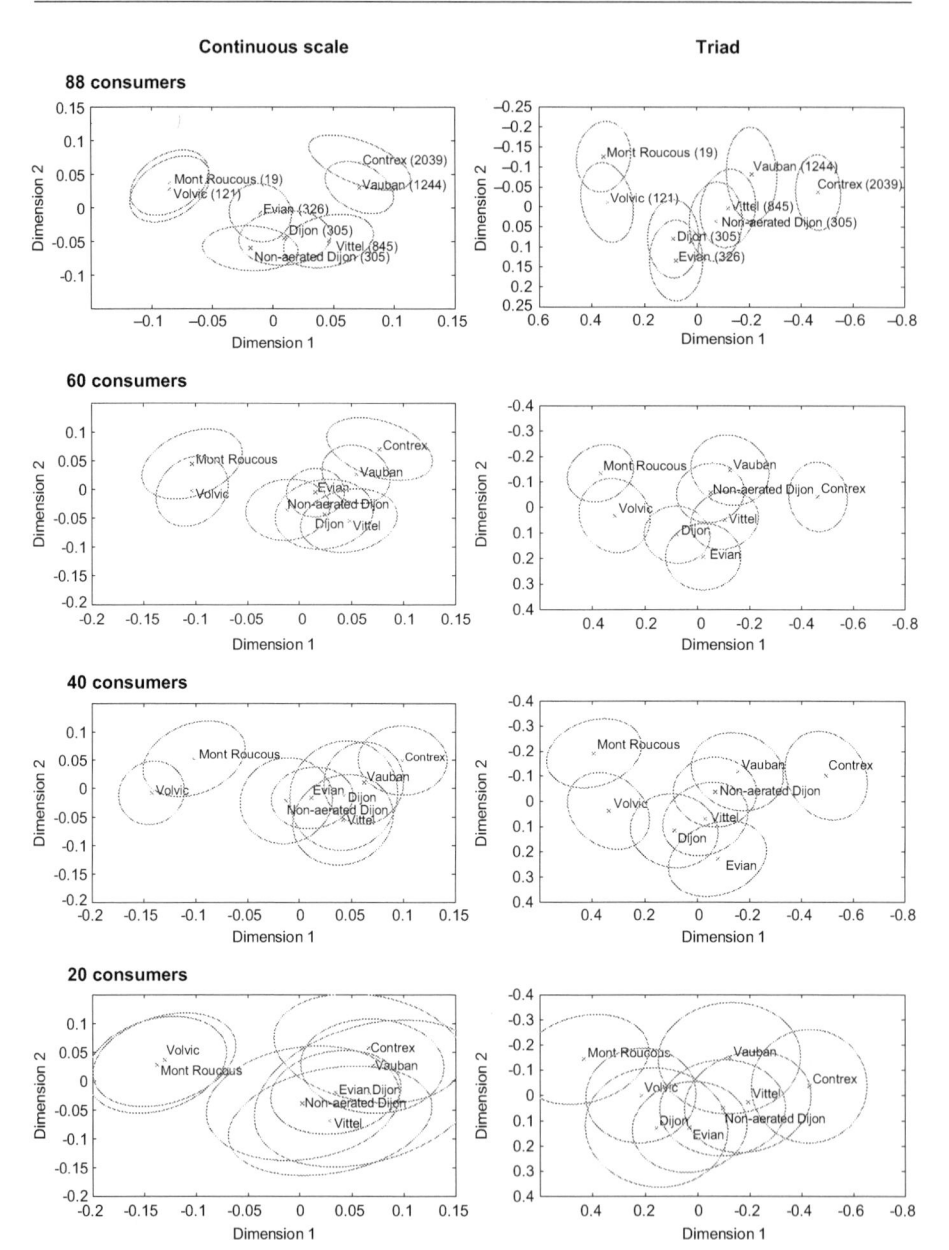

Figure 10.5 Maps obtained with continuous scale and "triad-PSP" for 88, 60, 40 and 20 consumers (1000 bootstrap samples).

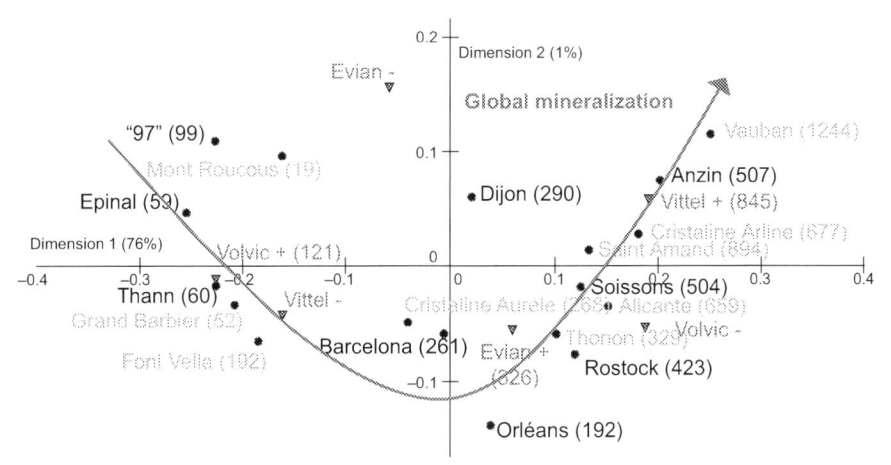

Polarized sensory positioning (PSP) as a sensory profiling technique   223

Figure 10.6 CA of aggregated data. On this graph, a water sample is close to a pole (Evian, Volvic, Vittel) if the co-occurrence sample/pole is high. Pole "+" means "resembles the most" and pole "−" means "resembles the least."

Three hundred and fifty-four consumers tasted nine out of 18 water samples (from all over the world) in a single trial (at a water congress). The methodology used was "triad" PSP and waters were tasted according to an incomplete block design. The purpose was to aggregate these incomplete data and to draw a common map for all the participants. Results are shown in Fig. 10.6.

The resulting map is still driven by the global mineralization of water samples. Waters with equivalent mineral contents are often close. Nevertheless, this kind of aggregation from incomplete blocks is quite common in our sensory field. It would be more interesting to aggregate data from several PSP studies. In addition, it would also be interesting to compare the results obtained with the incomplete block presentation versus the full block presentation.

10.5   Discussion of the choice of the poles

In such a methodology that essentially relies on comparison with predefined poles, the choice of poles is critical. Although there is no rule about how to choose poles, some advice can be given. Both the number and the characteristics of the poles are to be considered.

First, there is no clear rule about the number of poles needed in a PSP trial. Nevertheless, from a geometric point of view, a minimum of three "point to point" distances is necessary to locate a sample in a 2-dimension map. Thus, we recommend a minimum of three poles. If you increase the number of poles, results could be more accurate. Nevertheless, more poles imply more tastings and more comparisons.

Second, if you want to aggregate data from several studies, your poles must be stable over time, which is a first choice criterion. If you are not sure these poles will be available for future trials, you must at least know their "recipe" in order to reproduce them.

A first easy case is when you want to compare your products to several "references" on the market. Choose them as poles. This is what has been done with the water example, where Volvic, Evian and Vittel are very famous French brands. For marketing reasons, it was very interesting for Lyonnaise des Eaux to compare the waters they provide with these famous brands. Indeed, we (consumers) certainly evaluate products with references in mind and compare products with those "personal poles." This could reinforce the idea that poles could be selected according to consumers' point of view.

If this is not the case, but if you already have knowledge or expertise about the sensory space derived from your product set, use it. You can, for example, process a classification of your products on the basis of a quantitative analysis or a free sorting task. After classification, you can choose a prototype of each cluster as the best representative of this cluster (the closest to the barycentre, for example).

If you do not have any priori information about your product set, you will feel uncomfortable with the choice of your poles. Nevertheless, recent studies (manuscript in preparation) show that if we consider several "kinds" of products in the sensory space, there is no real importance as to the choice of the product in each "kind." This is consistent with what de Saldamando *et al.* (2013) observed. What is more, you can imagine an iterative sequence in the choice of your poles. If you see that samples lie outside your sensory space (very dissimilar to the poles), maybe you miss a pole. In the contrary, if two of your poles remain very close to each other in all your studies, maybe you can remove one of them.

10.6 Conclusion

PSP seems to be a promising methodology. For the first time it offers clear perspectives for routine sensory analysis of water samples. PSP methodology yielded promising results and should ultimately enable the sensory characteristics of a given product to be determined as well as limiting the number of samples to taste. In addition, PSP allows data aggregation.

The consistency criterion adopted for water samples only relies on the structure of the sensory space studied. Other criteria should be defined, such as repeatability, discrimination power of products (via ANOVAs and confidence ellipses) or others. These considerations will be the topic of future work.

There still remain a substantial number of questions to improve and generalize the use of PSP philosophy: the way data are collected (structure of the continuous scale), the way data are processed (application of the MAM), the influence of training on PSP results, of adding verbatim to the maps and so on. There is a clear need for further methodological research on these topics.

Nevertheless, PSP can be applied with confidence to any product space where its use seems to be relevant. PSP has already been used in cosmetics (Chréa *et al.*, 2011), aromas and beverages with success. Other PSP methodologies have already been envisaged, such as "Flash PSP" (Teillet *et al.*, 2013), which is a mix of Flash Profile (Dairou and Sieffermann, 2002) and PSP approaches. Enabling the aggregation of single (new) products and a quantification on several descriptors, Flash PSP can be helpful in a development process where new products are frequently developed.

PSP can also be adapted in a consumer-orientated perspective in order to relate perceptions to behaviours and emotions. In this view, one could imagine using pictures, ambiances, etc., as poles. Perspectives for the use of the PSP approach are extremely diverse.

References

Ares, G., de Saldamando, L., Vidal, L., Antúnez, L., Giménez, A. and Varela, P. (2013). Polarized projective mapping: comparison with polarized sensory positioning approaches. *Food Quality and Preference,* **28**(2), 510–518.

Brockhoff, P.B., Schlich, P. and Skovgaard, I. (2012). Accounting for scaling differences in sensory profile data: improved mixed model analysis of variance. *Presented at the Pangborn Sensory Science Symposium,* Toronto, 2011. Manuscript in preparation.

Busing, F.M.T.A., Groenen, P.J.F. and Heiser, W.J. (2005). Avoiding degeneracy in multidimensional unfolding by penalizing on the coefficient of variation. *Psychometrika,* **70**, 49–76.

Chréa, C., Teillet, E., Navarro, S. and Mougin, D. (2011). Application of the polarized sensory positioning in the cosmetic area. *Presented at the Pangborn Sensory Science Symposium,* Toronto, 2011. Manuscript in preparation.

Dairou, V. and Sieffermann, J.-M. (2002). A comparison of 14 jams characterized by conventional profile and a quick original method, the Flash Profile. *Journal of Food Science,* **67**, 826–834.

de Saldamando, L., Delgado, J., Herencia, P., Giménez, A. and Ares, G. (2013). Polarized sensory positioning: Do conclusions depend on the poles? *Food Quality and Preference,* **29**(1), 25–32.

Efron, B. (1979). Bootstrap methods: another look at the Jackknife. *The Annals of Statistics,* **7**(1), 1–26.

Green, P.E., Carmone, F.J. and Smith, S.M. (1989). *Multidimensional Scaling: Concepts and Applications.* Boston: Allyn and Bacon.

MacRae, A.W., Howgate, P. and Geelhoed, E.N. (1990). Assessing the similarity of odours by sorting and by triadic comparison. *Chemical Senses,* **15**, 661–699.

Teillet, E. (2009). Perception, préférence et comportement des consommateurs vis-à-vis d'eaux embouteillées et d'eaux du robinet. Thèse de 3ème cycle, Dijon, université de Bourgogne.

Teillet, E., Urbano, C., Cordelle, S. and Schlich, P. (2010a). Consumer perception and preference of bottled and tap waters. *Journal of Sensory Studies,* **25**(3), 463–480.

Teillet, E., Schlich, P., Urbano, C., Cordelle, S. and Guichard, E. (2010b). Sensory methodologies and the taste of water. *Food Quality and Preference,* **21**(8), 967–976.

Teillet, E., Petit, C. and Delarue, J. (2013). Combining PSP and Flash Profile... Why? How does it work? *Presented at the Pangborn Sensory Science Symposium,* Rio, 2013. Manuscript in preparation.

Check-all-that-apply (CATA) questions with consumers in practice: experimental considerations and impact on outcome

11

G. Ares[1], S.R. Jaeger[2]
[1]Universidad de la República, Montevideo, Uruguay; [2]The New Zealand Institute for Plant & Food Research Limited, Auckland, New Zealand

11.1 Introduction

A check-all-that-apply (CATA) question is a versatile multiple choice question in which respondents are presented with a list of words or phrases and asked to select all the options they consider appropriate. This question format has been extensively used in marketing research and is popular because it reduces participant response burden (Driesener and Romaniuk, 2006; Rasinki *et al.*, 1994; Smyth *et al.*, 2006).

CATA questions have been recently introduced to sensory and consumer science to obtain information about consumers' perception of products (Adams *et al.*, 2007). Although the method has been previously used with trained assessors (Campo *et al.*, 2010; Le Fur *et al.*, 2003; McCloskey *et al.*, 1996), its popularity has markedly increased for product sensory characterization with consumers (Varela and Ares, 2012). In this approach, consumers are presented with a set of products and a CATA question to characterize them. Consumers are asked to try the products and to answer the CATA question by selecting all the terms that they consider appropriate to describe each of the samples, without any constraint on the number of attributes that can be selected. The list of words or phrases in the CATA question usually include exclusively sensory characteristics of the product (Fig. 11.1a) but can also include hedonic terms, as well as terms related to non-sensory characteristics, such as usage occasions, product positioning and emotions (Fig. 11.1b) (Ares and Jaeger, 2013; Parente *et al.*, 2011; Piqueras-Fiszman and Jaeger, 2014; Plaehn, 2012).

Selecting terms from a list has been claimed to be an easy and intuitive task for consumers, which requires less cognitive effort than other attribute-based methodologies such as just-about-right or intensity scales (Adams *et al.*, 2007). Also, it has also been said that they lead to a more spontaneous evaluation than scales or forced-choice questions (Smyth *et al.*, 2006).

(a)
Please, check all the words or phrases which best describe this product:

☐	Sweet	☐	Bitter
☐	Bland	☐	Dry
☐	Sour	☐	Firm
☐	Chewy	☐	Crunchy
☐	Juicy	☐	Mealy
☐	Floral	☐	Soft
☐	Hard	☐	Off flavour

(b)
Please, check all the phrases that apply to describe the drink you have just tried:

☐	Good for nutrition	☐	Sour
☐	Active	☐	Energetic
☐	Orange flavour	☐	Good for gratification
☐	For the whole familiy	☐	Perfect for dieting
☐	It is a healthy option	☐	Enthusiastic
☐	Sweet	☐	It is the best way to start the morning
☐	Good to go along with meals	☐	Good for refreshing and hydrating
☐	Makes meals special	☐	Off flavour
☐	Calm	☐	Peaceful
☐	Bitter	☐	Perfect when practicing sports

Figure 11.1 Examples of CATA questions including: (a) sensory and (b) both sensory and non-sensory terms.

Despite their recent application for sensory characterization, CATA questions have already been used in consumer research for characterizing a wide range of products. Examples include snacks (Adams et al., 2007), apple and strawberry cultivars (Ares and Jaeger, 2013), crackers, potato crisps and beer (Jaeger et al., 2013b), ice-cream (Dooley et al., 2010), milk desserts (Ares et al., 2010a), orange-flavoured powdered drinks (Ares et al., 2011) whole grain breads (Meyners et al., 2013), citrus-flavoured sodas (Plaehn, 2012), and cosmetic products (Parente et al., 2010).

The application of CATA questions has been reported to be a quick alternative to gather information about consumer perception of the sensory characteristics of food products, providing similar information to that obtained using descriptive analysis with trained assessors (Ares et al., 2010a; Bruzzone et al., 2012; Dooley et al., 2010). Additionally, across four studies, Jaeger et al. (2013) showed high test-retest reliability of sensory product characterizations with CATA questions with consumers. These authors showed that product configurations and conclusions regarding similarities and differences among samples of different product categories were stable across sessions. Therefore, despite being easy and quick for participants, CATA responses are reliable. Overall, CATA questions have great potential for exploring consumer perception in both academic and industrial applications.

11.2 Implementation of check-all-that-apply (CATA) questions

CATA questions are a simple alternative for consumer-based sensory characterization. However, several methodological issues related to how CATA questions are implemented can strongly affect the results. The key aspects that should be taken into account when using this methodology are discussed in the following sections and recommendations are provided.

11.2.1 Design of the questionnaire

11.2.1.1 Type and number of terms

The selection of the list of words or phrases included in the CATA question is one of the main challenges for the implementation of the methodology. Terms included in the CATA questions should be easy for consumers to understand, and preferably related to the vocabulary they commonly use for describing the products. Input for generating the list of terms can be gained from published literature or results from previous qualitative consumer studies. Besides, terms related to product sensory characteristics can be selected based on the descriptors used by trained assessor panels to characterize the products, but care must be taken to ensure that consumers easily understand the selected terms.

If the objective of the study is to obtain a consumer-based description of the sensory characteristics of the products, the list of terms should only include sensory attributes. However, both sensory and non-sensory attributes can be included in the list of terms in those studies in which the relationship between the sensory characteristics of the products and consumers' emotional or conceptual associations are to be explored (Piqueras-Fiszman and Jaeger, 2014).

In consumer research, CATA questions are usually composed of 10 to 40 terms (Ares *et al.*, 2013b, 2014a, b; Dooley *et al.*, 2010; Jaeger *et al.*, 2013a; Lee *et al.*, 2013; Parente *et al.*, 2010). Results reported by Ares *et al.* (2013b) suggest that there may be an optimum range for the number of terms in a CATA question. Using short lists of terms can encourage consumers to use them all, decreasing their ability to discriminate among samples. On the other hand, large lists of terms can encourage consumers to use satisficing strategies, making them choose the first alternatives from the list, without thinking carefully about the product's sensory characteristics (Krosnick and Alwin, 1987; Rasinski *et al.*, 1994).

11.2.1.2 Order of the terms

One of the main drawbacks that have been reported when using CATA questions in marketing and survey research is that respondents rarely engage in deep processing; instead, they rely on primacy for minimizing cognitive effort (Krosnick, 1999; Sudman and Bradburn, 1992). Therefore, respondents may select the terms that easily catch their attention and that are easily found within the list of response options (Krosnick, 1999;

Krosnick and Alwin, 1987). For this reason, the layout of the list of terms plays a crucial role in consumer responses. Terms positioned at the beginning of the list are more easily found and more frequently selected than those located at the end of the list.

Hence, the order in which the terms are presented has been consistently reported to bias responses to CATA questions in marketing and survey research (Rasinski et al.,1994; Smyth et al., 2006) and also for the application of CATA questions for sensory characterization of food products (Ares and Jaeger, 2013; Castura, 2009; Lee et al., 2013). Castura (2009) reported that terms located at first positions within a block of terms increased the number of selections from 2.6% to 5.9%, and that the frequency of selection increased 10–20% when terms were located in the first row and first column. These results suggest that including the terms in a fixed order for all consumers can strongly bias results and should be avoided. Ares and Jaeger (2013) also reported that the order in which the terms are included in the CATA question strongly affected their frequency of use, and further that term order also influenced conclusions regarding similarities and differences among samples. When comparing results obtained using two CATA ballots which included the terms in different order, no significant differences among samples were identified for the terms not very sweet, no strawberry flavour, juicy, intense red colour and regular shape for one of the ballots, whereas according to the other ballot differences were significant.

Between-subjects randomization of the order in which the terms are presented in the CATA question has been suggested as a useful strategy to minimize the influence of primacy bias on consumer responses in marketing research (Jacoby, 1984; Krosnick, 1992). This implies that each assessor receives the CATA question with the terms in a different order. However, when CATA questions are used for sensory characterization, participants answer the question more than once. Thus, patterns of primacy bias are expected to appear within each participant as the test progresses, i.e. he/she may select more frequently the terms that easily catch their attention within the list of options. For this reason, to minimize this bias, randomization should also occur among participants, implying that the order in which the terms are presented in the CATA question should be modified from sample to sample for each assessor (Ares et al., 2013b). However, it is unclear yet if this type of randomization significantly increases response burden for participants.

Within-participant randomization of the order in which the terms are included in the CATA question can be performed using the Williams Latin square design (Williams, 1949). When collecting data using computer-based systems, randomization can be easily introduced in most available software. The implementation of randomization is equally possible when working with paper ballots. This can be done using Microsoft Office mail merge functionality. The within-participant randomization design for terms is obtained from an external source and placed in an Excel file identifying participants, samples, sample order, terms and term order. This file is linked to a Microsoft Office Word document that links to the Excel file. When properly configured, ballots will be printed identifying it with respect to participant and sample. Three-digit random codes can be specified to identify samples. Alternatively, a simplified strategy could be to consider a limited number of ballot versions and include them in balanced order for each assessor using a Williams Latin square design.

11.2.1.3 Influence of CATA questions on hedonic scores

CATA questions can be used concurrently with hedonic scores with the aim of understanding consumer preferences and identifying recommendations for product reformulation (Stone and Sidel, 2004). However, including questions about specific sensory characteristics can be a source of bias on hedonic scores (Popper *et al.*, 2004; Prescott *et al.*, 2011).

CATA questions are not thought to be cognitively demanding and do not encourage respondents to strongly focus their attention on each of the terms included in the list of options (Krosnick, 1999; Sudman and Bradburn, 1992). Therefore, when used together with hedonic scales, CATA questions could have a smaller effect than other attribute-based methodologies, such as just-about-right or intensity scales (Adams *et al.*, 2007). Jaeger *et al.* (2013b) reported weak and transient evidence of bias of hedonic scores when concurrently using CATA questions across a wide range of product categories (beer, fresh fruit, tea, flavoured water, crackers, savoury dips), suggesting the suitability of CATA questions for concurrent elicitation of consumers' sensory and hedonic responses to food products.

Regarding the position of the CATA question with respect to the hedonic question, although empirical evidence has not shown a strong effect (Ares and Jaeger, 2013; Jaeger *et al.*, 2013), placing the hedonic question before the CATA question is common practice.

11.2.1.4 Consideration of an ideal product

One of the main aims of new product development is to identify the sensory attributes that drive consumer preferences and the characteristics of the product that maximizes consumer liking (Lagrange and Norback, 1987). The Ideal Profile method has been developed for identifying the consumer ideal product (van Trijp *et al.*, 2007). Consumers are asked to rate attribute intensity for a set of samples and for their ideal product, using unstructured scales. Considering that it can be difficult and unintuitive for consumers to rate the ideal intensity of a large set of attributes using scales, Ares *et al.* (2014a) suggested asking consumers to answer a CATA question to describe the sensory characteristics of their ideal product. Using this approach, sensory characteristics of consumer ideal product at the aggregate level and for consumer segments with different preference patterns can be identified, which enables the identification of drivers of liking based exclusively on consumer perception. Besides, penalty analysis based on the comparison of consumer perception of the samples and their ideal product can be used to gather information about the impact of deviation from the ideal on liking scores (Ares *et al.*, 2014a).

11.2.2 Number of products

The number of samples to be used for sensory characterization using CATA questions typically ranges from 1 to 12, depending on the specific aim of the study and the sensory characteristics of the samples. One of the advantages of CATA questions, compared to methodologies such as sorting or projective mapping, is that they can be used to gather information about the sensory characteristics of small sample sets or to evaluate large sample sets in different sessions, due to the fact that the evaluation is monadic. When sensory spaces are to be generated, at least five samples have to be included in the study (Williams *et al.*, 2011).

CATA questions can be for performing internal preference mapping by projecting consumers' descriptions into the preference space. Besides, Ares *et al.* (2010a) and Dooley *et al.* (2010) used CATA questions to generate a consumer-based sensory space for external preference mapping. In these situations, consumers should evaluate at least six samples (Lawless and Heymann, 2010).

Samples are presented in monadic sequence, coded with three-digit random numbers, following a balanced rotation order (Williams Latin square design) to avoid presentation order and carry-over bias. Hence, best practice requires the use of experimental designs to minimize sample presentation order bias and within-participant randomization of CATA terms. Typically, both designs will be based on Willams Latin square designs but be different in order to reflect the actual number of samples and terms used. Designs should be developed to take into account the number of consumers in the study.

11.2.3 Number of consumers

The number of consumers used for product sensory characterization using CATA question ranges from 50 to 100 (Ares *et al.*, 2010a; Dooley *et al.*, 2010; Plaehn, 2012). Ares *et al.* (2014b) evaluated the influence of the number of consumers on the stability of sample and descriptor configurations obtained using CATA questions with a bootstrapping resampling approach. Considering that spatial configurations can be regarded as stable if similar results are obtained in repeated experiments performed under the same conditions (Blancher *et al.*, 2012), Ares *et al.* (2014b) obtained a large number of random subsets of different numbers of assessors from the original data of 13 consumer studies using bootstrapping. For each random subset, the sample configuration was obtained and the correlation with the reference configuration (that obtained with all the assessors) was calculated through the RV coefficient. Results suggested that when working with noticeably different samples, 60–80 consumers can be regarded as a reasonable compromise to obtain stable sample and descriptor configurations. However, research is still needed to study how the degree of difference among samples affects the minimum number of consumers necessary to reach stable configurations. Also, these numbers of consumers needed may change depending on the size of differences among samples, tentatively increasing if sample differences are small.

Considering that CATA questions are usually included in hedonic tests (Adams *et al.*, 2007; Ares *et al.*, 2010a; Dooley *et al.*, 2010; Jaeger *et al.*, 2013b), the minimum number of consumers necessary for obtaining reliable overall liking scores should be also considered when selecting the minimum number of consumers to be included in the study. For this reason, when CATA questions are elicited concurrently with overall liking scores, the usual number of consumers considered in hedonic tests (100–120) (Hough *et al.*, 2006; Lawless and Heymann, 2010; Mammasse and Schlich, 2014; Moskowitz, 1997) seems appropriate.

11.3 Analysis of data from CATA questions

Data from CATA questions consist of binary data which indicate if each consumer has selected each of the terms for describing each of the samples included in the study. An example of a typical data matrix, for a study with five samples and 60 consumers, is shown in Table 11.1. In the matrix, each column corresponds to a term of the CATA question. Using this data matrix, different approaches can be used for data analysis, which will be discussed in the following sections.

11.3.1 Summary data

The relevance of each term included in the CATA question for describing each sample is determined by calculating its frequency of use. Data are usually summarized using contingency tables which contain the number of consumers who selected each term for describing each sample (Meyners et al., 2013). Data can be displayed using counts or percentages, but the latter is more common.

11.3.2 Differences among samples

Cochran's Q test is used to evaluate if consumers detected significant differences among samples for each of the terms from the CATA question. Cochran's Q test is a

Table 11.1 **Example of the data matrix used for entering data from check-all-that-apply (CATA) questions**

Consumer	Sample	Sweet	Sour	...	Bitter	Thick
1	A	1	0	...	0	1
1	B	0	0	...	0	1
1	C	0	1	...	0	1
1	D	0	1	...	0	0
1	E	1	0	...	0	0
2	A	1	1	...	0	1
2	B	1	1	...	1	1
2	C	0	1	...	0	1
2	D	0	1	...	0	1
2	E	1	0	...	0	0
...	...	...	...	...	...	...
60	A	1	0	...	0	1
60	B	0	0	...	0	1
60	C	0	1	...	1	1
60	D	0	1	...	0	0
60	E	0	0	...	0	0

1 indicates that the consumer selected the term for describing the sample, while 0 indicates that the term was not selected. Data shown without randomization of terms and samples to more clearly visualize the structure.

Table 11.2 Example of the data matrix used for analysing data from one term of a check-all-that-apply (CATA) question using Cochran's Q test

Consumer	Sample 1	Sample 2	....	Sample x
1	1	0	...	0
2	1	0	...	0
...	...	...	...	...
n	0	1	...	1

Each cell indicates if the term was mentioned or not (1/0 respectively).

non-parametric statistical test, which is used in the analysis of two-way randomized block designs to check whether k treatments (i.e. samples in the test) have identical effects, when the response variable is binary (Manoukian, 1986). For each term of the CATA question, a data matrix is created containing samples in the columns, consumers in rows and where each cell indicates if the term was mentioned or not to describe each sample (1/0 respectively) (Table 11.2). According to Tate and Brown (1970), Cochran's Q test can be used if the number of consumers times the number of products is higher than 24, while Meyners and Castura (2014) indicated that at least 15 assessors should be used when comparing 2 or 3 samples. However, for external validity a larger number of consumers, as mentioned earlier (Section 11.2.3), may be needed. Post hoc pairwise comparisons can be carried out using the sign test (Meyners *et al.*, 2013).

11.3.3 Sample and term configurations

Correspondence analysis (CA) can be used to obtain a sensory map of the samples and the CATA terms. This map enables visualization of the similarities and differences between them, as well as their main sensory characteristics.

CA is a statistical method which enables simple and rapid visualization of the rows and columns of two-way contingency tables as points in a low-dimensional space (Greenacre, 2007). Similarly to Principal Component Analysis, CA projects the data from the contingency table onto orthogonal dimensions that sequentially represent as much of the variation of the experimental data as possible (Abdi and Williams, 2010). The positions of the points corresponding to the rows and columns in the dimensions of the space are consistent with their associations in the contingency table.

Although classical CA is based using χ^2-distances, Meyners *et al.* (2013) suggested using Hellinger distance, which is commonly used to quantify the similarity between two probability distributions (Cuadras and Cuadras, 2006), to minimize the influence of infrequently selected CATA terms on sample and terms configurations.

11.3.4 Penalty analysis

When consumers are asked to describe the samples and their ideal product, penalty analysis can be used to determine the drop in overall liking associated with a deviation from the ideal for each attribute included in the CATA question. Ares *et al.* (2014a) suggested a dummy variable approach to describe if an attribute was used to describe the product as in the ideal product (0) or differently (1).

For each attribute the percentage of consumers who used an attribute differently for describing a focal product and the ideal is determined, as well as the mean change in liking associated with that deviation from the ideal. A Kruskal–Wallis test can be performed considering each CATA variable as an independent variable and overall liking as a dependent variable, in order to determine if deviation from the ideal for each attribute has caused a significant decrease in overall liking (Plaehn, 2012).

Partial least squares (PLS) regression can be used to estimate the weight of the deviation from the ideal of each term from the CATA question, following a similar approach to that proposed by Xiong and Meullenet (2006) for just-about-right scales. A PLS model is calculated considering liking scores as dependent variables, and the dummy variables indicating if consumers described the product different from their ideal as regressors. Only attributes considered as deviating from the ideal for at least 20% of the consumers are usually considered, as suggested by Xiong and Meullenet (2006) and Plaehn (2012). This model also provides information about the maximum potential improvement on overall liking for a focal product, which is calculated as the difference between the model's intercept and actual mean liking score.

Meyners *et al.* (2013) extended the proposal of Ares *et al.* (2014a) by looking at the way in which the samples differed from the ideal product. These authors took into account whether the attribute was checked for the ideal product but not for the sample, checked for the sample but not for the ideal product, checked for both, or not checked for either. By looking at the decrease in liking of each of these alternatives, it is possible to classify attributes into "must-have" attributes (those which are present in the ideal and not in the sample and cause a large drop in liking), "to be avoided" attributes (which are present in the sample but not in the ideal and cause a large drop in liking) and "nice to have" (those which are present in the ideal and not in the sample and cause a small drop in liking). The benefit is greater input to guide product optimization efforts.

11.4 Case study: application of CATA questions for sensory characterization of plain yoghurt

To illustrate the data analysis approaches discussed above, an example of CATA study with 60 consumers. Six commercial samples of plain yoghurt (Samples A to F) available in the Uruguayan marketplace were evaluated. Twenty grams of each sample were served to assessors at 10°C in closed, odourless plastic containers labelled with three-

Sample No._____

How much do you like this yoghurt? ☐ ☐ ☐ ☐ ☐ ☐ ☐ ☐ ☐
Dislike very much   Like very much

Check all the terms that you consider appropriate to describe this yoghurt:

☐ Smooth   ☐ Firm   ☐ Heterogeneous
☐ Viscous   ☐ Creamy   ☐ Sweet
☐ Homogeneous   ☐ Sour   ☐ Fluid
☐ Liquid   ☐ Cream flavour   ☐ Milky flavour
☐ Thick   ☐ Off-flavour   ☐ Lumpy
☐ Gelatinous   ☐ Consistency   ☐ Aftertaste

Check all the terms that you consider appropriate to describe your IDEAL yoghurt:

☐ Smooth   ☐ Firm   ☐ Heterogeneous
☐ Viscous   ☐ Creamy   ☐ Sweet
☐ Homogeneous   ☐ Sour   ☐ Fluid
☐ Liquid   ☐ Cream flavour   ☐ Milky flavour
☐ Thick   ☐ Off-flavour   ☐ Lumpy
☐ Gelatinous   ☐ Consistency   ☐ Aftertaste

Figure 11.2 Example of the evaluation sheet used to evaluate the sensory characteristics of six yoghurt samples using a CATA question.

digit random numbers. Samples were presented monadically following a complete block design balanced for carry-over and position effects (Williams Latin square). Still mineral water was used for rinsing between samples, but was not mandatory.

Consumers were asked to try each of the samples and to score their overall liking using a horizontally labelled nine-point hedonic scale anchored at "dislike very much" (1) and "like very much" (9). After rating their overall liking, they completed a CATA question with 18 terms related to sensory characteristics of yoghurts. The terms were selected based on previous qualitative consumer studies (Ares *et al.*, 2008, 2010b): smooth, viscous, homogeneous, liquid, thick, gelatinous, firm, creamy, sour, cream flavour, off-flavour, consistency, heterogeneous, sweet, fluid, milky flavour, lumpy, aftertaste. After completing the CATA question for describing each sample, consumers were also asked to check all the terms they considered appropriate to describe their ideal yoghurt. The presentation order of the terms was randomized between and within participants. An example of the CATA question used in the study is shown in Fig. 11.2.

The frequency of use of each of the terms of the CATA question to describe the six yoghurt samples, as well as the ideal yoghurt, is shown in Table 11.3. The ideal yoghurt was described by the terms smooth, homogeneous, creamy, consistency and sweet, which indicates that these were the main drivers of liking for this type of product, as expected (Bayarri *et al.*, 2011; Pohjanheimo and Sandell, 2009).

Table 11.3 Number of consumers who used each of the terms of the CATA question to describe six yoghurt samples and the ideal product

Term	Samples						
	A	B	C	D	E	F	Ideal
Smooth***	35	32	28	26	10	2	40
Viscous***	14	10	4	7	2	1	4
Homogeneous**	14	13	18	18	15	4	29
Liquid***	0	6	31	8	37	33	3
Thick***	37	16	1	5	0	5	13
Gelatinous ns	6	4	0	3	0	3	3
Firm***	22	9	3	7	1	0	6
Creamy***	26	28	9	25	2	1	41
Sour **	9	11	5	5	15	18	10
Cream flavour***	25	12	5	11	0	2	18
Off-flavour***	6	7	12	14	27	29	1
Consistent***	22	10	3	7	0	1	20
Heterogeneous***	5	8	2	6	2	15	3
Sweet***	18	24	17	20	0	1	43
Fluid***	0	3	14	5	14	10	3
Milky flavour*	16	20	11	10	9	7	12
Lumpy*	6	4	1	2	0	6	0
Aftertaste**	4	3	11	8	15	13	0

***Indicates significant differences among samples according to Cochran's Q test at $p \leq 0.001$.
**Indicates significant differences at $p \leq 0.01$.
*Indicates significant differences at $p \leq 0.05$.
nsIndicates no significant differences ($p > 0.05$).
The ideal product was not included in Cochran's Q test.

According to Cochran's Q test, significant differences ($p \leq 0.05$) in the frequency with which 17 of the 18 terms of the CATA question were used to describe the yoghurt samples, suggested that consumers perceived differences in the sensory characteristics of the evaluated yoghurts.

Figure 11.3 shows the representation of the samples and terms in the first two coordinates of the CA performed on the frequency table using χ^2-distances. Together, the first and second dimensions explained 90.7% of the variance of the experimental data. The distance between the points corresponding to the samples is a measure of their similarity. Therefore, it can be concluded that three groups of samples with similar sensory characteristics were identified: Samples A, B and D, located at negative values of the first dimension; Sample C located at intermediate values of the first dimension and positive values of the second dimension; and finally Samples E and F, located at positive values of the first dimension (Fig. 11.3).

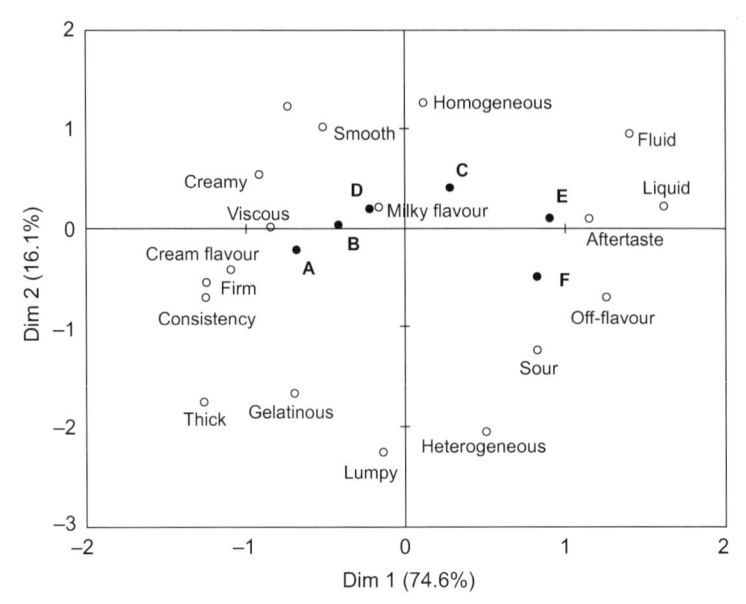

Figure 11.3 Sample and term representation in the first and second coordinates of the CA performed on the frequency of use of the terms of the CATA questions for the evaluation of six yoghurt samples.

Although the relative position of the rows and columns in the dimensions of the CA are not directly comparable and should be interpreted with care (Greenacre, 2007), some conclusions regarding general associations between the samples and the terms can be drawn. As shown in Fig. 11.3, Samples E and F were associated with the terms sour, off-flavour, aftertaste, liquid and fluid; while Samples A, B and D were mainly described with the terms creamy, viscous, cream flavour, firm and consistency.

Figure 11.4 shows mean drop in overall liking scores as a function of the proportion of consumers that checked an attribute for describing Sample D differently than for describing their ideal yoghurt. As shown, penalty analysis enabled the identification of directions for product improvement for each of the samples. In the case of Sample D, the attributes with the highest mean drop and deviation from the ideal were creamy, smooth and sweet. By looking at Table 11.3 and comparing the frequency of use of the terms for Sample D and the ideal yoghurt, it can be concluded that it is necessary to increase the creaminess, sweetness and smoothness to bring Sample D closer to consumers' ideal.

A PLS 1 regression model calculated overall liking and dependent variable and the dummy variables indicating if each consumer used the term identically for describing the sample and their ideal product as independent variables. Figure 11.5 shows regression coefficients for one of the six yoghurt samples, Sample A. Deviation from the ideal significantly affected overall liking for only a subset of attributes. Overall liking scores significantly decreased when the terms viscous, homogeneous, creamy, cream flavour, off-flavour, heterogeneous, sweet and lumpy deviated from the ideal.

CATA questions with consumers in practice 239

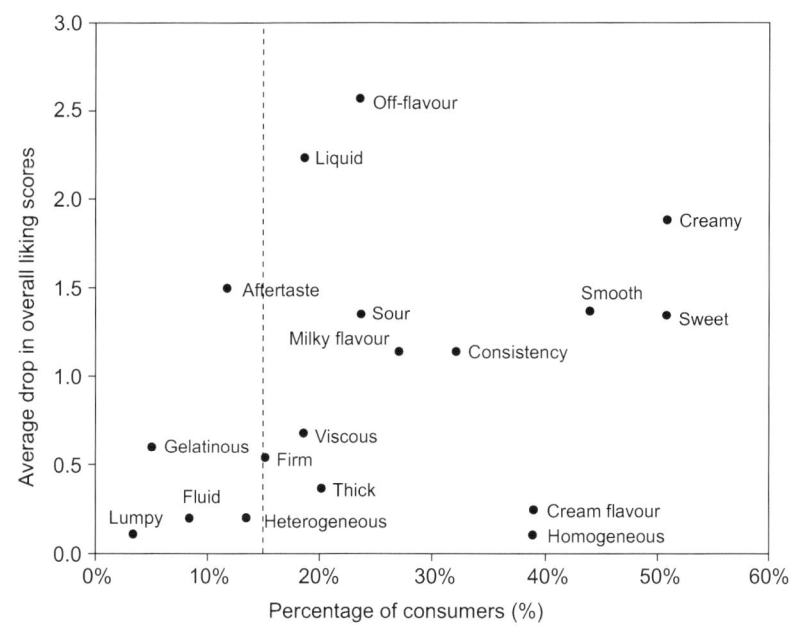

Figure 11.4 Mean drop in overall liking as a function of the percentage of consumers that checked an attribute differently than for the ideal product for one of the six yoghurt samples (Sample D).

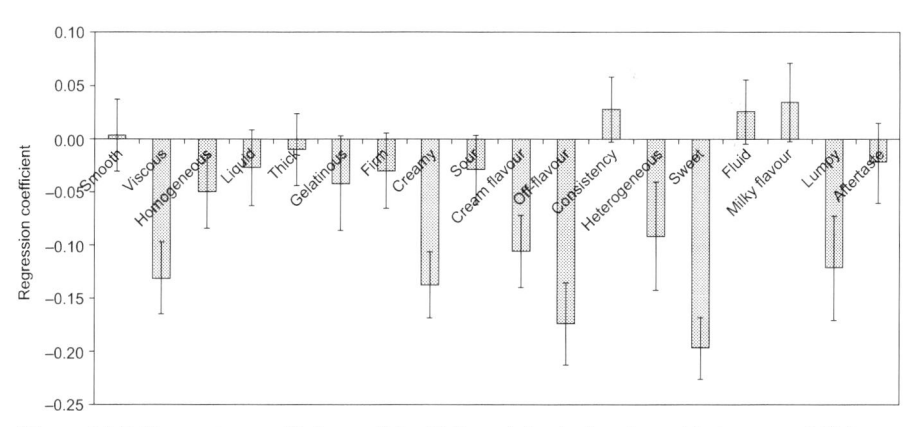

Figure 11.5 Regression coefficients of the PLS model calculated considering overall liking, and dependent variable and the dummy variable indicating if each consumer used the term identically for describing the sample and their ideal product as independent variables, for one of the six yoghurt samples (Sample A).

The attributes with the highest regression coefficients in the PLS model would be the priorities for product reformulation. Considering consumers' description of Sample A and the ideal product (Table 11.3), and taking into account the percentage of consumers who considered that the product deviated from the ideal, the priority for reformulation would be increasing sweetener concentration to increase sweetness perception.

11.5 Pros, cons and opportunities of the application of CATA questions

CATA questions are a simple and versatile tool for collecting information about consumer perception of the sensory and non-sensory characteristics of products. CATA questions are for uncovering consumers' emotional and conceptual associations of food products, and for studying how these associations are shaped by sensory and non-sensory characteristics.

When used for sensory characterization, this methodology has been reported to be quick, simple and not tedious for consumers (Ares *et al.*, 2011, 2013b), providing valid and reproducible information (Ares *et al.*, 2010a; Bruzzone *et al.*, 2012; Dooley *et al.*, 2010; Jaeger *et al.*, 2013a). However, it is important to stress that sensory product characterizations obtained using CATA questions cannot be regarded as a replacement for classical descriptive analysis with trained assessors. The latter methodology will always be more accurate, due to the fact that assessors are extensively trained in the identification and quantification of precisely defined sensory attributes. Considering that CATA questions do not encourage participants to engage in deep processing, there are several aspects related to how the methodology is implemented that deserve further exploration. In particular, it is necessary to further investigate how the design of the list of terms affects consumer processing and results from sensory characterization of products with different sensory complexity.

When used in marketing and survey research, one of the alternatives that have been reported to discourage satisficing response strategies and increase consumer attention is to ask them to answer yes/no to each one of the attributes included in the CATA question (Rasinski *et al.*,1994; Smyth *et al.*, 2006), as shown in Fig. 11.6. This strategy has already been implemented for sensory characterization by Ennis and Ennis (2013). However, considering that forced-choice CATA questions can be more tedious and time-consuming, research should be carried out comparing the usual application of CATA questions for sensory characterization with yes/no or applies/does not apply questions for each attribute, in terms of validity and reproducibility.

Other variants of CATA questions can also been implemented, such as asking assessors to select a predetermined number of response options (Campo *et al.*, 2010) or to combine CATA questions with intensity scales. Reinbach *et al.* (2014) asked participants to answer yes/no for seven overall flavour attributes of beer and to indicate the intensity of the selected flavours using a horizontal 15-point intensity scale anchored with "very weak" and "very strong." When comparing the usual CATA questions with CATA with intensity for sensory characterization of eight beers, Reinbach *et al.* (2014) reported high agreement between sample configurations and

Please indicate if each of the terms Applies or Does not apply for describing this chocolate

	Applies	Does not apply
Greasy	☐	☐
Burnt	☐	☐
Caramel	☐	☐
Vanilla	☐	☐
Salty	☐	☐
Soft	☐	☐
Off-flavour	☐	☐
Sweet	☐	☐
Hard	☐	☐
Bitter	☐	☐
Milky flavour	☐	☐
Acid	☐	☐

Figure 11.6 Example of forced-choice version of CATA questions.

perceived product differences, suggesting that discrimination among samples was not greatly enhanced by performing CATA-rating.

It is important to stress that the binary response format does not allow for a direct measurement of the intensity of the evaluated sensory attributes, which could hinder detailed descriptions and discrimination between products that have similar profiles in terms of their characteristic sensory attributes. In this sense, further research about the discriminative ability of CATA questions when working with highly similar products is necessary.

Furthermore, research is still needed to determine guidelines for the selection of terms to be included in the response options of CATA questions, particularly regarding the type and number of terms, as well as their layout on the evaluation sheet.

Although CATA questions are gaining popularity for sensory characterization by consumers, the approach is versatile and could also be used by trained panellists. Some challenges exist, however, including the minimum number of assessors required for statistical analysis. Yet, it may also present opportunities for sensory characterization of a large number of samples and/or using a large number of attributes. Hence, the application would be different from describing sensory analysis in the traditional sense, yet yield sensory characterization of products that are complex, such as wine, may be candidates for such application of CATA questions. This was exemplified by Campo *et al.* (2008), who used a list of 113 terms but forced assessors to select only the five most discriminant. Alternatively, CATA questions could be used when screening large number of product variants for selection of a subset to include in further detailed sensory profiling.

11.6 Conclusions

CATA questions are gaining popularity in sensory and consumer science, and seem an adequate option when a consumer-based sensory characterization is needed or in those situations where there is not enough time or resources to train a sensory panel. This methodology provides a rapid approach to obtaining information about the sensory characteristics of products directly from consumers. The task is simple and can be quickly and easily performed without the need of previous training. As with descriptive analysis, the methodology provides a description of the most relevant sensory characteristics of the products. However, it should be taken into account that CATA questions do not provide a direct measurement of the intensity of the evaluated sensory characteristics.

To support the development of CATA questions for sensory characterization with consumers, the current chapter has described the methodology and key aspects of its implementation. Despite the apparent ease with which CATA data can be collected and analysed, a number of important principles must be followed in application to ensure quality of data obtained and its value in guiding product innovation efforts. The most important of these principles have been described here, and this chapter constitutes a hands-on guide to CATA questions.

Acknowledgments

The authors are indebted to Comisión Sectorial de Investigación Científica (Universidad de la República, Uruguay) and to the New Zealand Institute for Plant and Food Research Ltd (Auckland, New Zealand).

References

Abdi, H. and Williams, L.J. (2010), "Correspondence analysis," In Salkind NJ, Dougherty DM, and Frey B, *Encyclopedia of Research Design,* Thousand Oaks, Sage, 267–278.
Adams, J., Williams, A., Lancaster, B. and Foley, M. (2007), "Advantages and uses of check-all-that-apply response compared to traditional scaling of attributes for salty snacks," Poster Presented at *7th Pangborn Sensory Science Symposium,* 12–16 August 2007. Minneapolis, USA.
Ares, G., Barreiro, C., Deliza, R., Giménez, A. and Gámbaro, A. (2010a), "Application of a check-all-that-apply question to the development of chocolate milk desserts," *J Sensory Stud,* **25**, 67–86.
Ares, G., Giménez, A. and Bruzzone, F. (2010), "Identifying consumers' texture vocabulary of milk desserts. Application of a check-all-that-apply question and free listing," *Braz*

J Food Technol, **6** SENSIBER, 19–21 de agosto de 2010, p. 98–105, DOI: 10.4260/BJFT201114E000112.

Ares, G., Giménez, A. and Gámbaro, A. (2008), "Understanding consumers' perception of conventional and functional yogurts using word association and hard laddering," *Food Qual Prefer*, **19**, 636–643.

Ares, G., Varela, P., Rado, G. and Giménez, A. (2011), "Are consumer profiling techniques equivalent for some product categories? The case of orange flavoured powdered drinks," *Int J Food Sci Technol*, **46**, 1600–1608.

Ares, G. and Jaeger, S.R. (2013), "Check-all-that-apply questions: Influence of attribute order on sensory product characterization," *Food Qual Prefer*, **28**, 141–153.

Ares, G., Jaeger, S.R., Bava, C.M., Chheang, S.L., Jin, D., Giménez, A., Vidal, L., Fiszman, S.M. and Varela, P. (2013b), "CATA questions for sensory product characterization: Raising awareness of biases," *Food Qual Prefer*, **30**, 114–127.

Ares, G., Dauber, C., Fernández, E., Giménez, A. and Varela, P. (2014a), "Penalty analysis based on CATA questions to identify drivers of liking and directions for product reformulation," *Food Qual Prefer*, **32**, 65–76.

Ares, G., Tárrega, A., Izquierdo, L. and Jaeger, S.R. (2014b), "Investigation of the number of consumers necessary to obtain stable sample and descriptor configurations from check-all-that-apply (CATA) questions," *Food Qual Prefer*, **31**, 135–141.

Bayarri, S., Carbonell, I., Barrios, E. X. and Costell, E. (2011), "Impact of sensory differences on consumer acceptability of yoghurt and yoghurt-like products," *Int Dairy J*, **21**, 111–118.

Blancher, G., Clavier, B., Egoroff, C., Duineveld, K. and Parcon, J. (2012), "A method to investigate the stability of a sorting map," *Food Qual Prefer*, **23**, 36–43.

Bruzzone, F., Ares, G. and Giménez, A. (2012), "Consumers 'texture perception of milk desserts. II – Comparison with trained assessors' data," *J Texture Stud*, **43**, 214–226

Campo, E., Do, B. V., Ferreira, V. and Valentin, D. (2008), "Aroma properties of young Spanish monovarietal white wines: A study using sorting task, list of terms and frequency of citation," *Aust J Grape Wine Res*, **14**, 104–115.

Campo, E., Ballester, J., Langlois, J., Dacremont, C. and Valentin, D. (2010), "Comparison of conventional descriptive analysis and a citation frequency-based descriptive method for odor profiling: An application to Burgundy Pinot noir wines," *Food Qual Prefer*, **21**, 44–55.

Castura, J.C. (2009), "Do panellists donkey vote in sensory choose-all-that-apply questions?" Oral presentation at *8th Pangborn Sensory Science Symposium*, 26–30 July. Florence, Italy.

Cuadras, C.M. and Cuadras, D. (2006). A parametric approach to correspondence analysis, *Linear Algebra Appl*, **417**, 64–74.

Dooley, L., Lee, Y.S. and Meullenet, J.F. (2010), "The application of check-all-that-apply (CATA) consumer profiling to preference mapping of vanilla ice cream and its comparison to classical external preference mapping," *Food Qual Prefer*, **21**, 394–401.

Driesener, C. and Romaniuk, J. (2006), "Comparing methods of brand image measurement," *Int J Market Res*, **48**, 681–698.

Ennis, D.M. and Ennis, J.M. (2013), "Analysis and Thurstonian scaling of applicability scores," *J Sensory Stud*, **28**, 188–193.

Greenacre, M. (2007), *Correspondence Analysis in Practice*, Boca Raton, Chapman and Hall/CRC.

Hough, G., Wakeling, I., Mucci, A., Chambers IV, E., Méndez Gallardo, I and Alves, L.R. (2006), "Number of consumers necessary for sensory acceptability tests," *Food Qual Prefer*, **17**, 522–526.

Jacoby, L.L. (1984), "Incidental vs. intentional retrieval: Remembering and awareness as separate issues." In Squire LR and Butters N, *Neuropsychology of Memory,* New York, Guilford Press.

Jaeger, S.R., Chheang, S.L., Yin, L., Bava, C.M., Giménez, A., Vidal, L. and Ares, G. (2013a), "Check-all-that-apply (CATA) responses elicited by consumers: Within-assessor reproducibility and stability of sensory product characterizations," *Food Qual Prefer*, **30**, 56–67.

Jaeger, S.R., Giacalone, D., Roigard, C.M., Pineau, B., Vidal, L., Giménez, A., Frøst, M.B. and Ares, G. (2013b), "Investigation of bias of hedonic scores when co-eliciting product attribute information using CATA questions," *Food Qual Prefer*, **30**, 242–249.

Krosnick, J.A. (1992), "The impact of cognitive sophistication and attitude importance on response-order and question-order effects," In Schwarz N and Sudman S, *Context Effects in Social and Psychological Research,* New York, Springer-Verlag, 203–218.

Krosnick, J.A. (1999), "Survey research," *Annu Rev Psychol*, **50**, 537–567.

Krosnick, J.A. and Alwin, D.F. (1987), "An evaluation of a cognitive theory of response order effects in survey measurement," *Public Opin Quart*, **51**, 201–219.

Lagrange, V. and Norback, J.P. (1987), "Product optimization and the acceptor set size," *J Sensory Stud*, **2**, 119–136.

Lawless, H.T. and Heymann, H. (2010), *Sensory Evaluation of Food. Principles and Practices.* New York, Springer.

Le Fur, Y., Mercurio, V., Moio, I., Blanquet, J. and Meunier, J.M. (2003), "A new approach to examine the relationships between sensory and gas chromatography olfactometry data using generalized procrustes analysis applied to six French Chardonnay wines," *J Agric Food Chem*, **51**, 443–452.

Lee, Y., Findlay, C. and Meullenet, J.F. (2013), "Experimental consideration for the use of check-all-that-apply questions to describe the sensory properties of orange juices," *Int J Food Sci Technol*, **48**, 215–219.

Mammasse, N., and Schlich, P. (2014), "Adequate number of consumers in a liking test. Insights from resampling in seven studies," *Food Qual Prefer*, **31**, 124–128.

Manoukian, E.B. (1986), *Mathematical Nonparametric Statistics,* New York, Gordon & Breach.

McCloskey, I.P., Sylvan, M. and Arrhenius, S.P. (1996), "Descriptive analysis for wine quality experts determining appellations by Chardonnay wine aroma," *J Sensory Stud*, **11**, 49–67.

Meyners, M. and Castura. J.C. (2014), "Check-all-that apply questions." In Varela P and Ares G, *Novel Techniques in Sensory Characterization and Consumer Profiling,* Boca Raton, Woodhead Publishing Limited.

Meyners, M., Castura, J.C. and Carr B.T. (2013), "Existing and new approaches for the analysis of CATA data," *Food Qual Prefer*, **30**, 309–319.

Moskowitz, H.R. (1997), " Base size in product testing: A psychophysical viewpoint and analysis," *Food Qual Prefer*, **8**, 247–255.

Parente, M.E., Ares, G. and Manzoni, A.V. (2010), "Application of two consumer profiling techniques to cosmetic emulsions," *J Sensory Stud*, **25**, 685–705.

Parente, M.E., Manzoni, A.V. and Ares, G. (2011), "External preference mapping of commercial antiaging creams based on consumers' responses to a check-all-that-apply question," *J Sensory Stud*, **26**, 158–166.

Piqueras-Fiszman, B. and Jaeger, S.R. (2014), "The impact of evoked consumption contexts and appropriateness on emotion responses," *Food Qual Prefer*, **32**, 277–288.

Plaehn, D. (2012), "CATA penalty/reward," *Food Qual Prefer*, **24**, 141–152.

Pohjanheimo, T. and Sandell, M. (2009), "Explaining the liking for drinking yoghurt: The role of sensory quality, food choice motives, health concern and product information," *Int Dairy J*, **19**, 459–466.

Popper, R., Rosenstock, W., Schraidt, M. and Kroll, B.J. (2004), "The effect of attribute questions on overall liking ratings," *Food Qual Prefer*, **15**, 853–858.

Prescott, J., Lee, S.M. and Kim, K. (2011), "Analytic approaches to evaluation modify hedonic responses," *Food Qual Prefer*, **22**, 391–393.

Rasinski, K.A., Mingay, D. and Bradburn, N.M. (1994), "Do respondents really mark all that apply on self-administered questions?," *Public Opin Quart*, **58**, 400–408.

Reinbach, H.C., Giacalone, D., Machado Ribeiro, L., Bredie, W.L.B. and Frøst, M.B. (2014), "Comparison of three sensory profiling methods based on consumer perception: CATA, CATA with intensity and Napping®," *Food Qual Prefer*, **32**, 160–166.

Smyth, J.D., Dillman, D.A., Christian, L.M. and Stern, M.J. (2006), "Comparing check-all and forced-choice question formats in web surveys," *Public Opin Quart*, **70**, 66–77.

Stone, H. and Sidel, J.L. (2004), *Sensory Evaluation Practices*, San Diego, Academic Press.

Sudman, S. and Bradburn, N.M. (1992), *Asking Questions*, San Francisco, Jossey-Bass.

Tate, M.W. and Brown, S.M. (1970), "Note on the Cochran Q test," *J Am Stat Assoc*, **65**, 155–160.

Van Trijp, H.C., Punter, P.H., Mickartz, F. and Kruithof, L. (2007), "The quest for the ideal product: Comparing different methods and approaches," *Food Qual Prefer*, **18**, 729–740.

Varela, P. and Ares, G. (2012), "Sensory profiling, the blurred line between sensory and consumer science. A review of novel methods for product characterization," *Food Res Int*, **48**, 893–908.

Williams, E.J. (1949), "Experimental designs for the estimation of residual effects of treatments," *Austral J Sci Res Ser A*, **2**, 149–168.

Williams, A., Carr, B.T. and Popper, R. (2011), "Exploring analysis options for check-all-that-apply (CATA) questions," Poster at *9th Rose-Marie Sensory Science Symposium*, 4–8 September 2011, Toronto, Canada.

Xiong, R. and Meullenet, J.F. (2006), "A PLS dummy variable approach to assess the impact of JAR attributes on liking," *Food Qual Prefer*, **17**, 188–198.

Open-ended questions in sensory testing practice

12

B. Piqueras-Fiszman
Wageningen University and Research Centre, Wageningen, The Netherlands

12.1 Introduction

> "When I use a word," Humpty Dumpty said, in rather a scornful tone, "it means just what I choose it to mean – neither more nor less."
> "The question is," said Alice, "whether you can make words mean so many different things."
> "The question is," said Humpty Dumpty, "which is to be master – that's all."
>
> Carroll, L. (1871) *Through the Looking Glass*, Chapter 6.

12.1.1 Overview

Free, spontaneous, idiosyncratic, easy, and fast. That is how responses to open-ended questions are, and these are the main aspects that attract researchers in sensory and consumer research to involve them more and more in their methodological routines. Open-ended questions have passed from being used merely as a possible support to explain other quantitative measures where, in many cases, the response is optional, to being part of the main task and contributing to the results as much as other measures. This chapter will begin by briefly describing this evolution and the main methodologies in which open-ended questions are currently being applied in our field, and the formats used (i.e., sentences or words, limited or unlimited).

Researchers should consider all the pros and cons of the entire procedure in order to make an informed decision about the choice of methodology. Next, these will be discussed while more specific issues are brought up as different case studies are presented throughout the chapter. And, existing out there are a number of similar questions/methods that share some of the benefits. In which specific situations do open-ended questions such as "win the battle," provide insights that no other method could? These situations will be described.

The sections that follow will explain the steps to be followed when using open-ended questions: from the instructions given and the collection of responses, through pre-processing the data, to its analysis and interpretation. At each of these stages, the particular features to consider for their potential applications will be highlighted.

Finally, what does the future hold? With the number of social networks, blogs, and so on available in which consumers spontaneously comment about products (without having the feeling that they are participating in a research study), an obvious question that arises is whether consumers really provide useful information that sensory science research can capitalise on. While the potential to fulfil consumer research/marketing purposes is clear, this chapter will question if this information could be exploited for product development and optimisation. Is extracting consumers' opinions from these public sources worth the effort? This section will put these topics on the table. Also, what can be done in the future to enhance the advantages of including open-ended questions, while diminishing the inconveniences?

At the end of the chapter, the key points seen throughout the chapter will be summarised.

12.1.2 Formats from the traditional approach to the novel methods

In sensory and consumer science, different methods have been used by researchers to measure consumers' perceptions and/or attitudes towards a set of products. Traditionally, key sensory drivers of liking have been studied, with techniques such as preference maps, that use objective assessments of product characteristics from trained panels (collected normally by means of rating scales), and liking ratings gathered from consumers. That is, the resulting information has often relied mainly on closed questions, while open-ended questions have been used to uncover and probe consumers' reasons for liking and facilitating final decision making (Lawless and Heymann, 2010). These questions would be normally positioned just after the liking question in a format like: "What did you like about the product?" (where no answers or any checklist of alternatives is provided), or "Is there anything else you would like to say?" consumers being able to express themselves freely in their own words. Moreover, consumers' responses to these open-ended questions are often not analysed in depth, but rather are used as a possible support for other quantitative questions, or to reduce the respondents' frustration by allowing them to explain (or "dump") their responses to other questions (Looker *et al.*, 1989; Sorensen, 1984).

However, this landscape is changing, and several rapid cost-effective methods for sensory characterisation have been recently developed and have been reported to be a good option for quickly gathering information about the sensory characteristics of food products. Many of these novel methods capitalise on the advantages of open-ended questions to uncover the perception of consumers (and depending on the aim of the project, also professional panels) at much earlier stages of the research process. In these methods, another format of these questions is being used, which basically consists in asking respondents to freely list a number of terms (limited or unlimited) that describe their perception of a product (sensory descriptors), and/or whatever they associate with it (general associations).

This elicitation technique can be the core of the task (often together with a liking, or buying intention question), being totally free, for instance, using phrases such as: "Please, write down all the words, images, associations, feelings, emotions and sensations that come to your mind when you see the following image of a plain yoghurt label," (as in Piqueras-Fiszman *et al.*, 2011), or allowing respondents to write down anything that would explain why they gave particular liking scores or to express whatever crossed their mind (ten Kleij and Musters, 2003; Ares *et al.*, 2010 limited subjects' explanations to four words). Symoneaux et al. (2012) conducted a study in which the untrained consumer panellists had the option of expressing freely what they liked and disliked about each of the products given, while Varela et al. (2014) asked their participants the same question (in a compulsory way) just after they had ranked the samples according to their preference. This task can also be limited to asking for sensory attributes, or even from a specific sensory category (e.g., only textures, or odours). For instance, Lawrence et al. (2013) asked their panel to "freely describe the perceived odours using as many words or expressions as you consider necessary."

Other recent methods use this elicitation technique as a complement to other tasks which are slightly more demanding than simply asking about liking. Examples of these methods are: rating the self-elicited attributes on scales (Free-Choice Profiling, Narain et al., 2004; Williams and Langron, 1984), ranking according to the perceived intensity of each one's own attributes (Flash Profiling, Delarue and Sieffermann, 2004; Sieffermann, 2000), and labelled sorting and *nap*ping according to perceived similarities or differences between the samples (e.g., Free Sorting, Faye et al., 2006; Popper and Heymann, 1996; and Ultra-Flash Profiling (UFP), Perrin and Pagès, 2009; Perrin et al., 2008). Previous chapters have already described these methodologies in detail, so here they will be briefly mentioned in as far as the focus remains on the open-ended question part of the task.

Their inclusion as one of the principal elements of the task certainly implies blurring the limit between sensory science and other fields, such as consumer psychology and sociology, where qualitative methods such as open-ended questions originate, and offers much more flexible and spontaneous insights from consumers (Piqueras-Fiszman et al., in press; Varela and Ares, 2012). This chapter will be more focused on these recent applications of open-ended questions, rather than on their traditional use.

12.2 General pros and cons of open-ended questions

In this section the main pros and cons of these questions will be discussed. As the chapter develops, more specific points will be brought depending on the type of open question used (the traditional type, or the term-elicitation approach used in more novel methods).

12.2.1 The advantages

The obvious main advantage of open-ended questions has already been mentioned: the spontaneity of the answers (unbiased) in the consumer's own words. Many researchers contend that by allowing consumers to respond freely, the question is better able to measure their salient concerns than in the close-ended format (which forces them to rate a set of given attributes or to choose among a fixed set of responses; see Geer, 1988; Kelley, 1983). This advantage is particularly relevant in relation to monadic sequential tests since the biases possibly introduced by the attributes would be avoided. In addition, when used in the traditional way (at the end of other quantitative/

closed questions) they may explain some of the responses given. If respondents feel limited by the closed questions and are not given the chance to "explain" themselves, a dumping effect might be observed, resulting in an artificial enhancement of scores on other scales (Clark and Lawless, 1994; Lawless and Clark, 1992) or on the frequency of selected terms from checklist questions.

Open-ended answers contain very rich information that may complement quantitative responses, but, in addition, they may provide the researcher with unpredicted insights. Importantly, these types of question are much more natural to respondents, who are more likely to be answering open-ended questions several times per day than coming across close-ended questions. Who is ever asked: "On a scale from 1 to 9 how much did you enjoy your holidays?" or "Tell me, from 1 to 9, how tasty my roast was"? (The probable response being: "What? Hmm. Well, I enjoyed it!"). Open questions do not require as much effort and reflection as closed questions do, a fact that also makes open questions easier to understand and to respond.

12.2.2 The catches

While the advantages of including open-ended questions in a sensory test make it, from what has been seen up to this point of the chapter, a simple and attractive way to go forward, criticisms have also been made against them. Among these is the belief that some respondents fail to answer these kinds of questions, not because they do not have views on the issue but because they are not articulate enough to put forth an answer (Geer, 1988). Possible reasons for this in our field are that they often have difficulties in verbalising the nuances of a product's sensory characteristics (Elmore *et al.*, 1999), or that they do not know exactly why they like a product. In addition, in the traditional approach, since normally these questions are not forced (and are seen purely as a support for other questions), respondents might just skip them to finish as soon as possible. To avoid "drop-outs," it is convenient that the test is carried out with a member of the research team present, encouraging them to complete the task or providing them with a previous unrelated example to inspire their possibly blocked minds.

Other disadvantages come at later stages of the procedure, namely at the time to analyse and interpret the terms or comments collected. As in any other qualitative research methods, the tougher part is to code and tabulate the responses. Since these are in the consumers' own words, they may be often ambiguous (mostly in self-administered tasks in which the researcher is not necessarily present or able to clarify them with the participant). In cross-cultural studies there is the additional problem of the translation of the terms used to describe the product and how they are being used. This issue makes the treatment of the responses (stemming and lemmatisation) even more difficult and time-consuming (it will be further discussed in Section 12.4). Furthermore, it is also legitimate to raise the same issue in generational studies. Therefore, the researchers leading the study should be somewhat acquainted with semantic analysis and, when encountering ambiguity in a participant's response, they should know how to look globally at the participant's demographic information and other given responses to try to make sense of it. This process, as one can imagine, can be very time-consuming (which translates to cost increases) and requires a great deal of experience and skill to avoid reaching misleading conclusions. Further down in the process is the analysis of the "clean" data, which is not as straightforward as when analysing close-ended questions. These issues will be covered more in detail in the next sections.

12.3 When open-ended questions are appropriate

There are different views about the usefulness (and worth) of open-ended questions (Lawless and Heymann, 2010; Stone and Sidel, 2004). As overviewed in the previous section, the skilful experimenter should carefully consider the research questions and analyse the pros and cons to make an informed decision of the procedure to follow. Open-ended questions are highly recommended in the following situations:

1. When it is important (if not essential) to collect people's opinions (or descriptions or associations) in *their own words*. What words do they use to describe their perception or experience with something? For instance, when a company has launched a new product, it could be valuable to ask an open-ended question about it to see what words the target consumers use to describe it. In addition, in case these allow for general associations, the responses could have marketing applications too. Remember also that these terms are likely to be the most salient characteristics of the product as perceived by the consumer.
2. When it is necessary to know the spontaneous reasoning behind consumers' (or untrained panels') perceptions of the sensory, hedonic and/or emotional differences or similarities between products. In this case, the researcher would not only want to measure how the participants perceive the products, but also to collect the genuine reasons for these perceptual differences, which will differ from one person to another.
3. When the aim is to discover something new. The research team might think that all the important possible attributes of a product could be provided by themselves. But, by allowing subjects to provide their own terms, ideas that had not been considered before might come up, and they might even suggest attributes that could be included in future studies' closed-ended questions.
4. When the data have to be collected in a faster way. In comparison with traditional procedures involving intensity ratings or conventional profiling, including open-ended questions can reduce the time of data collection significantly. Lawrence *et al.* (2013) reported that when using the free comments method, a group of wine professionals took an average of 1 h 15 min to evaluate 31 wines, whereas when the same wines were evaluated by another group of wine experts using the classical profiling method an average of 1 h 50 min was required to complete the task.
5. To simplify the task for the subjects (especially for particular groups of consumers). The free comments method is easy to perform; subjects do not need to understand the use of any types of scales nor "translate" their perception into a score. The ease of the technique is important when dealing with particular groups of subjects, such as the elderly, since they tend to become fatigued easily which consequently could affect their performance of the task. Piqueras-Fiszman *et al.*'s study (2011) involved older consumers in a free word association task which they had to complete by responding to a set of five yoghurt labels. While most of the elderly completed the task in a straightforward way (and they reported its easiness), many of them had doubts on what to write, and some others were blocked. A similar

situation would be expected from children. To avoid this situation when dealing with these particular age cohorts, a member of the research team should try to be present, encouraging them and providing examples to avoid blockages.
6. To compare different uses of a language between groups of consumers. Responses from open-ended questions can also be relevant to compare the spontaneous descriptions provided by different age groups or nationalities, and even how they differ regarding the number of terms used. Different groups of people use particular words and expressions that would not show up if they were given the same checklists, as in check-all-that-apply (CATA) tasks, or the same list of attributes to rate from.

12.4 Processing the answers: from raw to clean data

Regardless of the type of open-ended question, cleaning the data is the most tedious part of the process. This can be considered as "closing" the open-ended question, and is referred to as "post-coding" (Lebart et al., 1998). Nevertheless, it is important not to assume that these coded responses provide the same information as when obtained from closed-ended questions. Since this process has to be performed in a manual, qualitative way, it can be very time-consuming (imagine having eight products and 100 respondents providing an average of four terms for each product, that is 3200 terms to go through! And if they are allowed to provide sentences, this number can easily increase × 10, at least).

While the information extracted from these questions (textual data) can be extremely rich, dealing with large amounts of data can make it difficult for the researcher/s coding it to stay consistent and keep an overview of what is going on (ten Kleij and Musters, 2003). This is particularly the case for tasks in which consumers are asked to freely provide their thoughts about the set of stimuli. In this section, examples from tasks in which consumers are asked to provide short expressions or terms will be discussed. The reader interested in responses from, for instance, focus groups, where a text corpus is obtained (and several statistical units have to be derived from it), is referred to the following literature which covers in more detail the initial procedure of text segmentation and word coding (Dransfield et al., 2004; Lebart et al., 1998). In any case, the main steps would usually involve (Bécue-Bertaut et al., 2008):

1. A spelling correction.
2. A lemmatisation step which converts every word into its standardised form called a *lemma* (i.e., a dictionary entry such as infinitive verbs, singular form of nouns and a specific singular form for adjectives, in the case of languages which adjectives have gender) and also identifies the grammatical category of every word. Although certain lemmas can change their meaning depending on the context, or be influenced by the adverbs used, it is convenient to conserve every lemma (word) as a separate unit (Bécue-Bertaut et al., 2008).
3. Definition of a stoplist, which includes words that are considered meaningless or irrelevant in the study. In the framework of preference studies, the stoplist commonly consists of prepositions, articles, pronouns, conjunctions, possessives and demonstrative adjectives (ten Kleij and Musters, 2003).
4. Selection of a frequency threshold, since the comparison of lexical profiles is only meaningful if the words have a minimum frequency (Lebart et al., 1998).

Table 12.1 Example of how the table of raw data could look

S	Product 1	L1	Product 2	L2	...	Product p	Lp
1	Fresh, natural, fatty	5	Organic, healthy, digest	6	...	Sweet, fruity, female	6
2	Traditional, thick, healthy	7	Herbal, no additives, probiotic	6	...	Strawberry, low-fat, diet	8
...	...	...	...	...	...	...	...
n	Wellness, dense, full-fat	7	Easy to digest, organic, bland	4	...	Diet, sweet, fruits, healthy	8

The subjects (S) are arranged in rows and their individual terms are tabulated for each product, in columns. Other measures provided by each subject, such as liking (L) in this example, are also included in the table next to the corresponding product.

Table 12.1 shows a fictional example of how a table of raw data might look (taken from Piqueras-Fiszman et al.'s 2011 data). In this case, it represents data from a free word elicitation (or word association task), where liking scores for each yoghurt product were collected as well (note that, as mentioned previously, several other measures could be included in the table, but this is a simplified example).

What to do next? First, the words and/or the sentences provided have to be carefully interpreted and the words grouped together if they belong to the same category. Coding sentences implies taking word combinations into account (not to mention sarcasm, and so on) and breaking them into the words which best describe the meaning of the sentence. For example, the following sentence from a respondent: "It seems fresh, not very sticky, firm" could be firstly broken down as: "fresh," "not sticky" and "firm." Now, if another respondent says: "It is smooth and thick," we have to make the difficult choice of whether to group the first respondent's "not sticky" with the second's "smooth" into a single category or not (plus, are we sure they are referring to the same textural property?). The same occurs with "firm" and "thick." Also, some terms, such as "fresh" may apply to different sensory categories; in this case, "fresh" could mean "chilled" or "new/recently produced".[1] Therefore, the researcher has to go through all the terms and determine the most appropriate meaning. When we are not sure if two words refer to the same meaning, at this stage it is more convenient to keep them in separate groups. It is important to avoid disregarding synonyms that are elicited with a lower frequency, which would bias the results (Guerrero et al., 2010); however, stemming (grouping derivatives of the same word; e.g., sweet and sweetness) is a more straightforward task. The process of classification should ideally be performed by more than one researcher familiar with the language (and culture) in

---

[1] Kostov et al. (2013) have recently proposed a methodology based on multiple factor analysis for contingency tables (MFACT), in which all the words are kept (i.e., without filtering or grouping), and that detects those consensual words that really have the same meaning for most of the panellists describing the same product. In principle, this basically would solve, in an automatic way, the problem of having to decide whether one panellist is associating the same meaning to "fresh" as another panellist.

Table 12.2 **Example showing how individual terms can be grouped into category terms**

Category terms	Examples of individual terms that have been grouped together
Delicious	Delicious, nice taste, yummy, I like it…
Disgusting	Disgusting, awful, yuck, I don't like it, unpleasant taste…
Sweet	Sweet, very sweet, sugary…
Thick	Thick, consistent, dense…
Healthy	Health, healthy, wellness, good for you…
Bland	Bland, not very tasty, dull, no flavour…
Traditional	Traditional, homemade, like my Granma's, typical, as always…
…	…
t (Diet)	Diet, light, low-calories, zero, no fat, slimming…

Table 12.3 **The matrix (contingency table) showing how often occurrences of terms are used to describe each of the p number of products**

Product	Terms								Liking (mean)
	Delicious	Disgusting	Sweet	Thick	Healthy	Traditional	…	t (Diet)	
1	56	17	18	42	30	6	…	2	5.8
2	13	44	16	2	39	3	…	26	3.4
…	…	…	…	…	…	…	…	…	…
p	47	7	60	9	40	0	…	50	6.2

Other tables of complementary data can be juxtaposed too (as for the liking mean ratings shown here) to include them in the same analysis.

which the data have been collected. A good practice is performing it by triangulation (Modell, 2005) having three researchers considering word synonymy as determined by a dictionary and personal interpretation (Guerrero et al., 2010; Piqueras-Fiszman et al., 2011). After individually evaluating the data, agreement between their classifications has to be reached. Examples of category terms at this stage are shown in Table 12.2.

At this stage, the data have been cleaned from stopwords (e.g., the, a, etc.), spelling mistakes have been corrected, synonyms and stem-words have been grouped under one unique category term and there should be no ambiguity about what a category term is referring to (e.g., there could be two terms for "light" that should be considered: "light colour" and "light texture"). A contingency table (products in rows and terms in columns) should be created showing how often a category term has been used to describe each of the products (see Table 12.3). But we could still easily have

Raw data (individual terms, sentences, etc.) ⇨ Cleaning: Delete irrelevant stop-words, correct mistakes, stemming ⇨ Categorisation by synonyms + cut-off frequency ⇨ Reduced category terms ⇨ Generic categorisation ⇨ Dimensions (5–15)

Contingency table ⇅ Statistical analyses

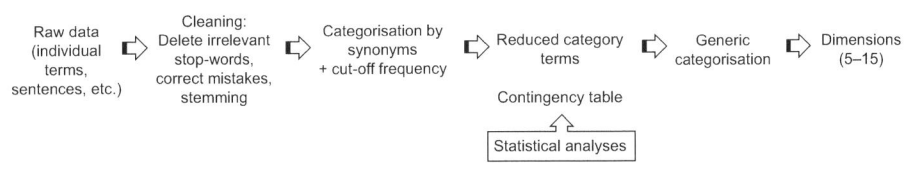

Figure 12.1 Diagram of the typical steps usually followed to process open-ended responses.

more than 100 terms, which is an unmanageable number; therefore, a frequency cut-off must be established in order to avoid including noise in the analyses. For this, a value of 5 or 10% is considered appropriate (depending on the number of words collected). Frequencies in each category are determined by counting the number of times that a word has been used to describe each product (occurrences), so if we had 100 consumers, and we determined a value of 10%, only category terms having more than ten terms across products would be considered for further analysis (for instance, in the example in Table 12.3, terms like "traditional" would be dropped out from the succeeding analysis steps).

Summarising, this long pre-processing part of the procedure (illustrated in Fig. 12.1) calls into question whether open-ended questions really make the methodologies in which they are included more "rapid" (if we do not limit this appellation to the stage of data collection). Certainly, depending on whether the questions are limited (to sensory descriptors, say) or free (unlimited), the size of the study (subjects and samples), and on the experience of the researchers managing the information, the timing of processing the data can vary greatly, from 5–8 h up to around 40 working hours (e.g., Galmarini *et al.*, 2013); if the task has been performed in a verbal or handwritten way, add at least 8 h to typewrite the responses first!

There are many text mining software packages out there that extract the words from unstructured texts (like sentences) and transform the information to make it accessible to the various data mining algorithms (Feldman and Sanger, 2007; Ghani *et al.*, 2006). However, these automated processes return an objective transformation of the text into numbers (numeric coding), without necessarily reflecting the correct meaning of the expressions or terms, which is the most valuable insight. Therefore, personal interpretation, mostly in technical fields, is recommended, at least in this pre-processing stage. Next, let us see briefly how these textual data are usually analysed, though going through the statistical details is not under the scope of this chapter; the interested reader is referred to the literature cited to explore the analyses more in depth.

12.5 Analysing the data: getting valuable outcomes from different applications

There are several available tests to perform to squeeze the most out of the data, which depend to a great extent on the entire dataset that has been collected for the purpose of the study. For instance, we may have liking or willingness to try mean ratings, the X and Y positions of the samples in a UFP task, and also datasets composed of responses from different groups of consumers that we may want to compare. The following sections will not only describe how to analyse and interpret the textual data from basic studies, but will also describe the analysis steps for those potential applications in which open-ended questions may have been included, providing unique valuable information.

12.5.1 Finding significant differences between the products

In sensory and consumer science studies, one of the common basic goals is to explore how the products differ in terms of the measures collected. Usually Chi-square is calculated to evaluate differences in the consumers' perception of the products. This can be first done globally, to test the independence between the rows and columns of the contingency table, and then per cell to evaluate if each category term is able to detect significant differences in the perception of the labels shown (Symoneaux *et al.*, 2012).

12.5.2 Perceptual maps based on open-ended questions

Let us recall that the contingency table crosses the modalities of two categorical variables (here, products and words). In other words, the rows and columns are of the same "nature", and lie at the same distance. However, in principal component analysis (PCA), this is not the case since the rows and columns are fundamentally different, and lie at two different distances.

Therefore, correspondence analysis (CA) should be used. CA is a method for representing data from a two-dimensional contingency table into a space so that the results can be visually examined in a map (Greenacre, 1984; ten Kleij and Musters, 2003). Similarly to any factorial approach, both the cloud of rows and the cloud of columns are closely linked, due to their relation of duality. However, in CA, this duality is even more apparent, as the rows and columns in contingency tables are fundamentally the same (and it is for this reason that the rows and the columns tend to be represented together on the same map). This is actually possible since both clouds share the same total inertia, as well as the same projected inertia on each dimension of same rank. Moreover, in CA, the relation between rows and columns is such that each row (product) is at the barycentre of the columns (words) assigned with a certain weight (which depends on the frequency of elicitation) and vice versa. This property (often referred to as barycentric, or pseudo-barycentric) is used to interpret both the position of one product in relation to all the words, and the position of one word in relation to all the products. However, it is not possible to draw conclusions about the distance between a row and a column, so this association has to be checked in the contingency table (see Husson *et al.*, 2011 for more detailed information).

An example of a CA representation of terms and products is shown in Fig. 12.2. Continuing with the same example described above, this map, together with its contingency table, would illustrate which terms were associated with each yoghurt product. For instance, the first dimension was positively correlated to terms related

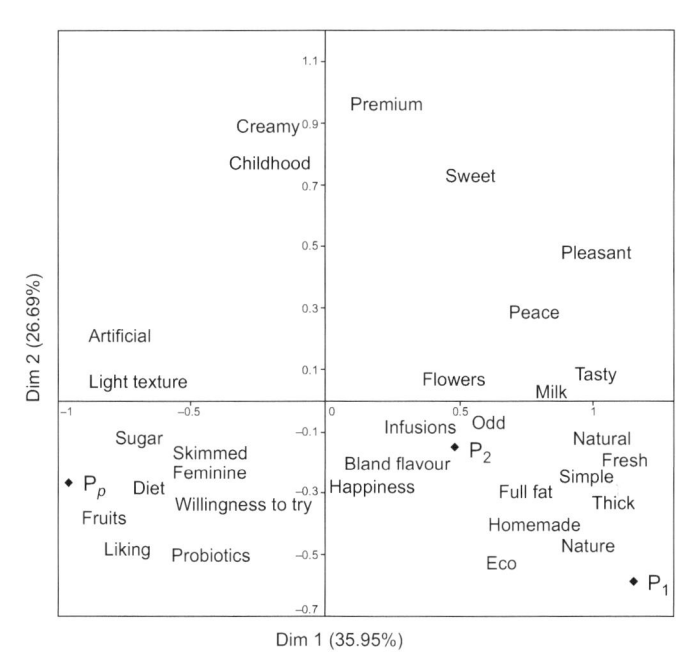

Figure 12.2 CA representing which category terms were particularly associated with each yoghurt product.

to nature, freshness, tastiness, full-fat content and naturalness. All these terms were much related to Product 1, situated in the same area of the map. This picture indicates that these subjects associated signs of freshness and nature to homemade products, which are usually rich in fat content. Product 2 was strictly related to flowers and infusions, but it was also regarded as odd (note that in the country where this study was performed there were no such yoghurt products). Product p (to keep the labelling for this example consistent throughout the chapter) was mainly related to diet, skimmed, fruits and women, which means that this product successfully conveyed those messages to these consumers, leading to a positive hedonic appraisal.

Other variables can be included in the analysis as supplementary variables (such as liking) as in external preference mapping. In fact, it has been demonstrated by a number of authors (e.g., Ares *et al.*, 2008; Symoneaux *et al.*, 2013; ten Kleij and Musters, 2003) that the results from open-ended questions are very similar to those obtained with traditional preference mapping techniques. The clear benefit of using open-ended questions with consumers in these methodologies is that all the data (liking or buying intention, and the sensory elements) come from the same individuals and the description is in their own language. In addition, what might be salient or important to them might not be significant for a trained panel. On the other hand, some sensory terms that consumers provide are somewhat vague and might not be very helpful for product developers. Therefore, it is suggested that consumers' open-ended responses are used as a complementary tool or when access to trained panels is not possible (although for marketing purposes, this alternative could be more convenient).

12.5.3 Comparing open-ended responses from different groups of subjects

As already mentioned, one of the most worthy reasons to use open-ended questions is to collect responses that, not only provide the consumers' spontaneous thoughts, but that are also expressed in a style that reflects their idiosyncrasy (culture, age, knowledge about the product, etc.). Therefore, a potential application of open-ended questions is to compare the descriptions or associations that different groups of consumers may provide. In this case, when open-ended responses have been collected from more than one group of subjects, most probably one might want to include several datasets (which will often consist of a mix of quantitative and frequency data) as active variables in a single factor analysis, instead of performing a factor analysis on one set of variables and considering the rest as supplementary variables which are simply projected into the space. The resulting map will allow visualisation of the consensus of the different datasets. This can be done by combining MFACT, an extension of multiple factor analysis (MFA) for dealing with multiple frequency tables (Bécue-Bertaut and Pagès, 2004), and classical MFA (Abdi et al., 2013; Escofier and Pagès, 1994) so that quantitative and/or categorical variables may be taken into account simultaneously as well (Bécue-Bertaut and Pagès, 2008).

Several researchers have used this analysis to compare open-ended responses obtained from subjects of different countries in cross-cultural studies (e.g. Ares et al., 2011; Bécue-Bertaut and Lê, 2011; Galmarini et al., 2013), which has allowed them to identify the terms that consumers use in their day-to-day life to describe products, finding separately likes and dislikes regardless of the different languages. Now, as mentioned previously, the interpretation of the words, even if they are limited to sensory attributes, has to be performed by a native speaker of the language in which the data have been collected and who is, preferably, familiar with the culture of the country. Then, since usually English is chosen as the consensual language in which to compare the responses and in which to show the results, the terms have to be translated by a professional who is, again, preferably acquainted with both cultures. If this possibility is not available (e.g., for economic reasons), then the researchers (who should be at least fluent in the two languages and have experience with the type of lexicon in question) can perform the translation, consulting related publications or databases, or other experts in the field, in the case of not knowing the most appropriate translation for a term. In Ares et al.'s study (2011) we compared consumers' associations from Spain and from Uruguay, and even though the language is Spanish for both countries, we were surprised at the differences found in the use of Spanish for terms with the same meaning!

Piqueras-Fiszman et al.'s (2011) study, of which some of the data have been shown throughout the chapter as examples, compared the free word associations from elderly to those provided by young consumers using MFACT as well. The results highlighted the difference in the associations that elderly and youngsters had of the labels, some of which would not have come up if a pre-determined list had been provided to them.

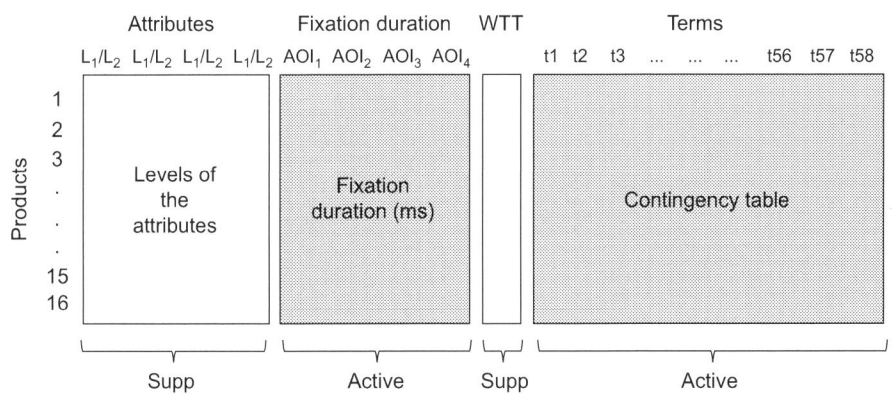

Open-ended questions in sensory testing practice 259

Figure 12.3 Structure of the juxtaposed tables of frequency, categorical and quantitative datasets from eye-tracking and word association responses. The products (1–16) are arranged in rows, while the attribute variations (in levels, two for each attribute), the fixation duration (in ms, for each area of interest (AOI) defined), the willingness to try (WTT) mean ratings, and the terms ($t = 58$) are arranged in columns. Since the interest relied on exploring the relationships between the associations and the attentional data of each AOI, only these two tables were considered as active in the MFA.

12.5.4 Combining spontaneous responses from different modalities

Piqueras-Fiszman *et al.* (2013) capitalised on the spontaneity of open-ended responses (by means of free word association) to combine the results with another unconstrained technique: eye-tracking. The idea behind this study was to capture consumers' thoughts while they were watching a set of jam products individually, which varied in four attributes. The data obtained were extremely rich, since they combined the total fixation duration for each area of interest defined (measured in ms), the frequency with which specific words were mentioned (contingency table), willingness to try ratings and the levels of the variations. This mixture of datasets was analysed using MFA as shown in Fig. 12.3.

12.5.5 Identifying the reasons responsible for the similarities/differences among samples in projecting mapping tasks

Open-ended responses are also extremely useful to use in projective mapping methods, since subjects freely provide terms that would help identify the underlying reasons for the similarities and differences perceived among the products. This technique is known as UFP, mentioned earlier, and the data obtained are usually analysed using MFA (Pagès, 2005; Perrin *et al.*, 2008) considering the coordinates of each assessor as a separate group of two un-standardised variables. The comments given for each of the samples are counted across consumers and a contingency table is built (as explained before) and considered as a set of supplementary variables (the

N	Dimension	Frequency of elicitation (%)		N	Dimension	Frequency of elicitation (%)
1	Memory associations	25.7		1	Individual's personal associations	32.4
2	Emotions	6.7				
3	Sensory	15.1		2	Sensory	15.1
4	Style of product	13.0		3	Style of product	13.0
5	Naturalness	10.0		4	Naturalness	10.0
6	Hedonic liking	4.9		5	Hedonic liking	7.0
7	Hedonic disliking	2.1				
8	Health	20.2		6	Health	22.3
9	Fat content	2.1				

Figure 12.4 The extent to which the array of category terms is narrowed down depends on the number of dimensions that the researcher might want to highlight. For instance, if showing the frequencies of liking and disliking individually is more relevant, then the two dimensions are kept separate.

correlation coefficients with the MFA factors are calculated and the variables are represented, but they do not participate in the construction of these factors; Pagès, 2005). This analysis provides a consensus representation of the samples, a representation of the descriptions provided by the panel, and also a representation of the members of the panel.

12.5.6 Narrowing the spectrum of terms into broader dimensions

If the elicitation task is free (not limited to sensory descriptors), very likely the array of terms (category terms) is still too wide. This is okay in order to have a nice perceptual representation on a map but, ideally, we would want to further reduce the information, facilitating the interpretation of the results obtained. To achieve this, the terms created in the previous stage should be grouped into more generic categories (or dimensions); this can be done again by triangulation until the three researchers have obtained a reduced manageable number of dimensions by consensus. Again, personal interpretation is needed, although since the data are already clean, this process can be automated by word substitution. Figure 12.4 presents an example of two possible ways in which certain category terms can be grouped in narrower or broader dimensions. There is not a rule of thumb to create these dimensions, as long as the researcher feels they represent and summarise appropriately the range of terms collected. Then the frequency of elicitation can be calculated for each dimension to get a clear, short and sweet picture of the types of terms that the subjects used.

12.5.7 Exploring semantic networks

Finally, it may also be interesting to explore the wealth of the data in terms of the semantic relationships and association between the words without any preliminary coding or other intervention. This can be done by considering the co-occurrence of certain pairs of words within a response in a survey. The similarity between words is derived from the proximity between their distributions (or lexical profiles) within the corpus, and the most relevant co-occurrences can be described by using a weighted undirected graph linking the lexical units. For more details about semantic networks see Bécue-Bertaut and Lebart (1998).

Having reviewed all the stages of the main methodologies based on open-ended questions, a hopefully helpful table of the advantages and disadvantages at each step is provided in Table 12.4 to summarise the key points.

12.6 Future trends and social media

As everybody must have heard at least once already, we are living in the so-called Internet and Information Age where online data usually grow in an exponentially explosive fashion. Consumers nowadays feel the need to express their thoughts about pretty much everything and publish them in social networks (e.g., Facebook, Twitter, Google+), blogs, forums and many other online sources that probably this author is not even aware of. Consequently, since some 10 years ago, marketers have harnessed the power of online resources to investigate their consumers without them even realising it.

Nowadays, there are a number of software packages tailored especially to explore each of these networks and extract the juicy information that companies are looking for (Ghiassi *et al.*, 2013; Li and Wu, 2010).. While it is not the aim of this section to provide an exhaustive review on the topics discussed in social media, a question that constantly springs to mind is whether consumers really provide useful information that sensory science research can capitalise on. The potential to marketers is clear: consumers do express their thoughts about brands and their experiences with products; but a step further from that and closer to sensory science, and the power of using this tool gets diminished. Unless a question is sneaked in by a company's page on Facebook, for instance, consumers very likely will not spontaneously describe the sensory properties of a product, but will perhaps indicate how much they are enjoying it (or hating it). Platforms such as Twitter (which is mainly used to publicise the outbreak of new stories, complaints and major public communications) have been used by researchers of different fields for a wide variety of purposes. Twitter has a data gathering tool (Twitter Application Programming Interface (API)) that allows researchers to capture tweets, their location, day and posting time, which makes it quite attractive. The majority of studies in these areas of Twitter research employ sentiment analysis approaches to identify and evaluate the opinions of users expressed in their tweets (see Abbasi *et al.*, 2008); however, several researchers have reported that the

Table 12.4 Summary of advantages and disadvantages of open-ended questions throughout the different stages

Stage	Advantages	Disadvantages
Task performance	It is fast (it requires a couple of seconds per product in word association tasks). • Cost saving. Instructions are simple and easy to understand, so it does not require constant supervision. It does not require training. It is practical to combine with other implicit tasks (such as eye-tracking), since the spontaneity of both tasks is maintained.	Some subjects might get blocked (mostly elderly people). • An example can be provided to "break the ice".
Pre-processing	Text mining software can extract the words from texts and export them into Excel-type files. Text processing software can facilitate the grouping of words and calculate frequencies easily once the cleaning part has been performed.	If the responses are oral (to gain spontaneity), transcribing them is required. Their responses might be ambiguous or difficult to understand. The cleaning and personal interpretation of the responses has to be done manually and this process can be very time-consuming. • If this process is done by several researchers, the duration can be reduced. It requires more than one researcher, at least fluent in the language of the data and acquainted with the subject's culture, to interpret the responses appropriately.
Analysis	More and more data mining and other statistical techniques allow for automated analysis and representations that make the visualisation of the textual data easy to understand and attractive.	
Outcome	Allows the discovery of new unexpected insights.	Not as precise as other conventional quantitative methods. • Can be used as a lexicon generator step in the conventional methods or in cases where access to a professional panel is not possible.

Table 12.4 Continued

Stage	Advantages	Disadvantages
	It provides similar results when compared to classical preference mapping or profiling procedures. The responses are spontaneous, which would represent the consumer's more salient thoughts about the products and/or the dominant sensations they experience. Expressed in the consumers' own words, which reflect their idiosyncrasy. • Ideal to compare responses from different groups of consumers.	Compared to classical preference mapping or profiling procedures, it does not identify sensory nuances between products as clearly.

language used in the tweets makes it difficult for them to determine sentiment, and some have had 80% of the tweets collected expressing no sentiment at all (Jansen et al., 2009). Another limitation is the amount of tweets that can be retrieved; on occasion, several weeks of collection have to pass to be able to collect a reasonable sample (c.15 000 tweets depending on the terms searched), which then have to be filtered to retain the ones that bring some relevant insight. Is it really relevant for us to know that someone is eating a chocolate bar while going to San Francisco, for instance? Furthermore, the process that follows after retrieving the targeted amount of tweets is definitely not shorter or easier than with the methods described throughout this chapter (see Ghiassi et al., 2013 for a review), and knowledge on text/data mining is required.

In brief, it could be said that although the possibility of using the trendiest social platforms in our research sounds exciting, one should not forget about how and why people use those media. While for marketing implications they are being successfully exploited, the applications to product development and optimisation are not currently clear and should still be explored in the future.

In any case, hopefully in the future, software packages will be developed to help even the non tech-savvy sensory researcher to extract valuable information from open-ended questions (and if the task is recorded orally, why not also transcribe it accurately) without missing the meaning of the expressions. Unfortunately, this idea is currently utopian, but there are fast advances being made in the area of computer intelligence regarding text and data mining, so time will tell.

Other ways to look at the future of open-ended questions in our field is by casting our attention over other unobtrusive measures that, in combination, would help explain what consumers are perceiving when those descriptors or association come to their minds. For example, Piqueras-Fiszman et al. (2013) used eye-tracking in what probably is the first study combining these two spontaneous techniques, demonstrating the additional benefits of combining them. Also, tablets are being more and more used to collect data from consumers in the marketing field. In the rapid sensory methodologies described here they would be useful since they include voice recognition (ideal to collect the responses orally) and could easily combine it with other computerised tasks, as they are currently being performed in sensory labs.

12.7 Conclusions

Throughout this chapter the formats and the strengths and weaknesses at the different stages of methodologies based on open-ended questions have been overviewed. While the responses obtained are unique, spontaneous and verbalised by the consumers themselves, interpreting them correctly and efficiently can be challenging. At the end of the day, the decision of which method to use relies upon the research team, who should analyse the timing and budget available at each stage and focus on the aims of the study in question. In addition, the major potential applications that could benefit from the inclusion of these questions have been discussed, together with what seems to be the directions of open-ended questions in the future. From the use of social media to the combination of open-ended questions with other spontaneous techniques and technologies which would make the data collection (and results) more natural and insightful, some easier alternative ideas to approach the spontaneous thoughts of large groups of consumers have been suggested.

The question of whether open-ended questions could be considered a rapid method is somewhat complex, since several points of view could be adopted depending on the stage of the procedure one is focusing on. In the sensory and consumer research domain, open-ended questions are most commonly used with consumers who are not required to be trained and who do not need to meet any specific criteria (related to the type of question itself) to be able to participate. That said, the research team might want to count with a panel fluent in the language of the survey and who have good verbalisation skills. Regarding the task itself, it is easily understood by consumers; the responses are spontaneous and normally aim for the consumers' subjective appraisal of the products. Taken together, these characteristics make open-ended questions to be considered as a "rapid" method with which to collect consumers' unique responses. In contrast, depending on the desired depth of treatment of the raw data, the interpretation and coding can be very time-consuming, and therefore make it "slow."

In terms of its application, open-ended questions can be the main task in a study, or serve as a complement to other more structured sensory tasks. Are they more efficient than Descriptive Analysis? In the former case, the nature and purpose of the questions and responses are simply not comparable to those obtained with Descriptive Analysis. In the latter case, the efficiency of the method would depend to a great extent on the task in which the questions are being included.

References

Abbasi, A., Chen, H. and Salem, A. (2008) "Sentiment analysis in multiple languages: Feature selection for opinion classification in web forums," *ACM Transactions Infor Syst*, **26** (3), 1–34. DOI: 10.1145/1361684.1361685.

Abdi, H., Williams, L. J. and Valentin, D. (2013) "Multiple factor analysis: Principal component analysis for multitable and multiblock data sets," *Wiley Interdiscip Rev Comput Stat*, **5**, 149–179. DOI: 10.1002/wics.1246.

Ares, G., Giménez, A., Barreiro, C. and Gámbaro, A. (2010) "Use of an open-ended question to identify drivers of liking of milk desserts. Comparison with preference mapping techniques," *Food Qual Prefer*, **21**, 286–294. DOI: 10.1016/j.foodqual.2009.05.006.

Ares, G., Piqueras-Fiszman, B., Varela, P., Morant, R., Martín López, A. and Fiszman, S. (2011) "Food labels: Do consumers perceive what semiotics want to convey?" *Food Qual Prefer*, **22**, 689–698. DOI: 10.1016/j.foodqual.2011.05.006.

Bécue-Bertaut, M., Alvarez-Esteban, R. and Pages, J. (2008) "Rating of products through scores and free-text assertions: Comparing and combining both," *Food Qual Prefer*, **19**, 122–134. DOI: 10.1016/j.foodqual.2007.07.006.

Bécue-Bertaut, M. and Lebart, L. (1998) "Clustering of texts using semantic graphs. application to open-ended questions in surveys." In *Data Science, Classification, and Related Methods* (Eds. C. Hayashi, K. Yajima, H.-H. Bock, N. Ohsumi, Y. Tanaka and Y. Baba), 480–487. Springer-Verlag, Tokyo.

Bécue-Bertaut, M. and Pagès, J. (2004) "A principal axes method for comparing contingency tables: MFACT," *Comput Stat Data An*, **45**, 481–503. DOI: 10.1016/S0167-9473(03)00003-3.

Bécue-Bertaut, M. and Pagès, J. (2008) "Multiple factor analysis and clustering of a mixture of quantitative, categorical and frequency data," *Comput Stat Data An*, **52**, 3255–3268. DOI: 10.1016/j.csda.2007.09.023.

Clark, C. C. and Lawless, H. T. (1994) "Limiting response alternatives in time-intensity scaling: An examination of the halo-dumping effect," *Chem Senses*, **19**, 538–594. DOI: 10.1093/chemse/19.6.583.

Delarue, J. and Sieffermann, J.-M. (2004) "Sensory mapping using Flash profile. Comparison with a conventional descriptive method for the evaluation of the flavour of fruit dairy products," *Food Qual Prefer*, **15**, 383–392. DOI: 10.1016/S0950-3293(03)00085-5.

Dransfield, E., Morrot, G., Martin, J. F. and Ngapo, T. M. (2004) "The application of a text clustering statistical analysis to aid the interpretation of focus group interviews," *Food Qual Prefer*, **15**, 477–488. DOI: 10.1016/j.foodqual.2003.08.004.

Elmore, J. R., Heymann, H., Johnson, J. and Hewett, J. E. (1999) "Preference mapping: Relating acceptance of 'creaminess' to a descriptive sensory map of a semi-solid," *Food Qual Prefer*, **10**, 465–475. DOI: 10.1016/S0950-3293(99)00046-4.

Escofier, B. and Pagès, J. (1994) "Multiple factor analysis (AFMULT package)," *Comput Stat Data An*, **18**, 121–140. DOI: 10.1016/0167-9473(94)90135-X.

Faye, P., Brémaud, D., Teillet, E., Courcoux, P., Giboreau, A. and Nicod, H. (2006) "An alternative to external preference mapping based on consumer perceptive mapping," *Food Qual Prefer*, **17**, 604–614. DOI: 10.1016/j.foodqual.2006.05.006.

Feldman, R. and Sanger, J. (2007) *The Text Mining Handbook: Advanced Approaches in Analyzing Unstructured Data.* Cambridge University Press, New York.

Galmarini, M. V., Symoneaux, R., Chollet, S. and Zamora, M. C. (2013) "Understanding apple consumers' expectations in terms of likes and dislikes: Use of comment analysis in a cross-cultural study," *Appetite*, **62**, 27–36. DOI: 10.1016/j.appet.2012.11.006.

Geer, J. G. (1988) "What do open-ended questions measure?" *Public Opin Quart*, **52**, 365–371. DOI: 10.1086/269113.

Ghani, R., Probst, K., Liu, Y., Krema, M. and Fano, A. (2006), "Text mining for product attribute extraction," *ACM SIGKDD Explorations Newsletter*, **8**(1), 41–48. DOI: 10.1145/1147234.1147241.

Ghiassi, M., Skinner, J. and Zimbra, D. (2013) "Twitter brand sentiment analysis: A hybrid system using n-gram analysis and dynamic artificial neural network" *Expert Syst Appl*, **40**, 6266–6282. DOI: 10.1016/j.eswa.2013.05.057.

Guerrero, L., Claret, A., Verbeke,W., Enderli, G., Zakowska-Biemans, S.,Vanhonacker, F., Issanchou, S, Sajdakowskad, M., Signe Granlie, B., Scalvedif, L., Contelf, M. and Hersleth, M. (2010) "Perception of traditional food products in six European regions using free word association," *Food Qual Prefer*, **21**, 225–233. DOI: 10.1016/j.foodqual.2009.06.003.

Husson, F., Lê, S. and Pagès, J. (2011) *Exploring Multivariate Analysis by Example Using R*. Boca Raton, CRC Press.

Jansen, B., Zhang, M., Sobel, K. and Chowdury, A. (2009) "Twitter power: Tweets as electronic word of mouth," *J Am Soc for Infor Sci Technol*, **60**, 2169–2188. DOI: 10.1002/asi.v60:11.

Kelley, S. (1983) *Interpreting Elections*. Princeton, Princeton University Press.

Kostov, B., Bécue-Bertaut, M. and Husson, F. (2014) "An original methodology for the analysis and interpretation of word-count based methods: Multiple factor analysis for contingency tables complemented by consensual words," *Food Qual Prefer*, **32A**, 35–40. DOI: 10.1016/j.foodqual.2013.06.009.

Lawless, H. T. and Clark, C. C. (1992) "Psychological biases in time-intensity scaling," *Food Technol*, **46**, 81–90.

Lawless, H. T. and Heymann, H. (2010), *Sensory Evaluation of Food. Principles and Practices*. Second Edition. Springer, New York.

Lawrence, G., Symoneaux, R., Maitre, I., Brossaud, F., Maestrojuan, M. and Mehinagic, E. (2013) "Using the Free comments method for sensory characterisation of Cabernet Franc wines: Comparison with classical profiling in a professional context," *Food Qual Prefer*, **30**, 145–155. DOI: 10.1016/j.foodqual.2013.04.005.

Lebart, L., Salem, A. and Berry, E. (1998), *Exploring Textual Data*. Dordrecht, Kluwer.

Li, N. and Wu, D. D. (2010) "Using text mining and sentiment analysis for online forums hotspot detection and forecast," *Decis Supp Syst*, **48**, 354–368.

Looker, E. D., Denton, M. A. and Davis, C. K. (1989) "Bridging the gap: Incorporating qualitative data into quantitative analyses," *Soc Sci Res*, **18**(4), 313–330. DOI: 10.1016/0049-089X(89)90011-2.

Modell, S. (2005) "Triangulation between case study and survey methods in management accounting research: An assessment of validity implications," *Manag Accounting Res*, **16**, 231–254.

Narain, C., Paterson, A. and Reid, E. (2004) "Free choice and conventional profiling of commercial black filter coffees to explore consumer perception of character," *Food Qual Prefer*, **15**, 31–41. DOI: 10.1016/S0950-3293(03)00020-X.

Pagès, J. (2005) "Collection and analysis of perceived product inter-distances using multiple factor analysis: Application to the study of 10 white wines from the Loire Valley," *Food Qual Prefer*, **16**, 642–649. DOI: 10.1016/j.foodqual.2005.01.006.

Perrin, L. and Pagès, J. (2009) "Construction of a product space from the ultra-flash profiling method: Application to 10 red wines from the Loire Valley," *J Sens Stud*, **24**, 373–395. DOI: 10.1111/j.1745-459X.2009.00216.x.

Perrin, L., Symoneaux, R., Maître, I., Asselin, C., Jourjon, F. and Pagès, J. (2008) "Comparison of three sensory methods for use with the Napping® procedure: Case of ten wines from Loire Valley," *Food Qual Prefer*, **19**, 1–11. DOI: 10.1016/j.foodqual.2007.06.005.

Piqueras-Fiszman, B., Ares, G. and Varela, P. (2011) "Semiotics and perception: Do labels convey the same messages to older and younger consumers?" *J Sens Stud*, **26**, 197–208. DOI: 10.1111/j.1745-459X.2011.00336.x.

Piqueras-Fiszman, B., Ares, G. and Varela, P. (in press), "An introduction to sensory evaluation techniques." in Nollet L and Toldrá F, *Handbook of Food Analysis. Third edition*, Boca Ratón, CRC Press.

Piqueras-Fiszman, B., Velasco, C., Salgado, A. and Spence, C. (2013) "Combining eye tracking and word association in order to relate attentional, cognitive, and affective information processing of the multisensory attributes of food packaging," *Food Qual Prefer*, **28**, 328–338. DOI: 10.1016/j.foodqual.2012.10.006.

Popper, P. and Heymann, H. (1996), "Analyzing differences among products and panelists by multidimensional scaling." In Naes T and Risvik E, *Multivariate Analysis of Data in Sensory Science,* Amsterdam, Elsevier, 159–184.

Sieffermann, J.-M. (2000) "Le profil Flash. Un outil rapide et innovant d'évaluation sensorielle descriptive." In Tec and Doc Paris, *AGORAL 2000 – XIIèmes Rencotres,* Paris, Lavoisier, 335–340.

Sorensen, H. (1984) *Consumer Taste Test Surveys. A Manual*. Corbett, Sorensen Associates.

Stone, H. and Sidel, J. L. (2004) *Sensory Evaluation Practices. Third Edition*. San Diego, Elsevier Academic.

Symoneaux, R., Galmarini, M. V. and Mehinagic, E. (2012) "Comment analysis of consumer's likes and dislikes as an alternative tool to preference mapping. A case study on apples," *Food Qual Prefer*, **24**, 59–66. DOI: 10.1016/j.foodqual.2011.08.013.

ten Kleij, F. and Musters, P. A. D. (2003) "Text analysis of open-ended survey responses: A complementary method to preference mapping," *Food Qual Prefer*, **14**, 43–52. DOI: 10.1016/S0950-3293(02)00011-3.

Varela, P. and Ares, G. (2012) "Sensory profiling, the blurred line between sensory and consumer science. A review of novel methods for product characterization," *Food Res Int*, **48**, 893–908. DOI: 10.1016/j.foodres.2012.06.037.

Varela, P., Beltrán, J. and Fiszman, S. (2014) "An alternative way to uncover drivers of coffee liking: Preference mapping based on consumers' preference ranking and open comments," *Food Qual Prefer*, **32B**, 152–159. DOI: 10.1016/j.foodqual.2013.03.004.

Williams, A. A and Langron, S. P. (1984) "The use of free choice profiling for the evaluation of commercial ports," *J Sci Food Agr*, **35**, 558–568. DOI: 10.1002/9780470385036.ch6a.

Temporal dominance of sensations (TDS) as a sensory profiling technique

13

N. Pineau[1], P. Schilch[2]
[1]Nestlé Research Center (Nestec Ltd), Lausanne, Switzerland; [2]Centre des Sciences du Goût et de l'Alimentation, Dijon, France

13.1 Introduction

Temporal dominance of sensations (TDS) is a relatively recent methodology in the sensory field that gives the opportunity to describe the evolution of the dominant sensory perceptions during the tasting of a food product. Practically, several subjects are asked to identify continuously over a given period of time the dominant perception among a list of sensory attributes. A TDS evaluation results in a sequence of dominant perceptions over the tasting period (e.g. from intake to swallowing of a piece of chocolate).

At a very early stage, Ep Köster, back in 1999 at CSGA (Centre de Sciences du Gout et de l'Alimentation, Dijon, France), dreamed about a "harmonium of sensations". He had in mind something like a piano on which each key would be a sensory attribute and, like a pianist, the sensory subject could play the melody of the product. This idea was somehow a response to a thought from Rose Mary Pangborn stated 40 years ago: "If an appropriate method could be developed, a modification of the so-called time-intensity study (Neilson, 1958) could be used to establish the temporal and sequential changes in apparent taste intensity of both compounds in a mixture" (Pangborn and Chrisp,1964).

A first attempt to build and test such a harmonium was made at CSGA in partnership with Fromagerie Bel and Danone in the early 2000s, but it was very difficult for the subjects to play with several keys/sensory attributes simultaneously. The use of the harmonium was basically reduced to one key at a time, like a pianist playing only with one finger. It was therefore decided to simplify the method by displaying the keys/sensory attributes on a computer screen and ask the subject to successively select the dominant perception: TDS was born!

The first visualization and analysis of TDS data were presented 3 years later at the Pangborn symposium (Pineau *et al.*, 2003) and started to be used by several companies. This method is nowadays well established in the sensory domain and has been successfully applied to many product categories.

This chapter first introduces the TDS method and its specificities. The focus is then on how properly to set a TDS experiment and train the panel, before digging into the analysis part. The chapter ends with some examples of applications that have contributed to the development of the method, and with the most recent innovations around TDS and alternative dynamic methods.

13.2 Overview of temporal dominance of sensations (TDS)

TDS is a sensory methodology aiming at recording the evolution of the dominant sensory perceptions of a product along the tasting period. The TDS computerized system shows the subject the entire list of attributes on a computer screen (Fig. 13.1). The subject is asked to click on the start button as soon as the product is into his/her mouth, then to consider which of the attributes is perceived as dominant and to select it on the screen. Thereafter, each time the subject feels the dominant perception has changed, he/she selects the new dominant attribute, until the perception ends. Along the testing of one product, the subject is free to select an attribute several times. Conversely, another attribute may not be selected at all. During the tasting, the computer records the sequence of dominant attributes.

To summarize TDS data and get a descriptive picture of each product, the most common representation is the TDS curve. The procedure (Section 13.4.2) considers each attribute separately. For each point of time, the proportion of evaluations (subject × replication) for which the given attribute was assessed as dominant is computed. These proportions are smoothed over time (e.g. using moving average or a more sophisticated algorithm such as SAS/TRANSREG) and displayed as curves of the evolution of the dominance rate for each attribute. In the example of Fig. 13.2, the

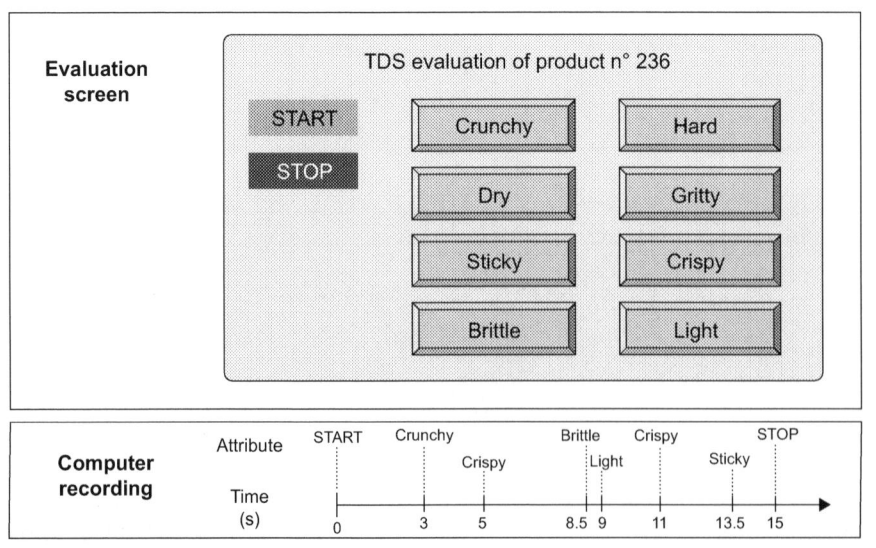

Figure 13.1 Example of TDS computer screen with buttons.

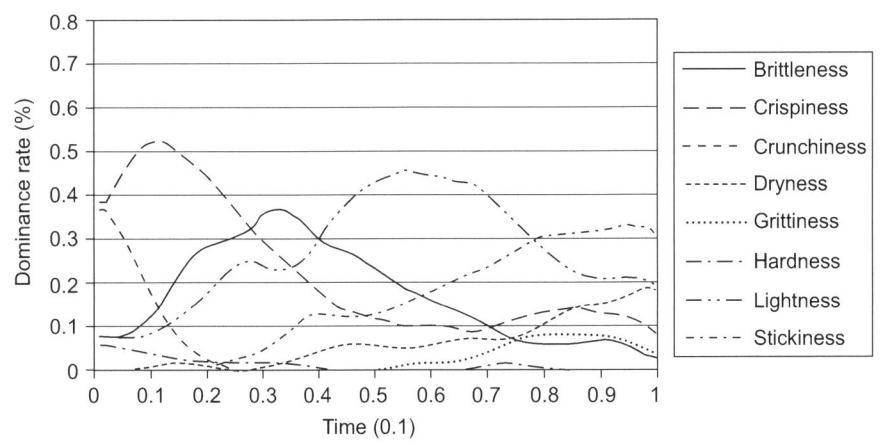

Figure 13.2 Example of a set of TDS curves on breakfast cereal products.
Source: Adapted from Lenfant *et al.* (2009).

product can be described as dominated by the crispy sensation at the beginning of the sequence, then brittle, light and finally sticky.

TDS is often compared to Time–Intensity (TI), the historical method for dynamic sensory measures, because both are time related measures. But these two methods actually do not target the same needs. TI is dedicated to the evolution of the intensity of one single sensory perception over time (e.g. the sweet intensity of an intense sweetener over time), whereas TDS is a multi-attribute method aimed at evidencing the sequence of dominant perceptions along tasting. Therefore, results from both methods cannot strictly be compared and the choice of the methodology depends on the question to be addressed (TI: quantitative evaluation of one sensory perception; TDS: evolution of the dominant quality/sensory perception along tasting) rather than on the superiority of one method versus another. In some cases, these methods can be used conjointly to characterize the different aspects of the products and evaluate if both temporal information give consistent results (see Section 13.8.2).

Among the temporal methods, TDS can be classified as a rapid method since it has the possibility to record temporal information on several sensory attributes during the same evaluation, whereas with TI, as many evaluations as the number of attributes are needed (i.e. potentially ten times more evaluations if considering ten attributes). The nature of collected information is however different (TI: intensity profile over time; TDS: selection of the dominant attribute over time).

In comparison with Descriptive Analysis, TDS can also be seen as a rapid method because it does not require any training step on the scoring of the intensity on a scale (still some exceptions, see Section 13.3.3). Indeed, TDS focuses on the choice of a quality/attribute among a list and not on the evaluation of the intensity of each attribute. This can save several training sessions and therefore accelerate the evaluation process.

13.3 TDS experiment and panel training

This section introduces the concept of dominance and details the most important steps to correctly set a TDS experiment, from the definition of the protocol to the training of the panel and the collection of measurement data.

13.3.1 Concept of dominance

The TDS methodology entirely relies on the capacity of the subjects to select the dominant perception among a list of sensory attributes. The definition of a dominant attribute is therefore naturally a key element. In the literature, several definitions have been given. It was either defined as the sensation "popping-up" (Pineau et al., 2009), or the sensation that "triggers the most your attention" (Le Reverend et al., 2008; Lenfant et al., 2009) or "the most intense" sensation (Labbe et al., 2009). In a recent paper, Saint-Eve et al. (2011) compared these three definitions and concluded that "the most intense" refers only to one aspect of the perception (intensity) whereas "popping-up" and "triggers the most your attention" had a broader scope (intensity, but also sensory contrasts along tasting), which better corresponds to the notion of dominance. From the panel performance perspective, the use of "the most intense" definition tends to evidence larger difference between products, but the same differences are still evidenced using the other definitions and the sequences of perceptions captured are richer.

From a general point of view, we can see that "the most intense" definition should be used when the focus is only the intensity, but in the majority of the cases a broader definition like "popping-up" or "triggers the most your attention" is more suitable. Practically, the concept of dominance can be introduced as such:

- The dominant attribute is the one which triggers the most your attention at a given time.
- It is likely to be the one with a rising intensity, allowing you to suddenly perceive it.
- But the dominant attribute does not have to be very or the most intense in the product.

To illustrate this definition, one can think about a piece of chocolate with the rising perception of an aroma (say coffee aroma) after several seconds. It is possible that the coffee aroma is not as strong as the chocolate aroma, but coffee aroma is still perceived as "dominant" because this is the sensory sensation that triggers most of your attention.

During panel training, this concept can also be illustrated through sounds, the attributes being the different instruments of a band, as presented by Lannuzel and Rogeaux (2007). For the subjects, it is generally easy to understand that the most triggering/dominant perception is not always the loudest one, but the one bringing the biggest change in the melody.

13.3.2 Attribute list

The selection of the attributes is critical and more important than for the Descriptive Analysis because the attributes are not evaluated independently (one by one).

Throughout a TDS evaluation, subjects have to continuously make a choice between several attributes to determine the sequence of dominant sensations. This decision task imposes that the attribute list is exhaustive (all potential dominant sensory perceptions must be there) and remains short enough to be handled by the subjects. Based on historical data on 21 TDS studies, Pineau *et al.* (2012) showed that long attribute lists (> 12 terms) do not seem to be used in an optimal way by the subjects, since all the attributes are not used (Fig. 13.3). It is therefore recommended to keep a relatively short list of attributes, say about ten, to ensure good data quality. The authors also emphasize spending enough time during the training on the definition and the validation of the attribute list with the panel, to ensure good reactivity of the subjects during the evaluation.

To determine the appropriate list of attributes, the glossary usually used for Descriptive Analysis can be used as a starting point to initiate the discussion with the subjects, but the attribute list for TDS is almost always different from the one used for Descriptive Analysis (at least, generally shorter). If the TDS evaluation is split into several phases (e.g. one TDS before swallowing and one TDS after swallowing), the attribute list can be different for each phase as long as there is no interest in the direct comparison of the TDS profiles across phases.

Saint-Eve *et al.* (2011) investigated the question of the quality of the attribute list by proposing to different panels with the same level of training either an optimal list of nine attributes (based on previous work on the same product area), or a reduced list of seven attributes with important missing attributes, or an extended list of 11 attributes with two additional but non-essential terms. Results showed that the use of a list missing dominant attributes led to a biased description of the products. Subjects missing an attribute have actually no other choice than reporting their selection on another term, which dramatically modifies the sequence of perceptions. On the other hand, it seems it does not harm to have some additional attributes since they were almost not used.

For the panel leader, it means it is important to spend enough time on the selection and the definition of the attributes, so that the subjects are very comfortable in identifying the sensory perceptions and can be quick in selecting the sensation they perceive as dominant during the evaluation. In this way, the number of attributes should also not be too large, say about ten as mentioned before.

13.3.3 Type of scale

For historical reasons, intensity has been generally recorded in the first TDS studies because the first TDS software (developed by Fizz, Biosystemes) was based on an evolution of a Descriptive Analysis module. This is, however, not recommended because it mixes up two different cognitive processes: the selection of a dominant attribute (qualitative task) and the intensity scoring (quantitative task). The use of buttons is therefore better aligned with the primary task of the subject, to identify the dominant attribute/sensation. The use of buttons is an easier alternative and was proven to be as efficient as the historical solution and simpler to handle by the subjects (Saint-Eve *et al.*, 2011). Most of the time, it is therefore more suitable to use buttons rather than intensity scales.

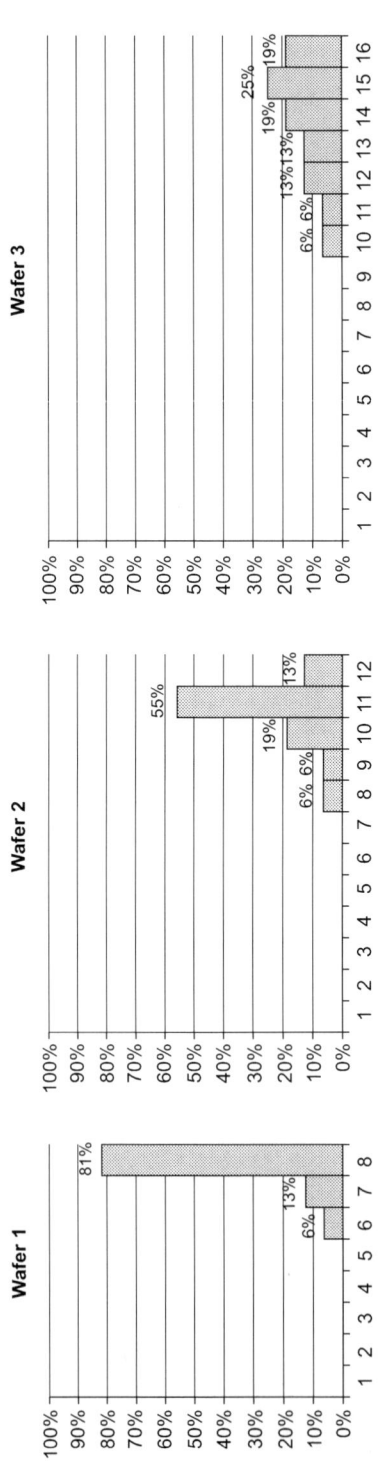

Figure 13.3 Distribution of the number of attributes selected at least once across all product evaluations. The percentage of subjects using at least 90% of the attributes in the list decreases from 81% (with a list of eight attributes) to 69% (list of 12 attributes) and finally to 44% (list of 16 attributes).

Source: From Pineau *et al.* (2012).

13.3.4 Products

The number of products to be used in a TDS session does not differ from regular (say Descriptive Analysis) sessions. Factors to account for are the same as for any other sensory methodology (see ISO norm 8589 for more details).

13.3.5 Tasting protocol

In order to ensure good reliability of the collected data and homogeneity among subject responses, it is worth defining a tasting protocol according to the product to be tested. In every case, one TDS evaluation of a food product corresponds to the eating of one spoonful, mouthful or sip of the product. For a solid food product it can be simply to ask the subject to record dominance from intake to swallowing (Pineau et al., 2009 on cheese, Lenfant et al., 2009 on breakfast cereals). If the aftertaste period is also of interest, it is possible to run one session with an instruction about the time to swallow. It is also possible to run a second TDS evaluation focusing on the aftertaste right after the tasting period. This second approach facilitates the analysis of both periods separately. For beverages, protocols are generally more elaborated. For example, Pessina (2006) proposed a detailed tasting protocol for the evaluation of wine (Table 13.1) to ensure good alignment across subjects. TDS has also been used to study the dominance of sounds in cars. In this case, the protocol was simply to listen to the sound sample but the sound samples were recorded in the same way to get comparable TDS profiles (hand break, door clapping, gear sounds at the same time in the different sequences).

Additionally, Pineau et al. (2012) observed that subjects can be biased by the order of the attributes in the list: attributes at the top of the list tended to be selected earlier in the evaluation than attributes at the bottom of the list. So, even if this attribute position effect is generally small compared to the differences among attributes, it is advised to balance the attribute order across subjects to reduce the potential bias due to the attribute position. In this way, each subject will have a different attribute order from the other subjects, but the attribute order will remain the same for a given subject for all product evaluations to facilitate the learning of the list and the reaction time to select the dominant attribute during evaluations.

Table 13.1 Protocol for the sensory evaluation of one sip of wine

Time (s)	Instruction
0	Put the wine (7.5 mL) in mouth and start to score
12	Inhale air three times and continue to score
20	Move the wine in mouth three times and swallow the wine
50	Stop to score

Source: From Pessina (2006).

13.3.6 Panel size

Since the TDS method aims at describing rather complex phenomena (sequence of dominant perceptions over time), published papers (Table 13.2) have generally used more subjects and replicates than for Descriptive Analysis (ISO 8586: 12 subjects, no replicate).

Pineau *et al.* (2009) investigated the question of the number of subjects by using three TDS datasets with large number of evaluations (data1: 43 evaluations = 43 different subjects; data2: 50 evaluations = 25 subjects × 2 replicates; data3: 48 evaluations = 16 subjects × 3 replicates) and simulating results for reduced panels. This study showed that a reduction to about 30 evaluations gives similar conclusions as the large dataset, but this number should slightly increase when the number of subjects decreases (and is compensated by a higher number of replicates). As a rule of thumb, it is therefore proposed to refer to the values presented in Table 13.3 to define the number of subjects and replicates to be used in TDS studies.

13.3.7 Panel training

The panel training for TDS has several specificities compared to other methods. First, the list of attributes is a critical point (see Section 13.3.2) so it is important to spend enough time on the selection of the attributes. Second, a TDS evaluation is generally not about the rating of a score for a selection of attributes, but it is a task choice. It is therefore important to focus on the differentiation between attributes (quality of the perceptions) rather than on the intensity of the perception for each attribute (quantitative measure). As proposed by Lannuzel and Rogeaux (2007), the concept of dominance can also be illustrated through sounds, which also has the advantage of limiting subject fatigue compared to food product tasting. Third, the subjects should be trained on the computerized system to feel comfortable with the use of the list. Last, to ensure the panel is adequately trained before the measurement sessions, the performance should be evaluated. Several publications propose methodologies to evaluate product discrimination, but only a few investigate specifically the question of the performances and propose methodologies to assess performance at different levels (discrimination and agreement, at panel or subject levels, considering the whole sequence or per time point, etc.). This broad and important topic will be detailed specifically at the end of the data analysis section in this chapter (Section 3.4).

Saint-Eve *et al.* (2011) investigated the effect of the number of training sessions on panel performance to discriminate the products. They evidenced that longer training on the definition/identification of the attributes (from three to five sessions of 1 h) increases product discrimination and richness of the dominance sequence, but that three sessions were already enough to catch major differences. So, as for any sensory method, a minimum number of training sessions is required, but the longer the training, the better the panel performance and the richer the results.

Table 13.2 **Non-exhaustive list of TDS publications on applications and corresponding numbers of attributes, products, subjects, replicates and evaluations per product**

Reference	Attributes	Products	Panellists	Replicates	Evaluation per product
Pessina et al. (2005)	9	5	30	2	60
Le Révérend et al. (2008)	6	6	12	2	24
Labbe et al. (2009)	5	9	43	1	43
Lenfant et al. (2009)	8	6	25	2	50
Meillon et al. (2009)	10	8	16	3	48
Pineau et al. (2009)	10	5	16	4	64
Meillon et al. (2010)	8	5	8	3	24
Teillet et al. (2010)	8	13	15	3	48
Barron et al. (2012)	11	7	16	2	32
Dinnella et al. (2012)	4 – 8	10	13	3	39
Ng et al. (2012)	9	11	11	3	33
Albert et al. (2012)	7	6	9	3	27
Bouteille et al. (2013)	11	13	16	2	32
Bruzzone et al. (2013)	8	8	10	2	20
Laguna et al. (2013)	7	6	13	3	39
Paulsen et al. (2013)	8	6	9	3	27
Hutchings et al. (2014)	9	4	20	3	60
Rosenthal and Share (2014)	6	3	15	2	30
Varela et al. (2014)	6	6	14	3	42

Table 13.3 **Proposal for minimum number of panellists required according to the number of replicates**

Number of replicates	Minimum number of panellists	Minimum number of evaluations
1	30	30
2	16	32
3	12	36
4	10	40

13.3.8 Panel behaviour

In order to build knowledge about the way the subjects use the attribute list during a TDS experiment, a database was built in collaboration between the Nestlé Research Center (Lausanne, Switzerland) and CSGA (Dijon, France) to collect TDS data representative of various conditions. The database currently contains 21 very different studies in terms of number and types of products, number and types of attributes and panel size. The analysis of these data (Pineau *et al.*, 2012) improved the knowledge of subject behaviour during TDS experiments, which could be useful in giving anchor points to the panel leader starting with this methodology.

Individual evaluations generally last between 20 and 40 s, depending on the type of product (products tasted in mouth and swallowed). For medium (> 5) to long (> 10) attributes lists, subjects are switched from one dominant attribute to another on average every 5.5 s, whatever the duration of the product evaluation. For short attributes lists (< 5), the average duration of attribute selection increased around 8 s. For TDS studies lasting several minutes (e.g. when investigating long lasting sensations), the time between two consecutive selections is much longer. The number of different attributes selected during one evaluation to characterize the sequence of dominant perceptions was around 4. And surprisingly, this number was relatively stable, regardless of the number of attributes in the list.

13.3.9 Data acquisition

Today, almost all sensory software programs offer the possibility to collect TDS data, and most have functionalities for data analyses. If it is not possible to have access to any of them, which could happen if the panel leader is out of his/her facilities, it is still possible to do TDS "by hand" simply by recording on a paper the sequence of the attributes (or codes related to each attribute to right them faster) and the time at which the dominance occurs using a stop watch. However, there are now many solutions to do "deported" TDS, i.e. sessions that can be run at home on a computer with an internet connection, on a tablet or a smartphone (e.g. with TimeSens, Fizz, EyeQuestion or Compusens).

13.4 Data analysis: representation of the sequence

This section describes how to visualize TDS data through curves of the evolution of the dominant sensations over time.

13.4.1 Sequence at individual level

The raw data collected by the acquisition tool are sequences of attributes selected over the tasting period, eventually with associated intensities. For each evaluation, this sequence can be represented as a horizontal bar with segments of different lengths to show the succession of selection of the dominant attributes. Figure 13.4 also displays how the data can be coded into a matrix. These individual dominance duration sequences can be visually compared across subjects or products.

13.4.2 Sequence at panel level: the TDS curves

To summarize TDS data and get a descriptive picture of each product, the most common representation is the TDS curve.

The procedure (Fig. 13.5) considers each attribute separately. For each point of time, the proportion of evaluations (subject × replication) for which the given attribute was assessed as dominant is computed. These proportions are smoothed over time (e.g. using moving average or more sophisticated algorithm like spline regression) and displayed as curves of the evolution of the dominance rate for each attribute. Numerically, this computation can also be seen as a simple averaging of the individual unfolded matrices (bottom part of Fig. 13.4) over the panel, which results in an attribute × time point matrix with values between 0% and 100% in each cell. In the dummy example of Fig. 13.5, the product can be described as dominated by the pasty perception at the beginning of the sequence, then sweet, sticky and sweet again.

13.4.3 For and against standardized TDS curves

Considering raw TDS data, the horizontal axis of a chart with TDS curves represents the time along the tasting in seconds. For a given attribute curve, high dominance proportions indicate a consensus among subjects to quote a specific attribute as the dominant one at a given time. However, an issue can arise when the tasting periods are different across subjects. Lenfant et al. (2009) explained that, typically for products that need to be masticated, the mastication behaviour, and in particular mastication duration until first swallowing, differed from one subject to the other, so time scales of sensory perceptions differed as well. In order to align mastication duration across subjects for the computation of the TDS curves, it is proposed to standardize the data of each subject according to individual mastication durations. After standardization, the x-axis of the TDS curves did not represent real time anymore but the mastication period from first scoring ($x = 0$) to swallowing ($x = 1$), as illustrated in Fig. 13.6.

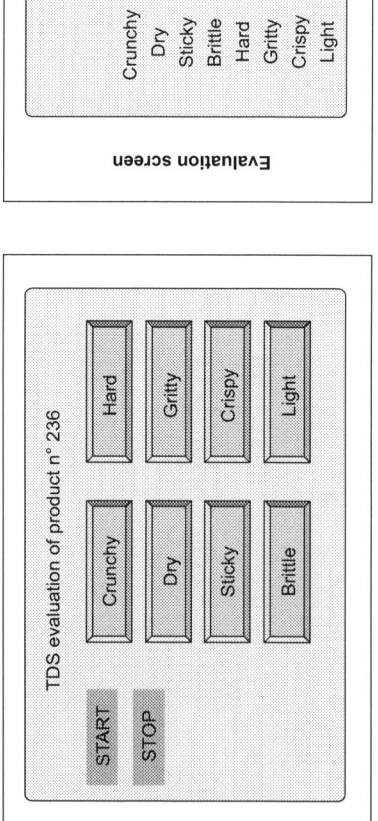

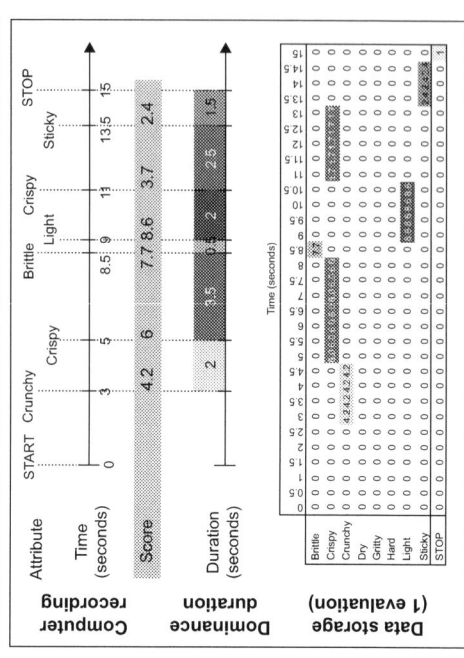

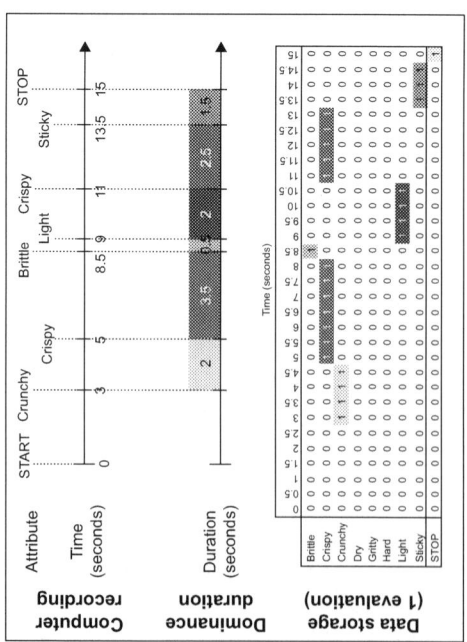

Figure 13.4 Handling of TDS data without or with intensity scores.

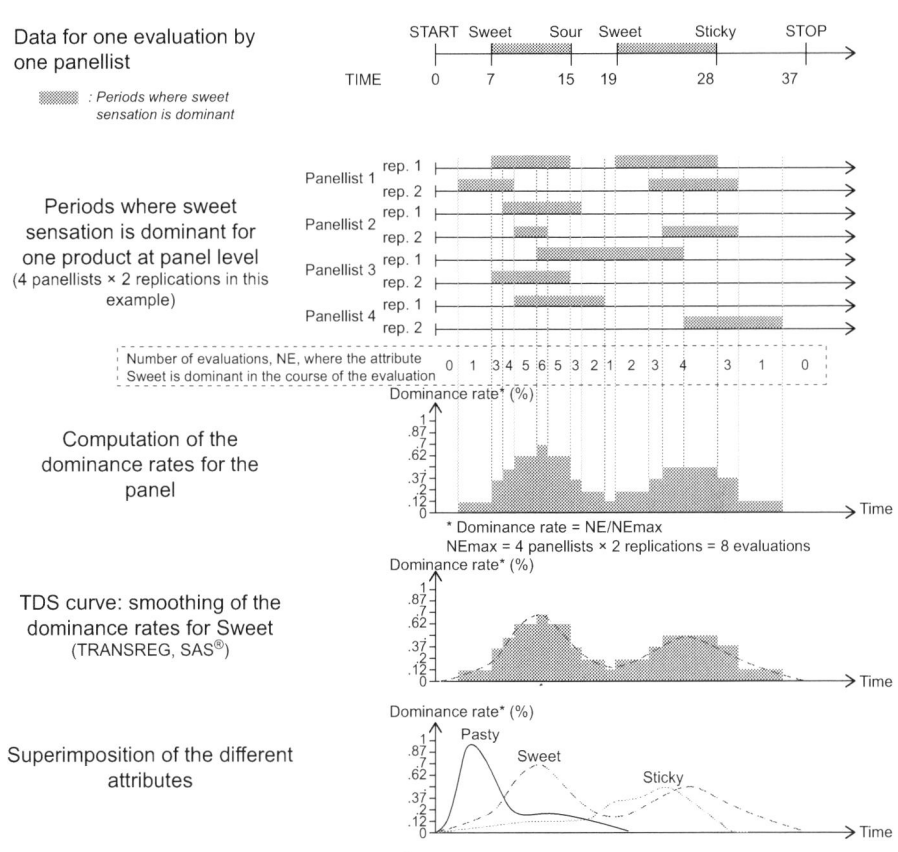

Temporal dominance of sensations (TDS) as a sensory profiling technique 281

Figure 13.5 Data processing to build TDS curves.
Source: From Pineau *et al.* (2009).

The authors justify the necessity for standardization as follows: "When data are not standardized, summing the two subjects data does not lead to any consensus about the perception A3 because there is no overlap between the times for which this attribute is perceived as being dominant for the two subjects. But standardization, by taking into account the duration until first swallowing for each subject, shows that the two subjects agree on the period where A3 is perceived as dominant, i.e. the end of the mastication." Standardized and non-standardized TDS curves are then compared for six wheat flakes. Figure 13.7 displays the corresponding TDS curves for one of the products and show higher levels of consensus with standardized data. For example, for crispiness maximum dominance rate moves from 40% to 53% when data are standardized. We can also see that it was difficult to interpret the non-standardized TDS curve for stickiness, whereas the standardized curves clearly evidence a rising of the stickiness at the end of the mastication.

However, by standardizing the data, it is not possible to have access to the real time anymore. To overcome this issue, Lenfant *et al.* (2009) proposed to analyse specifically the duration of the product evaluations (see Section 3.3.3).

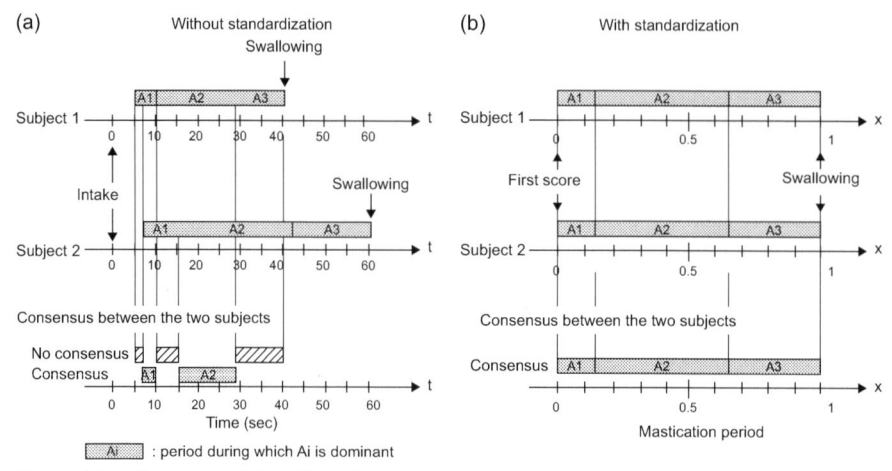

Figure 13.6 Example of the effect of data standardization through the comparison of TDS sequences between two subjects for data without standardization (a) and data with standardization (b) to account for the mastication time duration, considering three attributes: A1, A2 and A3.
Source: From Lenfant *et al.* (2009).

The choice of the representation (standardized or not) mainly depends on the protocol of the study. If the tasting protocol is precise (e.g. Table 13.1), with several anchor points referring to specific actions (intake, inhale air, move the product, swallow), data standardization is of less use because the protocol "naturally" standardizes the data. If the period of time of an evaluation can vary greatly, because products or subjects can induce different tasting durations (typically when mastication is involved), data standardization is recommended.

For the sake of simplicity, the next paragraphs of the data analysis section (Section 13.5) will consider only standardized data, even though the analyses presented can generally be applied to both kinds. The reader can refer to the corresponding publications for more details.

13.4.4 TDS curves and significance limit

In order to get more insight into the TDS graphical display, Pineau *et al.* (2009) propose to draw two additional lines on the TDS curves. The first one, called "chance level" (dotted line in Fig. 13.8), is the dominance rate that an attribute can obtain by chance. Its value, p, is equal to $1/n_a$, n_a being the number of attributes. The second one, called "significance level" (plain line in Fig. 13.8), is the minimum value for this proportion to be considered as significantly higher than p. It is calculated using a binomial test, with p the success rate by chance, n the number of trials, i.e. the number of evaluations (subject × replications) and α the risk to wrongly detect a proportion as significant (generally 5%). As an example, assuming a TDS experiment with a list of eight attributes (n_a), i.e. $p = 1/8$, and 25 subjects evaluating each product twice

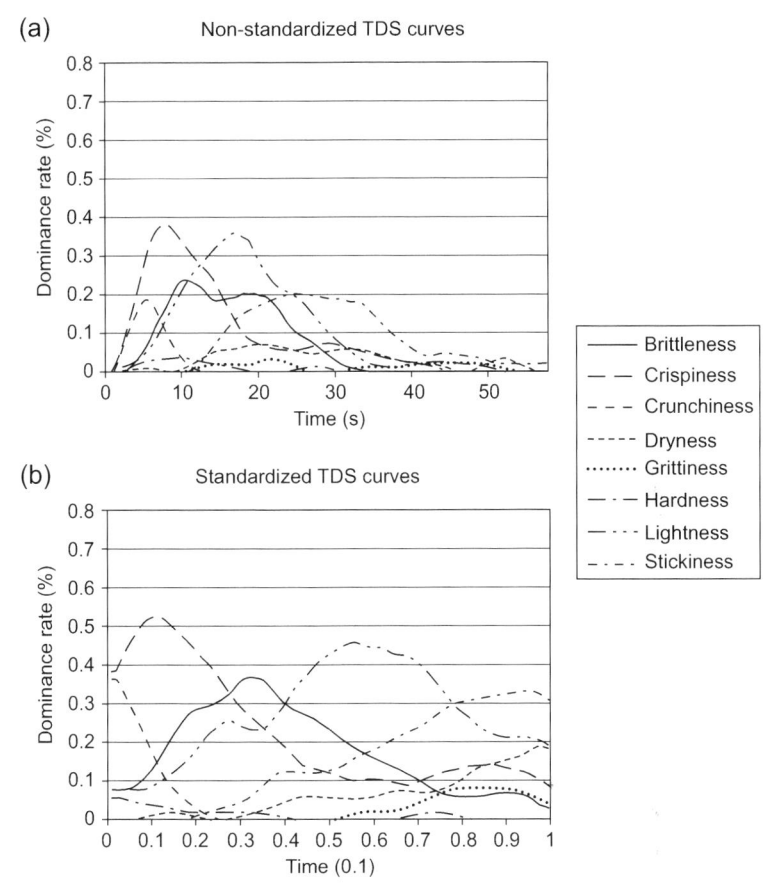

Figure 13.7 (a) Non-standardized vs (b) standardized TDS curves for texture evaluation of a breakfast cereal sample.

($n = 25 \times 2 = 50$), the minimum number of successes to be significantly above p for an α risk level of 5% is 11 (result from the binomial test). So in this case, the significance limit is set to 11/50 = 0.22. Practically, it means that dominance rates above 22% can be considered as significant. On the Fig. 13.8, we can therefore conclude that the product is significantly crispy and crunchy at first, then brittle, light and sticky before swallowing.

A normal approximation of the test can be used if $np(1 - p)$ is greater than 5 (Rosner, 1995). For example, considering again an attribute list of ten attributes, i.e. $p = 1/10$, the minimum number of observations should be $n = 5/(0.1 \times (1 - 0.1)) = 56$. In this case, the significance limit can be calculated according to Equation [13.1]:

$$P_s = P_0 + 1.645\sqrt{\frac{P_0(1-P_0)}{n}} \quad [13.1]$$

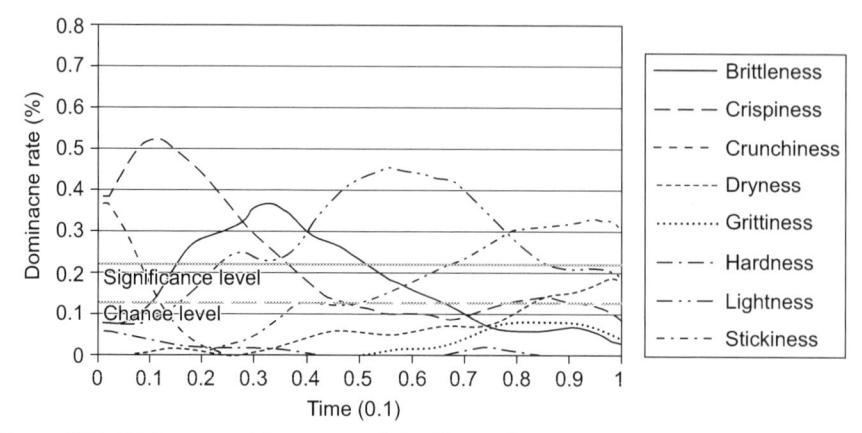

Figure 13.8 TDS curves with chance and significance levels.

where P_s: lowest significant proportion value ($a = 0.05$) at any point in time for a TDS curve, n: number of subject × replication.

Assuming a panel of 56 subjects evaluating products according to an attribute list of ten terms, the results of the calculation of the significance limit would be respectively about 18% and 17% for the binomial and normal approximation approaches, which are very close results.

13.5 Data analysis: representation of the product space

This section presents several ways of representing the product space accounting for the temporal nature of the TDS signal.

13.5.1 Individual parameters

To summarize TDS information, it is possible to extract parameters from individual evaluations to get an overview of the product space. The simplest parameter consists in recording the number of selection of each attribute for each product, whatever the time at which it is selected. Data can be collected in a product × attribute contingency table in which each cell contains the selection counts. A correspondence analysis can then be done based on this matrix to map the products and the attributes, and visualize their level of association. An alternative analysis consists in running a principal component analysis (PCA) based on the covariance matrix. In this case, the selection counts are simply considered as an indicator of the amplitude of the dominance of an attribute for a given product, and the biplot provides a descriptive view of the products' positioning in the attributes' space according to their dominance. Neither mapping proposed above accounts for the temporal nature of TDS data, but each provides a global/integrated and comparative view of the products. These

Temporal dominance of sensations (TDS) as a sensory profiling technique

maps can also be visually compared with the PCA from the results of a Descriptive Analysis to evidence similarity/differences between the approaches. Even though TDS and Descriptive Analysis target different needs/aspects, it is interesting to understand if the product structure remains the same or if some major differences are highlighted.

13.5.2 Parameters from TDS curves

Parameters can also be directly extracted from the TDS curves. Pineau *et al.* (2009) used three parameters in order to compare TDS curves to TI curves:

- V_{max}: maximum value for dominance rate (or average intensity for the TI curves)
- T_{max}: time to reach V_{max} from the beginning of the tasting
- D_{max}: duration period when dominance rate (or intensity) is higher than $0.9 \times V_{max}$.

Many other parameters used for TI curves or adapted from TI curves were also used for TDS curves (Fig. 13.9), such as the area under the curve (AUC), the AUC above the significance limit (area above significance: AAS), the first (respectively the last) time at which an attribute reaches the significance limit (Tfirst and Tlast) and the duration of significant dominance (Duration).

For each parameter, it is possible to gather data into a product × attribute matrix and run a PCA to display the product configuration and the relation between the attributes. The choice between a correlation and a covariance PCA depends on the parameter and the interpretation of the data. For instance, the AUC parameter represents somehow the strength of the consensus between the subjects to select one attribute as dominant. Attributes with high variability in AUC values among products are key drivers of product differences. Attributes with low variability do not evidence any difference between products, and generally correspond to seldom used attributes. Therefore, it is more likely that the user wants to give more weight to the attributes with more variability and will use a covariance PCA. For other parameters, like Tfirst, it is often observed that attributes more likely to appear at the beginning of a sequence (e.g. crispy when evaluating a wafer) have smaller variability than attributes more likely to appear later

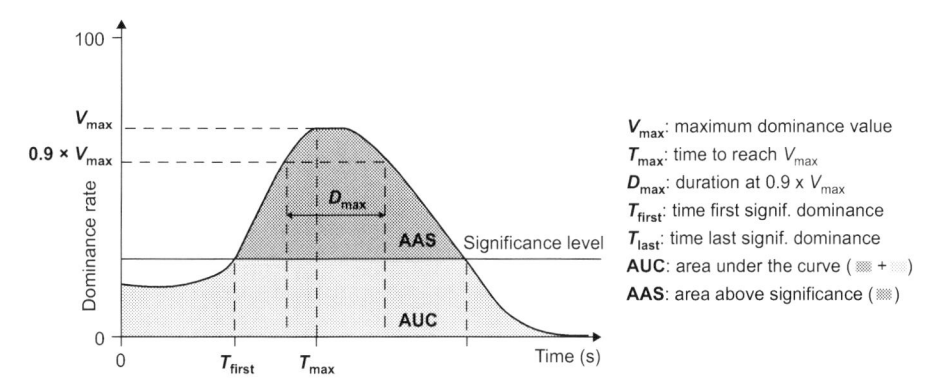

Figure 13.9 Potential parameters that can be extracted from TDS curves.

(e.g. sticky, still for a wafer). This might result in differences of parameter variability among attributes because of this "time position" effect. In this case, the user will be more likely to use a correlation PCA to give equal weight to the attributes.

13.5.3 TDS trajectory mapping

To visualize the evolution of the product sequences in the sensory space (called trajectory), Lenfant et al. (2009) proposed to run covariance PCA based on the dominance rates at different time points. The variables are the sensory attributes. The individuals are the samples at 11 equally spaced time points in order to have time points representing 0%, 10%, 20%, ..., 100% of the mastication period. The observations are the dominance rates. Since this study was based on six products and eight attributes, the data matrix for the PCA contains 66 individuals (6 × 11) and eight variables. The PCA biplot (based on the first two principal components) displays the 66 individuals and the eight variables. Sensory trajectories are shown by linking the 11 points of time corresponding to a same product: the first score (t0) is the beginning of a sensory trajectory and the end point (t100) corresponds to the last score before swallowing (Fig. 13.10).

On this example focusing on the texture of the products, the trajectory PCA exhibits both a global pattern of the product category and specificities among products. The global pattern can be visualized on the first axis, showing all products going from left to right, i.e. starting either hard, crackly or crispy, and finishing either light, sticky or gritty. This result indicates that, for this product category, there is a natural hierarchy of the perceptions ensuring that crackly, if perceived as dominant, will always be perceived at the beginning, whereas sticky will always be perceived at the end of the

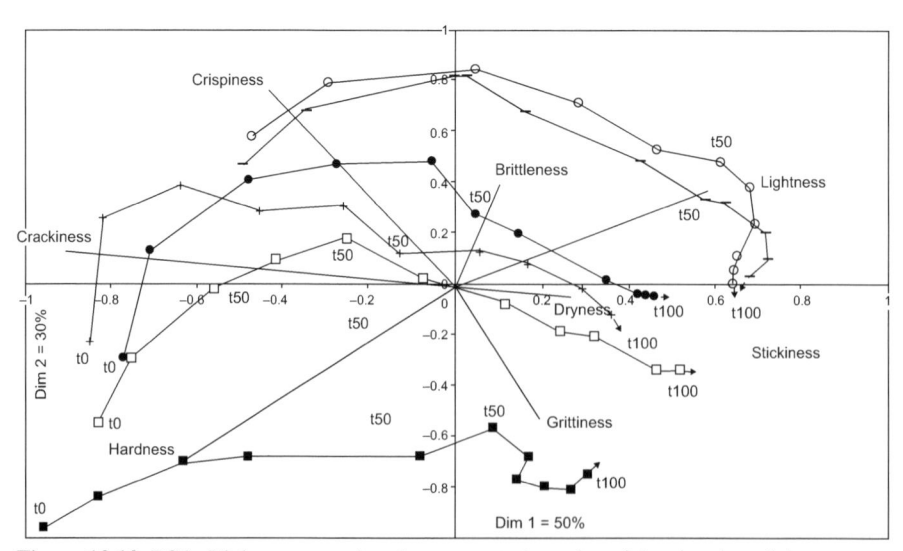

Figure 13.10 PCA. Biplot representing the texture trajectories of the six wheat flakes over the mastication period (□Wheat flakes A, ■ Wheat flakes B, – Wheat flakes C, ○ Wheat flakes D, ● Wheat flakes E and + Wheat flakes F).

mastication period. The specificities among products are evidenced on the second axis, showing three groups of products. Wheat flakes B is dominated by hardness at the beginning and grittiness at the end, which is not the case for the other products. In the middle, wheat flakes A, E and F start hard and crackly, then turn into crispy dominance and end in the sticky direction. Wheat flakes C and D have a crispy dominance right after the intake that evolves into brittleness and lightness and finally into some stickiness before swallowing.

This methodology was successfully applied in this context focusing on the texture of the products. It is also possible to use it with flavour or aroma attributes but the patterns were less obvious than for texture, making the trajectory map less easy to read.

13.6 Data analysis: comparison between products

This section introduces two approaches to compare products based on TDS data.

13.6.1 TDS difference curves

To compare the sequences of two products, it is possible to draw the difference of the TDS curves for these two products. As described by Pineau *et al.* (2009), it consists in displaying the p attribute curves of the differences of the dominance rates. These differences are plotted only when significantly different from zero, highlighting the differences between the products over time. The significance of the difference between two proportions is calculated for each time point and attribute using a McNemar test. If the number of observations is large enough (see Section 13.4.4), the normal approximation of the comparison of two binomial proportions can also be used (Kiemele *et al.*, 1995), as displayed in Equation [13.2]:

$$P_{t\,\text{diff}} = 1.96\sqrt{(1/n_1 + 1/n_2)P_{\text{moy}\,t}\left(1 - P_{\text{moy}\,t}\right)} \quad \text{where}$$

$$P_{\text{moy}\,t} = \frac{P_{1t}n_1 + P_{2t}n_2}{n_1 + n_2}$$

[13.2]

where $P_{t\,\text{diff}}$ is the least significant difference ($\alpha = 0.05$) for proportion difference at time t; P_{1t} is the proportion for product 1 at time t; P_{2t} is the proportion for product 2 at time t; n_1 is the number of subjects × replication having rated product 1; n_2 is the number of subjects × replication having rated product 2.

Note that this approach does not account for any correction for multiplicity of tests even though the number of tests is very high (number of time points × number of attributes, which represents generally hundreds of tests). The results should therefore simply be considered as a descriptive view of the most probable reasons for product differences, and not as a valid inference test to prove the significance of product differences.

The example presented in Pineau *et al.* (2009) shows that it is easy to evidence the specificities of the two products (Fig. 13.11). Product FC is more pasty at the

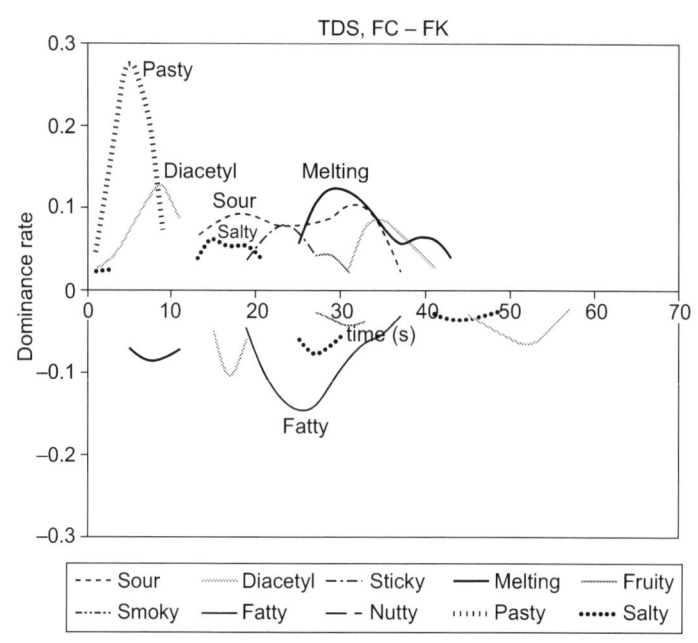

Figure 13.11 TDS graphical representation for the difference between products FC and FK. *Source*: From Pineau *et al.* (2009).

beginning, then more diacetyl, sour and melting, whereas FK is mainly perceived as fattier.

To get an idea of the specificity of one product compared to the other products of the test, it is also possible to calculate an average TDS profile of the other products by calculating the mean of the dominance proportion for each attribute × time point over the products, and using this average TDS profile to compare the remaining product to the others. As for the difference between two products, the curves of the difference will exhibit the specificities of one product versus the other ones.

13.6.2 Difference testing (randomization test approach)

Analyses presented earlier in this chapter aimed at giving a descriptive view of the TDS data and evidencing the main differences among products. In order to provide a framework to run statistical tests to compare the overall product sequences, and also to compare specific attributes or specific time points, Meyners and Pineau (2010) proposed an approach based on randomization tests. Details about this approach will not be discussed here, but the reader is encouraged to go through the original paper for more information. Shortly, this approach, as the one presented in the section above (Section 3.3.1), enables testing for the difference between two products for each attribute and time point. But, on top of this, it is also possible to test for the overall difference between two products considering all time points and attributes, and also for subsets of the full sequence, i.e. one attribute (for all time points) or one time point (for all attributes).

Temporal dominance of sensations (TDS) as a sensory profiling technique 289

To summarize the results for each pairwise comparison, the authors proposed a representation (example in Fig. 13.12) with time points in chronological order given on the horizontal axis, and attributes on the vertical axis. At each location defined by a time point and an attribute, a colour code is proposed to indicate if the dominance rates of the two products are different, and if so to indicate the product with the highest dominance rate (dark grey for product 1, light grey for product 2). In addition, attribute names are followed by a circle coloured in the same way to indicate if one product has significantly higher dominance rates than the other on the whole period of time. The circle can also be half dark grey and half light grey in case there were indeed differences but in different directions (e.g. for a given attribute, product 1 has higher dominance rate at the beginning of the sequence, but lower dominance rate at the end of the sequence). An additional row is added to indicate whether the differences at a certain time point between the products were larger than the threshold. Finally, product names in the title are coloured differently (dark and light grey) when the two products were found different at all.

As an example of interpretation of such a graph (Fig. 13.12), the title with different colours for each product name shows that product WF D is significantly different from product WF E (same colour if there is no significant overall difference). This overall difference can be explained by the higher dominance of brittleness and melting in product WF D (attributes with dark grey circle) and the higher dominance of crunchiness, grittiness and hardness for product WF E (attributes with light grey circle). For each attribute, more details about the period of the sequence when the products were different are given by the dark grey and light grey dots. For example, the difference in brittleness appears only at the beginning of the sequence. The dominance of crispiness is also different between the two products but not only in one direction (circle with two colours). Actually, crispiness dominance was higher in WF

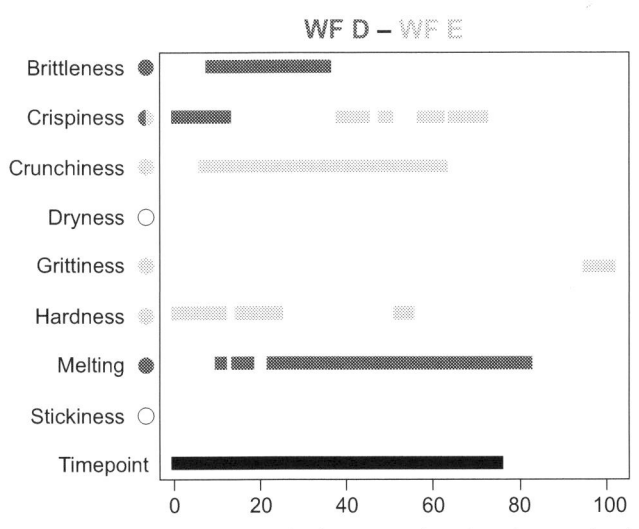

Figure 13.12 Representation of product pairwise comparison based on randomization tests. *Source*: From Meyners and Pineau (2010).

D at the beginning of the sequence (dark grey dots), but became higher in WF E in the middle of the sequence (light grey dots).

Using this representation, it is easy to display several pairwise comparisons on the same page to get an overview of the most critical differences between product pairs.

13.6.3 Analysis of the duration of the evaluations

When the evaluation protocol is not standardized, the duration of an evaluation can vary from one product to another (and/or from one subject to another). In this case, it is useful to compare the evaluation durations between products to know whether some products generally need a longer duration to be evaluated than others. To do so, the 2-way ANOVA model usually used for Descriptive Analysis data (product as fixed effect, subject as random effect, with interaction in case there is replicates) can be simply applied to the "evaluation duration" variable. A multiple comparison test can also be conducted to identify which products have significantly different evaluation durations from which other products.

The analysis of the evaluation duration provides complementary information. In particular, if TDS data are standardized (see Section 13.4.3), information about evaluation duration is lost, but this complementary analysis enables filling this gap. This is the case in the paper of Lenfant et al. (2009), for which the evaluation/mastication duration of the six wafer products were compared (Fig. 13.13) and evidenced long evaluation duration for product WF B (34.1 s) than for products WF C (29.9 s) and WF D (30.6 s).

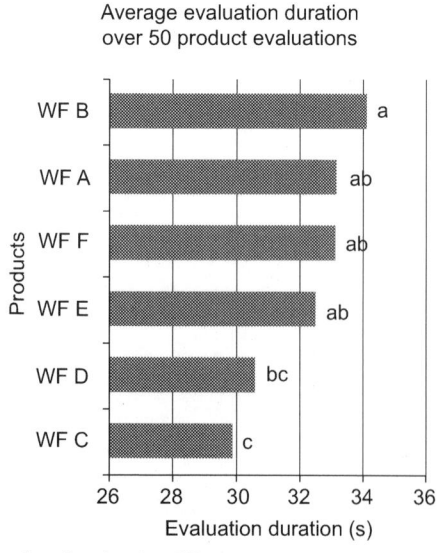

Figure 13.13 Evaluation duration for the different products averaged across subjects. Different letters account for significant difference ($\alpha = 0.05$) according to the Newman–Keuls test.
Source: Adapted from Lenfant et al. (2009).

13.7 Panel performance

This section details a way to proceed for the evaluation of the performance of a sensory panel in the TDS framework. It is proposed to use a dedicated testing protocol to better target the question of the panel performance, then to pre-process the data to fit with the performance context and finally to analyse the data according to a specific method to determine if the panel is performing well, highlight the good performing panellists and/or attributes and quickly identify the main points to be improved.

13.7.1 Protocol and data pre-treatment

Even if panel performance should be the first step before going for product evaluation, it is presented at the end of the data analysis part, since it relies on the product difference test concepts presented before. But contrary to product differences, the focus of the panel performance is to assess the ability of the panel to give consistent product differences across subjects rather than evidencing product differences as such. For this reason, Lepage *et al.* (2014), proposed a simple a priori protocol to check panel performance prior to product evaluation. This is based on the evaluation of a limited number of samples (say four) representative of the product space to be evaluated. Because of the nature of a TDS evaluation (selection of attributes as dominant or not, i.e. 0/1 data), the subjects are supposed to test all products at least in triplicate. In addition, to ensure that subjects' results are fully comparable and not biased by any differences in the product presentation order, the same presentation order is given to all subjects for all replicates.

Regarding the data analysis, the authors propose first a data pre-treatment to account for the singular nature of TDS data in the framework of panel performance. They consider the data in a summarized way by aggregating individual responses in a reduced number of time intervals rather than considering hundreds of time points (Fig. 13.14). This approach fits better with panel performance, because it will give a summarized overview of the performance rather than a check of the subject performance at every single time point of the product evaluation. Typically, considering three time intervals of equal size, this approach will result in panel performance evaluation at the beginning (first-third), in the middle (second-third) and at the end (third third) of the product evaluation. An overall performance measure over the whole evaluation period can also be derived from these three periods.

13.7.2 Performance indexes

Lepage *et al.* (2014) proposed four performance indexes in line with typical figures for panel performance (ISO 8586), namely the discrimination ability and the agreement, both at panel and subject levels. Even though repeatability is also generally part of the performance indicators, it is actually already embedded in the other indicators (Pineau, 2006). For instance, subject discrimination is declared as significant if product differences are large enough compared to individual non-repeatability (F-test), so

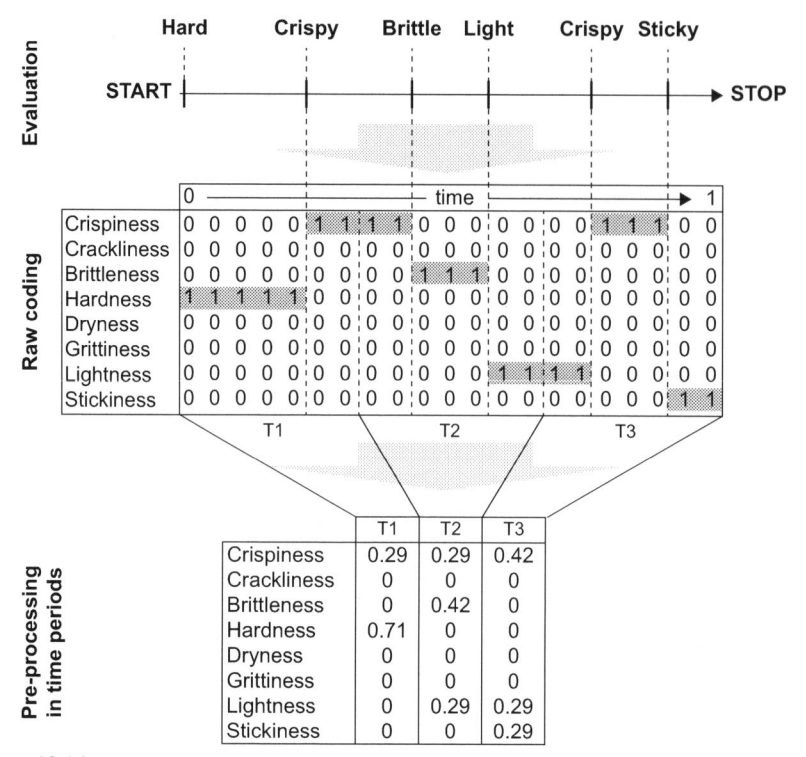

Figure 13.14 Data pre-processing for panel performance assessment.

	T1	T2	T3
Crispiness	0.29	0.29	0.42
Crackliness	0	0	0
Brittleness	0	0.42	0
Hardness	0.71	0	0
Dryness	0	0	0
Grittiness	0	0	0
Lightness	0	0.29	0.29
Stickiness	0	0	0.29

if the non-repeatability is too large compared to the size of the product differences, the subject will not be declared as significantly discriminating the products because of the non-repeatability. Therefore, to focus only on a reduced number of key performance indexes, nothing is proposed for the repeatability as such in their paper.

The four performance indexes are calculated at four different levels:

- For each time period and each attribute
- For each time period
- For each attribute
- Globally over all time periods and attributes

The calculation of each index is derived from the usual sum of square decomposition used in ANOVA, but applied to dominance rates. For instance, the discrimination index at panel level is simply the sum of square of the product effect (see the original paper for more details on the other indexes). However, the authors do not follow the F-test approach to test the significance for these indexes, since TDS data (or the residuals from any standard ANOVA model) are not normally distributed. They rather follow the permutation test approach proposed by Meyners and Pineau (2010) and Meyners (2011) and extend it to the scope of their indexes. The reader can refer to the original paper for more details about the testing procedure.

Temporal dominance of sensations (TDS) as a sensory profiling technique

Panel/Subject diagnostic

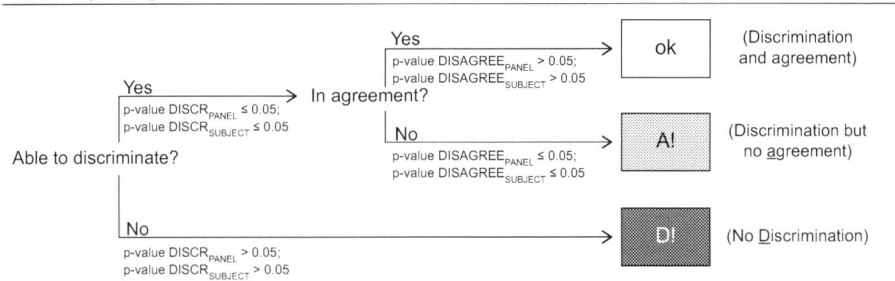

Figure 13.15 Sequential approach for panel and subject performance diagnostic.
Source: From Lepage *et al.* (2014)

The authors summarize the significance testing of the four indexes by proposing a sequential approach to building a diagnostic that focuses only on the key elements to be checked, both at panel and subject levels (Fig. 13.15).

Results at panel and subject levels, for each time period or globally, are summarized in a single attribute × subject table. In the example provided in Fig. 13.16, results can be interpreted as follows. The panel of 16 subjects is able to discriminate the four samples, overall and for eight out of nine attributes. Attribute particle is not discriminating because it is hardly ever elicited (i.e. maximum frequency of elicitation = 12% with only one out of 16 subjects being discriminating). The panel leader could therefore consider discarding this attribute from the list. The performance results over time further reveal that the panel is able to discriminate the four samples at each time point but according to varying attributes. Since products are never sticky or melting at the beginning and no more compact, hard, aerated or brittle at the end of mastication, subjects do not use these attributes at these time points. Consequently, the average number of elicited attributes (4.3) is much lower when considering the three time intervals (2.3, 2.2 and 1.6, respectively). In this case, the panel leader should be reassured that subjects are using the attributes in a very logical manner, which is certainly also part of panel performance. The performance results by subject reveal that one subject (S02) is not discriminating products overall (only one attribute is ok). Four additional subjects (S06, S14, S11 and S03) are discriminating products overall, but less than four attributes is ok (i.e. they use less than half of the attributes correctly). Among the 11 subjects who discriminate products overall and on more than half of the attributes, one subject (S08) is not in agreement. As a consequence, the panel leader should foresee corrective actions for 6 out of 16 subjects. The performance results by subjects and attributes help identifying these corrective actions. Subject S02 is so far from performing well that he should either be excluded or retrained from scratch. Subjects S06, S11, S14 and S03 use fewer attributes than average; they seem to not use melting, brittle, aerated and fatty properly and should therefore be retrained on these specific attributes. Finally, Subject S08 is not in overall agreement with the panel because he disagrees on attributes compact and crunchy. This subject should be specifically retrained on these terms. This is also the subject with the shortest average evaluation time (11.7 s, panel average = 20.7 s), which might explain some

	PANEL				SUBJECT (whole time period only)														
	All	t1	t2	t3		S07	S15	S04	S09	S01	S10	S12	S05	S08	S13	S16	S03	S11	S14	S06	S02
Average eval time (s)	20.7	6.9	6.9	6.9		28.0	24.3	31.2	23.9	23.1	14.4	14.7	18.3	11.7	25.5	26.5	20.7	13.1	19.8	22.1	15.0
n Att Sel/eval	4.3	2.3	2.2	1.6		3.9	3.6	7.9	3.9	5.3	3.4	4.3	4.1	4.3	4.3	6.5	3.3	2.9	3.8	4.7	3.0

	% max freq	All	t1	t2	t3		S07	S15	S04	S09	S01	S10	S12	S05	S08	S13	S16	S03	S11	S14	S06	S02	n subj ok
All	—	ok	ok	ok	ok		ok	ok	ok	ok	ok	ok	ok	ok	ok	ok	ok	ok	ok	ok	ok	Di	1.3
Sticky	73%	ok	Di	ok	ok		ok	ok	ok	ok	ok	ok	ok	ok	Al	ok	ok	ok	ok	ok	ok	Di	13
Compact	59%	ok	ok	ok	Di		ok	ok	ok	ok	ok	ok	ok	ok	Al	ok	Di	ok	ok	ok	ok	Di	12
Fatty	56%	ok	ok	ok	ok		ok	ok	ok	ok	ok	ok	Di	ok	Al	ok	ok	Al	Al	Di	ok	Di	11
Crunchy	80%	ok	ok	ok	ok		ok	ok	ok	ok	ok	ok	ok	ok	Al	ok	ok	Al	ok	Di	ok	Di	11
Hard	88%	ok	ok	ok	Di		ok	ok	ok	Di	ok	ok	ok	ok	ok	ok	ok	Di	ok	Di	Al	Di	11
Aerated	67%	ok	ok	Di	Di		ok	ok	ok	ok	ok	ok	ok	ok	Di	Di	ok	Di	Al	ok	ok	Di	10
Brittle	25%	ok	ok	Di	Di		ok	ok	Di	ok	ok	Di	ok	Di	ok	ok	Di	Di	ok	Di	Di	Di	9
Melting	45%	ok	ok	ok	Di		ok	ok	Di	ok	ok	Al	ok	ok	ok	ok	ok	Di	ok	ok	Di	Al	6
Particle	12%	Di	Di	Di	Di		Di	Di	ok	ok	Di	Di	ok	Di	Di	Di	ok	Di	Di	Di	Di	Di	1
n Att Ok		8	5	5	4		8	8	7	7	6	6	6	5	5	5	5	4	4	4	3	1	

Figure 13.16 Example of TDS performance figure. Left part of the figure presents panel performance; right part of the figure presents subject performance. Attributes are sorted according to n subj ok; subjects are sorted according to n att ok. Average eval time (s): average duration of an evaluation. n Att Sel/eval: average number of attribute selection per evaluation. % max freq: maximum frequency of selection of an attribute over time across products (extracted from TDS curves). n Att/subj ok: number of attribute/subject with the diagnostic "ok".

Source: From Lepage *et al.* (2014).

differences in the TDS evaluation. The panel leader could mention this difference and encourage the subject to taste the products more slowly.

This example illustrates that such a panel performance tool can very directly help a panel leader to take corrective actions (i.e. retrain specific subjects on specific attributes). It is therefore a very useful tool at the end of the training phase in order to define if the panel can enter the main study or if it needs some further training.

Dinella *et al.* (2013) also proposed a methodology to assess product differences and monitor subject discrimination ability in the TDS context according to a stepwise approach. The first step consists in the data visualization of the TDS curves to identify attributes, product pairs and/or subjects that are worthy of investigation. Then, an ANOVA approach is proposed to assess product differences. Compared to the permutation approach described above, the latter is based on more assumptions that might not fit perfectly with TDS data, but the authors confirm the validity of the approach since the distributions of the residuals fit with their assumption in the cases selected according to their first step.

13.8 Some applications

This section presents two examples of applications in different contexts: wine samples and car industry.

13.8.1 TDS method validation based on wine samples

To validate in practice the ability of a panel to describe the temporality of dominant perceptions during a TDS evaluation and to use several kinds of sensory attribute (e.g. taste and flavour) in the same evaluation, Pessina *et al.* (2005) manipulated a white wine to produce different expected dominance sequences, as presented in Table 13.4.

Each sample was tasted by a trained panel of 30 people in three different conditions:

Table 13.4 **Product description of manipulated white wine in Pessina *et al.* (2005)**

Sample	Description
BaseW	Italian BaseW (Falanghina) classical vinification (2004)
FruitW	BaseW with yeasts that produce esters and spiked with banana, apple and apricot flavours (Symrise)
FloralW	60% BaseW + 40% Italian Muscat Wine vinified in the same place and according to the same technology
DriedFrW	Italian BaseW (Falanghina) classical vinification (2003) with "sur-matured" grapes
WoodW	90% BaseW + 10% BaseW 4 months in new casks

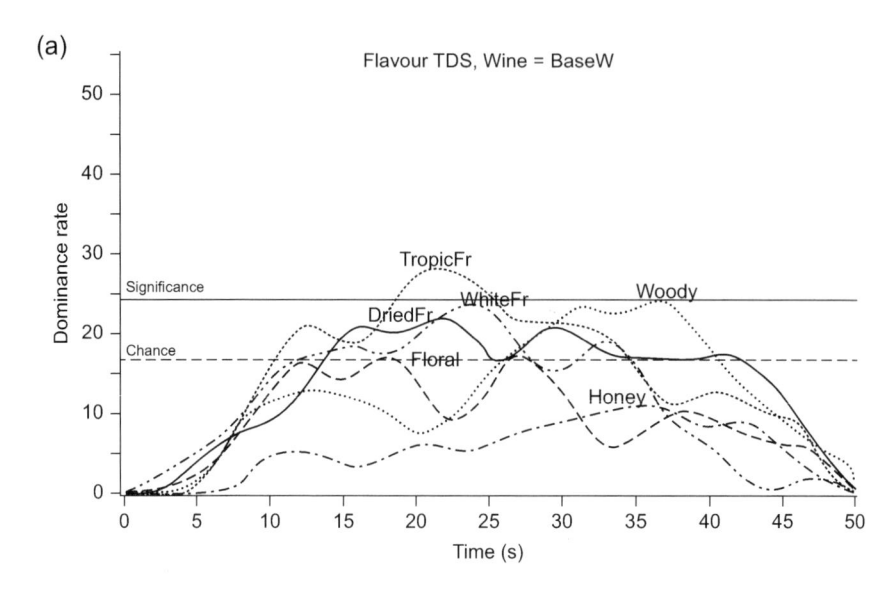

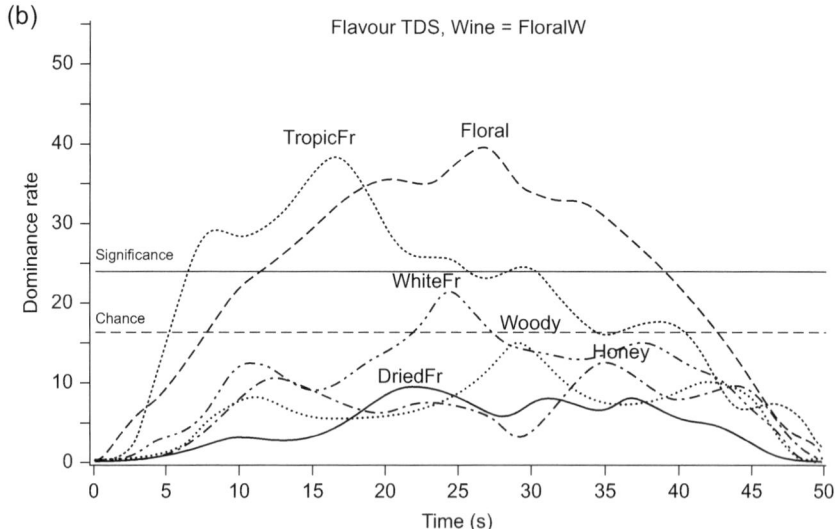

Figure 13.17 TDS curves for (a) a base wine (BaseW) vs (b) a blend of the BaseW and a more floral wine (FloralW).
Source: From Pessina *et al.* (2005).

- TDS evaluation on taste attributes only (three attributes)
- TDS evaluation on flavour attributes only (six attributes)
- TDS evaluation on taste and flavour attributes (nine attributes)

Results show the ability of the panel to evidence the expected dominance of sensations for the different wines, as exemplified with the product FloralW in Fig. 13.17.

Compared to the base wine, FloralW was dominated by the tropical fruit perception for the first half of the tasting and the by the floral note, which is consistent with wine expert knowledge on these products. The TDS methodology was therefore sensitive enough to provide the expected sensory differences in terms of sequences of dominant perceptions.

In addition, this experiment showed a good agreement of the results between TDS evaluation on taste attributes only, flavour attributes only and the combination of taste and flavour attributes (Fig. 13.18). This demonstrates subject ability to cope with different attribute categories in the same evaluation.

13.8.2 Temporal differences not captured with other methods

Pessina *et al.* (2005) investigated the difference between TI and TDS based on red wines spiked with different basic tastes (sweet, sour, bitter and astringent). As exemplified by the results for the wine spiked on sweetness (Fig. 13.19), the TI method is able to show a higher sweetness perception than for the other sensory attributes, but there is no temporal difference. Using the TDS method on the same wine clearly shows that sweetness is mainly perceived at the beginning of the tasting and that astringency dominates in the second part of the tasting. TDS methodology is therefore able to highlight temporal patterns that TI cannot detect. Focusing on the attribute astringent, these results also showed that a dominant attribute is not necessarily the most intense dominance, which is a practical proof of intensity and dominance being different concepts (see also Sections 13.3.1 and 13.3.3). Similar results were observed in a study on cheese product (Pineau *et al.*, 2003) in which the temporality of perceptions was evidenced with TDS methodology and not with TI. The discrepancy between intensity and dominance was highlighted here as well.

13.8.3 TDS application in cars

Egoroff *et al.* (2007) presented an application of TDS for the car industry. In this context, automotive sound sequences were recorded for 11 different cars while driving for the same period of time (one minute) and with the same protocol. These sequences were then played with headphones to ten subjects specialized in acoustic studies. Along the sound sequence, they had to identify the dominant sounds among a list of ten attributes generated by the panel. To illustrate the results, TDS curves for two different cars (a compact vehicle and a full-size vehicle) are presented in Fig. 13.20. Results evidenced different sequences of dominances depending on the car, such as the noise of the wind ("FF" attribute) that is more dominant in the compact car (less sound proof) than in the full-size car. More generally, this study highlighted that some attributes are specific to a part of the ride, e.g. the engine noise is dominant during the different acceleration situations, whereas the aerodynamic noise is dominant during the stationary situation. The authors expect to use these data to better understand the part of the ride that focuses customer attention.

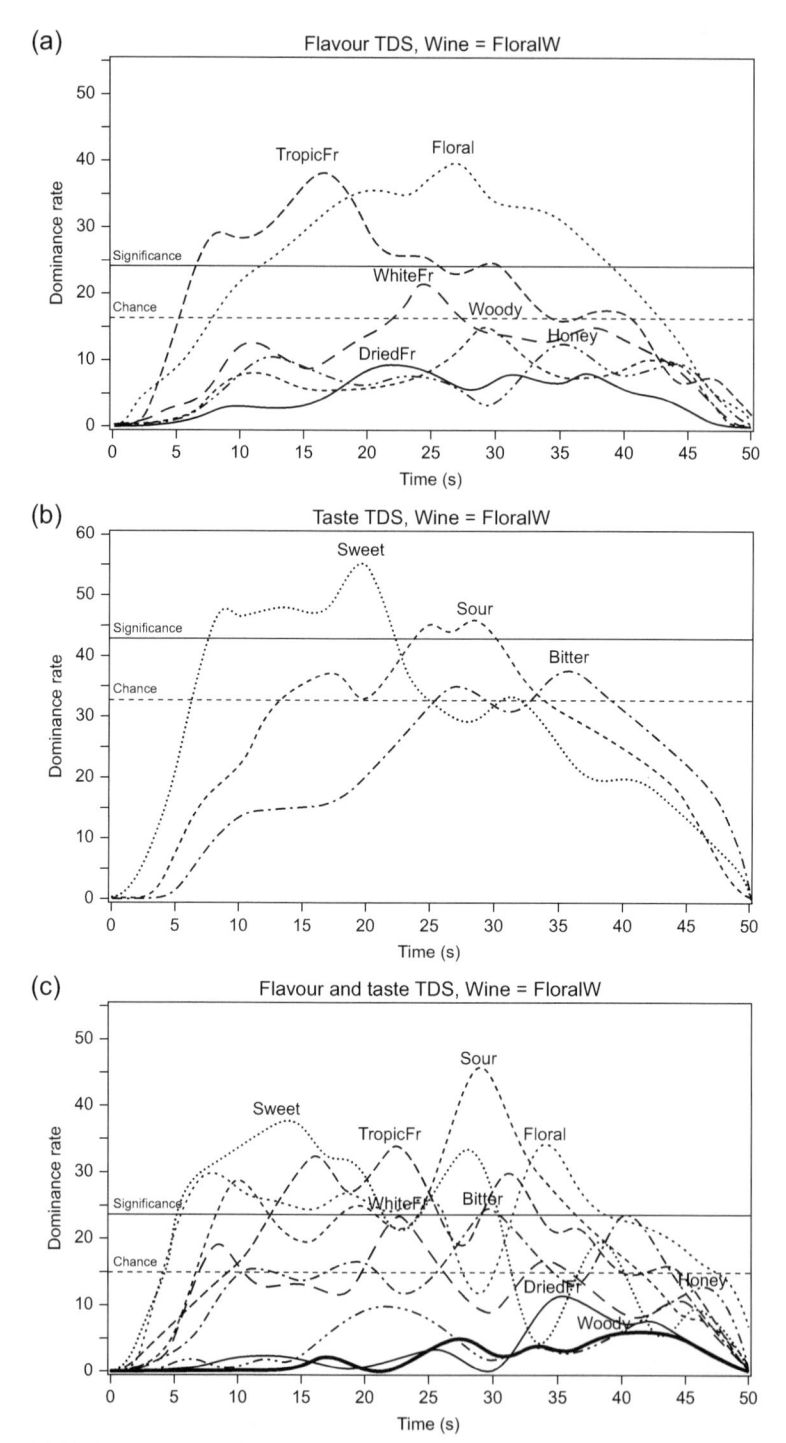

Figure 13.18 Comparison of TDS results on the same product (FloralW) for three sets of attributes ((a) flavour only, (b) taste only and (c) flavour + taste).
Source: From Pessina *et al*. (2005).

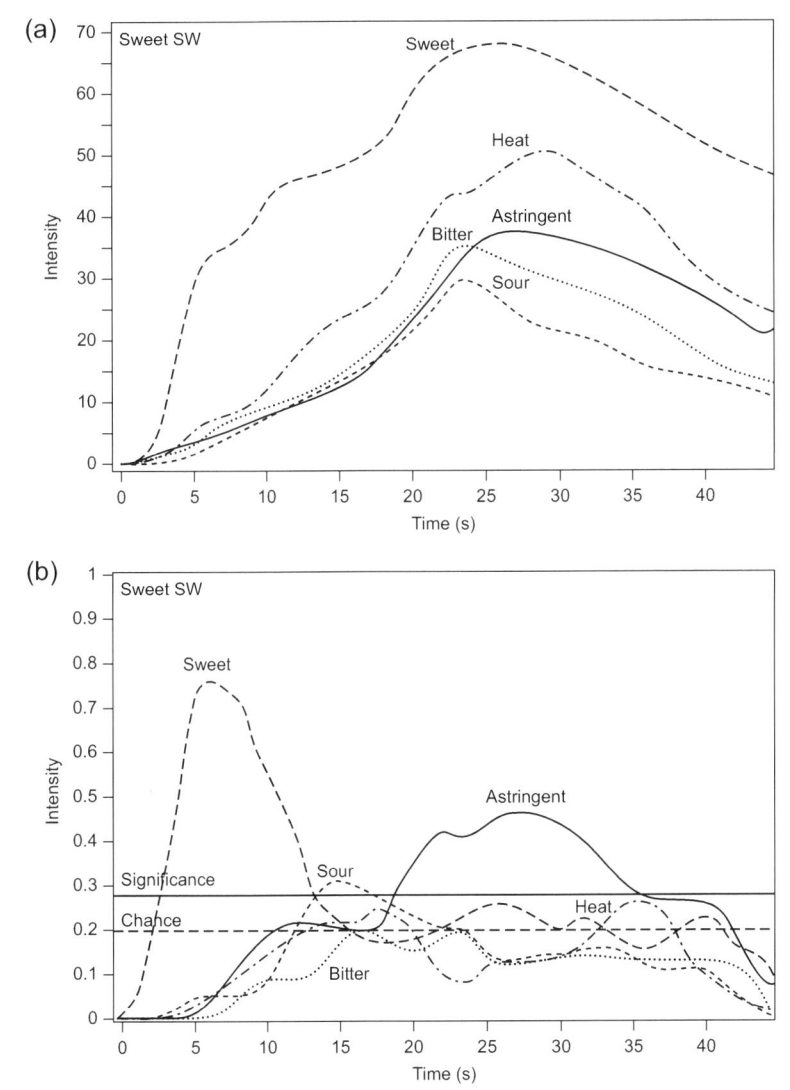

Figure 13.19 (a) TI results vs (b) TDS results on the same product. The sequence of the perceptions is highlighted with TDS but not with TI.
Source: From Pessina *et al.* (2005).

13.9 Future trends in TDS

This section investigates two potential areas for future developments related to the TDS methodology: the use of TDS in multi-sip or multi-bite context to cope with full portion size experiments, and the use of TDS with naïve consumers and the evaluation of the relation between TDS data and (temporal) liking.

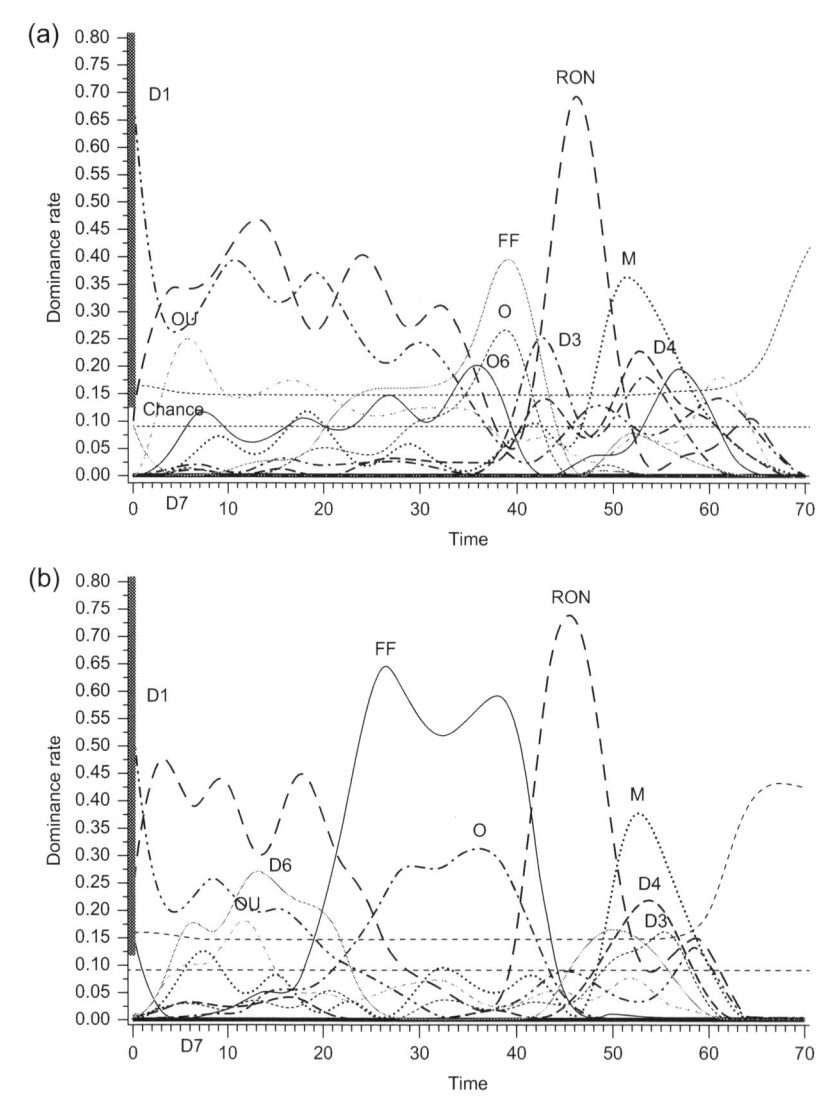

Figure 13.20 TDS curves for a full-size vehicle (a) and a compact vehicle (b). Letters represent different attributes, see text.
Source: From Egoroff *et al.* (2007).

13.9.1 Multi-bit or multi-sip TDS

Following the same concept as TDS, simplified versions were proposed to focus on the most important dominant perceptions of a product, meanwhile extending the period of recording to more than one single bite or one single sip. In this way, Pecore, Rathjen-Nowak and Tamminen (2011), proposed the temporal order of sensations (TOS), which consists in recording, for each spoonful (of a meat-based product in their case), the first, second and third dominant attributes. Time is therefore not recorded and the number of dominant attributes is limited to three. Data recorded over three spoonfuls and the aftertaste perception exhibited large differences between the meat products. In particular, it was possible to observe a delay of the spiciness dominance between two products.

To simplify the approach even further, Dugas *et al.* (2012) proposed to record only the most dominant attribute for each sip of an espresso cup. The protocol was designed so that each coffee cup had the same volume and the subjects were trained to consistently drink the full coffee cup in seven sips. To be able to display a large number of product sequences on the same page and rapidly evidencing the specificity of each espresso, the authors also proposed a condensed presentation of the sequences focusing on the significant part only (Fig. 13.21). As for classical TDS curves, the vertical axis represents dominance rates, but these ones are truncated above the significance limit. If none of the attributes is significantly dominant, the value is 0. Otherwise, all significant attributes sum up for each sip (horizontal axis). High vertical axis values indicate a high consensus of the panel to describe the dominance of the product. And each colour surface refers to one attribute to visually compare sequences. In the

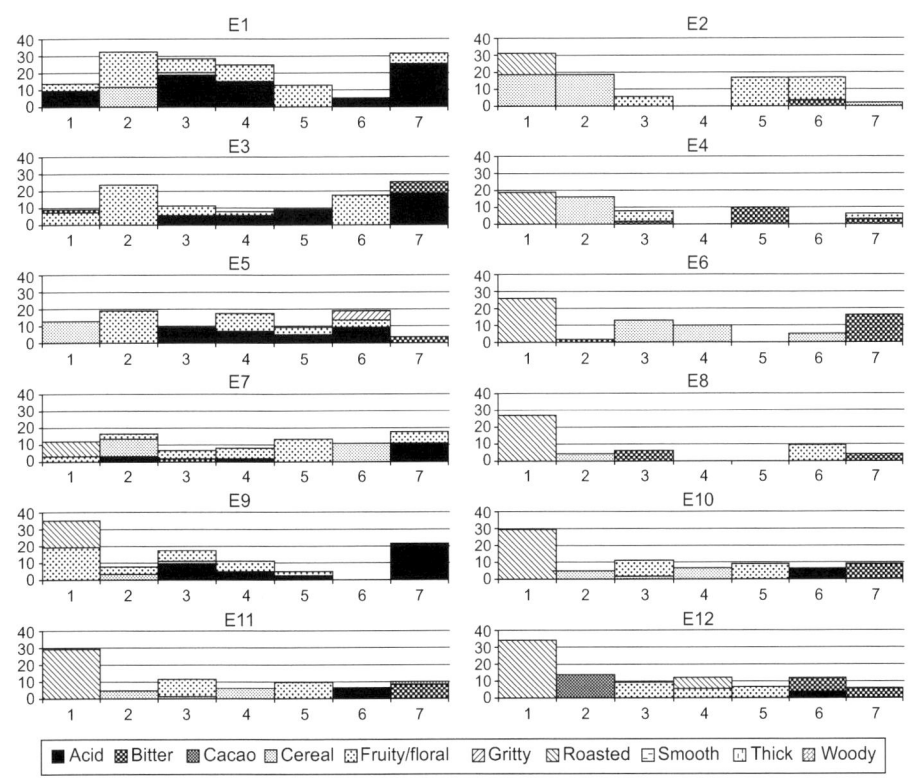

Figure 13.21 Simplified TDS data representations for 12 Nespresso products.
Source: Adapted from Dugas *et al.* (2012).

example of Fig. 13.21, subject agreement is globally better for some products (e.g. E1) than for others (e.g. E8). In addition, clear differences are observed among product sequences, e.g. E1 is dominated by fruity/floral and acid during almost the whole product evaluation, whereas E2 is fruity/floral only at the end and starts with cereal and/or roasted notes for the first sips.

TDS was recently used for applications considering more than one single mouthful or one single sip, and it has even been used for the evaluation of the combination of wine and cheese (Nilsen *et al.*, 2011). The same team also used this approach to the combination of sauce and salmon.

Given all these recent applications, it seems the TDS method is well adapted to the evaluation of complex products, and not only for one spoonful or one sip but also for the consumption of an entire portion size.

13.9.2 TDS with consumers

Since TDS can be seen as a relatively simple task (selection of one attribute/quality in a list over time, i.e. yes/No answer) compared to DA (scoring on a scale for which reference points should generally be learnt), it has recently been tested with naïve consumers instead of trained subjects. Depending on the applications, the subjects had explained to them the list of attributes (Saint-Eve *et al.*, 2011), or even simply the list of the attributes without any precise definition of the terms (Schlich, 2013). In both cases, the consumer panel was at least able to pick the same major product differences as a trained panel, showing the ability of the consumers and/or the methodology to give a reliable sequence of dominances even without training. Saint-Eve *et al.* (2011), however, showed that the trained panel was able to highlight more precise differences and richer sequences than the consumer panel. In Schlich (2013), the results were very close to each other with both panels. These examples tend to show the high potential of the methodology to get a fast descriptive picture of the products, even for untrained people. It is, however, recommended to give a short definition of each term to the consumers prior to the test to ensure that everybody is talking the same language.

13.10 Conclusion

The conclusion aims at highlighting the main differences between the TDS methodology and more classical methodologies such as Descriptive Analysis. It also explains why TDS can be seen as a rapid method and the potential future trends related to the further development of this methodology.

13.10.1 TDS: an established method

The TDS methodology is now an established method in the sensory field to characterize the temporality of the sensations during the tasting of a spoonful or a sip of a product. Statistical methodologies have been developed to characterize product sequences, map product differences, compare them and test the significance of their difference. A panel performance methodology has also been developed to assess the validity of a trained panel.

13.10.2 Comparison with Descriptive Analysis

Compared to the usual Descriptive Analysis, TDS brings a time dimension that can be useful in discovering dynamic/temporal differences between products that cannot be identified by Descriptive Analysis. But more importantly, the biggest difference between both approaches is a matter of mindset. Descriptive Analysis provides a quantification of the intensity of each sensory dimension/attribute in a kind of "chirurgical" way. TDS gives a temporal overview of the most striking product characteristics along the tasting considering the product as a whole. This is a fundamental conceptual difference regarding the way the product is considered, and can bring another angle of view on the product. We can also hypothesize that this fundamental difference makes the TDS methodology closer to the real consumer experience than Descriptive Analysis. This is one of the reasons why some papers were recently published to investigate the potential of the TDS method with consumers and the relation with (temporal) liking (Sudre *et al.*, 2012; Thomas *et al.*, 2014).

13.10.3 TDS as a rapid method

In the framework of the temporal sensory methods, TDS can be considered as a rapid method compared to the historical TI method, since TI requires as many evaluations as the number of attributes to be tested whereas they are all assessed in the same evaluation with TDS method. But it is important to remember that these methods do not target the same needs, as explained in introduction.

Compared to the usual Descriptive Analysis, the answer is actually two-fold. If TDS is used to get a detailed picture of product temporality and rich sequences of sensations, the training of a TDS expert panel is generally slightly shorter than the training of a Descriptive Analysis panel. More in detail, a bit more time is needed with TDS to emphasize the qualitative differences among attributes but there is no need to train on the scoring as long as the user chose buttons instead of intensity scales to record dominance. This can globally save several training sessions on the intensity scoring and thereby accelerate the evaluation process. But TDS can also be used to get a quick overview of the product space. In this case, it has been highlighted in two papers that untrained people can still catch the most important product differences (Albert *et al.*, 2014; Saint-Eve *et al.*, 2011). In this regard, there is at least potential to use TDS as a rapid method, meaning with low level of training, or even with untrained people. This however requires good a priori knowledge of the product category (typically, previous sensory experiments on the same product category) from the panel leader to be able to define a suitable attribute list without or with low input of the panel.

13.10.4 TDS in the future

Beyond the original method, many alternative applications are also growing to adapt the TDS methodology based on the concept of the dominance to longer evaluation periods, such as the evaluation of an entire portion size (multi-sip or multi-bite), the evaluation of product associations (salmon and sauce, cheese and wine) or the assessment of some products by consumer panels. We can therefore imagine that other extensions might arise over the coming years to cope with other specific problems, such as the dominant sensory sensations throughout an entire meal, or the dominant perceptions (or even emotions) during the use of a multi-serving product such as a breakfast cereal box, from the first servings at the top of the box to the last servings at the bottom of the box with e.g. more broken pieces that could displease the consumer. This kind of sensory experiment could also be coupled with liking rating over time, to better identify the critical events/time points influencing the overall liking of a product and better understand the making process to build liking. To "follow" the trained subject or the consumer over several days of use of the product, remote acquisition devices will be required, such as watches, smart phones and tablets. Commercial data acquisition solutions are already available in some software and will probably soon be developed in most of the sensory and consumer software on the market. In this dynamic framework, suitable protocols and dedicated statistical methodologies to analyse and interpret these new kinds of data will have to be developed.

References

Albert, A., Salvador, A., Schlich, P. and Fiszman, S. (2012). Comparison between temporal dominance of sensations (TDS) and key-attribute sensory profiling for evaluating solid food with contrasting textural layers: Fish sticks. *Food Quality and Preference*, **24**(1), 111–118.

Barron, D., Pineau, N., Matthey-Doret, W., Ali, S., Sudre, J., Germain, J. C., Kolodziejczyk, E., Pollien, P., Labbe, D., Jarisch, C., Dugas, V., Hartmann, C. and Folmer, B. (2012). Impact of crema on the aroma release and the in-mouth sensory perception of espresso coffee. *Food and Function*, **3**(9), 923–930.

Bouteille, R., Cordelle, S., Laval, C., Tournier, C., Lecanu, B., This, H. and Schlich, P. (2013). Sensory exploration of the freshness sensation in plain yoghurts and yoghurt-like products. *Food Quality and Preference*, **30**(2), 282–292.

Bruzzone, F., Ares, G. and Gimenez, A. (2013). Temporal aspects of yoghurt texture perception. *International Dairy Journal*, **29**(2), 124–134.

Dinnella, C., Masi, C., Zoboli, G. and Monteleone, E. (2012). Sensory functionality of extra-virgin olive oil in vegetable foods assessed by temporal dominance of sensations and descriptive analysis. *Food Quality and Preference*, **26**(2), 141–150.

Dinnella, C., Masi, C., Naes, T. and Monteleone, E. (2013). A new approach in TDS data analysis: A case study on sweetened coffee. *Food Quality and Preference*, **30**(1), 33–46.

Dugas, V., Pineau, N. and Folmer, B. (2012). Evaluating whole cup experience in gourmet espresso coffee by using dynamic methods. *Eurosens Symposium,* Bern, Switzerland.

Hutchings, S. C., Foster, K. D., Grigor, J. M. V., Bronlund, J. E. and Morgenstern, M. P. (2014). Temporal dominance of sensations: A comparison between younger and older subjects for the perception of food texture. *Food Quality and Preference*, **31**, 106–115.

ISO 8586, Sensory analysis – General guidelines for the selection, training and monitoring of selected and expert assessors.

ISO 8589, Sensory analysis – General guidance for the design of test rooms.

Kiemele, J., Schmidt, S. and Berdine, R. (1995). *Basic Statistics, Tools for Continuous Improvement* (4th edn.). Colorado Springs, USA: Air Academy Press.

Labbe, D., Schlich, P., Pineau, N., Gilbert, F. and Martin, N. (2009). Temporal dominance of sensations and sensory profiling: A comparative study. *Food Quality and Preference*, **20**, 216–221.

Laguna, L., Varela, P., Salvador, A. and Fiszman, S. (2013). A new sensory tool to analyse the oral trajectory of biscuits with different fat and fibre contents. *Food Research International*, **51**(2), 544–553.

Lannuzel, C. and Rogeaux, M. (2007). *How to speed up Temporal Dominance of Sensations Training*. Seventh Pangborn symposium, p 84.

Lenfant, F., Loret, C., Pineau, N., Hartmann, C. and Martin, N. (2009). Perception of oral food breakdown: The concept of sensory trajectory. *Appetite*, **52**, 659–667.

Lepage, M., Neville, T., Rytz, A., Schlich, P., Martin, N. and Pineau, N. (2014). Panel performance for Temporal Dominance of Sensations. *Food Quality and Preference*, **38**, 24–29.

Le Révérend, F. M., Hidrio, C., Fernandes, A. and Aubry, V. (2008). Comparison between temporal dominance of sensations and time intensity results. *Food Quality and Preference*, **19**, 174–178.

Meillon, S., Urbano, C. and Schlich, P. (2009). Contribution of the temporal dominance of sensations method to the sensory description of subtle differences in partially dealcoholized red wines. *Food Quality and Preference*, **20**, 490–499.

Meillon, S., Viala, D., Medel, M., Urbano, C., Guillot, G. and Schlich, P. (2010). Impact of partial alcohol reduction in Syrah wine on perceived complexity and temporality of sensations and link with preference. *Food Quality and Preference*, **21**(7), 732–740.

Meyners, M. (2011). Panel and panellist agreement for product comparisons in studies of Temporal Dominance of Sensations. *Food Quality and Preference*, **22**(4), 365–370.

Meyners, M. and Pineau, N. (2010). Statistical inference for temporal dominance of sensations (TDS) data. *Food Quality and Preference*, **21**(7), 805–814.

Neilson, A. J. (1958). Time-intensity studies. In Nieman, C. (ed.), *Flavor Research and Food Acceptance*, pp. 88–93. New York: Reinhold Publ. Corp.

Ng, M., Lawlor, J. B., Chandra, S., Chaya, C., Hewson, L. and Hort, J. (2012). Using quantitative descriptive analysis and temporal dominance of sensations analysis as complementary methods for profiling commercial blackcurrant squashes. *Food Quality and Preference*, **25**(2), 121–134.

Pangborn, R.M. and Chrisp, R.B. (1964) Taste interrelationships VI: Sucrose, sodium chloride, and citric acid in canned tomato juice. *Journal of Food Science*, **29**, 490–498.

Paulsen, M. T., Naes, T., Ueland, O., Rukke, E.-O. and Hersleth, M. (2013). Preference mapping of salmon-sauce combinations: The influence of temporal properties. *Food Quality and Preference*, **27**(2), 120–127.

Pecore, Rathjen-Nowak and Tamminen (2011). *Tenth Pangborn Symposium*, Toronto, Canada.

Pessina, R., Boivin, L., Moio, L. and Schlich, P. (2005). Application of TDS to taste and flavor in wine. *6th Pangborn Sensory Science Symposium*, Harrogate, UK.

Pessina, R. (2006). Temporal dominance of taste and flavour of wine. Thesis, ENSBANA, Dijon/Università degli studi, Foggia. Available from: http://www.sudoc.fr/109217985.

Pineau, N., Cordelle, S. and Schlich, P. (2003). Temporal dominance of sensations: A new technique to record several sensory attributes simultaneously over time. *Fifth Pangborn, Symposium* (p. 121), 20–24 July.

Pineau, N., Schlich, P., Cordelle, S., Mathonnière, C., Issanchou, S., Imbert, A., Rogeaux, M., Etiévant, P. and Köster, E.. (2009). Temporal Dominance of Sensations: Construction of the TDS curves and comparison with time–intensity. *Food Quality and Preference*, **20**, 450–455.

Pineau, N., Goupil de Bouillé, A., Lepage, M., Lenfant, F., Schlich, P., Martin, N. and Rytz, A. (2012). Temporal Dominance of Sensations: What is a good attribute list? *Food Quality and Preference*, **26**, 159–165.

Pineau, N., Glérum, A., Robin, T., Thémans, M., Martin, N., Bierlaire, M. and Rytz, A. (2009b). TDS: Is it possible to reduce panel size without changing the quality of the results? *Eighth Pangborn Sensory Science Symposium*, 26–30 July, O5.3.

Rosenthal, J. and Share, C. (2014). Temporal dominance of sensations of peanuts and peanut products in relation to Hutchings and Lillford's "breakdown path". *Food Quality and Preference*, **32**, 311–316.

Rosner, B. (1995). *Fundamentals of Biostatistics* (4th edn.). Chapters 4–7. Belmont: Duxbury Press.

Saint-Eve, A., Lenfant, F., Teillet, E., Pineau, N. and Martin, N. (2011). Impact of panel training, attribute list, type of response and dominance definition on TDS response. 9th Pangborn sensory symposium, Toronto, Canada, P1.9.06.

Schlich, P. (2004). L'analyse en variables canoniques des données de profil sensoriels. *Eighth European Symposium of Agro-Industry and Statistical Methods*, Rennes.

Sudre, J., Pineau, N., Loret, C. and Martin, N. (2012). Comparison of methods to monitor liking of food during consumption. *Food Quality and Preference*, **24**(1), 179–189.

Teillet, E., Schlich, P., Urbano, C., Cordelle, S. and Guichard, E. (2010). Sensory methodologies and the taste of water. *Food Quality and Preference*, **21**(8), 967–976.

Thomas, A., Visalli, M., Cordelle, S. and Schlich, P. (2014). *Temporal Drivers of Liking*. Food Quality and Preference, In Press, Available online 21 March 2014.

Varela, P., Pintor, A. and Fiszman, S. (2014). How hydrocolloids affect the temporal oral perception of ice cream. *Food Hydrocolloids*, **36**, 220–228.

Ideal profiling as a sensory profiling technique

14

T. Worch[1], P.H. Punter[2]
[1]Qi Statistics, Ruscombe, Reading, UK; [2]OP&P Product Research, Utrecht, the Netherlands

14.1 Introduction

In sensory science, the common practice is to conduct hedonic and descriptive analysis separately. Hedonic analysis is conducted with consumers who evaluate the products in terms of liking (Thomson, 1988), whereas descriptive analysis is traditionally conducted with expert, or trained, panellists who generate sensory profiles of the products (Stone *et al.*, 1974; Stone and Sidel, 2004; Meilgaard *et al.*, 2007).

Although these two types of data – sensory profiles and hedonic ratings for the same set of products – contain valuable information for product developers, they are not very actionable in isolation. However, once combined, this information becomes really powerful. If product developers gain insight into how the sensory perception relates to the appreciation of the products, they can try to apply this information to improve the products (Moskowitz and Sidel, 1971; Moskowitz *et al.*, 1977). More particularly, if one can tell product developers which sensory characteristics (and at which intensity) are influencing consumers' appreciation, or why some products are more liked than others in terms of sensory attributes, R&D can start to create new potentially successful products (product development), or improve existing products (product optimization).

In order to provide this information to product developers, several statistical tools have been developed that combine expert profile data with hedonic consumer data – for example, internal and external preference mapping techniques (Carroll, 1972; Danzart, 2009), and derivatives such as PrefMFA (Worch, 2013), Consumer Preference Analysis (CPA) (Lê *et al.*, 2006), Landscape Segmentation Analysis (LSA) (Ennis, 2005), Euclidean Distance Ideal Point Modelling (EDIPM) (Meullenet *et al.*, 2008), Bayesian Sensory Model Integrated with Characteristics (BaSic) (Nestrud, 2012), etc. Although the core of these methods differs, they all share the common underlying (simplified) assumption that for each interviewed consumer an ideal product exists for which liking is maximized (Booth and Conner, 1990). Hence, the aim of these methodologies is to provide drivers of liking (which attributes play a possible role in consumers liking) and to define an optimally liked product that can be used as a reference to match. Such an optimal product corresponds to the ideal product.

These statistical tools are widely used in practice, and their success does not need to be proven. However, such methods rely on statistical models, and a large sample set is required to get sufficient degrees of freedom when these complex models are fitted. Moreover, when the link between the two blocks of data is weak, the subsequent models might show limitations (low goodness of fit).

An alternative to statistical modelling is to directly ask consumers to make the link between what they perceive and what they like. This is possible since it has been shown that consumers are able to rate intensities for an imaginary product (Booth et al., 1987). One possibility is the use of the Just About Right scale (JAR) (Rothman and Parker, 2009) with which consumers are asked to rate the difference between the perceived intensities and their internal ideal intensities for a predefined list of attributes by saying whether the perceived intensity is *just about right*, *too little/much too little* or *too much/much too much*. The classical analysis for this type of data is called penalty analysis (Meullenet et al., 2007). It consists in calculating the difference in liking for each product and each attribute between JAR and each non-JAR category (*much too little* and *too little* being grouped together, and *much too much* and *too much* being grouped together).

This type of test is frequently used by practitioners, as it provides relevant information about consumers' perceptions and expectations of the products. In this way, the practitioners and product developers are getting information from the main actors in the market, namely the consumers.

This analysis is performed for each product separately and can be seen as a diagnostic approach, since it helps the user and product developers in highlighting the attributes that would need improvement in each product. However, this approach has some drawbacks: (1) What does the level *just about right* mean exactly to consumers? (Gacula et al., 2007). (2) When a deviation from ideal is detected (e.g. a product is too sweet), it is unclear how far from the ideal it is (and hence how much it should be modified). (3) If a product is modified according to one sensory attribute (e.g. the sweetness intensity is decreased), it is not possible to evaluate the reaction of the consumers who first said that the product was JAR in sweetness (will they still find the product *JAR*? Or will it be *too little*?).

In recent years, the use of consumers in sensory tasks has gained more and more interest. This is particularly true since it has been shown on different occasions that relevant descriptive information can be obtained with consumers. For example, several studies have shown that sensory profiles obtained with consumers match profiles obtained with trained panels (Moskowitz, 1996; Husson et al., 2001; Worch et al., 2010a).

If consumers can profile products, and if consumers can rate the products relative to their own ideal (as with the JAR scale), then we could ask consumers to rate their ideal directly on a predefined list of attributes. This is the essence of the Ideal Profile Method (IPM) (Worch et al., 2013) that is presented here. As will be shown, the data obtained from this approach are directly actionable for product development and/or product improvement (Hoggan, 1975). However, since these particular data are obtained from consumers who are rating a fictive ideal product, they require particular attention from the practitioners.

The concepts, as well as the corresponding step-by-step methodology for the analysis of the IPM data (called the Ideal Profile Analysis (IPA)), will be presented. Finally, the advantages/inconveniences of the IPM and its practical use compared to other methods (such as the Preference Mapping or JAR scale) will be discussed.

14.2 Principle and properties of the Ideal Profile Method (IPM)

14.2.1 The IPM, in practice

In the IPM, consumers are asked to rate both the perceived and ideal intensities of products on a predefined list of sensory attributes. In this case, P tested products will yield P sensory profiles and P ideal profiles per consumer. Additionally, the consumers rate the products on overall liking. In this sense, the IPM can be seen as a mix between a classical profiling task (such as QDA®, except that it is performed by consumers) and a JAR task in which the ideal intensities are asked explicitly.

In practice, two successive questions are asked for each attribute, first a sensory rating question and second an ideal rating question. For example, if the first question asked of the consumer is "How sweet is this sample", the following question is "What is your ideal level of sweetness for this sample" (Fig. 14.1).

In this example, the consumers are asked to rate both their perceived and ideal intensities on a 100 mm line scale, with anchor points at 10 and 90. No restrictions concerning the scale used are given, except that it should be identical for both the perceived and ideal intensities. Additionally, the consumers are asked to rate the products on liking. Specific liking questions can also be asked. For hedonic questions, a 9-point category scale is used.

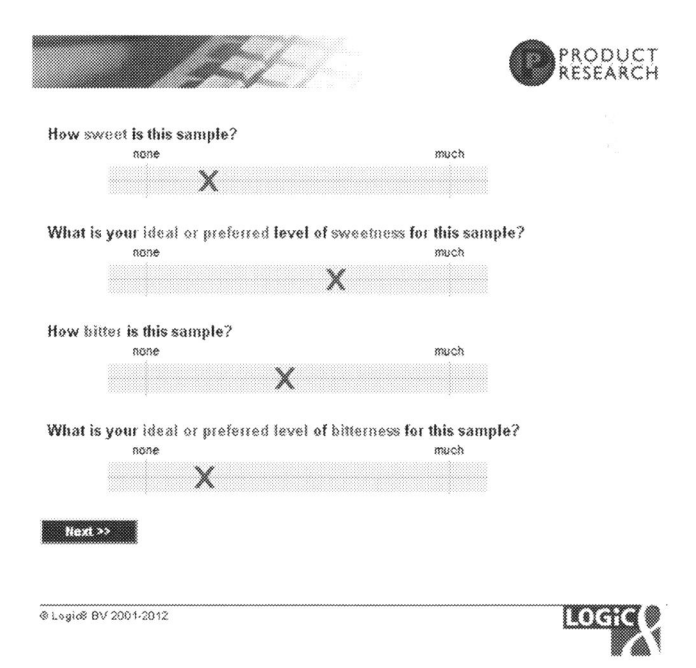

Figure 14.1 Extract of questionnaire for the IPM in the computer based software (EyeQuestion, www.logic8.com).

Table 14.1 **Presentation of the data obtained from consumer** j

	Attr.1	...	Attr. a	...	Attr. A	Ideal attr. 1	...	Ideal attr. a	...	Ideal attr. A	Liking
Product 1											
...											
Product p			y_{jpa}					z_{jpa}			h_{jp}
...											
Product P											

On the left side, the sensory profiles of the products; in the middle, the ideal profiles; on the right side, the vector of hedonic scores. Here, j indicates the consumer, p the product and a the attribute.

Table 14.2 **Description of the data acquisition in the different methods**

	Respondents	Perceived intensity	Liking scores	Ideal Profiles	Deviation (perceived–ideal)
QDA®	Experts or trained panellists	Measured	–	Estimated	Estimated
CLT	Consumers	–	Measured	Estimated	Estimated
JAR	Consumers	–	Measured	–	Measured
IPM	Consumers	Measured	Measured	Measured	Calculated

At the end of the test, each consumer has thus generated three blocks of data (Table 14.1): the sensory profiles of the products (noted y_{jpa}), the ideal profiles (noted z_{jpa}) and the vector of hedonic scores (noted h_{jp}).

It can already be noticed that, for each consumer, the difference between the perceived intensity scores and the ideal scores is akin to JAR measurement. If, for a consumer j, a product p and an attribute a, the difference $y_{jpa} - z_{jpa}$ is close to 0, the product is *just about right*. If this difference is positive, the product is *too much*. Similarly, if the difference is negative, the product is *too little*. But compared to the JAR scale, in the case of the IPM, the intensity of the difference is also provided.

A more detailed comparison of the ins and outs of the IPM compared to other methods (JAR and the classical procedure for PrefMap involving Quantitative Descriptive Analysis (QDA) from experts and Central Location Test (CLT) with consumers) is presented in Table 14.2 (Van Trijp *et al.*, 2007).

14.2.2 Variants of the IPM

The methodology of the IPM in its present version has been used by OP&P Product Research (Utrecht, The Netherlands; www.opp.nl) in the past 20 years. The particularity of this approach is that both perceived and ideal intensity are rated for each product.

Some practitioners think that this approach is too demanding for consumers, and suggest that the ideal should be asked for only once, either at the beginning or at the end of the test. Such a procedure was used in the past (Moskowitz, 1972; Cooper *et al.*, 1989). Although no proof has been given, Szczesniak *et al.* (1975) claimed that whether the ideal is asked for once at the beginning or at the end does not make much difference. In our experience, if the practitioner decides to ask for the ideal only once, it would be more suitable to ask for it at the end, because the consumers would have references to rate the ideal against (the set of products they tested). However, such an approach has consequences, since the ideal ratings are affected by the product tested. Further research needs to be done on this topic to evaluate the relevance of repeating the ideal measurements.

More fundamentally, asking for the ideal intensity only once implies that consumers refer to a more general (or abstract) ideal product. In practice, consumers do have a notion of what their ideal should be, based on their experience and memory. Just as with the JAR procedure, we are interested to know if a particular attribute (say sweetness), for a specific product, and at a specific time should be stronger or weaker. Hence, it seems more adequate to ask for the ideal intensities for every product in conjunction with the intensity rating.

Also in our experience, participants in our tests have never complained about having to rate both the perceived and ideal intensity for each attribute for each product. And besides providing much more information on the reason why each product is liked (or disliked), the repetition of the ideal measurement brings variability within consumers of their ideal ratings. This variability appears to be very useful, statistically.

14.2.3 Properties of the IPM

Since the IPM is intended to be performed with consumers, training would be inappropriate. We thus recommend not to train the assessors, neither on the products, the attributes, nor the method. This should avoid biases in consumers' responses primarily as a result of an increased knowledge or familiarity with the product set. Besides, on the practical side, training would imply longer, or even additional, sessions. It would thus limit the feasibility of the method with consumers. In conventional descriptive analysis, training aims at reducing noise in the measurement by increasing the panellists' accuracy (better understanding and detection of the attributes, better use of the scale, etc.). The absence of training in IPM should thus be compensated in order to ensure a good sensitivity at the panel level. This can be done by increasing the sample size, i.e. the number of assessors in the panel. In fact, as in most of the quantitative hedonic tests, the sample size is much larger in the case of the IPM than in classical profiling. In the literature, it has been stated that a minimum of 100 consumers is required (Moskowitz, 1997).

From our experience, in consumer tests about 20% of the data cannot be explained, and hence are considered as random "noise". In Section 14.4.2, a methodology that allows identifying the inconsistent consumers is presented. Once identified, the user can decide to discard these particular consumers from the analysis, hence reducing the noise. However, we do not recommend this since the random noise will average out, and discarding consumers might have undesirable side effects, such as unbalancing the design.

The number of products that could be included in an IPM study is a frequently asked question.

Assuming sessions of 1 h, six to eight products can be assessed in one session (for each product 60–80 questions can be asked). There is no maximum to the number of products, since there can be more than one session. In practice, there will be seldom more than 24 products (with a maximum of six sessions over a 2 week period).

As in any sensory test with more than one product, the products should be presented in a balanced order, following an adequate experimental design (MacFie *et al.*, 1989).

Additionally, the absence of preliminary training limits the use of specific or technical attributes. The list of predefined attributes asked during the IPM test must include only non-technical terms that are understood by most respondents. Specific or technical terms should be explained.

Another way to reduce noise in the data consists in correcting the consumer data from the difference in the use of the scale. It is well known that not all customers are rating the products in a similar way: some are using the higher part of the scale while some others are using the lower part of the scale. In hedonic testing, this is often corrected by transforming hedonic scores into "preference" scores (in practice, the hedonic scores are centred for each consumer). In the IPM, such difference in the use of the scale is also observed both for the perceived and ideal ratings. The impact of differences in the use of scale is limited in the case of perceived intensities, as average data are used. However, this is not the case for ideal ratings, in which consumers' ideals are directly compared. In this latter, it is necessary to "correct" the ideal data from the different use of the scale.

In order to correct the individual ideal scores provided by a consumer j, the averaged perceived intensity this consumer j provided is subtracted from his/her ideal scores. Such a corrected ideal score is denoted as $\tilde{z}_{jpa}$ (Equation [14.1]).

$$\begin{cases} \tilde{z}_{jpa} = z_{jpa} - (1/P)\sum_{p=1}^{P} y_{jpa} \\ \tilde{z}_{jpa} = z_{jpa} - \overline{y}_{j.a} \end{cases} \qquad [14.1]$$

This procedure of correcting the ideal scores should be performed every time the ideal products are compared between consumers.

14.3 IPM, a tool for product development and product optimization

Although the IPM is mainly described as a tool for product optimization, it also can be very useful for product development. Indeed, the methodology that is described here can also be used to find the sensory characteristics of a product which are important to consumers. Once the sensory characteristics (and their optimal level) is defined, the corresponding product can be formulated and compared to already existing products for validation.

14.3.1 Definition of an ideal product

An online dictionary (source: online Oxford dictionary: www.oxforddictionaries.com) defines the term "ideal" as "satisfying one's conception of what is perfect; most suitable," "representing an abstract or hypothetical optimum", "a standard or principle to be aimed at".

Through these definitions, it appears that the ideal can be seen as a standard or goal that would be highly satisfying (optimum). By adapting this definition to the sensory field for an ideal product, the two concepts – standard/goal and optimum – should be kept. This notion of ideal should hence be defined according to two dimensions: a sensory dimension (the characteristic of the product itself – a standard) and a hedonic dimension (the high overall liking score – an optimum).

Applied to sensory, we propose this definition for an ideal product: it is a product with particular sensory characteristics which would maximize liking. With this definition, both notions of standard/goal (i.e. product with particular sensory characteristics) and of optimum (i.e. would maximize liking) are kept.

14.3.2 A tool for optimization

Most of the methodologies that are used for product optimization share the (simplified) notion that an ideal product exists (in external preference mapping, multiple ideals might be found, even at the individual level). Obviously, the IPM shares this notion as well. A core assumption underlying this thinking is that there is a link between the liking score of a product and the distance between the perceived intensity and the ideal intensity for a given attribute. Under this assumption, the further a product deviates from its ideal, the lower the liking score. In JAR tasks, such an assumption is tested in the penalty analysis, in which the impact of each attribute not being JAR on overall liking (the product is more distant from its ideal for that attribute) is estimated through the calculation of the penalty.

In IPM, the liking score is considered as a weighted linear combination of the distances between the attributes' levels and their corresponding ideals (Equation [14.2]; van Trijp *et al.*, 2007; Worch *et al.*, 2010b).

$$h_{jp} = H_{jp} - \sum_{a=1}^{Att} b_a \left(z_{jpa} - y_{jpa} \right) \qquad [14.2]$$

In this formula, H_{jp} corresponds to the liking score of the ideal product (often considered as 9 on a 9-point hedonic scale) and b_a corresponds to the weight of an attribute on liking. If b_a is positive (respectively negative), the attribute a can be considered as a positive (respectively negative) driver of liking. And the higher is $|b_a|$, the more influential is attribute a on liking.

In IPM, the sensory profiles of the products, the information regarding the consumers' ideals, and the hedonic scores are collected directly for a set of products. Once the test is finished, the data are immediately actionable for product optimization and/or product development. The only missing parameter at this moment is the coefficient b_a, used to estimate the weight of each attribute on liking. Since the consumers have provided information on how they perceived the products and on how they liked them, the weights b_a can be computed.

The coefficients can be calculated either individually (i.e. for each consumer separately) or on an overall basis (i.e. at the panel level). Depending on the point of view adopted, the attribute weight can be obtained either by using the correlation coefficient measured between the liking scores and the intensity ratings for that attribute, or through multivariate regression such as PCR (regression on principal components) or PLS (partial least square regression).

Finally, once the weights b_a are estimated, guidance can be given to the product developers. The possible gain in liking can be estimated for each attribute if that attribute were at its ideal level. Such a measure somehow corresponds to the mean drop estimated in the penalty when the JAR scale is used. The attributes which would show the largest gain in liking if they were ideal are the most important ones to improve.

Since the information is directly measured, the distance between the actual level of each attribute and its corresponding ideal is available, and shows whether the intensity should be increased or decreased, and to what extent.

Beside its great advantage, it must be noted that this methodology presents a limitation, as it considers each attribute separately, without taking into consideration the collinearity between sensory attributes. Indeed, the optimization is often formulated as the potential improvement when improving one attribute, the other one being kept unchanged. Actually, such an assumption is not realistic as, for instance, changing the intensity of sweetness directly impacts the perception of bitterness and sourness. Moreover, it is not clear whether interactions between attributes are taken into consideration in the IPM task. When consumers provide their ideal intensity in sweetness, for instance, do they provide an ideal balance of sweetness, sourness and bitterness? Or do they provide an ideal intensity for each attribute with respect to the true perception of the others (i.e. ideal sweetness, if sourness and bitterness are at the intensity perceived; ideal sourness, if bitterness and sweetness are at the intensity perceived, etc.)?

For those reasons, the guidance on improvement provided by the IPM should be considered as a "path to follow," rather than a true recipe of the ideal product.

14.3.3 Sensory profile of the ideal of reference

In the previous formula (see Section 14.2), it is assumed that the optimization of a product requires comparing its sensory profile to the sensory profile of the ideal product used as reference to match. Before determining the sensory profile of the ideal product of reference, it is worth noticing that from a statistical point of view the analysis of ideal data is inversed compared to the analysis of sensory data. Indeed, with sensory data, the quality of an assessor (whether it is an expert panellist or a consumer) is measured through his/her ability in discriminating the products (large variance between products within consumer) and in being in agreement with the rest of the panel (low variance between consumer within product). When ideal data are considered, a low variance between consumers within product would suggest that they all share the same ideal. And a large variance between products within a consumer suggests that the ideal scores are dependent on the products evaluated – in other words, that consumers have multiple ideals. Based on this statement, it seems reasonable to expect that for ideal data, the variance between consumers within products can be large while the variance within consumer between products should be relatively small.

Additionally, like in any optimization procedure, it must be checked whether consumers should be segmented according to their preferences. The IPM is no exception to this rule. However, in the case of the IPM, consumers could be either segmented based on liking or based on their ideal profiles. From a practical point of view, since consumers can be very different in terms of ideals (some define their ideals within the sensory space, while some others are more extreme), it is more valuable to segment the consumers based on their liking. The classical approach in segmenting consumers could then be used (hierarchical vs non-hierarchical methods, using similarities or dissimilarities, involving the linkage of interest). Once homogeneous groups of consumers are defined, the optimization can then be performed for each group separately. In this case, one ideal product of reference is defined for each group (Cooper et al., 1989).

Since consumers provide ideals that can be very different, a common ideal should be defined. Companies are indeed not interested in using individual ideal products for product optimization, since formulating one product for each consumer would not be viable. So which ideal should be considered as reference?

The first (and simplest) solution consists in considering the sensory profile of the average ideal product as reference (Hoggan, 1975; Sczezesniak et al., 1975). This ideal corresponds to the average score for each attribute over all ideal ratings. This solution is the most common one in the literature, and it corresponds to a product that is liked by most consumers, without being necessarily optimal.

A second solution consists in defining as an ideal of reference the ideal product shared by a maximum of consumers. In this case, the ideal might not please all the consumers (in the absolute sense), but fully satisfies a majority of them. Such solution is the one provided by the Ideal Mapping (IdMap) (Worch et al., 2012c). Technically, the IdMap defines the ideal area for each consumer (by constructing ideal confidence ellipses using the variability within consumer of the ideal ratings) on the product space and searches for the location within the product space in which a maximum of individual ideal areas are overlapping. The sensory profile of the ideal product shared by a maximum of consumers is then used as reference to match in the product optimization.

Although the two solutions – average ideal product vs solution from the IdMap-proposed here – are different from the point of view adopted (the IdMap tends to target a smaller proportion of the population while the mean satisfies a maximum of consumers), in practice they can arrive at a similar result. This is particularly true when consumers all share an ideal that is quite close, or when the ideal products from the consumers are homogeneously distributed (the maximum of ideal points being in the centre of the clouds).

14.4 Additional valuable properties of the IPM

14.4.1 The IPM and the notion of multiple ideals

In most product optimization procedures which explicitly estimate ideal points (LSA, EDIPM, BaSic, etc.), it is assumed that each consumer is associating the product set to one unique ideal product. Although this assumption is reasonable in the case where the product set consists of very similar products, it is not necessarily correct when the products are more heterogeneous. Let us consider a test in which milk chocolates and dark chocolates are compared. In such case, it is fair to say that consumers might like both the milk and the dark chocolates, and hence have one ideal for the milk chocolates and one ideal for the dark chocolate. The optimization procedure should then associate the milk chocolates with the ideal of reference defined for the milk chocolates, and the dark chocolates with the ideal of reference defined for the dark chocolates.

To generalize, let us consider a product set made up of products belonging to the same category (e.g. the chocolate category). This category can be divided into latent subcategories[1], each latent subcategory being associated with one unique ideal (e.g. the milk chocolate subcategory, and the dark chocolate subcategory). If, for the given product set, the consumers associate all the products to one unique ideal, then the product set is made of one unique latent subcategory. However, if the consumers associate the product set to I ideals, then the product set is made of I latent subcategories.

In order to define the number of latent subcategories present in the product set, the variability of the ideal ratings between products within consumer is used (Worch and Ennis, 2013). This procedure is only possible if each consumer rates his/her ideal for each product. More precisely, the average ideal product is calculated for each product separately (a product set is made of P products yielding P averaged ideal products, the average being calculated over the consumers). In the case where all the average ideal products are sensibly the same, the majority of consumers' ideals are cueing one unique ideal. However, if a systematic shift is observed for one particular product,

---

[1] The notion of a subcategory does not relate directly to the one used in Marketing: we define here a subcategory of product as a subset of products, for which all the products (from that subset) are associated with the same ideal. Hence, this notion of subcategory (called latent subcategory) is linked to the notion of ideal, rather than to the notion of product.

the corresponding average ideal product is more distant from the others, meaning that it is cueing a different ideal. In this case, the product set is defined by more than one latent subcategory. In practice, the sensory profiles of the average ideals computed for each product can be compared visually (using spider plots) or statistically through confidence ellipses (constructed using the variability within consumers) and the Hotelling T^2 test.

The decision whether one or multiple ideals should be considered is a decision that should not only be based on statistical grounds, but also on the business arguments. If two or more latent subcategories exist, is it worth formulating more than one optimal product? Or should only one optimal product be formulated? Such a decision is obviously based on the size of the differences between the ideal of each latent subcategory, and on a business decision as to whether one or multiple products should be placed on the market.

14.4.2 Consistency of the ideal data

As mentioned in Section 14.2.3, the validity of the ideal data needs to be checked, since the data are obtained from consumers who are asked to rate a fictive product. For that reason, before any optimization, one has to make sure that the consumers are providing relevant information when rating their ideals.

Since the ideal product has both a sensory and a hedonic dimension – the ideal being defined as a product with particular sensory characteristics that would maximize liking – checking for the consistency of the ideal data should be done both from the sensory and the hedonic point of view. These consistencies are checked both at the consumer and at the panel level.

An ideal product provided by a consumer is considered consistent from the sensory point of view if the sensory profile associated with that ideal has similar sensory characteristics as the most liked product(s) (Worch *et al.*, 2012a). From an attribute point of view, this means that if a consumer said he/she appreciated more the products that he/she perceived as more intense for a certain attribute a, he/she should define his/her ideal level for that attribute a as rather intense. For example, if a consumer declared liking the products that are sweeter more (sweetness being a positive driver of liking, the sweeter the better), his/her ideal product should be defined as rather sweet. Inversely, if a consumer declared liking the products that are more bitter-less (bitterness being a negative driver of liking, so the less bitter the better), his/her ideal product should be defined as rather not bitter.

Statistically, the sensory consistency is assessed at the panel level by verifying that the ideal information provided by the consumers is making the link between the sensory and the hedonic description of the products. For a consistent panel, the projection as illustrative of the product p within the ideal space is located close to the ideal products of consumers who liked particularly p more. At the consumer level, consumer j is consistent if he/she provided high ideal ratings for positive drivers of liking and low ideal ratings for negative drivers of liking. More systematically, such an assertion can be evaluated by verifying that consumers provide high ideal ratings for attributes that are intense on the products they like most (and vice versa). To do so, the correlation is calculated across attributes between the corrected ideal scores of a consumer and the correlation coefficient defining drivers of liking. In the first case, the corrected ideal score of an attribute is positive if the consumer wants more of that attribute; in the second case, the correlation between the hedonic scores and the perceived intensity of an attribute is positive if the attribute is a driver of liking. For a consistent consumer, such correlation is expected to be high and positive.

By definition, a consumer should like his/her ideal more than any other product, if such ideal product happens to exist in the product set. In other words, the ideal data provided by consumers should correspond to products that would be more appreciated than any current product tested (Worch *et al.*, 2012b). Practically, this means that the ideal product should be associated with a liking score which should be larger than the liking scores given to the products tested. Since the liking score of the ideal product cannot be directly measured (such product does not physically exist), it must be estimated through statistical models. We thus define the estimated liking score associated to the ideal product as the *liking potential* of that ideal product.

Mathematically, the liking potential of the ideal product of consumer j can be estimated through the following steps: based on the liking scores and the perceived intensities j given to the products, a model explaining liking in function of the perception of the attributes can be generated. Such model can be obtained by PLS regression or PCR. This model is then applied to the sensory profile of the ideal product defined by j in order to estimate the liking potential of that ideal product. If consumer j is consistent, the liking potential thus obtained is larger than the liking scores provided by j to the products tested.

A summary of both the sensory and hedonic consistency is given in Fig. 14.2. In this figure, it can be seen that the consumer considered is consistent (both from a sensory and hedonic point of view) if the ideal intensity he/she provides is situated within the grey area (by considering that sweetness is a driver of liking here), i.e. in the area of maximum liking.

It is worth mentioning that the consistency of the ideal data as defined here is evaluated according to the liking scores. This point of view has been adopted since the vast majority consumers associate ideals with optimal product in terms of liking. Consumers that are providing their ideals based on a criterion other than liking will be considered arbitrary and inconsistent. Of course, there are cases when consumers can be very consistent in their description of their ideals without relating them to liking. For example, consumers caring about their health might define their ideals in terms of low fat, low sugar and/or low salt content consistently, although they say they prefer products with higher fat, sugar and/or salt content. This is a limitation of this approach that the user should be aware of. Additionally, since the ideal information is directly linked to liking, the optimization performed is done based on liking. Such procedure cannot be used easily to improve products in terms of costs, side benefits, etc.

A second limitation that it is important to point out is that in these procedures, only the linear relationship between attributes' level and liking scores are considered. Although in some cases quadratic effects would be a better fit, in our experience linear relationships are often sufficient to evaluate the consistency of the ideal products.

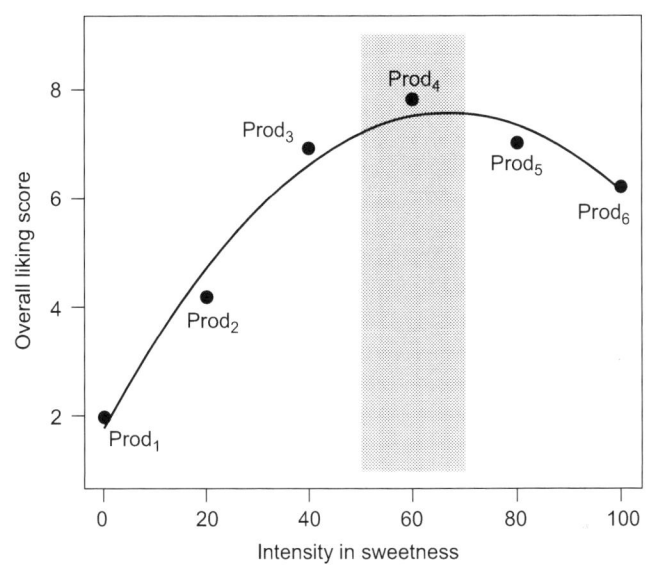

Figure 14.2 Summary of the consistency of ideal information provided by a consumer: the consumer can be considered consistent if he/she (roughly) provides an ideal intensity for sweetness within the grey area.

14.4.3 Guideline for the analysis of ideal data

As has been described previously, the analysis of ideal data should be done in four steps:

1. The *consistency* of the ideal data provided from consumers should be evaluated both from a *sensory* and *hedonic* point of view.
2. If the consumers are consistent in the description of their ideal, homogeneous groups of consumers (*clustering*) and of products (*single or multiple ideals*) should be defined.
3. For each homogeneous group of consumer and product, the *ideal product of reference* is then defined (either by using the mean, or through the IdMap).
4. Finally, the ideal product of reference thus defined is used to *guide product developers on improvement* by defining which (and how) attributes should be optimized in priority.

These four steps are the core of the IPA, the analysis of Ideal Profile data (Worch, 2012).

14.5 Illustration of the Ideal Profile Analysis (IPA)

The dataset used for illustration involves 13 sauces tested by 112 Dutch consumers using the IPM. The consumers were asked to rate each product on both perceived and ideal intensity for 23 attributes. A 100 mm line scale with anchor points at 10 and 90 was used. The consumers were also asked to rate the products on overall liking using a 9-point category scale.

To illustrate the principles described in this chapter, the methodology of the IPA is applied to the dataset presented here. This analysis starts with the assessment of the consistency of the ideal ratings provided by the consumers.

14.5.1 Consistency

14.5.1.1 Sensory consistency

Consumers are consistent from a sensory point of view if they rate their ideal product with similar sensory characteristics to the product they like most. The evaluation of sensory consistency (at the panel level) is done by evaluating whether the ideal information provided is making the link between the perception and the appreciation of the products. Such evaluation is done by double projection as supplementary of the sensory profiles (supplementary entities) and the hedonic scores (supplementary variables) within the ideal space (Fig. 14.3).

The Fig. 14.3 shows that consumers who described their ideal as less intense in herbs and spices (i.e. *32* and *52*) have their ideal close to Products 1, 4 and 12 (represented in grey in Fig. 14.3a). These consumers also liked the Products 1, 4 and 12 more than the rest of the panel (represented in grey in Fig. 14.3b). On the other hand, the consumers who described their ideal as more intense in herbs and spices described their ideal with similar characteristic Products 6 and 13. These consumers also appreciated Products 6 and 13 more than the rest of the panel.

Since consumers are defining their ideals with similar characteristics to the products they like more, the ideal is making the link between the perception and the appreciation of the products. Hence, it can be concluded that the panel of consumer is consistent from the sensory point of view.

14.5.1.2 Hedonic consistency

Consumers are consistent from a hedonic point of view if their ideals would be potentially more liked than the products tested. To evaluate the hedonic consistency of the ideal profiles, the liking potential of the ideal product needs to be estimated. To do so, a model expressing the liking scores in function of the perceived sensory characteristics of the products is defined for each consumer. Each individual model is then applied to the ideal ratings provided by the consumer considered. Finally, hedonic consistency is assessed by comparing the estimated liking potential of the average ideal products with the hedonic scores provided for the products tested. In order to facilitate the comparison, the liking potential is made relative to the liking scores obtained for the products.

Since the quality of fit of each individual model influences the estimation of the liking potential, the relative liking potential is represented in function of the R^2 of the individual model (Fig. 14.4).

In this example, the majority of consumers are associated with individual models which fit well the data (high R^2) and with relatively high liking potential (the majority

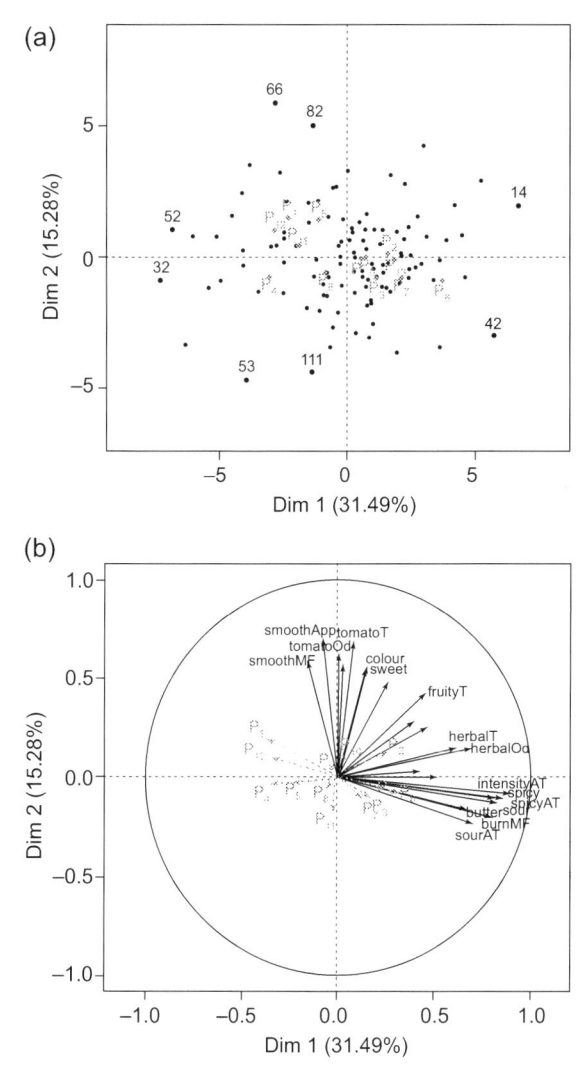

Ideal profiling as a sensory profiling technique 321

Figure 14.3 Evaluation of the sensory consistency of the ideal products at the panel level. Panel (a) represents the consumer ideal space with projection as illustrative of the products profiles (in grey). Panel (b) shows the corresponding variables representation with projection as illustrative of the hedonic scores (in grey).

of the relative liking potential are larger than 0.5 highlighted in the figure with the dotted line). In fact, only a few consumers have a negative liking potential. It can also be noted that some consumers are associated with individual models in which the goodness of fit is lower than 0.4. For those consumers, although the estimation of the liking potential is mainly positive, it is more difficult to conclude as to their consistency.

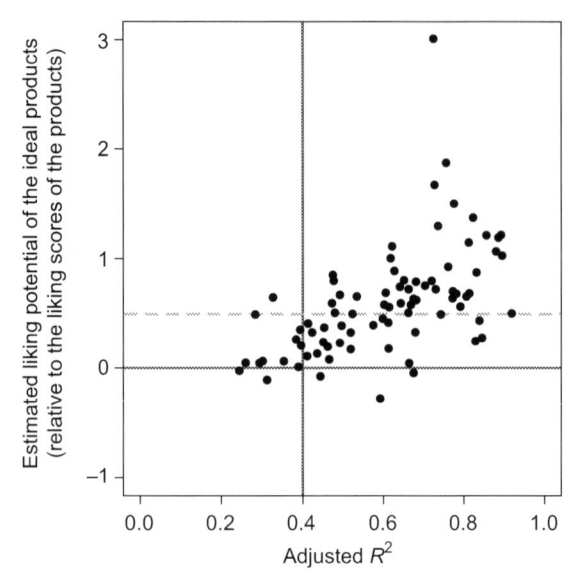

Figure 14.4 Assessment of the hedonic consistency of the consumers' ideals. The estimated liking potentials associated with the ideal products of each consumer (and made relative to the liking scores of the products tested) are represented as a function of the R^2 of the individual models.

Since consumers' ideals are associated with high estimated liking potential, their ideals would be potentially more liked than the products tested. The ideal information consumers provide would potentially improve the products tested. Hence, it can be concluded that the consumers are consistent from the hedonic point of view.

14.5.1.3 Conclusions

Most consumers are consistent both from the sensory and hedonic point of view when rating their ideal. The ideal information provided can be used to optimize the products tested. Still, it should be noted that the consistency has been evaluated according to the liking scores: consumers for whom it is concluded that they are not consistent might still be so in practice, although the statistics cannot prove it. This would be the case, for instance, if consumers are rating their ideal according to a criterion other than liking (e.g. health issues, etc.). Unfortunately, this statement cannot be verified here.

Since the ideal data are consistent, we are confident in using them for optimizing the products. However, before providing guidance on improvement, the homogeneity of the panel of consumers and of the product set need to be evaluated.

14.5.2 Homogeneous groups

14.5.2.1 Homogeneous groups of consumers (clustering)

Any optimization technique requires checking if segments of consumers exist (according to their preference). If segments exist but have not been taken into consideration,

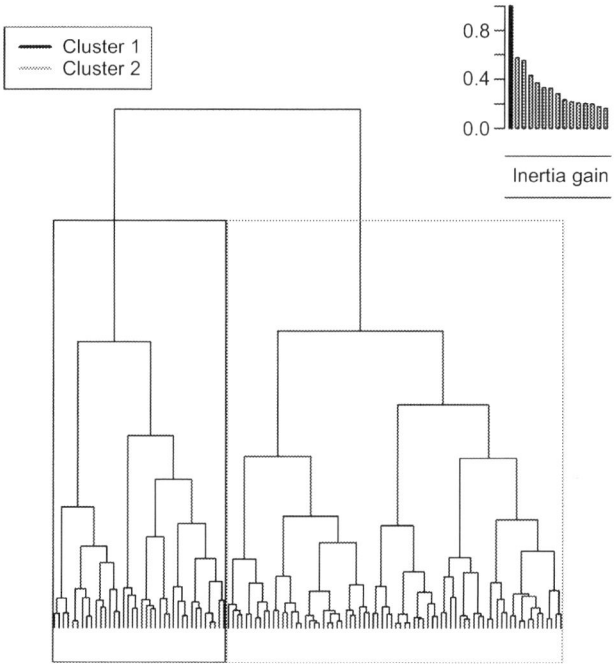

Figure 14.5 Dendrogram highlighting the two cluster solutions in the segmentation procedure.

there is a risk that the optimization is based on a hybrid product which combines characteristics of the optimum products from different segments. This hybrid product might not be appreciated by the majority of consumers, and hence would lead to product deterioration instead of product improvement.

In this example, the clustering technique deployed is hierarchical clustering using the Ward criterion (Husson *et al.*, 2010, 2011). Two segments of consumers are found: Cluster 1 (60 consumers) likes the Products 2, 6 and 13 more, while Cluster 2 (52 consumers) likes Products 1, 4, 8 and 12 more (Figs 14.5 and 14.6).

Since the panel size is limited (only 112 consumers took part in the test) the optimization will be performed on the entire panel of consumers. This approach is not optimum (one optimization procedure should be considered for each segment) but is convenient, as the goal of this chapter is to illustrate a methodology rather than improving the products from this study.

14.5.2.2 Homogeneous groups of products (single or multiple ideals)

As we have just seen, segmenting consumers according to their preference is of utmost importance in optimization procedures. Similarly, users should also ensure that the products tested all belong to the same subcategory of products, i.e. are all associated with one unique ideal. If this is not the case, the optimization procedure should be adapted by improving each product according to its adequate ideal product of reference.

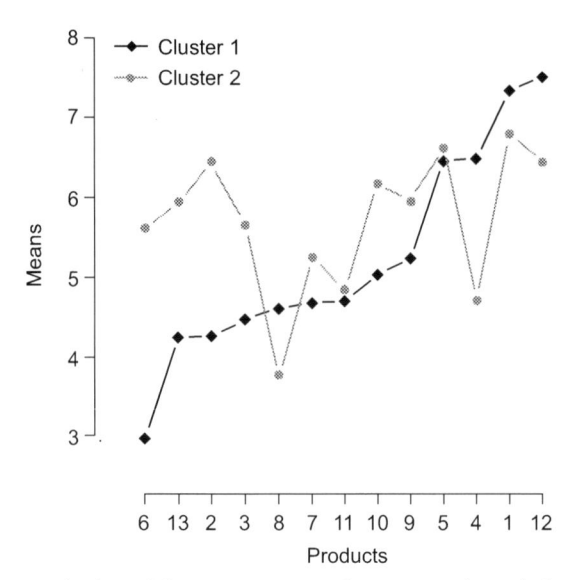

Figure 14.6 Characterization of the two segments of consumers through the average liking scores for each product.

The assessment of single or multiple ideals can be done using multivariate analysis by evaluating whether the average ideal products associated with each product tested are close or distant on the sensory space. Since the notion of close and distant is relative in multivariate analysis, confidence ellipses obtained by total bootstrap (using the variability between consumers of the ideal ratings) are plotted around each average ideal product in order to help in coming to a conclusion on the homogeneity of the product set. In this example, it can be seen that the confidence ellipses associated with the ideal products are all overlapping (Fig. 14.7): the entire product set is cueing one unique ideal.

The optimization of the entire product set is done according to one unique ideal product of reference.

14.5.3 Ideal of reference (IdMap)

The last step before performing the optimization consists of defining the ideal product that will be used as the reference to match. Two different procedures could be used: (1) the average ideal product for the adequate group of consumers and products; or (2) the ideal that is common to a maximum of consumers. We are adopting the second point of view. Such a solution is directly provided by the Ideal Mapping (IdMap). Fundamentally, the IdMap consists in associating each consumer with his/her ideal zone on the sensory map. From this map, the ideal area of the sensory space that is shared by the maximum number of consumers defines the ideal product of reference to match in the optimization procedure. This ideal product is located in the darkest area of the IdMap. In this example, the ideal product of reference is located close to Products 1, 8 and 11 (Fig. 14.8).

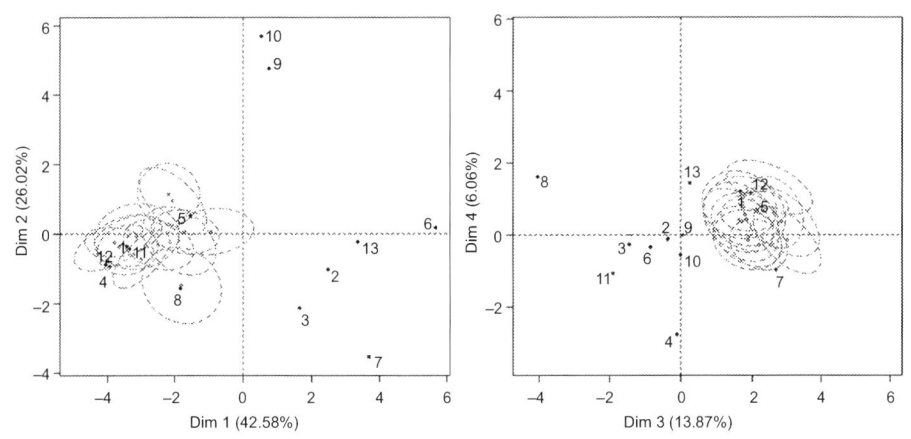

Ideal profiling as a sensory profiling technique 325

Figure 14.7 Assessment of a single or multiple ideals by multivariate analysis through the use of confidence ellipses. The results are given for the first four dimensions.

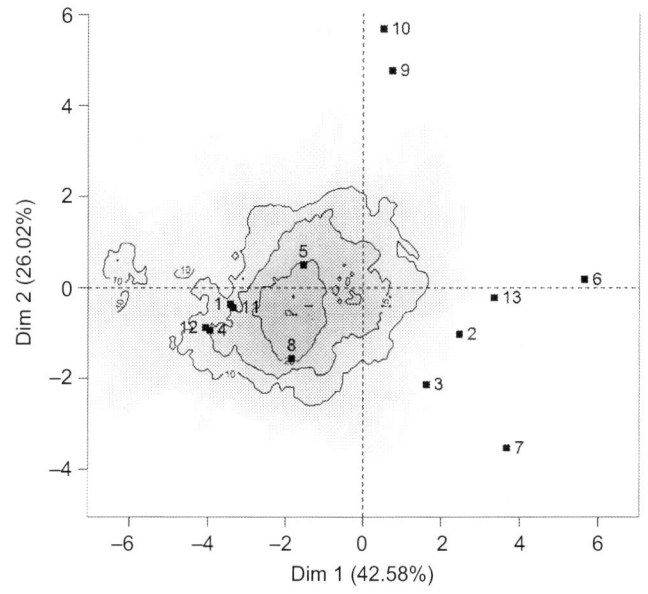

Figure 14.8 Solution of the IdMap defining the ideal of reference to match in the optimization procedure.

The sensory profile of the ideal product of reference is given in Fig. 14.9. In this figure, the sensory profile of the ideal product of reference is compared to the sensory profile of Product 1 (close to the ideal), Product 3 (further away from the ideal) and Product 9 (very far from the ideal).

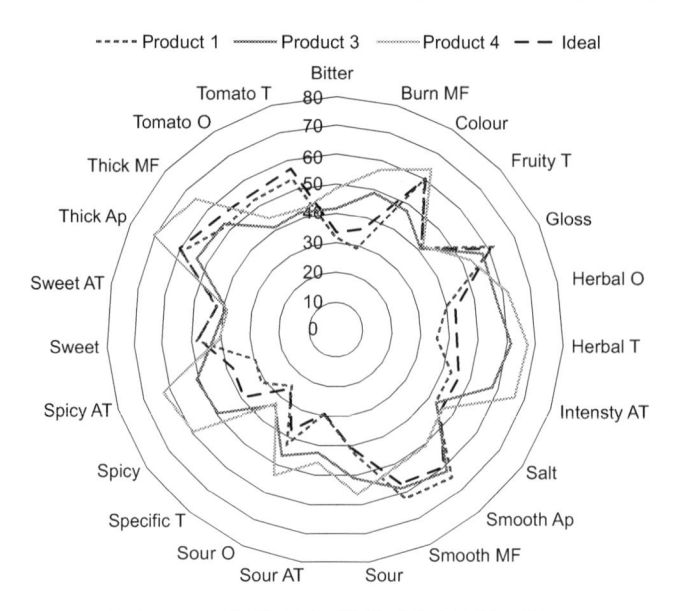

Figure 14.9 Comparison of the sensory characteristics of the ideal product of reference (obtained from the IdMap) to the sensory profiles of Products 1, 3 and 6.

14.5.4 Optimization

As in any optimization procedure, guiding the improvement of a product implies considering two pieces of information: (1) its deviation from the ideal of reference; and (2) weighting by the importance of each attribute on liking (see Equation [14.2]). In our example, the drivers of liking/disliking are provided by regression on principal component: the taste of tomato, the smoothness of the sauce, as well as the sweet/fruity taste are positive drivers, while the herbal taste, the spiciness and the bitter/sourness are drivers of disliking (Fig. 14.10).

Based on these drivers of liking and on the deviation of a product from the ideal of reference defined in Section 14.5.3, guidance on improvement can be provided. Figure 14.11 details the example of two particular products, i.e. Product 1 (closer to the ideal) and Product 3 (further away from the ideal).

Product 1 requires less optimization than Product 3. This result was as expected since Product 1 is closer to the ideal of reference than Product 3. For Product 1, the *spiciness* of the product (*taste, aftertaste* and *burnt mouth feel*) as well as the *herbal taste* should be slightly increased (black diamond on Fig. 14.11) to improve the product (the grey bars in Fig. 14.11 provide an estimation in percentage of the potential gain in liking if the product is at its optimum level for each attribute). This result may be surprising since these characteristics were defined as drivers of disliking (Fig. 14.10). However, it can be seen from the IdMap (Fig. 14.8) that the ideal product is less extreme on the first dimension than some products (e.g. Product 1), meaning

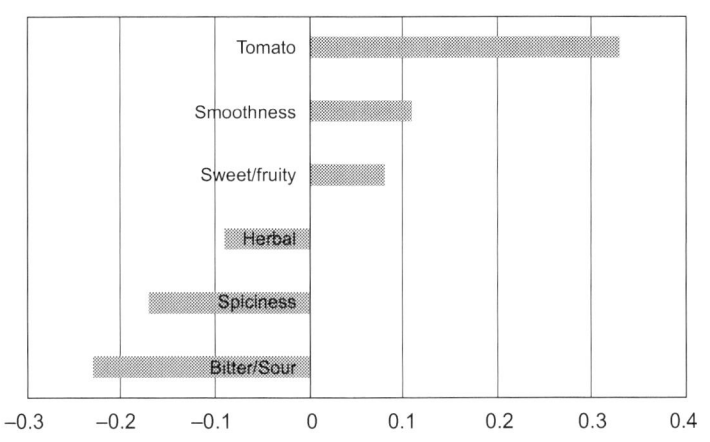

Figure 14.10 Drivers of liking/disliking obtained by regression on principal component. Here the regression weights are related to the PCA dimensions which are named according to their interpretation.

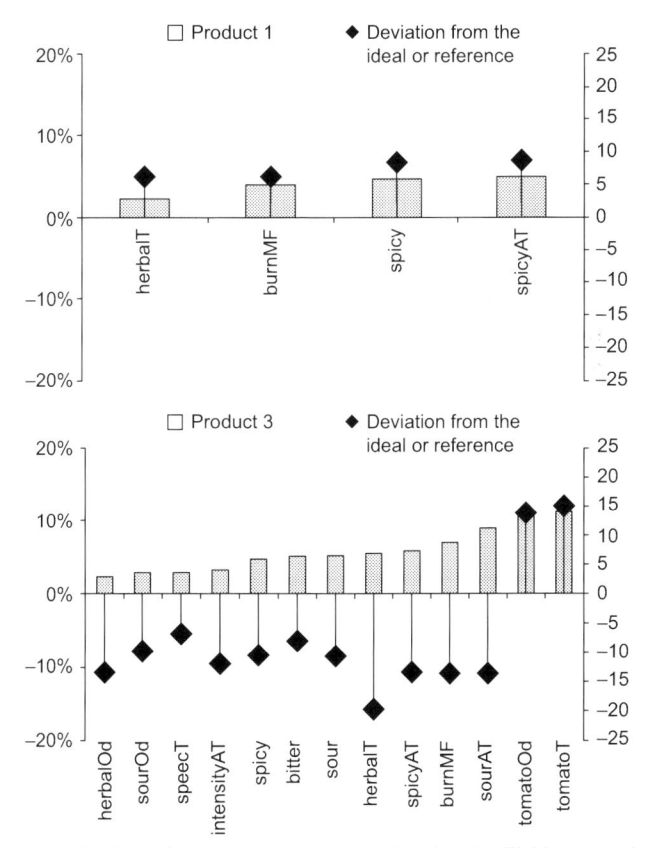

Figure 14.11 Optimization of Product 1 and Product 3 using the Fishbone method. In these plots, the grey histograms represent the potential gain in liking while the black diamonds represent the deviation of that product from the ideal product of reference.

that some attributes show saturation: too much spiciness has a negative effect on liking but absence of spiciness is not optimal either, suggesting that the right amount of spiciness should be dosed.

For Product 3, the main improvement concerns the *tomato* character, which should be increased (black diamond) to improve the products. Decreasing *sourness, spiciness* and *herbal taste* would also improve this product (although the effect on liking is less impacting).

As we have seen previously that Product 6 is even further away from the ideal, it is expected that the guidance for improvement of that product is even more important and involves more attributes than for Product 3. Such results were verified (but not shown here).

14.6 Conclusions

The IPM is a useful tool for product optimization and product development and can be applied to a large variety of products (food, beverages, cosmetics, etc.). Compared to other optimization tools (such as external preference mapping, or LSA for example), it has the advantage that a large variety of information is gathered directly from the same consumers. This large variety of data allows the user to:

1. Evaluate the consistency of the data (through the consistency).
   Although the statistics are different, this is in line with panel performance routines performed on expert data. Such a procedure cannot be performed for methodologies in which the ideal product is estimated statistically. And, as for JAR scale, no similar routine exists, to our knowledge.
2. Check the presence of eventual subcategories of products within the product set.
   In many cases, methodologies are making strong assumptions that consumers have one unique ideal. Although this assumption is often verified, situations where it is not the case also exist. In those latter cases, considering one unique ideal can be misleading. Again, this procedure is (to our knowledge) unique to the IPM and requires the consumers to rate their ideal for each product tested.
   It is worth saying that external preference mapping techniques and tasks involving JAR questions are able to deal with multiple ideals, either by estimating different acceptance areas in the product space for a given consumer (PrefMap) or by providing guidance on improvement for each product separately.
3. Study more deeply the variability of the ideal products between consumers.
   It would be a strong assumption to consider that consumers all share a similar ideal. Even within homogeneous clusters (in terms of liking) consumers do not all share the same ideal: although the same characteristics are highlighted, some will be more extreme than others in terms of ideal intensity for each attribute. This is often observed in penalty analysis involving JAR scale when clusters of consumers would still disagree for some attributes, a large proportion finding it *too much* while another large proportion of consumers finding it *too little*.
4. Guide more precisely on product optimization.

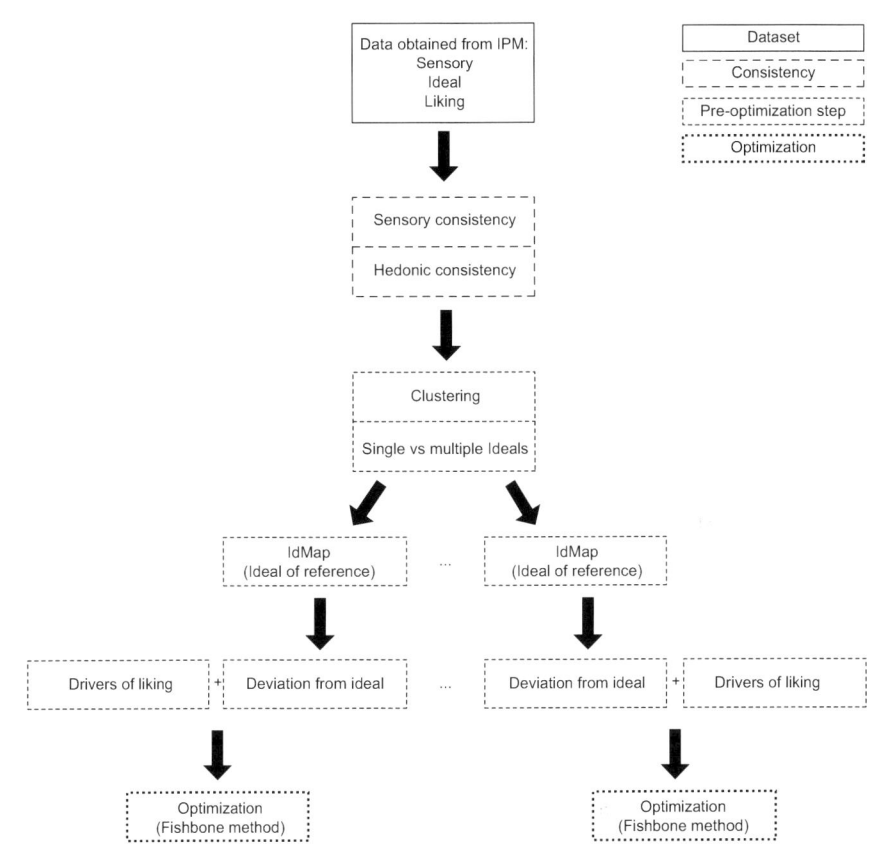

Ideal profiling as a sensory profiling technique 329

Figure 14.12 Schematic representation of the IPA.

With the IPM, since the ideal level of the product of reference is estimated (using the mean or the IdMap solution), the exact difference between the perception of a product and this ideal of reference can be computed. This deviation from the ideal hence provides both the direction of the change (i.e. increase or decrease the level of that attribute) and the intensity of change to apply (i.e. should the product be changed slightly or a lot on this attribute). Such information is not available in JAR scale, since in penalty analysis the degrees of difference are often omitted.

Such advantages are made possible thanks to the IPA (a summary of the methodology of the IPA is provided in Fig. 14.12).

However, gathering so much information has some drawbacks, and some users might be reluctant to use this method as they would find it too demanding for consumers. Moreover, since no preliminary training is performed, the questionnaire should be set up in a way that it is understandable by consumers. But thanks to the analyses developed for this type of data, it is possible to evaluate in each case how efficiently consumers are performing, hence giving more confidence in the conclusions drawn. Although most of the analyses presented here can be performed within your usual statistical software, users should bear in mind that preliminary data transformations might be required (e.g. correction of the ideal data). For those who are more interested, software packages including the IPA routines exist: a selected part of the routines is available in Senstools (www.senstools.com), while the entire IPA has been programmed in the R language (R Core Team, 2012) and made available within the SensoMineR package (Lê and Husson, 2008).

IPM can be seen as a rapid methodology for different reasons. First, it is a rapid method in terms of data acquisition, as it measures acceptance (i.e. liking), sensory perception and ideal intensities of a set of products, in one unique test. Second, in terms of analysis, it links consumers' liking directly to their perception, and their desires (through ideals), without requiring any particular statistical model. In terms of interpretation and guidance, it shows directly by which attributes products fail to satisfy the consumer and the effect on liking when these attributes reach ideal levels. Such outcomes do not provide a technical recipe for improvement, but give guidance to the developer that are directly actionable. Finally, IPM is time- (and effort-) saving: as it is performed by consumers, training sessions are not recommended here.

References

Booth, D.A. and Conner, M.T. (1990) "Characterisation and measurement of influences on food acceptability by analysis of choice differences: theory and practice," *Food Quality and Preference*, **2**, 75–85.
Booth, D.A., Conner, M.T. and Marie, S. (1987) "Sweetness and food selection: measurement of sweetners' effects on acceptance," in Dobbing J, *Sweetness*, London, Springer-Verlag.
Carroll, J.D. (1972) "Individual differences and multidimensional scaling," in Shepard R, Romney A and Nerloves S, *Multidimensional Scaling: Theory and Applications in the Behavioral Sciences*, New York, Academic Press, 105–155.
Cooper, H.R., Earle, M.D. and Triggs, C.M. (1989) "Ratios of ideals – a new twist to an old idea," in *Product Testing with Consumers for Research Guidance*, Philadelphia, ASTM STP 1035, 54–63.
Danzart, M. (2009) "Cartographie des préférences," in *SSHA Evaluation Sensorielle, Manuel Méthodologique*, Paris, Lavoisier, 443–450.
Ennis, D.M. (2005) "Analytic approaches to accounting for individual ideal point," *IFPress*, **82**, 2–3.
Gacula, M., Rutenbeck, S., Pollack, L., Resurreccion, A.V.A. and Moskowitz, H.R. (2007) "The just about right intensity scale: functional analyses and relation to hedonics," *Journal of Sensory Studies*, **22**, 194–211.
Hoggan, J. (1975) "New product development," *MBAA Technical Quarterly*, **12**, 81–86.
Husson, F., Josse, J. and Pagès, J. (2010) "Principal component methods – hierarchical clustering – partitional clustering: why would we need to choose for visualizing data?" *Technical Report – Agro Campus*, retrieved from http://www.agrocampus-ouest.fr/math002E.
Husson, F., Le Dien, S. and Pagès, J. (2001) "Which value can be granted to sensory profiles given by consumers? Methodology and results," *Food Quality and Preference,* **12**, 291–296.
Husson, F., Lê, S. and Pagès, J. (2011) "Clustering" in Husson F., Lê S. and Pagès J. *Exploratory Multivariate Analysis by Example Using R*, London, CRC Press, 169–204.

Lê, S. and Husson, F. (2008) "SensoMineR: a package for sensory data analysis," *Journal of Sensory Studies*, **23**, 14–25.

Lê, S., Husson, F. and Pagès, J. (2006) "Another look at sensory data: how to have your salmon and eat it, too!" *Food Quality and Preference*, **17**, 658–668.

MacFie, H.J., Bratchell, N., Greenhoff, K. and Vallis, L.V. (1989) "Designs to balance the effect of order of presentation and first-order carry-over effects in hall tests," *Journal of Sensory Studies*, **4**, 129–148.

Meilgaard, M.C., Carr, T. and Civille, G.V. (2007). *Sensory Evaluation Techniques*. London, CRC Press (Fourth Edition).

Meullenet, J.F., Lovely, C., Threlfall, R., Morris, J.R. and Striegler, R.K. (2008) "An ideal point density plot method for determining an optimal sensory profile for Muscadine grape juice," *Food Quality and Preference*, **19**, 210–219.

Meullenet, J.F., Xiong, R. and Findlay, C.J. (2007) *Multivariate and Probabilistic Analyses of Sensory Science Problems*, Ames, Blackwell Publishing, 208–210.

Moskowitz, H.R. (1972) "Subjective ideals and sensory optimization in evaluating perceptual dimensions in food," *Journal of Applied Psychology*, **56**, 60–66.

Moskowitz, H.R. (1996) "Experts versus Consumers: a comparison," *Journal of Sensory Studies*, **11**, 19–37.

Moskowitz, H.R. (1997) "Base size in product testing: a psychophysical viewpoint and analysis," *Food Quality and Preference*, **8**, 247–255.

Moskowitz, H.R. and Sidel, J.L. (1971) "Magnitude and hedonic scales of food acceptability," *Journal of Food Science*, **36**, 677–680.

Moskowitz, H.R., Stanley, D.W. and Chandler, J.W. (1977) "The eclipse method: optimizing product formulation through a consumer generated ideal sensory profile," *Canadian Institute of Food Science Technology Journal*, **10**, 161–168.

Nestrud, M. (2012) "Product landscaping the bayesian way: uncovering the evaluative dimensions of consumers," *Oral Presentation* at the *3rd Society of Sensory Professionals Conference*, 10–12 October, Jersey City, NJ, USA.

R Core Team (2012) "R: A language and environment for statistical computing," *R Foundation for Statistical Computing*, Vienna, Austria. URL: http://www.R-project.org/.

Rothman, L. and Parker, M. (2009) "Just-About-Right JAR Scales: Design, Usage, Benefits and Risks," *ASTM International*, Manual MNL-63-EB, USA.

Stone, H. and Sidel, J.L. (2004) *Sensory Evaluation Practices*. San Diego, Academic Press, 241–246.

Stone, H., Sidel, J., Oliver, S., Woosley, A. and Singleton, R.C. (1974) "Sensory evaluation by quantitative descriptive analysis," *Food Technology*, **28**, 24–34.

Szczesniak, A., Loew, B.J. and Skinner, E.Z. (1975) "Consumer texture profile technique," *Journal of Food Science*, **40**, 1253–1256.

Thomson, D.M.H. (1988). *Food Acceptability*. London, Elsevier.

Van Trijp, H.C., Punter, P.H., Mickartz, F. and Kruithof, L. (2007) "The quest for the ideal product: comparing different methods and approaches," *Food Quality and Preference*, **18**, 729–740.

Worch, T. (2012) "The Ideal Profile Analysis: From the validation to the statistical analysis of Ideal Profile data," PhD document, retrieved from www.opp.nl/uk/.

Worch, T. (2013) "PrefMFA, a solution taking the best of both internal and external preference mapping techniques," *Food Quality and Preference*, **30**, 181–191.

Worch, T., Dooley, L., Meullenet, J.F. and Punter, P.H. (2010b) "Comparison of PLS dummy variables and Fishbone method to determine optimal product characteristics from ideal profiles," *Food Quality and Preference*, **21**, 1077–1087.

Worch, T. and Ennis, J.M. (2013) "Investigating the single ideal assumption using Ideal Profile Method," *Food Quality and Preference*, **29**, 40–47.

Worch, T., Lê, S. and Punter, P. (2010a) "How reliable are the consumers? Comparison of sensory profiles from consumers and experts," *Food Quality and Preference,* **21**, 309–318.

Worch, T., Lê, S., Punter, P. and Pagès, J. (2012a) "Assessment of the consistency of ideal profiles according to non-ideal data for IPM," *Food Quality and Preference,* **24**, 99–110.

Worch, T., Lê, S., Punter, P. and Pagès, J. (2012b) "Extension of the consistency of the data obtained with the Ideal Profile Method: would the ideal products be more liked than the tested products?" *Food Quality and Preference,* **26**, 74–80.

Worch, T., Lê, S., Punter, P. and Pagès, J. (2012c) "Construction of an Ideal Map *IdMap* based on the ideal profiles obtained directly from consumers," *Food Quality and Preference,* **26**, 93–104.

Worch, T., Lê, S., Punter, P. and Pagès, J. (2013) "Ideal Profile Method (IPM): the ins and outs," *Food Quality and Preference*, **28**, 45–59.

Part Three

Applications in new product development and consumer research

Adoption and use of Flash Profiling in daily new product development: a testimonial

15

C. Petit, E. Vanzeveren

Puratos N.V, Groot-Bijgaarden, Belgium

15.1 Introduction

Developing sensory descriptive methodologies can be quite challenging in a highly competitive industrial environment, particularly for companies commercializing a large diversity of products and confronted with limited resources to allocate to sensory analysis. In this chapter, we will explain why/how we decided to implement Flash Profile for descriptive sensory analysis for new product development, and how this methodology can be easily and quickly implemented to answer a diversity of sensory issues. We will present the lessons we have learned from more than ten years of experience in the use of this method. More specifically, we will examine the advantages and limitations of the use of Flash Profile in an industrial context. We will then describe some solutions we have found to overcome these limitations, and discuss the need to develop or adopt new sensory descriptive methodologies.

15.2 Flash Profile as a starting point

15.2.1 Industrial context and challenges

Conventional descriptive analysis, such as Quantitative Descriptive Analysis (QDA)® or Spectrum, has been successfully implemented and used for years in many food companies as a standard measurement tool to assist product development. When we decided to implement sensory analysis methods in our company in 2001, we soon realized that we faced several constraints that made the implementation of such methodologies rather difficult in our situation.

First of all, as a producer of ingredients for the bakery, patisserie and chocolate sectors, we sell products to be used in a multitude of final applications: from bread to custard creams, from chocolate to glazes, from sponge cakes to fruit fillings. Also, we produce and sell our ingredients all around the world. We develop international recipes, but each country also develops its own local recipes for each application. For one specific product category, custard cream mixes for instance, we can have in our range several hundreds of references.

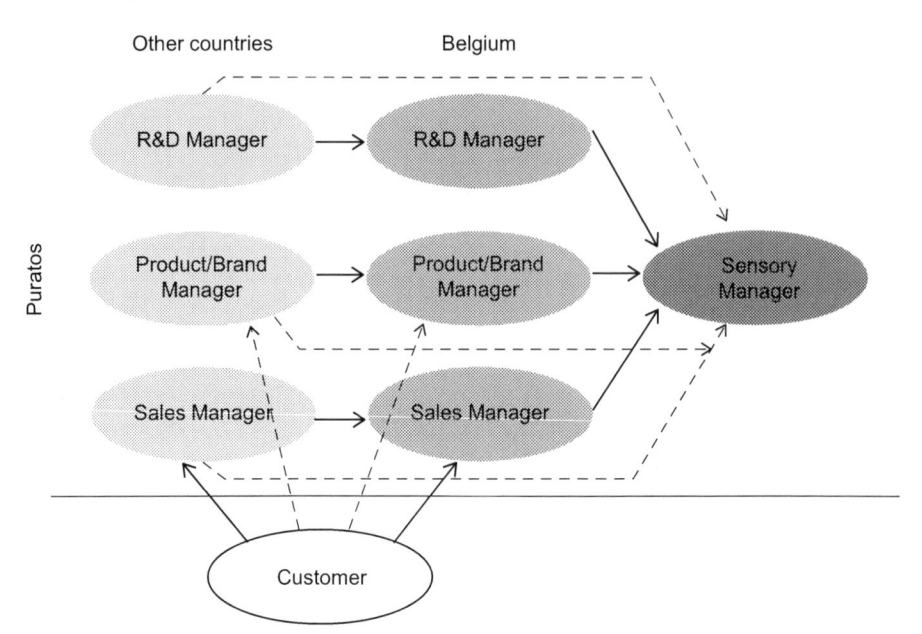

Figure 15.1 Diagram showing the origin of sensory descriptive requests within the Puratos group.

Secondly, Puratos is a business-to-business company: we are not selling directly to the final consumers, but to customers producing bakery, patisserie or chocolate end-products. In this context, we are confronted with a diversity of stakeholders and communication channels (Fig. 15.1). In addition, the development time of a product is very short in our business (on average one to two years). We thus need to be able to react very quickly to a sensory request, especially if it comes directly from one of our customers.

Thirdly, due to our limited resources at the time we started to implement sensory analysis methodologies, it was not possible to externalize the sensory panels: we had to begin the work with internal judges. Working with internal people certainly has certain advantages (motivation, implication of people, flexibility in organization), but it also has specific constraints: we need to ensure that there is always a pool of judges available for any test, and we are faced with a substantial turnover of judges (about 25% of the panel each year).

Consequently, using conventional descriptive analysis methodologies, which would require training panellists on all our applications to answer short-term requests, did not seem practically realistic. Moreover, our needs in terms of sensory description were primarily to highlight the main differences between our products, rather than to get a very fine sensory description of each product. From this perspective, conventional sensory descriptive techniques did not appear to be the optimal choice.

15.2.2 Advantages of Flash Profile in our context

Amongst the most recently developed sensory descriptive methods, Comparative Free Choice Profiling, also called Flash Profile, was particularly suitable for us, because it is based on a comparative evaluation procedure (Delarue and Sieffermann, 2004). In conventional profiling methodologies the products are evaluated in a sequential monadic way: the products are presented and each is evaluated by means of all the sensory descriptors. On the contrary, with Flash Profile the judges evaluate the whole sample set attribute by attribute: they receive all products simultaneously and have to rank them all on each attribute. Comparing and ranking products is considered to be a much easier task for assessors than giving an absolute intensity score for a single product. We indeed found it easy to train our internal judges on this evaluation procedure.

The second interest of Flash Profile for us was that it is based on free vocabulary, as in FCP (Williams and Langron, 1984): the attributes are selected by each judge individually to describe the main perceptible differences for a set of products. This perfectly met our need to focus on the main sensory differences between products. Free vocabulary was also an advantage considering the diversity of our product ranges: letting the judges free to generate their own descriptors allowed us to build up our knowledge of descriptive terms for all product categories. Also, by using individual descriptions we can combine different point of views for each evaluation: one judge may indeed focus on visual texture characteristics, while another one on mouthfeel properties, and a third on specific flavour notes. By including in our sensory panel internal judges with different backgrounds and different levels of expertise, we quickly enriched our sensory descriptive vocabulary for most of our products. It was also quite relevant in the context of an international company, where we have to work with judges using different languages.

Finally, a huge advantage of Flash Profile was its rapidity. After being trained on the evaluation procedure, the judges are directly ready to perform their first evaluation. Each evaluation session, typically for six to seven samples, takes about 30 min per judge. This includes the generation of a few sensory descriptors, followed by the comparative evaluation of the products on each descriptor. With between 8 and 12 judges per session, we can complete the sensory description of our target products in a few hours.

15.2.3 Internal development of sensory descriptive analysis

Thanks to Flash Profile, we have been able to develop sensory analysis internally with few resources and a tight schedule. Convincing our management was not difficult: without this methodology we would not have been able to implement sensory descriptive analysis at all. Starting with only ten volunteer judges in 2001, we are now managing in Belgium a panel of 40 internal expert judges, who can work on all product categories. In 2012 we conducted more than 100 sensory descriptive sessions in Belgium for new product development purposes; this included descriptive tests done at the headquarters and at two other Belgian sites (Fig. 15.2).

Sensory analysis is nowadays spreading out within Puratos Group, with several countries locally implementing the main methodologies following our internal guidelines. China and Brazil are now in the process of building up panels for sensory descriptive testing, following the same steps that we did in the past (Table 15.1). Using Flash Profile, they can quickly get concrete results in a context of very limited resources.

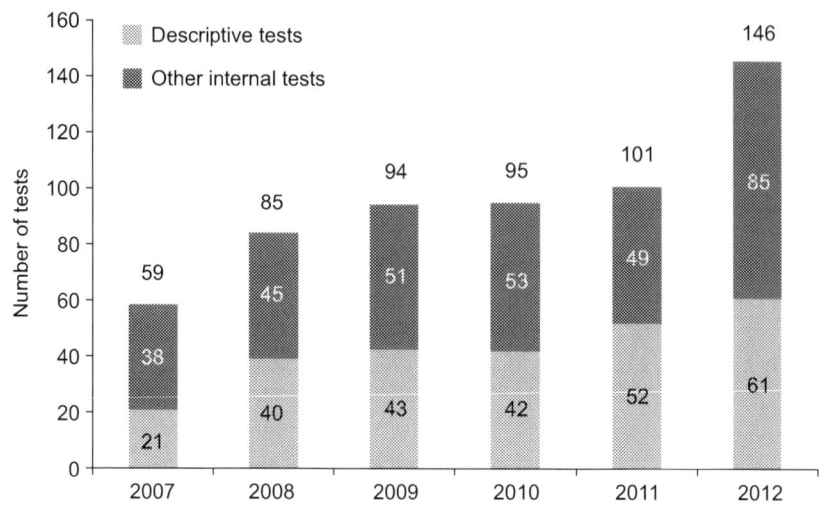

Figure 15.2 Number and type of internal sensory tests for product development (R&D) conducted at Puratos' Belgium headquarters between 2007 and 2012.

Table 15.1 Internal human resources for sensory analysis within Puratos group

Country/site	Sensory activities	Sensory panels
Belgium headquarters, Brussels	Discriminative and descriptive tests	70 judges for discriminative testing, 25 judges for descriptive testing
Belgium Erembodegem	Discriminative and descriptive tests	15 judges
Belgium Andenne	Discriminative tests	40 judges
Brazil	Discriminative and descriptive tests	Training on-going
USA	Discriminative tests	Training on-going
China	Discriminative and descriptive tests	Training on-going

15.3 Flash Profile as a reference methodology

15.3.1 Explorative studies and product benchmarking

Flash Profile has become our internal methodological reference for sensory descriptive analysis, and we use it in a diversity of situations requiring descriptive sensory analysis.

It can be of great help for explorative studies on a new or not-well known product category. In this case we will focus on the development of a specific sensory vocabulary, which is actually the first step of Flash Profile. For instance, we have recently explored the sensory characteristics of new chocolate glazes from our international range. This investigation has allowed us to determine on which sensory aspects the glazes differentiate, and has helped us to orientate further communication strategies.

For most of our projects linked to new product development, we want to investigate in detail the sensory proximities/differences perceived between samples. For this objective, we usually rely on sensory maps obtained from Flash Profile. After the judges have ranked the products on each of their selected descriptors, the sensory map is obtained by generalized Procrustean analysis (GPA) (Gower, 1975). Depending on the objective of the study, we can use the descriptive sensory map to compare several prototypes to an internal reference, or to compare several competitor samples to our own products for benchmarking.

When consumer data are also available on the same products, we can build up an External Preference Mapping model to visualize how consumer preferences are linked to the sensory characteristics of the products (Schlich, 1995). This tool is especially powerful in orientating further product development towards consumer appreciation. Figure 15.3 shows an example of Preference Mapping, relating the sensory characteristics of eight European toast breads from different countries to the preference of Spanish consumers for the same toast breads. We can see that consumers globally prefer softer and moister products. The Spanish bread (actually the brand of the market leader) is close to this profile, so it is well in line with the taste expectations of the Spanish consumers.

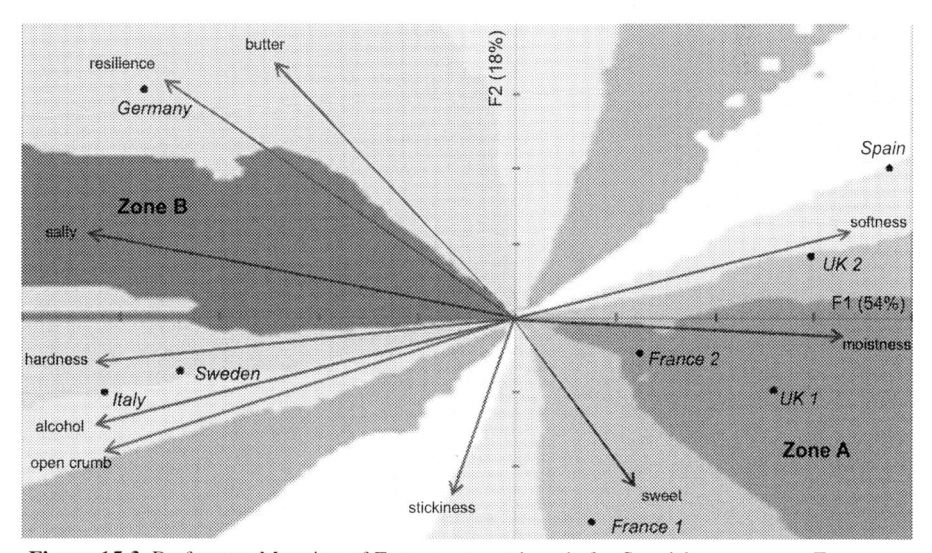

Figure 15.3 Preference Mapping of European toast breads for Spanish consumers. Zone A corresponds to higher consumer preferences, Zone B corresponds to lower consumer preferences.

Flash Profile can thus be useful in the early stages of a project, for instance to explore a range of products from the market, as well as during the course of the development process, to help fine-tuning prototypes to match consumers' expectations. As obtaining a sensory map requires mobilizing our resources for a short period of time, we regularly use Flash Profile at different stages during the same project. The sensory map can be a very valuable tool to orientate the development of new products.

15.3.2 Sensory description as a tool for communication with customers

The sensory descriptive map is also a powerful communication tool to our customers. As a business-to-business company, our challenge is not only to understand better the preferences and sensory expectations of the end-consumers, but also to communicate on these finding to our direct customers, who are selling the final products made with our ingredients. Using Flash Profile, we have developed several communication tools highly valued by our customers. Figure 15.4 represents one of these tools, called "What's your Texture?." It is based on a sensory map representing the differences in texture between soft breads, which is presented to the customer together with real demonstration samples. All samples are made using the same recipe, but with different texture improver solutions. It is possible to also include physical measurements into the sensory map, by adding these measurements as supplementary variables in the GPA. The objective of this tool is to show how we can adapt the texture of a soft bread to the needs and desires of the customer. By directly evaluating the samples

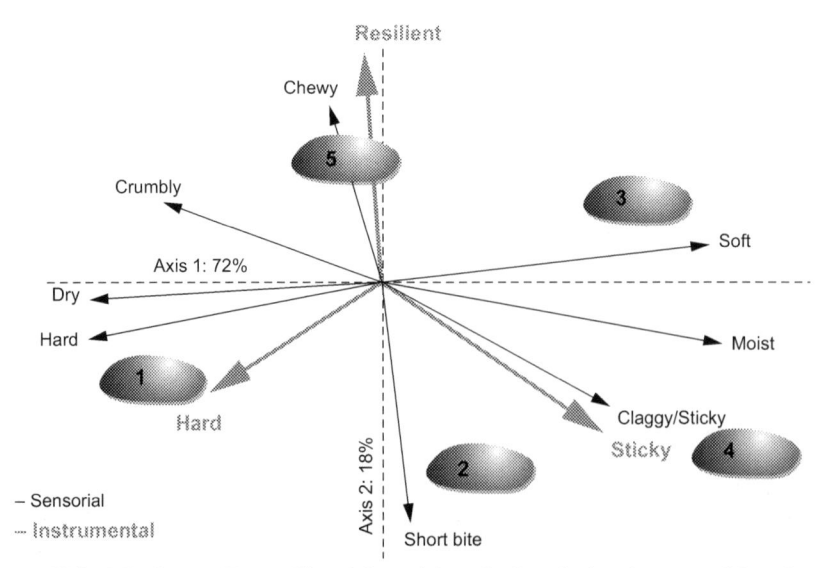

Figure 15.4 "What's your Texture?" tool for soft breads. Sample 1: reference soft bread without texture improver; Samples 2 to 5: soft breads with different texture improver solutions.

during the demonstration and referring to the "texture" map, we are sure we speak the same language as the customer and we understand the type of texture they want to get for their product. Similar tools have been created for different applications: for instance, the texture of donuts, or the flavour of pure chocolate.

Though they are not always easy to communicate to people who are not familiar with factorial analyses, such sensory maps encounter great success with most of our customers. Indeed, beyond focusing on the technical advantages of the products, it allows speaking about perceptions using a common vocabulary.

15.3.3 Development and maintenance of a sensory panel for Flash Profile

According to our experience, it is easier to select judges to participate in Flash Profile studies than in conventional descriptive analysis studies. Indeed, as discussed previously, getting used to the comparative evaluation procedure is rather intuitive and does not require an intensive training from the judges. This makes the task very accessible to people with a basic knowledge of sensory analysis techniques. Basically, we select our panel members based on their ability to describe several products from our range during an individual training session. However, we cannot ensure that all judges will at the beginning be efficient in all product categories. The main difficulty for them is then to develop a relevant and precise sensory vocabulary for several types of products and several sensory dimensions (texture, taste, flavour). Depending on the individuals and on their own frequency of participation, this process can take from a few weeks to several months.

Moreover, maintaining the expertise of the panel over time is a difficult task. Indeed, the frequency of evaluation requests for each type of product can vary a lot in the course of a year. Some products are evaluated every week, while other products are only evaluated every three months. For this reason, we have to regularly organize group training sessions focused on specific product ranges. The aim of these sessions is for the judges to train themselves on the main sensory descriptors relative to the product ranges they are less used to, or with which they feel less comfortable.

15.4 Limitations and perspectives in the use of Flash Profile

15.4.1 Current limitations of the methodology

Despite its enormous advantages, Flash Profile cannot be used to answer all our sensory issues, and we are now facing some limitations with this methodology.

First, we need to highlight the constraint of the number of products to be evaluated simultaneously by the judges. Comparing two products requires tasting each one of them at least once, but comparing N products together requires a much higher number of tastings. When evaluating baked goods (bread or cakes), which are quite satiating products, the perceived fullness can affect the evaluation. Also for products with very intense flavour, such as fruit fillings or pure chocolate, the judges can rapidly come to sensory saturation. Based on our experience, we have to limit the number of products to be evaluated in one tasting session to six or seven. In the case of pure chocolate, we lower this limit to five samples. This means that for each Flash Profile exercise we can evaluate a maximum of five to seven products. Most of the time, a preliminary selection step is needed to determine the set of products to be included in the sensory descriptive session. This selection is usually based on informal tasting with a few experienced judges and/or physical texture measurements.

A second limitation of the comparative evaluation methodology is that it is rather difficult to compare results from different sessions obtained at different times. Indeed, with Flash Profile the products are ranked in order of increasing intensity for each descriptor, but they are not given any absolute intensity scores on a scale. Contrary to conventional descriptive methodologies, it is thus much more difficult to correlate several sensory maps. Two maps can be qualitatively compared, but combining data into one map is not straightforward and will require specific calculation techniques. The free vocabulary also adds to this complexity and to the difficulty in comparing data from different exercises.

The fact that the evaluations are based on ranks and not on intensity scores can finally lead to difficulties in communicating the results: we can always say that Product B is softer than Product A, but "how much softer?" is a recurring question. It could only be answered by including standard products in the evaluations and by using intensity scales.

15.4.2 Perspectives for further development of descriptive sensory analysis

From this perspective, we are now evolving by introducing more conventional descriptive methodologies to complement our current practices. Our first challenge is to train our judges to give intensity scores to products, without having time to spend on specific out-of-project training sessions. With this purpose, we recently introduced a rank-rating evaluation procedure: the judges have to rank the products on a linear scale with labels from 0 to 10 (Ishii *et al.*, 2007). Products are still evaluated in a comparative way, but after ranking them the judges can also indicate the perceived distances between them on the scale. This last task is not mandatory, and the given scores are mostly not recorded. Even so, we noticed that most judges spontaneously used the digit-labels to position the products, and try to give this information about distances. Working with a rank-rating procedure seems to help the judges to learn more easily how to use intensity scales. It proved to be a relevant way to start training our judges on new products categories.

To go further with intensity scoring, we are now developing scoring cards specifically for the evaluation of the texture of some key product categories: soft bread, cake and custard creams. Our objective is to unify across all countries our evaluation procedures and standards. To do so, we followed the main steps of the QDA®

methodology: selection of internal references, set-up of a descriptor list by group discussion, definition of intensity scales with high and low reference points, training of the judges, and checking for individual repeatability and consensus. While this process is highly time consuming and monopolizes lots of resources, it is of real importance for the international development of sensory descriptive analysis within the company.

Also, we are currently working on a new descriptive methodology to combine the results of descriptive evaluations obtained at different times. As discussed in Section 15.4.1, the evaluation procedure in Flash Profile implies that we cannot directly combine results from evaluations at different times. By including into the sample set a few standards which are common to all evaluations, and by using a specific calculation module, we are now able to pool data from different evaluation sessions and to analyse them into one sensory map, even if we did not have the same judges in the different sessions (Teillet *et al.*, 2013).

15.5 Conclusion

Being a quick and effective method to describe the main sensory differences between products, Flash Profile has established itself as the reference sensory descriptive methodology in our highly constrained industrial environment. After several years of extensively using this methodology, we have been building up our sensory knowledge on most of our product ranges. We have developed a sensory vocabulary for R&D, and an efficient way to augment our product performances for our customers using sensory mapping.

From our point of view, Flash Profile cannot replace conventional descriptive analysis methodologies. As it does not allow a precise quantification of the intensity of sensory attributes, it does not seem suitable for tracking in time the evolution of the sensory properties of products. However, Flash Profile can be seen as a powerful complement or alternative to conventional sensory profiling techniques at the earlier stages of product development, for sensory benchmarking or rapid characterization of a product set. Its rapidity also offers an incontestable advantage over conventional methods in a business environment.

References

Delarue, J. and Sieffermann, J.M. (2004) "Sensory mapping using Flash profile. Comparison with a conventional descriptive method for the evaluation of the flavor of fruit dairy products." *Food Quality and Preference*, **15**(4), 383–392.

Gower, J.C. (1975) "Generalized procustes analysis" *Psychometrika*, **40**, 33–51.

Ishii, R., Chang, H.K. and O'Mahony, M. (2007) "A comparison of serial monadic and attribute-by-attribute protocols for simple descriptive analysis with untrained judges." *Food Quality and Preference*, **18**(2), 440–449.

Schlich, P. (1995) "Preference Mapping: Relating consumer preferences to sensory or instrumental measurements," in Etievant P and Schrier P, *Bioflavour*, INRA Dijon, 135–150.

Teillet, E., Petit, C. and Delarue, J. (2013) "Combining PSP and Flash Profile... Why? How does it work?," *10th Pangborn Sensory Science Symposium*, Brazil.

Williams, A.A. and Langron, S.P. (1984) "The use of free-choice profiling for the evaluation of commercial ports." *Journal of Science of Food and Agriculture*, **35**, 558–568.

Improving team tasting in the food industry 16

M. Rogeaux
Danone Nutricia Research, Palaiseau, France

16.1 Introduction: the ever-increasing importance of new tasting methods within the project teams

As companies, we are currently facing a new context in which project teams need to evaluate their products. This need is obvious, of course; it is something developers have been working on for some time. But the need to improve, standardize and formalize the approach appears to have become increasingly important, for three reasons:

- First of all, a product's organoleptic quality has become an obligation. As a result, in the majority of our projects, a sensory evaluation is required in different phases of a project, and this evaluation must be objective. Traditional sensory profile approaches remain relevant and accurate – but they are also time-consuming and costly, and fail to allow all evaluations to be performed. Some of the evaluations must be managed on a day-to-day basis by the project team using lighter methods that are managed by the project team independently (i.e. less complicated, less burdensome and quicker methods).
- Within project teams, there is also a need for better integration of the sensory information provided through tasting sessions. Thus, in order to allow projects to be managed more efficiently, we must be able to evaluate products within project teams. This then enables the results to be integrated immediately, and allows the products' sensory performances to be taken into consideration by the entire project team, which is then better aligned for the next step of the project.
- Lastly, the nature of the differences to be detected is becoming increasingly precise and subtle. Basic evaluation without a precise design is inadequate. The challenge is therefore to be able to set up new sensory approaches that will be able to detect differences between products.

Against this backdrop, it is easy to see the increasing need to implement the new tool "team tasting", which involves the internal project team in the sensory evaluation. This chapter will analyse how this tool can be used, according to the following plan:

1. Precise analysis of the concrete situations where this type of evaluation is appropriate;
2. Analyses of the opportunities and constraints linked to a project team evaluation;
3. Presentation of the methodology used;
4. Presentation of concrete results;
5. Prospects.

16.2 Precise analysis of the concrete situations where evaluation by team tasting is appropriate

A project team needs to be able to accurately characterize its products throughout the development process. From a practical point of view, it can be said that four relevant "needs" can be managed effectively through team tasting.

16.2.1 Need 1: basic characterization of the products

When the aim is to increase our knowledge of a category of product, we often need to evaluate the main differences between the products in a simple but structured way. This evaluation gives a first indication of the differences between three to ten products, and enables these differences to be taken into account by the project team in the early phases of a project. This evaluation can also serve as a screening process for selecting a sub-set of products, e.g., 10–20 products from a base of 60–80 products. It is important that the characterization is performed reliably using such characterization tools as expert profiling if a trained panel is available and, if the budget allows, alternatively faster tools such as sorting or flash.

16.2.2 Need 2: monitoring of improvements during development work

Development work is built around the modification of certain sensory properties to give the consumer a more satisfactory experience of the product. The product must comply with an approved specific brief before the development teams start working to improve it. The major approval stages are, of course, validated for sensory and consumer tests, but day-to-day, after each important process/formulation modification, it is important to check whether the tests are in line with the objectives. The products are therefore evaluated regularly to monitor whether the solutions implemented are moving in the right direction from a process and formulation point of view. If this need is met, it significantly reduces risk in project management.

16.2.3 Need 3: product comparison

When the intended aim is to change a formula, reduce costs or duplicate a formula, an evaluation is sought. Holistic tests (with same–different or triangular tests) can be used in the context of a similarity-based approach (beta risk). In practice, these tests can certainly be implemented, but at project team level they are impossible to manage because of insufficient participants. Rework of products is also difficult, because these tests do not provide any information on how products differ. We have therefore formalized an approach that can evaluate proximity between tests and a reference product.

16.2.4 Need 4: comparison of products on specific criteria

Sometimes, a project is focused solely on one or two sensory aspects. In this case, we attempt to accurately position the product's deviation with regard to these sensory properties.

16.3 Analysis of opportunities and constraints linked to project team evaluation

An internal "team tasting" evaluation strategy has certain important specific features which must be taken into consideration.

16.3.1 Opportunities

Firstly, there are some very strong opportunities presented by such an approach:

- *Very high motivation of the judges*: It is easy to motivate the project team. Each judge easily perceives his/her own contribution. The judges are also able to better monitor their tests and better evaluate their competitors.
- *Increased knowledge of the sensory methodology and the category*: Participation in the tasting team provides the taster with a context for the sensory evaluation. Participating in the evaluation allows the taster to understand from within the notion of objective evaluation, and enables an understanding of the advantages of this approach compared to profile and qualified-panel approaches. Repeated sessions and vocabulary-teaching approaches also provide knowledge on the product category.
- *Budgetary advantages*: There is no external budget with this method. It is therefore easy to introduce into project management. However, this advantage should not be overstated, as the project team's participation still represents a cost to the company.
- *Almost no prerequisites*: This type of evaluation is performed without any training, approval or performance phase. Therefore it can be quickly implemented with just a short learning session. We also believe it is possible to evaluate the products in a meeting room, although the use of a standardized room is recommended. In fact, this evaluation method can be used for any project in any country, if managed by a coordinator.

16.3.2 Constraints

The fact that the evaluation is performed via a project team presents certain constraints, however:

- *Type of judges and level of training*: It is by definition the project team who will evaluate the products. We must therefore comply with the following obligations:
  a. *No selection*: From a group dynamic point of view, selection is not possible.
  b. *Reduced staff*: There are often between six and ten members of staff and for certain projects we can extend the project team to around 15–20 people.
  c. *No, or only light, training*: It is impossible to provide the judges with real training. A specific learning module can be set up to quickly explain the sensory bases and the essential

recommendations for performing an evaluation in good conditions (duration: around 2 h for the judges).
- *Type of task*: The task must be simple and adapted to the target. We therefore favour comparison and ranking approaches, e.g., evaluation by criteria using a Check-All-That-Apply (CATA) approach, or a quantitative approach limited to very simple items with a five-point scale.
- *Need for communication warnings*: Sensory analysis has built its reputation on the reliability of its recommendations. We must, of course, work on and control the reliability of the acquisition mode and processing. However, under no circumstances can the quality of the data equal that of the profile data. Explanatory work must therefore be carried out to clearly differentiate this approach from more traditional approaches.
- *Coordination*: The session is coordinated by a person from the team. This coordinator must be trained to be capable of preparing, coordinating and exploiting the data correctly. One day's training is quite enough to achieve these objectives, often with coaching on group animation.
- The approaches implemented during such project evaluations must incorporate these specific requirements, managing the constraints and encouraging the opportunities.

16.4 An approach adapted to Danone's needs but integrated with the limits of the team tasting

In order to manage the team tasting evaluation, we have worked on establishing specific test designs. As regards the four needs detected (Section 16.2), we have structured and adapted four approaches for Danone based on around 10 years of trials and regular improvements. Each method has a common framework and its own design specificities.

16.4.1 The common framework

Six basic rules are established, which should be communicated to all teams systematically during basic training. They are of course applied systematically for all sessions.

- Blind evaluation with a three-figure coding
- Sample quality control
  - Systematic pre-evaluation of the products
  - Validation of the practical aspects: temperature management, product uniformity
- Individual evaluation without interaction with the group, followed by a group discussion and pooling session at the end of the test
- Controlled order of the samples
  - Avoid any bias linked to ordering or reporting
- Appropriate tasting conditions: room temperature, atmosphere
  - An evaluation in an ISO standardized room is an advantage, but should not be obligatory
- Protocol monitoring
  - Explanation of clear instructions.

Table 16.1 Use of team tasting depending on project need

Needs	Approach
Basic characterization of the products	Descriptive team tasting
Monitoring of improvements during development work	Descriptive team tasting
Product comparison	Comparative team tasting
Comparison of products on specific criteria	Ranking team tasting

16.4.2 Management of the needs

Three different approaches have been determined in relation to the management of the four needs, as shown in Table 16.1

16.4.2.1 Descriptive team tasting

Descriptive team tasting is based on a simple evaluation of products based on items and CATA. The approach is formalized in the plan in Table 16.2 and some concrete examples are described in the following text.

Practical example
Based on previous knowledge and the objective of the project, we have defined the item list showing which items need to be evaluated. First, a questionnaire was designed (Fig. 16.1). Based on the evaluation (15 judges), we can visualize the average data (Fig. 16.2a) and the number of citations of CATA items (Fig. 16.2b). In order to clarify differences, we can use a table (shown in Fig. 16.3) that categorizes the data by judge. We can also use a statistical analysis of the CATA items (Fig. 16.4). In order to highlight significant differences amongst the products, we use a letter code. For Red colour, for Trial 1, we have the letters bcD, which means that this value is more important than that for the other products (risk 5%). In the same way, for strawberry note (Straw), the letter D indicates that Trial 1 is more often quoted (risk 1%) than Trial 4.

We can conclude that Trial 1 is creamier, with a redder colour and more of a strawberry taste. Trial 2 appears less creamy and more pear-flavoured. Trial 4 appears less sweet.

16.4.2.2 Comparative team tasting

The objective of comparative team tasting is to determine whether a difference exists between one or more samples and each given descriptor. This methodology is an internal adaptation of the difference-from-control test (Meilgaard, 2000). The modality of comparison should be predefined: giving a quotation makes the test more difficult and is not recommended for non-expert assessors (O'Mahony, 1995; Kim and O'Mahony,

Table 16.2 Main characteristics of descriptive team tasting

Steps	Description
Session design and coordination	At the start of the test, check if the participants fully understand the terms. Each session: evaluation of 3–15 products. Sequential monadic evaluation of the products. With several evaluations, possible to evaluate: until 60 products…
Questionnaire	The questionnaire is made up of a combination of quantitative criteria: scales (4–8 depending on the studies) on simple aspects such as sugar, acid, etc., together with CATA description criteria such as aromatic score and specific texture. Example in Fig. 16.1
Pre-analysis	In order to sum up the opinion of the entire group, a mean calculation is performed. Example in Figs 16.2a and 16.2b Another with data centring for each individual to eliminate the impact of using a scale Example in Fig. 16.3
Exploitation of the data via consensus. *Possible if all the judges are present at the same time.*	Pooling and sharing the results. Discussion and validation of the main differences observed between the products (based on CATA and quantitative criteria) Search for a consensus situation to orientate the next steps of the project.
Exploitation of the data via statistics *Possible with a minimum of 15 data entries per product*	For this approach, the statistical approach is mainly focused on the CATA data, and via the Cochran's Q test Kruskal–Wallis test we attempt to reveal citation differences. Example in Fig. 16.4

1998). In the final design, the test is close to the "A/not-A test with sureness," but applies several criteria. The design of the "comparative team tasting" test is described in Table 16.3.

Practical example

Two trials, A and B, are to be compared to a standard. First, five items are defined in line with the project knowledge of the designer. The questionnaire is constructed as shown in Fig. 16.5.

Based on the evaluation (15 judges), we can draw the following conclusions (Figs 16.6 and 16.7): Trial A is perceived to have significantly fewer fruit pieces and to be less sweet than the reference. Trial B shows no significant difference from the reference. Therefore, B is probably a good proposal to further the project.

Improving team tasting in the food industry 351

Figure 16.1 Example of questionnaire – descriptive team tasting.

(a)

SCALE	Creamy	Sweet	Aroma
Trial1	3.62	3.85	2.54
Trial2	3.08	3.23	2.46
Trial3	1.69	3.31	3.08
Trial4	2.23	2.23	2.54

(b)

CATA	fruit_piece	red_color	green_aroma	red_aroma	straw	pear	apple	pineapple
Activia Strawberry	6	7	1	3	4	4	4	5
Yop Strawberry	5	1	2	5	4	3	5	4
Actimel Strawberry	0	7	7	5	2	6	3	2
Yoplait Strawberry	1	1	4	5	0	3	3	4

Figure 16.2 (a) Result of the scale evaluation – descriptive team tasting; (b) result of the CATA evaluation – descriptive team tasting.

	Creamy	Sweet	Aroma
Trial 1	0.96	0.69	−0.12
Trial 2	0.42	0.08	−0.19
Trial 3	−0.96	0.15	0.42
Trial 4	−0.42	−0.92	−0.12

Figure 16.3 Result of the scale analysis – descriptive team tasting with centred data average.

	fruit_piece	red_color	green_aroma	red_aroma	straw	pear	apple
Trial1(A)	8	12^{bcD}	4	8	8^d	7	9
Trial2(B)	8	5	4	6	5^d	3	8
Trial3(C)	4	7^d	8	6	4	9^{b,a}	5
Trial4(D)	2	1	4	5	0	3	3

Significant difference: A'<0.001, A<0.01, a<0.05, a'<0.1.

Figure 16.4 A result of the CATA analysis – descriptive team tasting, with Kruskal-Wallis test.

Table 16.3 **Description of the comparative team tasting design**

Steps	Description
Session design and coordination	Each session: evaluation of two to four products + reference evaluation Comparative evaluation of the products versus the reference • At the start of the test, check if the participants fully understand the terms • Evaluation of the reference • Then evaluation of the products + evaluation of the reference as an another product *For example, with three trials, we obtain four evaluations*: *Trial1/ref* *Trial2/ref* *Reference/ref* *Trial3/ref*
Questionnaire	Evaluation of four to eight items to evaluate the deviation from this reference (higher/equal/lower). The questionnaire contains four to eight terms that are very easy to evaluate Example of questionnaire is given in Fig. 16.5
Exploitation of the data via consensus *Possible if all the judges are present at the same time*	Pooling of the results Discussion and validation of the main differences observed between the products. Search for a consensus to direct the remainder of the project
Exploitation of the data via statistics *Possible with a minimum of 15 data entries per product*	Based on the theory of "the A/not-A test with sureness", cf Bi (2006) and Christensen et al. (2011), a specific processing method was built in order to evaluate the level of the difference with an appropriate indicator: d' association to the confidence interval. In this way, it makes it possible to calculate a d-value combined with an alpha value giving the probability of a deviation from the reference. An example of results is presented in Figs 16.6 and 16.7.

Improving team tasting in the food industry 353

Figure 16.5 Example of questionnaire – comparative team tasting.

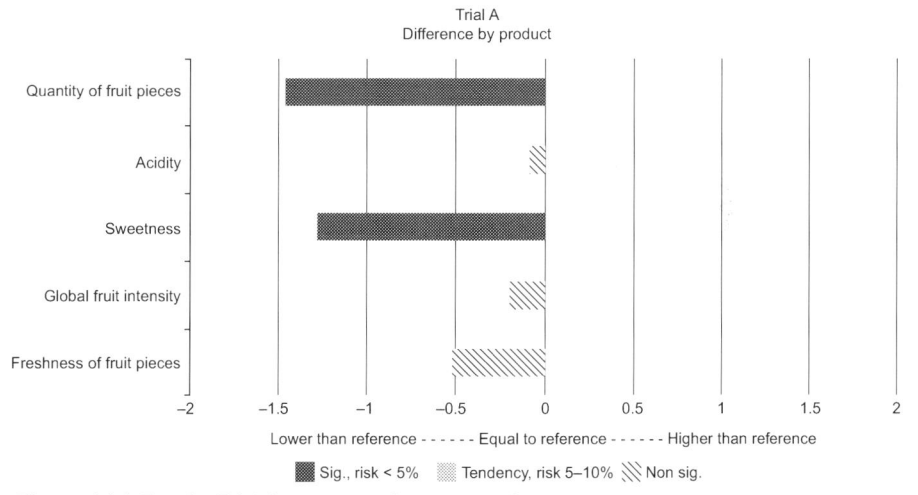

Figure 16.6 Results Trial A – comparative team tasting.

16.4.2.3 Ranking team tasting

The design of this test is simple, focusing on ranking the products. This is an easy task for non-trained judges, but does require some specific elements in terms of number of products and characteristics. The design is described in Table 16.4.

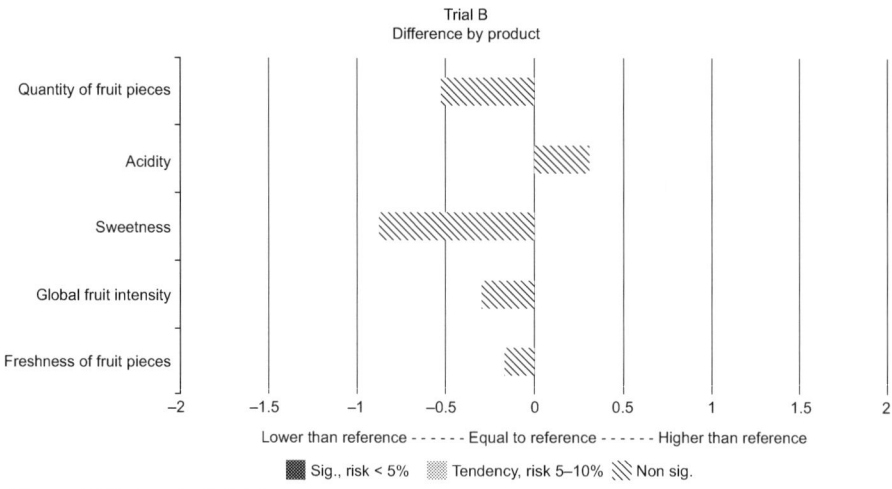

Figure 16.7 Results Trial B – comparative team tasting.

Table 16.4 Description of ranking team tasting

Steps	Description
Session design and coordination	Based on a comparative evaluation of all the products, each judge ranks the products on some items At the start of the test, we need to check if the participants fully understand the terms Each session: evaluation of three to six products, on two to four items
Questionnaire	The questionnaire contains two to four terms that are very easy to evaluate (we will check there are not too many) An example is shown in Fig. 16.8 We authorized the use of the same rank for several products
Pre-analysis	In order to sum up the opinion of the entire group, a mean rank calculation is performed The dispersion of the ranks can also be displayed
Exploitation of the data via consensus *Possible if all the judges are present at the same time*	Pooling of the results Discussion and validation of the main differences observed between the products Search for a consensus to direct the remainder of the project
Exploitation of the data via statistics *Possible with a minimum of 15 data entries per product*	Based on the ranks, a Friedman test is set up for each attribute

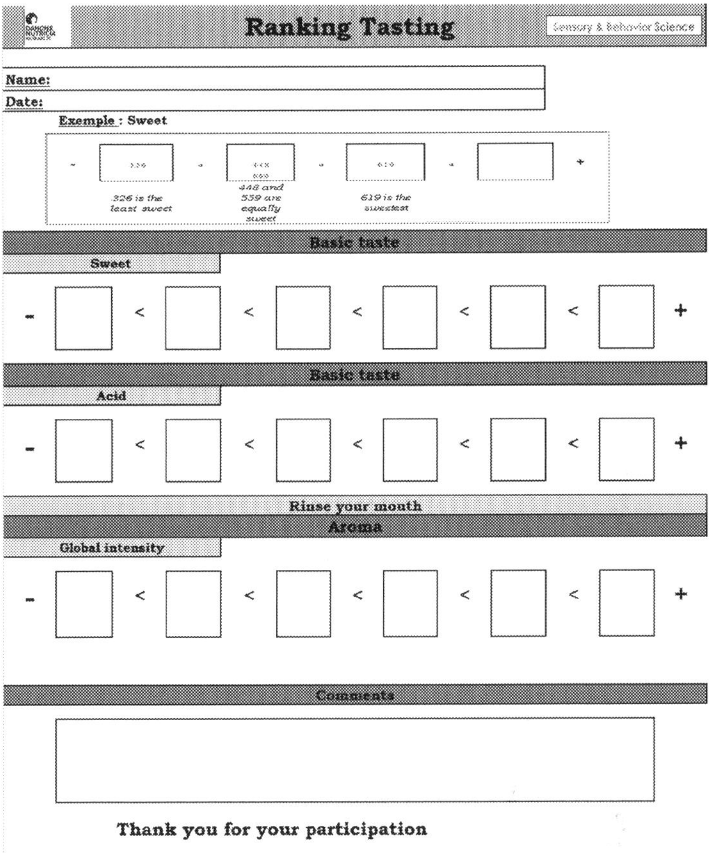

Improving team tasting in the food industry 355

Figure 16.8 Example of questionnaire – ranking team tasting.

Practical example

The project requires a focus on three aspects: citrus, intensity aroma and acidity. We want to compare four products according to these criteria, so we construct a questionnaire (Fig. 16.8). Based on the data process, we obtained the values shown in Fig. 16.9. We conclude that Trial Product 678 has more citrus taste and is more acidic than the others. On intensity, this product is also the strongest, but equal to Trial Product 125.

16.4.3 Management and exploitation of data

As indicated, we must always decide between two methods: analysis via consensus, or via a statistical approach. We consider this double process to be important, because it enables us to reach a conclusion in all situations by integrating the constraints for this type of test, i.e., generally low staff numbers, and we also avoid having to use a statistical data process with no adequate conditions. In order for the statistical test to

Ranking analysis results

Citrus

	A	B	C
Trial 056	1		
Trial 125		2.38	
Trial 111		2.62	
Trial 678			4

Aroma int

	A	B
Trial 111	1.59	
Trial 056	1.66	
Trial 125		2.94
Trial 678		3.81

Acid

	A	B	C
Trial 056	1.31		
Trial 111		2.38	
Trial 125		2.69	
Trial 678			3.62

Probability level : 5%

Figure 16.9 Results of ranking based on a Kruskal-Wallis analysis – ranking team tasting.

be robust, we can identify a threshold of 15 people to qualify studies for which a statistical approach could be used. This threshold is arrived at based on three elements:

- Analysis of the tasks the judge is asked to perform, and the associated statistical approach;
- Cochran and McNemar test, Kruskal-Wallis test, specific data process to calculate D prime with A/not-A test with sureness;
- Judges' low qualification level leads to a greater variability of responses.

For robustness, we structure the construction of the consensus, and guide the coordinator during training to ensure that he/she collects all of the opinions provided, that he/she does not give undue weight to any one participant, that he/she does not provide any information regarding the identity of the samples (and make sure that participants do not remove covers, tape, etc.) and then removes blindfolds until the end of the debriefing session and, most importantly, that the debriefing always ends with a conclusion and an action plan.

16.4.4 Factors for success to guarantee correct application of the method

All the factors quoted in Section 16.4.1 need to be followed carefully, but in order to obtain relevant results we need to master two important factors, as outlined below.

16.4.4.1 Importance of coordinator/teams

The coordinator guarantees the quality of the process monitoring. Regular follow-up work and discussions with the coordinators are important in order to maintain the quality of the approach and the quality of the interpretation (by consensus or statistical processing). This point is key, because the judges have limited knowledge and training, and their turnover is high.

16.4.4.2 Quality of the terms of the evaluation

The choice of terms included in the questionnaire is fundamental. They must be simple, relevant and easy to understand via references and/or definitions.

16.5 Implementation examples (common in R&D field)

With the help of two case studies, it is shown how this approach can be implemented in a project.

16.5.1 Case 1: characterize twelve products in order to select two

Context and objective:
- Launch of a new product on the Spanish market
- The project brief is to deliver a product that provides a fresh orange flavour, no off notes and low sweetness.
- After three months of work, 12 prototypes are formulated.
- Two prototypes out of 12 are to be selected in order to evaluate the products with sensory profiles, along with a consumer screening test.

Method chosen: Descriptive team tasting

Questionnaire: Seven sensory criteria are chosen to cover the main dimensions of variation within the trials, with of course a focus on the important dimensions for the project. There are four scale criteria, namely acid intensity, sweet intensity, global aroma intensity, orange intensity, and three CATAs, namely off note, fresh aroma and citrus note.

Note: For aroma note, CATA is often selected as it allows for an easier evaluation with untrained judges.

Session:
- R&D/marketing project team (16 people)
- One session of 1 h:

	Sweet	Intensity aroma	Intensity Orange	Acid
Trial1	1.15	0.49	0.2	0.28
Trial2	1.01	0.28	−0.38	0.06
Trial3	−1.13	0.63	0.91	−0.08
Trial4	−0.35	−0.94	−0.23	−0.08
Trial5	0.72	0.06	−0.09	−0.08
Trial6	0.58	−0.08	−0.52	−0.37
Trial7	−1.13	0.21	0.62	−0.08
Trial8	−0.85	−0.65	−0.52	0.35

Figure 16.10 Mean value on centred data (case 1).

	Off note	Fresh aroma	Citrus note
Trial1(A)	12^{CDfGH}	14^{BDeFH}	2
Trial2(B)	10^{CDGH}	2	4
Trial3(C)	0	14^{BDeFH}	14^{ABdEFH}
Trial4(D)	2	2	8^{ae}
Trial5(E)	10^{CdGH}	8^{bF}	2
Trial6(F)	6^{ogh}	0	4
Trial7(G)	0	10^{BdFh}	10^{AE}
Trial8(H)	0	2	6

Significant differences: A' < 0.001; A<0.01; a<0.05; a'<0.1

Figure 16.11 CATA analysis associated to Cochran and Mac Nemar test (case 1).

- 30 min acquisition
- 30 min debrief based on statistics.

Main results: Based on the results presented in Figures 16.10 and 16.11, we can conclude that the best trials in line with the objectives are Trials 3 and 7.

Conclusion: In the next step, the two trials are evaluated in sensory consumer tests. It is important to note that only when the stakes are very low we validate a product based only on team tasting. The session leader should always make this clear – team tasting does not replace classic sensory tools.

16.5.2 Case 2: compare two trials to a standard

Context and objective:

- Analyse the possibility of achieving the same sensory perception as the reference product, in the Japanese market, through two new processes (A and B).
- These new processes may impact the perceived texture and, indirectly, the perception of the sweet/acid balance and also the aroma perception and the aftertaste.
- The strategy is to pre-validate the prototypes with team tasting. The action standards defined are as follows:

Improving team tasting in the food industry 359

- If a difference is detected, reworking is required.
- If no differences are perceived, this conclusion should be validated with sensory checks.

Method chosen: Comparative team tasting

Questionnaire: Nine criteria to cover the main possible impacts:

- Texture: smoothness/thickness/stickiness
- Taste: sweetness/acidity/bitterness
- Aroma: global dairy notes/acetic acid/cheesy aroma

Session:
R&D project team (15 people)
One session of 45 min:

- 15 min acquisition
- 30 min debrief based on statistics

Main results: Based on the results presented in Figs 16.12 and 16.13, we can conclude that Process 2 should not be accepted for further analysis. To render it acceptable, we need to try to increase two criteria: thickness and stickiness. Process 1 is in line with our requirements, so we can validate this formula for further analysis. If possible, we will then try to understand why the product resulting from this process is less thick than the reference, and attempt to correct this.

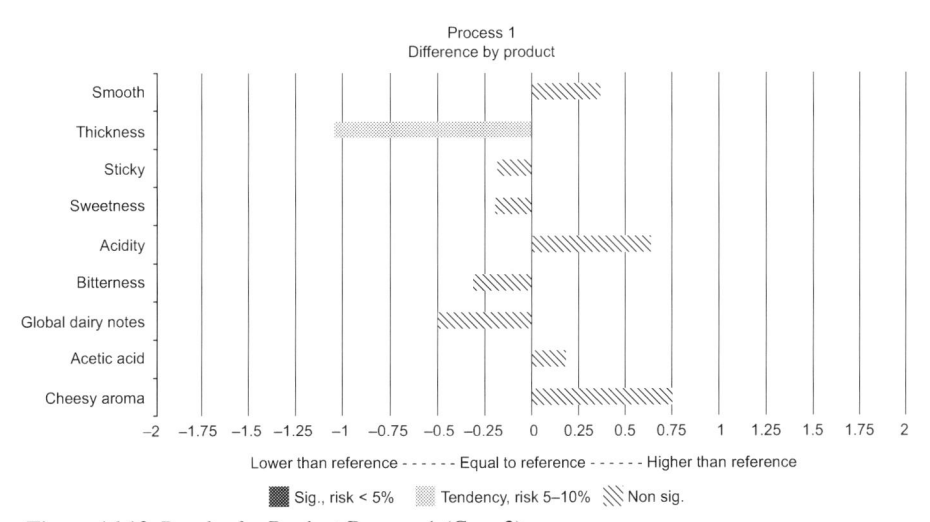

Figure 16.12 Results for Product Process 1 (Case 2).

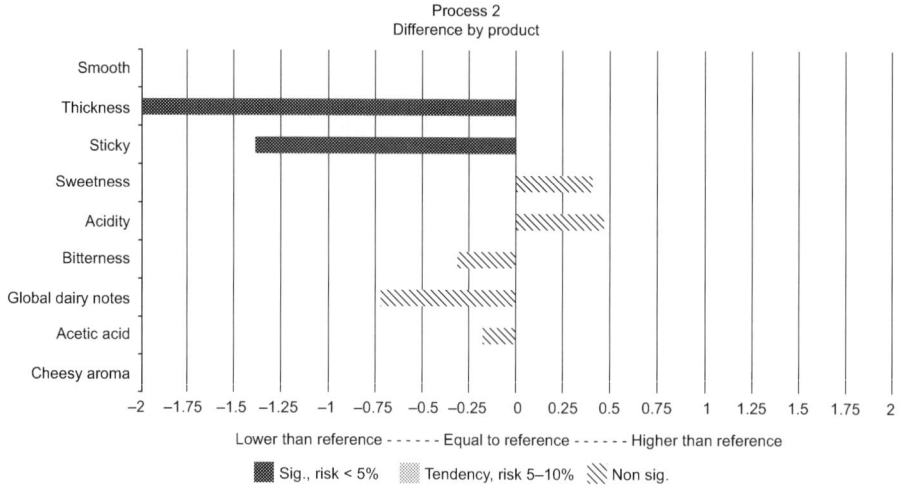

Figure 16.13 Results for Product Process 2 (Case 2).

Conclusion: As the next step, Process 2 will be evaluated with sensory tests.

16.6 Analysis and prospects

We have followed a simple approach that is "easy and pragmatic". In the analysis, it will be shown how to introduce the evaluation at the right way in an organisation. In prospect, the strategy will be to reinforce this simplicity.

16.6.1 Analysis

In order to obtain a complete overview of this approach, it is important to look at two points:

- How this method is perceived and used within the R&D team;
- How this method is perceived compared with the more classic tools.

16.6.1.1 How this method is perceived and used within the R&D team

This method has been used for several years within Danone. At first, we only used descriptive team tasting with scale and a consensual debrief. Given the importance of the evaluation, we have adapted the method to obtain better responses and so provide more relevant support. Recently, we have created a specific data process tool to facilitate and harmonize the test. This tool is now available to all the Danone R&D teams and is combined with training. On the whole, feedback has been very positive, and the method is constantly in deployment amongst the international R&D teams for the Danone dairy and water divisions. The expected impact has been achieved, and with the above-mentioned approaches the projects are better managed at sensory level. This frees up time for our classic "expert sensory panel" (indeed, often the team tasting provides information that is enough for a project follow-up). The expert panel can then be used for other applications.

It is also important to note some secondary impacts:

- In terms of product relationship, we have observed a new familiarity. The project teams are now evaluating products more by themselves and working on the sensory dimensions in a more objective way.
- Sometimes, in order to improve the quality of team members' descriptions, training and term qualification work is often requested and specific training activities are often set up.
- At the decision-making stage, concluding with a consensus task reinforces the mutual respect within the project team.

Often, a "hedonistic" tendency exists in the project teams, which it is important to avoid in order to maintain an objective description. To this end, attempts have been made to set up some consumer checks that are easy to process, non-expensive and quick. This has proved a challenge!

16.6.1.2 How this method is perceived compared with more classic tools

This type of tasting must be used correctly in addition to the other sensory approaches, and should never be seen as a substitute for them. This is now well understood. However, two actions should be maintained:

- First, constant training of the tasting team is required. This training should be focused on the relevant business issues or challenges.
- Secondly, we must also always ensure we communicate results in a specific way, and that we do not communicate team tasting and results obtained from expert panels and/or qualified panels in similar ways. Building specific software to process the data is of course well in line with this objective.

16.6.2 Prospects

The method defined above is well in line with our objectives: it is a quick, easy method that can be set up alone by the project team. One way to improve use would be to create software dedicated to the task of integrating acquisition and data processing. We are building an acquisition interface for PCs and tablets to enable simpler, faster and more reliable acquisition. This acquisition method facilitates consensus work via instant exchanges on the acquired data, and is combined with data processing.

References

Bi, J. and Ennis, D.M. (2001). Statistical models for the A-Not A method. *Journal of Sensory Studies*, **16**, 215–237.

Christensen, R.H.B., Cleaver, G. and Brockhoff, P.B. (2011). Statistical and Thurstonian models for the A-not protocol with and without sureness. *Food Quality and Preference*, **22**, 542–549.

Kim, K. and O'Mahony, M. (1998). A new approach to category scales of intensity i: traditional versus rank-rating. *Journal of Sensory Studies*, **13**, 241–249.

Meilgaard, M., Civille, C.V. and Carr, B.T. (2000). *Sensory Evaluation Techniques.* 2nd edn. CRC Press. 448 p.

O'Mahony, M. (1995). *Sensory Evaluation of Food: Statistical Methods and Procedures.* New York: Marcel Dekker, Inc.

Alternative methods of sensory testing: working with chefs, culinary professionals and brew masters

17

M.B. Frøst[1], D. Giacalone[2], K.K. Rasmussen[3]
[1]University of Copenhagen and Nordic Food Lab, Frederiksberg, Denmark;
[2]University of Copenhagen, Frederiksberg, Denmark; [3]University of California, Berkeley, CA, USA

17.1 Introduction

Chefs, brewers and related professionals occupy a particular role in society – they work with the transformation of some of the most basic matters of our lives. They are part highly skilled workers and part artists with a vision for the end result of their transformation. In general, they work with their senses and have a highly educated palate. However, in the hectic work environment of the restaurant kitchen or the brewery, there is little tradition for formal sensory tests. In the development lab, when prototyping new foods or beverages, there is time, but little money and tradition for sensory tests. To take sensory methods to this new territory outside the well-known facilities of the sensory laboratory is a terrifying task. Will the knowledgeable experts and educated palates ridicule the sensory profession? Will they understand the method, and the value it can give them in their work? Will they understand the results? Will they think the results are too obvious, and not worth spending the time on?

Those were some of the thoughts that rummaged around in the brain of the first author some years ago, when we first attempted to move fast sensory methods out of the sensory laboratory and into the real life of brewers and chefs.

17.2 Background: fast descriptive methods and persons with no formal sensory training in sensory tests

Descriptive analysis is known to produce detailed, robust and repeatable results, as documented by numerous scientific publications (for a review on the topic, see Dijksterhuis and Byrne, 2005; Murray *et al.*, 2001). However, it also has certain drawbacks. First, it is a very slow method, particularly because of the extended training phase. Second, it is a very expensive method. Maintaining a sensory panel is generally not affordable for small and medium-sized enterprises (SMEs) in the food industry, and can be a significant spending also for larger manufacturing companies. Third, it is possible that trained sensory panellists experience the product differently from the other professionals, such as chefs of brewers, or the sensory panellists may take into account sensory characteristics that may be irrelevant for the consumers (Ares *et al.*, 2010), thus providing high quality results but with lower external validity.

In order to address these drawbacks, a number of alternative descriptive methodologies have been proposed over the years, most of which require little or no training and are easily implementable with trained panellists, with other groups of professionals, and even with consumers. The idea that consumers (or generally untrained subjects) can be used for descriptive tasks – traditionally a highly controversial topic among sensory scholars (Moskowitz *et al.*, 2003) – is increasingly accepted due to three factors:

1. Strong evidence that untrained respondents, such as consumers, can provide valid and meaningful sensory (descriptive) information (e.g. Bruzzone *et al.*, 2012; Worch *et al.*, 2010);
2. Methodological developments that facilitate the collection and analyses of such responses;
3. A general consensus that a deeper consumer involvement at an early stage in the product development life-cycle is beneficial to the success of food products development (Grunert *et al.*, 2008; Stewart-Knox and Mitchell, 2003; Trijp *et al.*, 2008; van Kleef *et al.*, 2005).

A useful way to classify rapid descriptive methodologies, according to recent reviews on the topic (Valentin *et al.*, 2012; Varela and Ares, 2012), is the distinction between verbal-based methods and similarity-based methods. Verbal methods can be based on both monadic and concurrent evaluations of a number of products on individual sensory descriptors. Examples of this class of methodology are Free Choice Profiling (Williams and Langron, 1984), its Flash Profile variant (Dairou and Sieffermann, 2002) and check-all-that-apply (CATA) questionnaires (Adams *et al.*, 2007). Similarly to descriptive analysis, these methods produce a descriptive sensory profiling of the products, while bypassing the time-consuming steps of attribute and scaling alignment that is a key aspect of descriptive analysis (Valentin *et al.*, 2012).

In the second class of methods, similarity-based, respondents are presented with all products simultaneously, and give a global evaluation expressed as the perceived inter-product differences. These methods are sometimes defined "holistic," because they require the respondent to consider the product as a whole, unlike the "reductionist" verbal-based methods which require respondents to decompose the stimulus into multiple descriptors. In reality, some similarity-based methods may include a verbalization task, but this occurs only after the global evaluation, and the output is usually not used to build the perceptual space beyond simple correlational measures. The simplest (and probably best known) of these methods is the free sorting task (Lawless, 1989; Lawless *et al.*, 1995), which has been applied for sensory evaluation of beers, and is reportedly applicable with trained panellists and consumers alike (Chollet *et al.*, 2011). Another important similarity-based method is projective mapping (Risvik *et al.*, 1994), the first method to introduce the idea of expressing product differences as Euclidean differences by means of projection onto a two-dimensional space. Various adaptations and modifications of the original projective mapping technique have been proposed, the best known of which is napping® (Pagés, 2003, 2005). It is important to remark that this solution provides a sensory configuration that is not necessarily driven by the sensory variables with the strongest structure, but by those that are relatively more important for the respondent. Accordingly, some authors have observed that this method can be thought of as producing both quantitative and qualitative sensory information (Chollet et al., 2011). The terms napping and projective mapping are sometimes improperly used as synonyms in the literature; Appendix 1 provides a description of the differences between the two methods.

When working with expert tasters from different professions, such as brewers and chefs in these case studies, it may be necessary to change the location to accommodate their participation. Although there have been numerous systematic studies of fast sensory methods, and how the results fare compared to conventional descriptive analysis (see e.g. Dehlholm, 2012; Delarue and Sieffermann, 2004; Perrin et al., 2008), there has been no systematic evaluation of the effect of the test location. The effects can thus only be estimated based on sound sensory methodology considerations, and text books regarding this are good to consult (e.g. Lawless and Heymann, 2010). In general, the most important factors for a successful descriptive sensory profile is that the samples should be served blind, the respondents should not discuss their results and the respondents should carry out the task in a room suitable for sensory analysis, most often according to ISO-standards (ISO-8589:2007, 2007). The necessary considerations when carrying out descriptive analysis in a development kitchen, or a meeting room in a brewery, is thus to make sure that the respondents do not discuss the samples while they evaluate them, and that in general they disturb each other as little as possible. In general, chefs and food professionals like to discuss the food they test, and in particular, they discuss the quality of the samples. Silence during tasting is not to be assumed, so careful and repeated instructions need to be given. The effects of carrying out tests in less optimal physical environments are smaller than the effects of respondents discussing the samples during tasting. The main focus has been to go where the brewers and chefs are, and instruct them carefully to decrease bias from other respondents. We present the methods and results from two cases:

1. Brewers and beer novices assessing a set of Danish craft style beer
2. Exploring the world of spice blends and pastes with chefs and other food experts

The cases are presented and discussed individually, after a short description of the applied data analysis methods and the considerations in that connection.

17.3 Data analysis of projective descriptive methods

Data obtained by napping are usually analysed by multiple factor analysis (MFA), a multivariate data analytical technique that seeks the common structure between several blocks of variables (i.e. the individual panellists in a napping task) describing the same observations (the samples), see FactoMineR's website for further information (Husson et al., 2008; Lé and Husson, 2008; Lê et al., 2008). MFA can be thought of as a principal component analysis (PCA) in two steps. The main difference between them is that MFA takes into account individual differences, rather than averaging the data (Nestrud and Lawless, 2008).

MFA starts by computing an initial PCA on each individual matrix X_j, containing the sample coordinates for individual respondents, and subsequently transformed into a new matrix $X_{NEW\,j}$ such that:

$$X_{NEW\,j} = \frac{1}{\sqrt{\lambda_1^j}} \times X_j$$

where λ_1^j represent the first eigenvalue of the initial PCA of matrix X_j. The quantity $\sqrt{\lambda_1^j}$, called *first singular value* in MFA jargon, is basically a matrix equivalent of the standard deviation. This procedure corresponds to a normalization, i.e. the first eigenvalues of the transformed $X_{NEW\,j}$ matrices are all equal to 1 (see also Chapter 9 for additional considerations on the use of MFA to analyse napping data). That prevents the blocks with the largest variance from exerting an overwhelming influence: in napping, this means accounting for individual panellists' differences in the use of the projective space on the paper.

After this step, the data blocks are concatenated in a global data table on which a new PCA is run, i.e. by singular value decomposition of the matrix $X_{NEW} = \left[X_{NEW1} | X_{NEW2} | ... | X_{NEWJ} | \right]$. The descriptive data from Ultra-Flash Profiles (UFP) are usually treated as supplementary variables to the MFA on the napping coordinates. "Supplementary variables" means that UFP data are not used to construct the MFA model, but correlation coefficients of the UFP sensory descriptors are calculated and can be presented in the product space to aid the interpretation. It is important to remark that this solution provides a sensory configuration that is not necessarily driven by the sensory variables with the strongest structure, but by those that are relatively more important for the panellist[1]. Accordingly, some authors have observed that this method can be thought of as producing both quantitative and qualitative sensory information (Chollet et al., 2011). Although much positive can be said about the approach, it is not easy to use for non-statisticians, and certainly not understandable to people without good statistical command. These considerations are important when considering how to communicate the results to professionals from other walks of life, such as chefs and brewers.

For our data analysis we have kept simplicity and communication to non-sensory professionals as key points.

---

[1] According to the inventor of the method, this is the main advantage of napping over DA (Pagès, 2005). The latter produces (ultimately) a data matrix crossing products and descriptors. Such data, typically containing mean values over panellists, is then mean-centred column-wise and analysed by principal component analysis (PCA), either by giving identical weight to the same variables after a normalization procedure so that each descriptor gets the same variance (usually dividing the data by the sample standard deviation), or by keeping the weight of each descriptors proportional to its variance (unscaled PCA). Whichever solution one chooses, Pagès (2005) observes, the weights given to the descriptors do not necessarily correspond to the actual importance for the subject.

So for all data analysis we restrict ourselves to PCA on the raw position data, and downweigh the importance of the descriptors for this analysis and presentation to non-specialists. Similar to FactoMineR's analysis, the emphasis of the data analysis is on the positions. We consider them to be the most important because generally non-specialists need to understand the interrelationship between the tested samples. In addition, in our instructions we emphasize the positioning of the samples, and instruct panellists to add descriptors after placing the samples. In respondents' evaluation, it can then occur that samples that are perceived as very different, may share one or a few descriptors. Further, the argumentation for emphasis on positions is that with fewer trained respondents, the variation in descriptors increases. The type of descriptors that respondents use is partly affected by their background – the level of expertise they have with the products being evaluated, their particular cultural and professional background, etc. The focus on communication and thereby also simplicity in the data gives a number of good arguments to select PCA as the data analysis method[2]. For all data analysis we have used full cross validation, leaving one product out at a time. We consider both calibrated and validated explained variance to decide the optimal number of components for the individual data set. For presentation of the data, we restrict ourselves to score plots of validated variance for analysis of interrelationships between products, and use correlation loading plots to elucidate how different descriptors correlate to each other (Martens and Martens, 2001).

17.4 Case study 1: brewers and novices assessing beer

This study was hosted by the Danish craft brewery *Indslev*, which was at that time involved in a collaborative project with the University of Copenhagen, *Danish Microbrew* (see e.g. Giacalone, 2013). The panel consisted of eight brewers and nine other participants who we categorize as novices in terms of describing beer sensory characteristics. The panellists were on that occasion introduced to fast sensory methods, in particular napping. The purpose of this study was two-fold. The first aim was to assess whether napping, performed by relatively knowledgeable subjects but without formal sensory training, would succeed in discriminating between the beers and provide meaningful results for the brewers. The second aim was to compare beer novices and brewers with regards to their performance as napping panellists, with the working hypothesis that the latter group would be more consistent in discriminating and profiling the beers. The latter goal is not treated in depth here. Interested readers may consult Giacalone, Machado, and Frøst (2013). Nine Danish beers – seven commercially available and two experimental brews – were used as test stimuli

---

[2] At the Department of Food Science at the University of Copenhagen, a number of small animations and lectures with examples of multivariate data analysis concepts and explanations of the principles have been developed. These are broadcast on a YouTube channel ("YouTube channel: Quality And Technology"). This allows presentations of complex computations in a generally understandable form to a broader audience. We have used these videos on many occasions to teach non-statisticians about the background for the analysis and the resulting plots.

Table 17.1 The beer samples tested in case study 1 (see also Giacalone *et al.*, 2013)

Beer name	Brewery	Beer type*	Flavouring
Nutty	Ørbæk Bryggeri	Brown Ale	Walnuts
Fynsk Forår	Ørbæk Bryggeri	Pale Ale	Elderflower
Havre Stout	Bryggeri Skovlyst	Stout	Oat and rye
Classens Lise	Halsnæs Bryghus	Pale Ale	Chamomile and heather honey
Enebær Stout	Grauballe Bryghus	Stout	Juniper berries
Bøgebryg	Bryggeri Skovlyst	Amber Ale	Beech twigs
Oak Aged Cranberry Bastard	Hornbeer	Fruit Beer	Cranberries
Rosehip Beer	Experimental	Pale Lager	Rosehip powder[†]
Pine Beer	Experimental	Pale Lager	Pine needles extract[‡]

[*] Self-reported by the producer.
[†] "Organic rose hip powder" Coesam SA Laboratorios de Cosmetica. Concentration = 5% w/w.
[‡] "Pin Thyrol," Firmenich SA. Concentration = 0.00625% (6.25 µL/100 mL).

(Table 17.1). The selection was made in order to be representative of main trends among Danish craft brewers, and also because each of them contained a specific ingredient of interest (e.g. due to its sensory and/or functional properties).

Experimental set-up and the instructions given to the participants were according to Pagès (2003, 2005) with one exception: in the original napping method the task is completely holistic, whereas in this case panellists were specifically instructed to focus on taste and smell characteristics of the beers. This variant is sometimes referred to as *partial* napping (Dehlholm *et al.*, 2012) and was chosen in part because there were some clear differences between samples with regards to colour (which we were less interested in), and in part because this variant was shown to produce results more similar to those of conventional descriptive analysis (Dehlholm *et al.*, 2012). At the end of the napping task, participants also did a UFP (Perrin *et al.*, 2008) evaluation of the samples: i.e., they added the sensory descriptors they found appropriate to describe the samples, and wrote them down directly on the sheet.

17.5 Results and discussion of partial napping of beer

The 17 respondents generated a total of 325 words or combinations of words to describe the nine beer samples. However, when the words were grouped according to meaning, the number decreased to 136, with 66 of them being unique descriptors used by only one respondent for one beer. The consensus configuration obtained from PCA of the individual respondents' data is showed in Fig. 17.1. It shows product clusters and Fig. 17.2 gives a presentation of the underlying sensory properties. The

Alternative methods of sensory testing

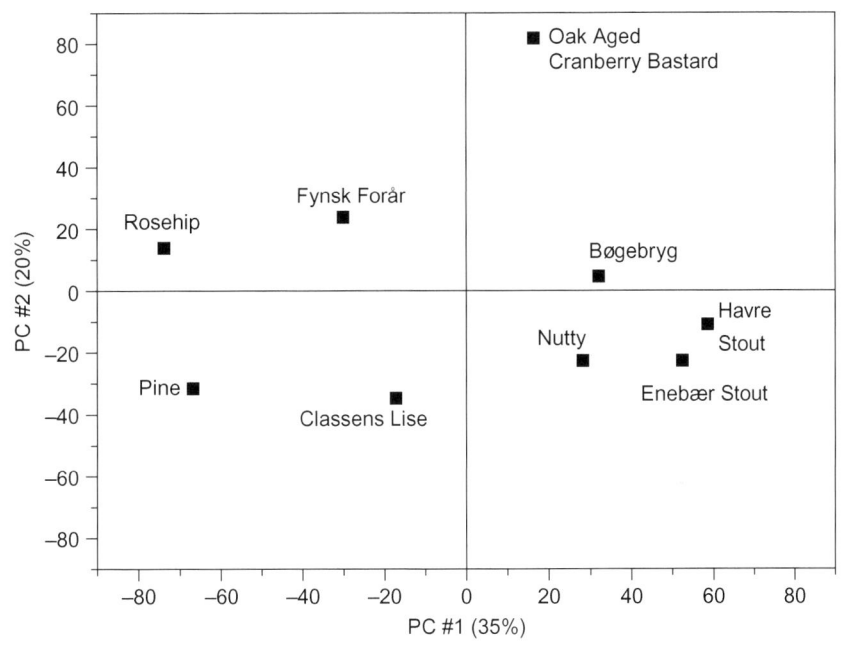

Figure 17.1 Score plot from PCA, Principal Components 1 and 2. Map of nine beers, showing the interrelationship between samples. See Table 17.1 for further information about samples.

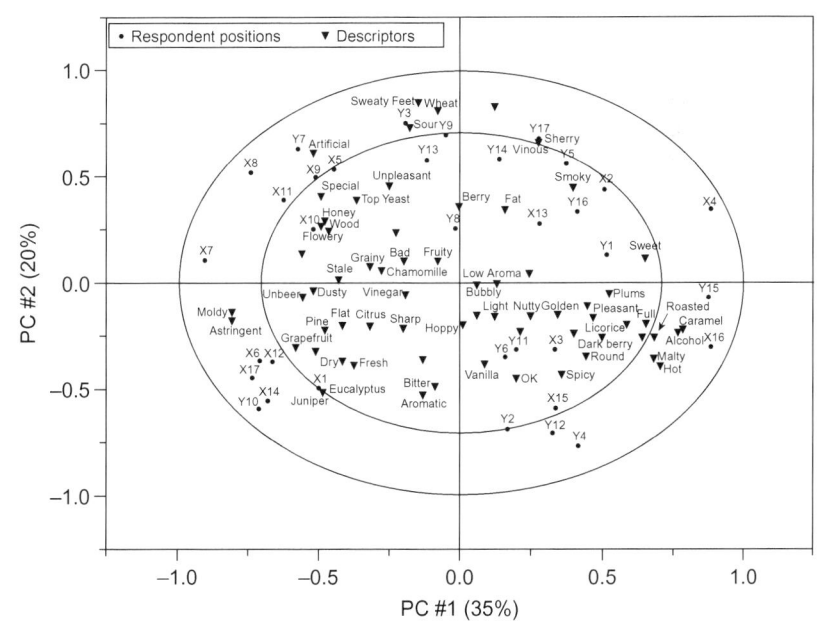

Figure 17.2 Correlation loading plot from PCA, Principal Components 1 and 2. Map of respondents' positioning (X and Y followed by respondent number ●) and descriptors (▼) used by respondents for sensory properties of the beer.

majority of the respondents positioned the beers according to either the style or the specific adjunct flavour addition, and accordingly the first principal component (PC1) separates lagers and pale ales from the four dark ales. The former group included "Bøgebryg", "Havre Stout", "Enebær Stout" and "Nutty". The correlation loading plot (Fig. 17.2) shows that they are primarily characterized by such descriptors as Sweet, Caramel, Alcohol, Roasted, Malty, Hot and Full, and by more specific characteristics such as Licorice, Dark Berries and Plums. All are descriptors which match the general beer style of these four samples. The correlation loading plot (Fig. 17.2) shows that descriptors in the opposite end of PC1 are the following: Astringent, and general woody descriptors, such as Pine, Woody, and citrus fruit characteristics such as Citrus and Grapefruit. The two beers in this group positioned in the most negative part of PC1 were the two experimental brews, Rose Hip and Pine Needle. Other sensory descriptors that were used to describe these two in particular were naturally Rosehip (four occurrences observed in raw data) and Pine, respectively (nine occurrences), but also Astringent and Dry. Fynsk Forår, a pale ale with added Elderflower, is positioned somewhat positive on PC2 and medium negative on PC1. It received descriptors such as Flowery and Elderflower, that are also positioned in that direction in the correlation loading plot. The second component indicates the unique characteristics of one sample (Oak Aged Cranberry Bastard). This is a fruit beer brewed with cranberries, which was often characterized as Sour (nine occurrences) and opposed to Bitter, a descriptor more often used to describe all of the other beers. Further, Oak Aged Cranberry Bastard is produced by spontaneous fermentation (as in some Belgian beer styles): this characteristic was perceived by some panellists, as the descriptors Wild yeast and Sweaty Feet were positively correlated with the second component.

Furthermore, we were interested in looking at the level of agreement of the individual respondents with the general consensus configuration. In line with our expectations, we found that the brewers had configurations more consistent with the consensus profile, suggesting that the degree of product expertise may increase panellists' reliability in a napping task (Giacalone *et al.*, 2013). In general, the product discrimination by napping and the sensory characterization obtained by UFP were clearly interpretable, in the sense that they matched our knowledge of the samples and the characteristics specified by the producers. The method was generally very well received by all panellists, who experienced it as a sort of tasting game. In addition, the brewers and other collaborators in the project understood the results well when they were presented with them. Considering its speed (the whole napping session was conducted within 30 minutes, including time for introducing the method), napping is very advantageous for exploratory purposes in early stages of product development. In a brewery setting, the first issues in a product development project is the design of the brew (e.g. deciding on a specific brewing style), and then the formulation (i.e. deciding on type and quantity of raw materials and processing). At this stage, the pilot plant can be used to develop initial prototypes that may be evaluated by napping, together with existing products or similar in-market alternatives, to obtain an overall description of the product space the experimental beer occupies, and provide a summarized description of the underlying sensory dimensions. Such a test can be used for rapid product screening, to gather feedback on the product and process specifications, and/or for vocabulary generation. In addition, it serves to document the sensory outcome of experimental brews in a systematic manner. If one is sufficiently acquainted with the method, a napping test can be fairly easily arranged, for example with available co-workers: a good solution, since they will be high in beer expertise and will be acquainted with the method, both of which are desirable characteristics for a napping panel. Our results with this particular test showed that the experts (brewers) were more in agreement in their evaluation of the sensory properties than the novices (Giacalone *et al.*, 2013). From this we infer that the product expertise they have, and the product language they share, makes them more suitable for fast projective sensory methods than general consumers, mainly because of their expertise in assessing the sensory properties, or more often the quality of the products in their field of expertise.

17.6 Case study 2: exploring the world of spice blends and pastes with chefs and other food experts

Our purpose was to capture the sensory differences in a large set of traditional spice blends and pastes using methods similar to napping® and UFP. Samples were chosen in order to include a unique variety of traditional blends from around the world in addition to several "New Nordic cuisine" samples, such as Juniper Ant Paste and Peaso. Consequently, the samples incompletely represent the very large differences that exist in mixes of different flavourful ingredients from a broad range of cultures. All samples were held frozen before being dispensed and brought to room temperature 45 minutes prior to testing to ensure consistency between trials. Samples are either liquid-based, oil-based, dry-based, dairy-based, or fermented. Samples, sample categories, and ingredients are listed in Table 17.2. A total of 26 subjects, including chefs, students majoring in a food-related field, and other food professionals participated in the study. Due to our large sample size the study would benefit from individuals experienced in tasting food, as they more quickly and accurately are able to identify and name the flavours in the mixes than novices would be. For this reason, we chose to draw from a pool of food knowledgeable persons.

The study included 29 different aromatic blends presented at the same time, which is quite a large number of samples. Thus, to reduce the likelihood of fatigue and inaccurate results, we were prompted to both recruit experienced subjects as well as make subject placement of samples and data quantification easier by increasing the size and including a grid on the napping® sheet, identifying this method as "Big Grid" napping. For this an A0-sized (84.1 by 118.9 cm) plastic-coated sheet was used. A grid of 60 by 90 cm with 2 cm squares was printed on the sheet. For ease of data collection, numbers representing the centimetre position of each line intersection from the lower left corner were added as labels outside the frame of the grid (see Fig. 17.3). The study were conducted at the Nordic Food Lab, located on a houseboat in Copenhagen, Denmark. Due to logistical reasons, respondents completed the study one person at a time. Before the experiment, a quick 5–10 minute instruction session was delivered

Table 17.2 Spice blends and the ingredients they contain

Category	Blend	Composition – ingredients
Liquid-based	BBQ Chipotle Blend	Chipotle, panela, apple vinegar
	Jerk paste	Onion, vinegar, scotch bonnet pepper, allspice, black pepper
	Juniper and Ant Paste	Juniper berry, thyme oil, ant, verbena, lemon thyme, woodruff
	Tomato/epazote	Tomato, epazote, piquin chilli
	Pickling blend	Coriander seed, vinegar, juniper berry, bay leaf
	Vihna d'alhos base	White wine, onion, paprika, garlic
Oil-based	Afro Bahian Base	Coconut milk, cassava, green pepper, cilantro
	Aji Panca Adobo	Aji panca, beer, grape seed oil, soy sauce
	Ligurian Pesto	Parmesan, pine nut, French beans, basil, potato, garlic
	Massaman	Shallot, lemongrass, galangal, garlic, red chile, shrimp paste, coriander seed, cumin, peppercorn, clove, cardamom
	Mole Negro	Guajillo chilli, mulato pasilla chilli, chille chipotle mora, sesame seed, peanut, almond, walnut, pecan, raisin, semi-sweet role, cinnamon, black pepper, clove, cumin, thyme, Mexican oregano, bay leaf
	Pipian	Pumpkin seed, tomato, corn, achiote, epazote, Mexican oregano
	Recado	Achiote, masa, cumin, pepper
	Salsa Verde	Capers, flat leaf parsley, olive oil, anchovy, lemon and rind
Dry-based	Berbere Mix	Coriander, clove, fenugreek, black pepper, cayenne, ginger, allspice, cumin, cardamom, cinnamon, nutmeg
	Dukkha	Hazelnut, cumin, sesame, coriander seed, black pepper
	Quatre Epice	Black pepper, cinnamon, clove, ginger, nutmeg
	Za'atar	Oregano, thyme, sumac, sesame seed, cumin
	Shichimi Togarashi	Szechuan peppercorn, buckwheat, koji, orange, garlic, nori, red chilli pepper, ginger, sesame seed
	Chinese Five-Spice	Star anise, Szechuan peppercorn, fennel seed, clove, coriander seed, cinnamon
	Panch Puran	Fenugreek seed, nigella, cumin seed, black mustard seed, fennel seed
	Garam Masala	White pepper, cinnamon, clove, cardamom, nutmeg, fennel seed, coriander seed, bay leaf

Alternative methods of sensory testing

Table 17.2 Continued

Category	Blend	Composition – ingredients
Dairy-based	Fresh Dill marinade	Crème fraiche, Dijon mustard, dill, honey, fennel seed
	Kadi	Yogurt (cow's milk), onion, Graham flour, ginger, cumin, mustard seed, turmeric powder
	Aji Escabeche/ Peanuts	Aji Escabeche, huacatay, peanut, milk, crème fraiche, fresh cheese
	Tikka Masala	Yogurt (cow's milk), onion, ginger, garlic, paprika, coriander seed, cumin, turmeric, cayenne, tomato
Fermented	NFL Fermented Bean	Black soy bean inoculated with *Aspergillus oryzae*, icing sugar, food molasses, white wine vinegar, black garlic
	Lacto Blueberry	Blueberry, sea salt
	Peaso	Yellow peas, buckwheat, koji, sea salt

and participants were not provided with any additional details about potential word suggestions to describe samples, but instead were instructed to determine their own criteria for placing and describing samples. Respondents did not discuss their evaluations with others during the assessment. To counteract sensory adaptation, respondents were provided with a small container of roasted coffee beans to sniff when their noses adapted to the aromas of the blended spices/pastes. Sample descriptions from respondents were highly varied, but by grouping descriptors with the same or very similar meaning, a consistent set of descriptive words was formulated to be used for further statistical analysis.

17.7 Results and discussion of spice blends and pastes

A visual representation of the results from the data analysis of the projective mapping of the 29 mixes/pastes is shown in Figs 17.4 and 17.5. From a culinary and sensory viewpoint, all samples are different from each other. For the present purpose, the most interesting part is the degree of agreement between individuals' maps of the differences. Consider the enormous memory load that is used to keep track of the sensory properties of so many samples during the projective mapping. The summed validated explained variance for the first two PCs is 37%. Analysis of the screen plot for both validated and calibrated variance (not shown) indicates that three may be the optimal number of components, but we limit the plots to only PCs 1–2. PC3 accounts for an additional 9% of validated variance, whereas the total calibrated variance is 30% after three PCs. A total of 335 descriptors were used to describe the different samples. Of these 121 were unique descriptors, used by only one panellist to describe one sample; in addition, the total number of descriptors that were used five times or less is 233. So the verbal response to the task varied greatly.

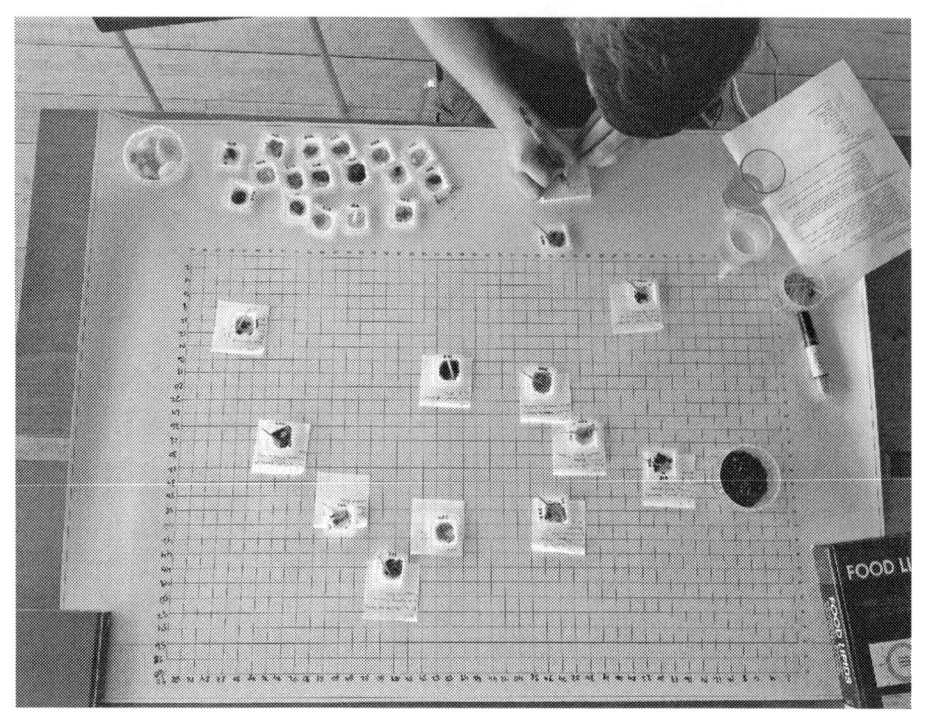

Figure 17.3 Respondent carrying out the task of positioning and describing 29 spice mixes and pastes. Note the size and the grid, identifying this method is "Big Grid" napping.

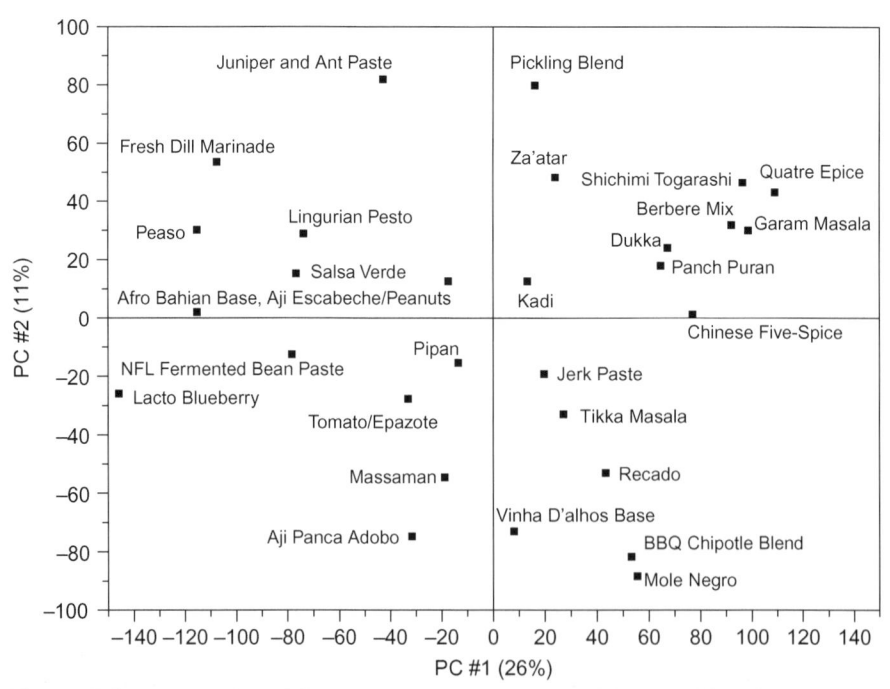

Figure 17.4 Score plot from PCA, Principal components 1 and 2. Map of 29 spice blends, showing the interrelationship between samples. See Table 17.2 for further information about samples.

Alternative methods of sensory testing 375

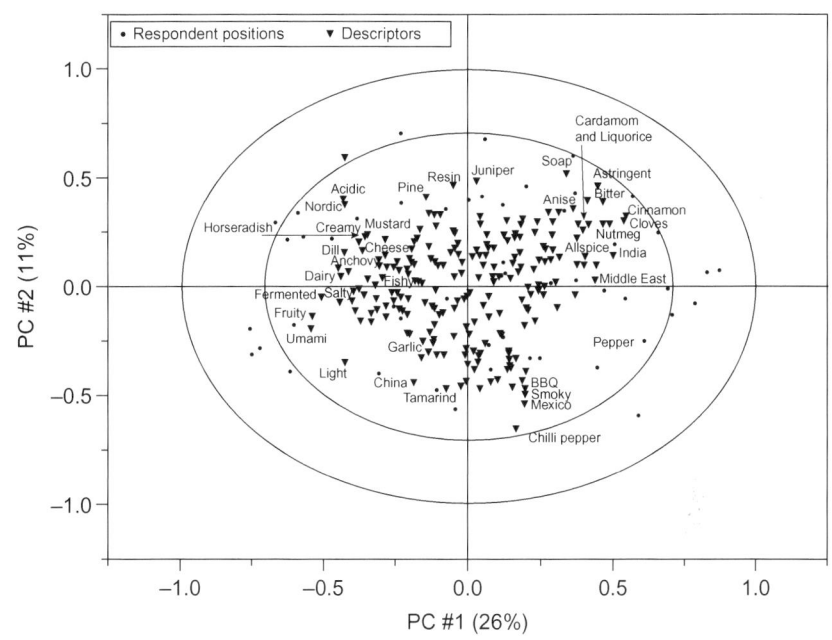

Figure 17.5 Correlation loading plot from PCA, Principal Components 1 and 2. Map of respondents' positioning (●, labels omitted for brevity and clarity of the figure) and descriptors (▼) used by respondents for sensory properties of the spice blends. For clarity only important descriptors are labelled.

Using data from 26 respondents with a large variation in geographical background and an unheard of 29 different potent samples to evaluate, the first question we pose is: are the results from the experiment comparable to other studies? From a range of studies with eight or more samples and many respondents the range of explained variance in the first two dimensions is reported as low as 22% (Torri et al., 2013, 12 wines described by 81 wine consumers) and up to as high as 62% (Reinbach et al., 2014, 8 beers with 55 respondents ranging from novices to brew masters; Ribeiro, 2011). At face value, the present results with 29 spice blend samples evaluated by 26 very diverse respondents from the food and restaurant trade appear comparable to other results. The litmus test is if the results make sense from a culinary and sensory point of view, i.e. can meaningful interpretations be made from the data? The overall observations from the data in terms of the interrelationship between the samples and their descriptions, are given in the following.

Samples positioned in the negative part of PC2 and in the positive part of PC1 (lower right part of Fig. 17.4), are Recado, BBQ Chipotle and Mole Negro, which all emerge from the southern part of North America or Central America. they are characterized with descriptors such as Smoky, Mexico, BBQ, Chilli Pepper and Pepper (Fig. 17.5). The samples positioned in the positive part of PC1 and PC2 (right and upper right part of Fig. 17.4), are Shichimi Togarashi, Quatre Epice, Berbere Mix, Garam Masala, and to some degree the group also encompasses Dukkha,

Panch Puran and Chinese Five-Spice. The geographical origin of those samples is well spread. However, they are all mixes of dried spices that are used as a basis for a wide range of dishes in their respective kitchens. The descriptors most often used to describe them are: Cinnamon, Cloves, Nutmeg, Bitter, Astringent, Allspice, Liquorice and Cardamom (Fig. 17.5). Samples positioned in the highest positive part of PC2 (top of Fig. 17.4) are Juniper Ant Paste and Pickling blend, samples that appear to be completely unrelated, but when the ingredients are studied (Table 17.2), they both contain Junipers, which is one of the descriptors characterizing these samples in addition to Resinous and Pine, both have characteristics that Juniper possesses. Samples in the lowest negative part of PC1 (left side of Fig. 17.4) share their origin. Peaso, Black bean paste and Lacto Blueberry originate from the Nordic Food Lab from the Noma restaurant, and are fermented flavour enhancers, with a significant umami taste component to them, which are also the descriptors most associated with them (Fig. 17.5). Afro Bahian Base is grouped with these samples. Bear in mind that the position of each sample is based solely on how the sample is positioned by individuals. The main descriptors for Afro Bahian Base are Coconut, Fresh, Cool and Cucumber (from raw data, not shown), that are also positioned in the direction (Fig. 17.5), albeit closer to the centre than many other descriptors. The downweighting of the descriptors in the analysis is the cause of this observation. In the upper vicinity of this group is Fresh dill marinade, a dairy-based Scandinavian concoction (see Table 17.2), used as a sauce or a marinade. The words associated with it are: Acidic, Nordic, Dill, Creamy, Mustard and Horseradish. These correspond to the ingredients, where the characteristic Horseradish originates from the same compound as mustard – allyl isothiocyanate. Closer to the centre in the same direction are Ligurian Pesto and Salsa Verde, both of them South European oil-based sauces, with a high content of umami-rich ingredients (parmesan and anchovies, respectively), and some pungent characteristics from Garlic – in the Ligurian Pesto, and from lemon with its rind in the Salsa Verde. Naturally the descriptors most often associated with these two samples are strongly related to this: Fishy, Anchovy and Salty for the Salsa and Cheese and Garlic for the Pesto, which are also located in that direction on the loading plots (Fig. 17.5). The samples positioned around zero on PC1 and the most negative part of PC2 are Aji Panca Adobo, Massaman and Vinhas d'Alhos. They originate from three different places: Ecuador/Peru, Thailand and Portugal, respectively. They also have very different compositions (Table 17.2), but by their overall perception in the set of mixes they are relatively similar. There are a few descriptors that they share, such as Salty, Garlic and Light (from raw data, not shown), but they are not located in that direction in the loading plot (Fig. 17.5), as they also characterize a number of other samples. Towards the centre of the plot are samples such as Aji Escabeche/Peanuts and Kadi, where both are dairy-based, and although they share a dairy base, the share only very few characteristics (Nutty and Curry, from raw data). In the lower part of the centre is Pipian, Jerk paste, Tomato/Epazote and Tikka Masala. In general, the samples in the centre of the figure are those that are less extreme with no specific characteristics being very dominant.

All in all, the interrelationship of the samples based on the consensus map from all 26 respondents is meaningful, when aligned with the culinary and gastronomic knowledge regarding those samples. However, as the respondents could freely list as many descriptors as they wanted for each sample, and there were no restrictions on which descriptors to use, the list was immense. Respondents' responses reflected their interest in food, and their experience in verbalizing the sensory information they processed. A majority of the respondents attempted to guess where the individual mixes/pastes came from, by listing countries or regions. A number of them responded with information about their suggested use for the spice mixes/pastes, e.g. *"very nice for cabbage and cake, good for Danish rolled meat sausage and liver pate,"* which was not included in the data analysis. Had we done so, the results from the descriptors would have been much more complex. Of the 335 different descriptors, 126 only occurred once, and a total of 221 were used five times or less. Respondents took between 45 minutes and 2 h to complete the projective map of the samples, and although most were delighted by the sheer variety in sensory properties they were exposed to, it was also an exhausting task to complete. In general, the feedback we collected was that it was a good experience, but there were a lot of samples to assess.

We have not presented the results to all of the participants at the time of the completion of the manuscript, but those to whom we have presented the results responded enthusiastically to the information. To many chefs this is a completely new method to organize their tastings, and the results can be understood by chefs in terms of their culinary organization of knowledge about what different spice blends and pastes can be used for, or as a reminder of what they have tasted on a particular occasion. The variation of the method we introduced – so-called Big Grid napping – was perfectly suited for a large number of samples. The use of a plastic-coated sheet of paper had the additional benefit that it could easily be cleaned with a wet cloth. It has proved very durable and can still be used more than a year later.

17.8 General discussion and recommendations

Initially there were two main questions with taking fast sensory methods out of the sensory laboratory, and into the domains of the professional kitchen or the brewery. One was if chefs and brewers would realize the benefits of the method to them, and the other was if they could understand the results of the data analysis. In general, brewers have a good enough understanding of the benefits of science to their field, so that they see the benefits of organized tasting, and systematic recording of the results of tastings. Chefs in general are less scientifically orientated than brewers, but in recent years the fields of science and gastronomy have benefitted from each other in several connections (see e.g. Risbo *et al.*, 2013). Chefs who have an interest in science appreciate the dual nature of Nordic Food Lab, with both a theoretical and a practical approach to food. Nordic Food Lab seeks to reconcile the two approaches, using modern and traditional methods to combine craft and science for delicious results. Using fast sensory methods with chefs and related professions, we have found that the organized and systematic collection of tasting notes is not overly complicated for the conditions in an experimental kitchen.

As mentioned in the introduction, there have been no systematic tests of the effect of changing the location from a sensory laboratory to different settings, such as a brewery or an experimental kitchen. However, for the practical purposes that we have explored in these case studies, we find that any effect of the location itself may be of a very small magnitude, if the general guidelines that panellists should not discuss their evaluation during sessions are followed. Having said that, it also remained a challenge for chefs and brewers to restrain themselves from voicing their opinions. In the case of the spice blends, panellists would generally do the evaluation alone, and were thus not able to disturb other participants, whereas with the brewers, everybody was seated in the same room and few utterances were made during the session.

The two case studies are very different in their nature. The case study with spice blends shows that it is possible to evaluate a very large pool of samples, and obtain meaningful results from the analysis. However, it is necessary to consider that the variety in the sensory properties in the sample set was very large. It is very likely that had there been only minor sensory differences, the results would not have been so clear. Although we did not systematically compare different variations of the method, we conclude that the new variation of the method that we presented, so-called Big Grid napping, facilitated evaluation of the large number of samples. Comparing the two cases, it is clear that the diverse backgrounds of the respondents for the spice blends/pastes increased both the variety in the descriptors, as well as the number of descriptors. We propose limitations to respondents' use of descriptors, both in terms of the type that they can use, restricting them to sensory descriptors and discouraging more general or holistic terms. However, there may be some applications where e.g. suggesting appropriate uses of a sample, can be beneficial. We also propose to limit the number of descriptors that a respondent can give to a sample. An upper limit of five descriptors appears to be a reasonable balance between restrictions and enough options to give a good description of a sample. An effect of limiting the number could be a more focused process in the selection of which descriptors best characterize a particular sample.

The variety in backgrounds of the respondents in both case studies is likely to have led to increased variety in the descriptors used. One of the differences previously found between different levels of expertise is the type of words used to describe the products. In another experiment with craft style beer, Ribeiro (2011) found that beer experts to a larger extent used specific sensory descriptors (e.g. Malty, Ester, Astringent), whereas less experienced beer drinkers used more abstract and integrated terms (e.g. Summer, Heavy, Youth). Differences in descriptors usage also highlight the apparent benefit of focusing the data analysis on the positions of the samples on the response sheet, and downweighting the importance of the descriptors. Descriptors should be regarding supplemental information that aids in the interpretation of the differences between samples. In an unpublished study of expert chocolatiers' and chefs' perceptions of premium chocolates, Magelund (2013) applied a procedure in which participants tasted the samples, and discussed descriptors that were suitable, prior to carrying out a projective mapping of the samples. The procedure provided more aligned use of descriptors among the respondents.

Successful communication of the results to non-sensory scientists requires careful attention to the knowledge that the receivers have. Multivariate data analysis is a concept that requires thorough explanation, and meticulous attention to the immediate feedback from the receiving audience. PCA can be explained to laymen. The analogy to a geographical map of a region is straightforward. The main purpose of a PCA is to reduce the complexity in data presentation, by extracting the main underlying phenomena and structure of a data set. It is easy to communicate that to laymen, since it is a common feature of much analysis we as humans do in many different activities – we tacitly extract what is important, and eliminate or put less attention to less important features[3].

References

Adams, J., Williams, S., Lancaster, B. and Foley, M. (2007). Advantages and uses of check-all-that-apply response compared to traditional scaling of attributes for salty snacks. In *7th Pangborn Sensory Science Symposium. Minneapolis, MN, USA*.

Ares, G., Deliza, R., Barreiro, C., Gimenez, A. and Gambaro, A. (2010). Comparison of two sensory profiling techniques based on consumer perception. *Food Quality and Preference*, **21**(4), 417–426.

Bruzzone, F., Ares, G. and Giménez, A. (2012). Consumers' texture perception of milk desserts. II – comparison with trained assessors' data. *Journal of Texture Studies*, **43**(3), 214–226.

Chollet, S., Lelièvre, M., Abdi, H. and Valentin, D. (2011). Sort and beer: Everything you wanted to know about the sorting task but did not dare to ask. *Food Quality and Preference*, **22**(6), 507–520.

Dairou, V. and Sieffermann, J. M. (2002). A comparison of 14 jams characterized by conventional profile and a quick original method, the flash profile. *Journal of Food Quality*, **67**(2), 826–834.

---

[3] At the University of Copenhagen we have implemented a short course aimed at practitioners working with product development in SMEs. The purpose is to teach them how to use fast sensory methods, both similarity-based and verbal-based sensory methods. In the teaching situation we use short animation films to teach PCA at a conceptual level, and the reception from the participants has been very positive. The considerations regarding data analysis have been how to make it operational and simple to use for laymen. The freeware PanelCheck is the most user-friendly free software, with very simple instructions necessary for successful operation of the program (see e.g. Tomic *et al.*, 2010). Although the software is designed for data from conventional descriptive analysis, it is possible to analyse data from projective mapping. The challenge is to provide guidelines for evaluation of the quality of the data and results that are both rigid and simple to understand. The experience in the Sensory Science Group at KU is that a necessary precaution when non-sensory scientists analyse the data is a strict focus on the explained variance. Although the range of explained variance in published studies can vary substantially, we recommend that it should be above 30% for the first two PCs. To keep the data analysis simple, we recommend a two-step procedure, with initial analysis of the positions only, and subsequent analysis of the positions and the descriptors (standardized data), and limit the analysis in the latter to plots that give information about which descriptors are attributed to different samples.

Dehlholm, C. (2012). *Descriptive Sensory Evaluations, Comparison and Applicability of Novel Rapid Methodologies.* PhD Thesis, University of Copenhagen.

Dehlholm, C., Brockhoff, P. B. and Bredie, W. L. P. (2012). Confidence ellipses: A variation based on parametric bootstrapping applicable on Multiple Factor Analysis results for rapid graphical evaluation. *Food Quality and Preference*, **26**(2), 278–280.

Delarue, J. and Sieffermann, J.-M. M. (2004). Sensory mapping using Flash profile. Comparison with a conventional descriptive method for the evaluation of the flavour of fruit dairy products. *Food Quality and Preference*, **15**(4), 383–392. doi:10.1016/S0950-3293(03)00085-5.

Dijksterhuis, G. B. and Byrne, D. V. (2005). Does the mind reflect the mouth? Sensory profiling and the future. *Critical Reviews in Food Science and Nutrition*, **45**(7–8), 527–534. doi:10.1080/10408690590907660.

Giacalone, D. (2013). *Consumers' Perception of Novel Beers.* PhD Thesis, University of Copenhagen.

Giacalone, D., Ribeiro, L. M. and Frøst, M. B. (2013). Consumer-based product profiling: Application of Partial Napping® for sensory characterization of specialty beers by novices and experts. *Journal of Food Products Marketing*, **19**, 201–218.

Grunert, K. G., Jensen, B. B., Sonne, A.-M., Brunsø, K., Byrne, D.V., Clausen, C., Friis, A., Holm, L., Hyldig, G., Kristensen, N. H., Lettl, C. and Scholderer, J. (2008). User-oriented innovation in the food sector: Relevant streams of research and an agenda for future work. *Trends in Food Science & Technology*, **19**(11), 590–602.

Husson, F., Josse, J., Lê, S. and Pagès, J. (2008, February 21). FactoMineR: Exploratory Multivariate Data Analysis with R. Retrieved February 05, 2014, from http://factominer.free.fr/index.html.

ISO-8589:2007. (2007). Sensory Analysis – General guidance for the design of test rooms. Geneva, Switzerland: International Organisation for Standardisation.

Lawless, H. T. (1989). Exploration of fragrance categories and ambiguous odors using multidimensional scaling and cluster analysis. *Chemical Senses*, **14**(3), 349–360.

Lawless, H. T. and Heymann, H. (2010). *Sensory Evaluation of Foods – Principles and Practices* (2nd, edn.). New York: Springer.

Lawless, H. T., Sheng, N. and Knoops, S. S. C. (1995). Multidimensional scaling data applied to cheese perception. *Food Quality and Preference*, **6**, 91–98.

Lé, S. and Husson, F. (2008). SensoMineR: A package for sensory data analysis. *Journal of Sensory Studies*, **23**(1), 14–25.

Lê, S., Josse, J. and Husson, F. (2008). FactoMineR: An R package for multivariate analysis. *Journal of Statistical Software*, **25**(1), 1–18.

Magelund, C. (2013). *Sensory Evaluation and Consumer Acceptance of New Premium Dark Chocolates.* MSc Thesis, University of Copenhagen.

Martens, H. and Martens, M. (2001). *Multivariate Analysis of Quality An Introduction.* Chichester: Wiley and Sons Ltd.

Morand, E. and Pagès, J. (2006). Procrustes multiple factor analysis to analyse the overall perception of food products. *Food Quality and Preference*, **17**(1–2), 36–42.

Moskowitz, H. R., Muñoz, A. M. and Gacula, M. C. (2003). *Viewpoints and Controversies in Sensory Science and Consumer Product Testing.* Trumbull, CT: Food & Nutrition Press, Inc.: Online version published in 2008: http://onlinelibrary.wiley.com/doi/10.1002/9780470385128.index/summary

Murray, J. M., Delahunty, C. M. and Baxter, I. A. (2001). Descriptive sensory analysis: Past, present and future. *Food Research International*, **34**(6), 461–471.

Nestrud, M. A. and Lawless, H. T. (2008). Perceptual mapping of citrus juices using projective mapping and profiling data from culinary professionals and consumers. *Food Quality and Preference*, **19**(4), 431–438. doi:10.1016/j.foodqual.2008.01.001

Pagès, J. (2003). Direct collection of sensory distances: Application to the evaluation of ten white wines of the Loire Valley. *Sciences Des Aliments*, **23**(5–6), 679–688.

Pagès, J. (2005). Collection and analysis of perceived product inter-distances using multiple factor analysis: Application to the study of 10 white wines from the Loire Valley. *Food Quality and Preference*, **16**(7), 642–649.

Perrin, L., Symoneaux, R., Maître, I., Asselin, C., Jourjon, F. and Pagès, J. (2008). Comparison of three sensory methods for use with the Napping® procedure: Case of ten wines from Loire valley. *Food Quality and Preference*, **19**(1), 1–11.

Reinbach, H. C., Giacalone, D., Ribeiro, L. M., Bredie, W. L. P. and Frøst, M. B. (2014). Comparison of three sensory profiling methods based on consumer perception: CATA, CATA with intensity and Napping®. *Food Quality and Preference*, **32**, 160–166.

Ribeiro, L. M. (2011). *Perception and Description of Premium Beers by Beer Experts, Novices and Enthusiasts*. MSc Thesis, University of Copenhagen.

Risbo, J., Mouritsen, O. G., Frøst, M. B., Evans, J. D. and Reade, B. (2013). Culinary science in Denmark: Molecular gastronomy and beyond. *Journal of Culinary Science and Technology*, **11**(2), 111–130.

Risvik, E., McEwan, J. A., Colwill, J. S., Rogers, R. and Lyon, D. H. (1994). Projective mapping: A tool for sensory analysis and consumer research. *Food Quality and Preference*, **5**, 263–269.

Stewart-Knox, B. and Mitchell, P. (2003). What separates the winners from the losers in new food product development? *Trends in Food Science & Technology*, **14**(1–2), 58–64.

Tomic, O., Luciano, G., Nilsen, A., Hyldig, G., Lorensen, K. and Næs, T. (2010). Analysing sensory panel performance in a proficiency test using the PanelCheck software. *European Food Research and Technology*, **230**, 497–511.

Torri, L., Dinnella, C., Recchia, A., Naes, T., Tuorila, H. and Monteleone, E. (2013). Projective Mapping for interpreting wine aroma differences as perceived by naïve and experienced assessors. *Food Quality and Preference*, **29**(1), 6–15.

Trijp, H. C. M. and van Kleef, E. (2008). Newness, value and new product performance. *Trends in Food Science & Technology*, **19**(11), 562–573.

Valentin, D., Chollet, S., Lelièvre, M. and Abdi, H. (2012). Quick and dirty but still pretty good: A review of new descriptive methods in food science. *International Journal of Food Science & Technology*, **47**(8), 1563–1578.

Van Kleef, E., van Trijp, H. C. M. and Luning, P. (2005). Consumer research in the early stages of new product development: A critical review of methods and techniques. *Food Quality and Preference*, **16**(3), 181–201.

Varela, P. and Ares, G. (2012). Sensory profiling, the blurred line between sensory and consumer science. A review of novel methods for product characterization. *Food Research International*, **48**(2), 893–908.

Williams, A. A. and Langron, S. P. (1984). The use of free-choice profiling for the evaluation of commerical ports. *Journal of the Science of Food and Agriculture*, **35**(5), 558–568.

Worch, T., Le, S. and Punter, P. (2010). How reliable are the consumers? Comparison of sensory profiles from consumers and experts. *Food Quality and Preference*, **21**(3), 309–318.

Appendix: Projective mapping versus napping (see also Chapter 9)

Projective mapping and napping are often improperly used as synonyms in the literature. It is more correct to state that napping is a specific case of projective mapping. While napping has a specified protocol (Pagès, 2005) with regards to materials, task instructions and data analysis, projective mapping is a more generic approach to sensory evaluations. Table 17.3 shows an overview of the differences between projective mapping and napping.

Although the issue of terminology may appear trivial, it is useful to bring this up because method users need to be aware that protocol modifications may alter the way panellists face the task and produce slightly different results. For example, as Dehlholm (2012) has shown, the shape and size of the frame affect the projection strategies the panellists adopt (particularly the use of the first and second dimension). Further, data analyses other than MFA on unscaled data may yield results that do not reflect individual panellist's use of the space (Morand and Pagès, 2006).

Table 17.3 **A schematic overview of the differences between projective mapping and napping**

	Projective mapping	Napping
Frame geometry and size	Rectangular (A4 or A3) or square (60 × 60)	Rectangular (60 × 40)
Frame look	Drawn axes, gridline, or blank	Blank
Data analyses	GPA, PCA, MDS-INDSCAL, STATIS	MFA on unscaled data

Sensory testing with flavourists: challenges and solutions

B. Veinand

Givaudan International SA, Kemptthal, Switzerland

18.1 Introduction

Herbs, spices and cooking techniques have been used to naturally flavour food for thousands of years. After the Industrial Revolution of the nineteenth century, food products began to be processed and preserved, in order to extend their shelf-life and enable their export to distant countries. These technologies have enabled the food industry to provide consumers with safe products, but have affected their intrinsic flavour quality. The flavour industry was born in order to compensate this loss of flavour, but nowadays the industry aims at much more. It aims at creating flavour solutions for food and beverages that will fully satisfy consumers by taking into account their likings, usages, attitudes, emotions and memories. A typical role of sensory testing within the flavour industry is to characterise the type of flavour that would appeal to consumers in a specific food product (e.g. would British consumers prefer a boiled-chicken or a roasted-chicken stock cube?). In order to inspire, guide and validate the efforts of the flavourists, sensory experts provide them with a standardised flavour description, which complements their individual expert point of view. In an international company such as Givaudan, flavourists are not necessarily native or living in the country for which their flavour is designed. They need support from sensory experts to obtain knowledge about the flavour directions that consumers experience in their market and about the sensory cues that would be liked or preferred in the future.

The success of the interaction between flavourists and sensory experts is intimately linked to their ability to communicate. Sensory experts must indeed understand the flavourists' ways of working, and flavourists must correctly interpret the sensory insights. The challenges of communication have been well described by Werber (1993):

> "Among what I think, what I want to say, what I think I say, what I say; what you want to hear, what you think you hear, what you hear; what you want to understand, what you think you understand, what you understand, there are ten possible difficulties to communicate. But let's try it though...."

The interaction between flavourists and sensory experts induces various challenges, due to their different fields of expertise and ways of working. Fortunately, the difficulties of communication can be overcome thanks to specific strategies which involve flavourists in sensory sessions. They can either be invited to attend sensory sessions as observers, to better understand how a sensory expert panel works, or be invited to actively participate in sensory sessions. In the latter case, rapid descriptive methodologies must be used instead of standard descriptive ones, in order to meet the specific flavourists' time constraints and ways of working, which will be described later in this chapter.

The usual positive outcome of these strategies enables highlighting the complementarity and synergetic interaction of the two types of expertise.

18.2 Roles and responsibilities

The roles and responsibilities of a flavourist and of a sensory expert within the flavour industry are very different. The former aims at creating flavour solutions to be applied in food bases, and the latter at studying the perception of the obtained flavoured food products.

18.2.1 Flavourists

18.2.1.1 Flavourists' role and responsibilities

In essence, flavourists are chemists. They use chemistry to mimic or modify the olfactory, gustatory and trigeminal properties of food products. They are responsible for creating flavours, according to customer needs and strategic goals, by using creativity, curiosity and passion. Mosciano (2006) described the role of the flavourist as an aspiring painter and a musician. Flavour creation is the fusion of science and artistry, as there are numerous possibilities for combining molecules to achieve a desired flavour perception. The way flavourists combine them is very personal. Working in multidisciplinary teams with food scientists, marketing, sales, sensory and analytical scientists, flavourists interpret all sorts of insights to create the right flavour in the right application for the right market.

18.2.1.2 Flavourists' main objectives

The main activities of a flavourist are three-fold. The first is to match an existing flavour, for example in the case of legislation changes. The second is to optimise an existing flavour, when for instance a customer aims at making his product preferred, in comparison to the current version or the main competition, in order to increase his market share. And the third is to create a completely new flavour – for example, this could be a new product development initiated by a customer in the context of a range extension or a new brand creation.

18.2.2 Three types of sensory experts

A wide range of methods is used in sensory science, from discrimination tests (usually to determine whether a significant difference exists between two or more products) to descriptive tests (to characterise the differences existing between products). This chapter focuses on descriptive sensory methodologies.

Within Givaudan, a typical descriptive sensory support is provided by three complementary types of sensory experts. Sensory expert panellists are in charge of expressing their flavour perceptions. Sensory panel leaders are in charge of recording the collected data and ensuring their quality. Finally, sensory scientists are in charge of analysing, interpreting and communicating the results to flavourists.

18.2.2.1 Sensory expert panellists' roles and responsibilities

The role of Givaudan sensory expert panellists is to describe as objectively and precisely as possible their perceptions of flavours applied in food products – flavour perception being here the complex set of olfactory (orthonasal and retronasal), taste, trigeminal and mouthfeel sensations perceived while eating. Sensory expert panellists obviously cannot ignore the appearance and the texture of the food products they taste, but they must not focus on these aspects. Red lights are used in the sensory booths to hide, as much as possible, the appearance differences between the food products and minimise the influence of these differences on the flavour perception. Literature indeed shows that colour can strongly influence the determination of odours, both qualitatively (Gilbert *et al.*, 1996; Stillman, 1993; Zellner and Whitten, 1999; Zellner *et al.*, 1991) and quantitatively (Kemp and Gilbert, 1997; Zellner and Kautz, 1990).

Conventional descriptive methods, such as quantitative descriptive analysis (QDA), usually imply a rather limited training of the assessors, which would not be sufficient in the flavour industry, as in-depth flavour description is a particularly challenging task. This is the reason why Givaudan relies on carefully selected and extensively trained sensory expert panellists. They are screened according to their sensitivity to a large diversity of molecules and flavour ingredients, their aptitude in expressing their sensations, and their ability to work in a team of around 15 people. After their selection, each panellist goes through an intensive training, mainly to learn the basics of sensory science, the Givaudan flavour language and the use of intensity scales. This initial training usually lasts for a year, to ensure high quality of future sensory data.

18.2.2.2 Sensory panel leaders' role and responsibilities

Sensory panel leaders provide sensory expert panellists with appropriate training in order to efficiently capture relevant sensory descriptive information. They are responsible for the maintenance of the quality of the data and the motivation of the sensory expert panellists by monitoring their performance and retraining them as often as necessary. For each project, sensory panel leaders adapt the process of evaluation according to the food product characteristics (i.e. temperature of consumption, palate cleansing methods, break duration, etc).

18.2.2.3 Sensory scientists' role and responsibilities

Most of the sensory scientists of Givaudan have a food science background. Based upon project objectives and sensory testing needs, they decide on adapted sensory approaches and collaborate with the cross-functional team to deliver appropriate results. They analyse, interpret and communicate the collected data in order to provide the flavourist with actionable insights.

18.2.2.4 Sensory experts' main objectives

For each flavourist's type of activity that was previously mentioned, sensory descriptive support can be provided to inspire, guide and validate their efforts. Below is a brief review of the various types of situations.

In the frame of a matching project, the sensory expert panellists would for instance describe the target to provide a starting point to the flavourist for his first match proposal development. They may also have to describe the differences between the target and already created match proposal(s), in order to determine what aspects of the targeted flavour profile have already been reached, or are still missing. The flavourist would then know where to concentrate his efforts in the matching process.

When an existing flavour must be optimised by a flavourist in order to make it better liked than that of a specific target (e.g. the current product), it is common practice to work with a set of several products, which may include products from the marketplace as well as a series of Givaudan prototypes. This ensures coverage of a diversity of flavour directions. Sensory expert panellists' job is then to accurately describe the flavour profiles of a group of products. Combining the flavour profile results with the results of a hedonic test performed with consumers (with the exact same set of products), as in the external preference mapping technique, allows providing sensory drivers of consumer acceptance to the flavourist (i.e. the descriptors that positively or negatively influence the hedonic appreciation of the consumers, and those having no influence). The flavourist would also get a prediction of the profiles of optimised flavour proposals (i.e. the intensities of all the descriptors that are predicted to be in combination of the most liked flavour) (Danzart *et al.*, 2004).

Finally, in the context of new flavour development, the sensory expert panellists would for example have to assess the flavour profile of selected market products, in order to highlight what flavour directions already exist and what the possible gaps are.

In an ideal situation, flavourists would intuitively understand outcomes provided by sensory experts. However, as their backgrounds, their fields of expertise and their ways of working are different, communication between these two functions is often challenging. In the following, I will present how these two types of experts differ, before presenting the solutions that we devised at Givaudan to improve the situation and make them work more efficiently together.

18.3 Different ways of working

In their respective positions, flavourists and sensory experts include frequent tasting sessions of food products, as well as descriptions of their associated flavour sensations. However, the ways they evaluate the food products, and how they express and quantify their perceptions, are very different. Acknowledging these differences is important in adapting communication and ensuring clear understanding between both communities.

18.3.1 Evaluation of the products

Technical tasting sessions organised by flavourists follow an empirical course, whereas sensory blind tasting sessions follow a standardised approach.

For the flavourists, the main goal of technical tasting sessions is to familiarise themselves with a specific product set or to evaluate the performance of their own flavour creations. In that respect, flavourists are fully aware of all the project details, such as the requester, the objectives, the targeted market, the product recipe, etc., which strongly influence their perceptions and their judgements.

On the other hand, sensory expert panellists are never informed about the project stakes when they describe flavour profiles. Blind tastings enable the collection of objective information. This avoids many psychological biases. An expectation error would consist, for instance, of having the preconceived idea that a vanilla ice-cream sold under a well-known brand would have a more intense vanilla flavour than a cheaper brand, even before having tasted them. A blind evaluation using plastic cups labelled with a three-digit code ensures the minimisation of any such preconceived idea, in order to objectively determine whether the intensity of vanilla flavour is different from one product to another.

Flavourists, however, are not used to the total absence of context, and are rarely aware of its possible impact on perception. They would thus expect technical tasting and a sensory evaluation to provide similar results, and they may be disturbed when faced with difference between the outcomes of these two types of evaluation. For instance, a recent sensory market study on various European best-sellers of vanilla icecreams highlighted the complete absence of vanilla flavour notes in some of the products coming from the same country. This insight was completely unexpected by the flavourist community, and especially relevant to the ones who develop vanilla icecream flavours for this particular country. They now know that they must focus on creamy and dairy notes instead of vanillin or vanilla bean notes to provide vanilla flavours which fit this specific market.

18.3.2 Deconstruction of the flavour perception

As most flavourists are chemists, they have an extensive knowledge of the numerous molecules and ingredients that can be combined to develop flavours. A typical part of their studies consist in learning an extensive list of flavour molecules, their corresponding smell and/or taste and their usage in flavour formulation. Therefore, they tend to use molecule names to describe their flavour perceptions, mentally analysing and deconstructing the formula they think is hidden behind this perception.

On the other hand, sensory experts do not provide ingredient information; they aim at determining the perception of the flavour directions that characterise the profile of a given product. Using the metaphor of painting, if a flavourist were a painter trying to copy an existing masterpiece by a famous impressionist – e.g. Claude Monet – he would have a specific approach. He would first try to find out which colour pigments were used to achieve this specific painting. He would then combine this information with his artistic ability to reproduce the whole painting, even if not all pigments are available to him.

Impressionists sought to express their perceptions of nature, rather than create exact representations. Similarly, flavourists need support from sensory expert panellists to provide them with a generalised human perception to complement their individual expert perceptions and thereby ensure a good reproduction of the painting. Sensory experts would give specific colour names and their associated intensity. For instance, the flavourist might perceive a small difference between his draft and the original version, based on an overall impression without always being able to understand on what aspects. Thanks to a comparison of both paintings, sensory experts would describe all the possible differences and let the flavourist know that he should slightly lighten the blue to get closer to the original painting. The flavourist would identify that the cobalt pigment is the one responsible for the blue colour and would slightly decrease its quantity.

For the sensory support to be helpful an effective translation must be ensured between what a sensory expert panellist means and what the flavourist understands (for instance what colour pigment corresponds to "bright purple"? Are there several pigments involved?). And this can only be achieved thanks to a strong collaboration between both functions.

18.3.3 Description of the flavour perception

As stated before, a flavourist is a highly trained expert in chemistry with an artistic mind. Therefore, flavourists often use their own language to describe their flavour perceptions, using ingredients and molecule names, such as "maltol" and "cis-3-hexenol," combined with more consumer- and artist-linked vocabulary such as "rich," "authentic" and "balanced." Their personal notes, taken while tasting, facilitate their creation process. As this process is usually individual, it does not need to be standardised or comprehensible to the whole flavourist community. In some specific cases, when flavourists seek advice from other flavourists, who by definition share a similar expertise, they can easily adjust their language and use only molecule and ingredient names to base their common understanding.

Flavourists' descriptions are fundamentally personal, and flavourists are indeed very efficient as long as they work individually. However, this contrasts with the language used by sensory expert panellists, which must be harmonised and understandable by a large number of people, starting with the sensory expert panellists themselves and including the stakeholders to whom the sensory insights will be communicated. This is particularly challenging since human beings are not naturally able to express their smell and taste perceptions as objectively and efficiently as they would their visual or auditory perceptions. Each person relies on his own expertise, culture, memories and sensitivity while describing a flavour. In addition, the concepts associated with flavour perception are not easily translated and understandable from one language to another.

To minimise these numerous barriers, Givaudan has developed a global flavour language called Sense It™. This corresponds to a language dictionary in which each descriptor is given a name (corresponding to an identified flavour direction), a definition (an accurate explanation of what the flavour direction's name stands for) and, more importantly, a physical reference illustrating each aroma, taste, trigeminal sensation and mouthfeel perception. References are single chemical substances, spices/extracts, ingredients, or flavours, which can be either smelled or tasted. Nowadays, Sense It™ contains more than 370 descriptors, with reference materials classified over 140 flavour lexicons (e.g. orange or chicken lexicon). Sense It™ language is not exhaustive; however, its large number of descriptors enables describing flavour perception for most of the product categories from the marketplace around the world. It is constantly being evaluated and adapted by a team of flavourists and sensory experts, according to specific needs and any evolutions of product characteristics on the market.

All Givaudan sensory expert panellists are trained, and constantly retrained, on Sense It™ language to ensure relevant sensory outcomes.

The usual technique used by the sensory panel leader to familiarise the sensory expert panellists to Sense It™ language is the organisation of individual smelling and tasting sessions of the references belonging to a lexicon of interest. The sensory expert panellists are invited to take notes of their personal memory associations to each stimulus, in order to better remember it (e.g. the reference "jammy strawberry" reminds me of my grandmother's jam). This individual step is followed by a group discussion where they share their individual notes. This step enables the sensory panel leader to check whether all the descriptors are understood and being used in the same way by all members of the group. Once the sensory expert panellists are familiarised with the lexicon, a quiz is organised wherein they must individually find the name of the references that are presented to them blind. This exercise enables the sensory panel leader to build individual training plans focused on the most problematic descriptors for each sensory expert panellist.

The challenge is that, whereas sensory expert panellists can take time to train themselves, flavourists cannot afford to spend such amount of their resources to correctly learn the complete Sense It™ language, or even the most important lexicons.

18.3.4 Quantification of the flavour description

For all the above mentioned objectives, it is important to be able to describe a flavour qualitatively (i.e. this chicken bouillon contains some pyrazine), but also quantitatively.

Using their own expertise and sensitivity, flavourists usually quantify their perceptions making use of categories (e.g. this chicken bouillon contains a huge amount of pyrazine) or of product comparison (e.g. this chicken bouillon contains much more pyrazine than the previous one). Givaudan sensory expert panellists are trained to quantify their perceptions using a 100 mm unstructured scale. Various practical exercises can be used by the sensory panel leader to train them. They can, for instance, practice scaling by estimating the amount of shading on various shapes (Meilgaard *et al.*, 2007). Flavourists are not trained in using scales of intensity. Therefore, they might sometimes misinterpret the meaning of intensity scores of sensory profiles.

18.3.5 Individualisation or standardisation of the flavour description

As previously mentioned, the work of flavourists is mostly individual; they base their judgements on their own expert perceptions during technical tastings. They are not used to dealing with standardised flavour descriptions obtained from a team of sensory expert panellists. Even though flavourists are highly trained, a certain part of subjectivity might subsist in their flavour evaluation, due to their individual liking and/or sensitivity. As a consequence, the results of sensory evaluation do not always match their own results. A simple example could be the perception of bitterness. Indeed, it is well known by the scientific community that sensitivity to bitterness widely differs from one individual to another, due to genetic differences (Delwiche *et al.*, 2001). A flavourist who is extremely sensitive to molecules that induce bitterness will often perceive a high intensity of bitterness in some food products and will focus his attention on this aspect. In parallel, a sensory profile of the same food products obtained by a group of sensory expert panellists with various sensitivities to bitterness might weight the resulting mean intensity of the descriptor bitter. In that specific case, it is crucial that sensory experts explain to the flavourist the possible reasons why the sensory results differ from his own perception.

After listing all the differences in the ways of working between flavourists and sensory experts (Table 18.1), it appears obvious that the communication between these two communities of experts is very challenging. From their language to their tasting approaches, everything differs. Strategies have been established in Givaudan to take advantage of these differences and obtain a synergetic effect from both types of expertise, which *in fine*, enables Givaudan to provide the right flavours in the right applications for the right markets.

18.4 Strategies to complement both types of expertise

The key to these strategies is to make flavourists attend sensory sessions, either as guests or as active participants, depending on the selected sensory methodology.

Various sensory methodologies are used in Givaudan to characterise the sensory flavour profiles of food products. In each project, the selection of the methodology depends on the objectives of the study, the number of products to be evaluated, their similarity to each other, the type and the precision of the information being looked for, and the time and resources that are available. The sensory scientist in charge of each project must carefully take all these parameters into account to select the descriptive method that will deliver the most appropriate results using an optimal amount of resources.

Table 18.1 Summary of the main differences between flavourists' and sensory experts' fields of expertise and ways of working

	Flavourist	Sensory expert
Roles and responsibilities	Creation of flavour solutions to be applied in food bases	Study of the perception of food products to support flavourists' work
Usual background	Chemist	Food Scientist
Ways of working		
Evaluation of products	Empirical technical tasting with context awareness	Blind standardised tasting
Deconstruction of flavour perception	Link between flavour perception and assumed flavour formula	Link between flavour perception and associated descriptors
Description of flavour perception	Molecules/ingredient names, combined with consumer and artist-linked verbatim	Standardised flavour language (Sense It™)
Quantification of the flavour description	Quantification by categories and ranking	Quantification with 100 mm unstructured scale
Individualisation or standardisation of the flavour description	Individual	Consensual

Whenever a flavourist needs accurate qualitative and quantitative information about the flavour profile of specific food products to efficiently guide his flavour creation process, the sensory scientist typically decides to run a quantitative flavour profiling (QFP) on this set of products. This is the type of sensory session that flavourists will be invited to attend as observers. In some other projects, flavourists might only need a general overview of the main flavour directions existing in a product set, to gain for instance general knowledge on a specific market. In that case, rapid descriptive methodologies, such as free sorting and flash profile, will be preferred. This is the type of methodology that flavourists will be invited to actively experience.

18.4.1 Sensory evaluation sessions observed by flavourists

18.4.1.1 Quantitative flavour profiling (QFP) methodology

The QFP has been developed at Givaudan (Stampanoni, 1993) as an enhancement of the QDA®. The QFP technique uses sensory expert panellists to identify and quantify the flavour perception of food products (i.e. ortho- and retro-nasal aroma, taste, trigeminal sensation and mouthfeel) using the Givaudan global flavour language Sense It™. The perception of the appearance and the texture of the food products is not measured. The basic steps of this method include: language generation, training of the sensory expert panellists, individual data collection of repeated measurements, use of linear scale and analysis of results by univariate and multivariate methods.

During the language generation session(s), sensory expert panellists blind taste every product of the set and determine which of the Sense It™ references are needed to describe all the flavour characteristics of the product space. This step is at first individually performed in the sensory booths to enable an optimal focus from the sensory expert panellists. It is eventually discussed by the team of sensory expert panellists, in order to achieve a consensual list of descriptors, which will be reviewed and finalised in the training step.

18.4.1.2   QFP step flavourists are invited to observe

Flavourists are invited to attend the language generation step of QFPs and observe how sensory expert panellists build the Sense It™ descriptor list. They are strongly advised to smell or taste every Sense It™ reference that is discussed within the session. They may communicate questions to the sensory panel leader in charge of the sensory test, who will then check the answers with the sensory expert panellists. Indeed, they do not actively participate in the discussion, in order to prevent any bias of the sensory expert panellists' perception.

18.4.1.3   Two-ways benefits

The involvement of the flavourists in the sensory language generation sessions brings benefit to both types of experts. The benefits to flavourists are listed first, followed by the benefits to the sensory experts.

After having been exposed to some language generation sessions, flavourists usually realise that this time spent at the beginning of a project enables them to eventually speed up their creation process. Thanks to this step, they understand much better Sense It™ descriptor meanings and therefore how products are perceived by the sensory expert panellists. They also see how much a blind tasting impacts the whole flavour perception. Consequently, the sensory outputs which will be given to them at the end of a sensory evaluation are much more trusted by the flavourist community, who will then use the results more efficiently and more frequently.

Observation of the sensory expert panellists' ways of working makes the flavourists also aware that each piece of sensory descriptive information must always be linked to the set of products that was evaluated within a sensory test. Sensory scores of intensity are relative to the context of the product space and must not be taken as absolute scores. Indeed, even if the contrast and convergence effects (Amerine *et al.*, 1965) are minimised by the balanced or randomised order of presentation of the samples in every sensory test, they are never totally eliminated.

Table 18.2 Givaudan Sense It™ descriptors usually selected to describe mushroom flavour profile, and their definition

Descriptor	Definition
Astringent	Drying mouthfeel typically associated with tasting tannin (in water), strong black tea, unripe banana, or young red wine
Bitter	Basic taste sensation associated with substances such as caffeine or quinine diluted in water
Earthy	Aroma associated with damp or mossy soil, similar to a musty cellar
Fatty-fried	Aroma associated with heated frying oil
Hay-like	Dried grass, tea and tobacco aroma associated with hay, similar to a hay stack in a field
HVP-like	Meaty, brothy and vegetable aroma associated with hydrolysed vegetable protein (HVP) diluted in water
Metallic	Aroma associated with iron or other metals in water
Mouldy	Sour and ammonia-like aroma associated with cheese products showing mould on their surface; similar to Camembert cheese
Mushroom	Fruity, earthy aroma associated with fresh button mushrooms
Salty	Basic taste sensation associated with table salt (NaCl) diluted in water
Sour	Basic taste sensation associated with acids in solution
Stale bread	Woody, hay-like aroma associated with old, stale bread
Truffle	Sulphury aroma associated with aged button mushrooms and truffles
Umami	Basic taste sensation associated with mono sodium glutamate (MSG), characterised by fullness of flavour in the mouth; often found in bouillons, soy sauce and mushrooms
Vegetables-mushroom	Aroma associated with mushrooms
Yeasty-savoury	Meaty, yeasty aroma and taste associated with yeast extract powder
Vegetables-potato	Vegetative, sulphurous and earthy aroma associated with cooked potatoes

In some cases, Sense It™ is not sufficient for sensory expert panellists to describe specific flavour characteristics. Listening to sensory expert panellists' issues, flavourists may quickly propose new physical references to fill Sense It™ language gaps and avoid any delays in the completion of the sensory test.

Taking the example of the mushroom Sense It™ lexicon, the diversity of the descriptors belonging to this lexicon is usually great enough to describe most of the savoury food products having standard mushroom aspects (see Table 18.2). However, if the objective of a project is to accurately describe the specificities of several mushroom species in food application (e.g. bouillons), Sense It™ is not precise enough. The help of a flavourist is therefore required to suggest new physical references to the sensory expert panellists in order to enable them to better express and quantify the differences between the mushroom flavour profiles. The sensory outputs would then describe, for example, what characterises an oyster mushroom and makes it different from a boletus.

18.4.2 Sensory evaluations performed by flavourists

Flavourists are sometimes requested to actively participate in sensory evaluation sessions. The main benefits of these sessions are three-fold: to increase the flavourists' knowledge of sensory methodologies, to help flavourists gain general knowledge of the market they work on, and to improve communication between flavourists and sensory experts by aligning their languages. However, it is important to understand that the aim of these invitations is not to replace sensory expert panellists by flavourists. Their generated data are under no circumstances utilised or communicated as sensory results.

18.4.2.1 Methodologies

As previously explained, flavourists are neither extensively trained in Givaudan Sense It™ global language, nor used to quantify their perception with linear scales. Adapted sensory methodologies, which take into account flavourists' ways of working, must therefore be selected. In this context the most frequently selected methodologies in Givaudan are flash profile and free sorting. Both methodologies are indeed adapted to flavourists' ways of working, as they are based on individually elicited attributes and do not imply the use of intensity rating scales. Free sorting is usually preferred to flash profile in the context of large product sets.

Flash profile is a technique developed by Sieffermann (2000) that combines free-choice profiling with a comparative evaluation of the product set. This method is divided into two consecutive steps. It first consists in asking the assessors to individually generate their own attributes to describe the product set. In the second step, assessors are asked to rank the samples according to each of those attributes. As assessors have the whole sample set in front of them, it forces them to focus on the differences they perceive and to generate discriminant attributes.

Flavourists are usually comfortable with this methodology, as the requested deconstruction of the flavour perception to discriminate products is not far from their way of working. They find the ranking task easy to perform, as long as the number of products to rank is not too large (the sensory scientist in charge of the project usually selects flash profile for studies of eight to ten products maximum).

As explained by Faye et al. (2004)' the free sorting task consists of grouping samples according to their similarities, and allows the evaluation of large sets of products. Givaudan assessors are usually asked to describe the groups they have formed or the products themselves, to associate vocabulary to the stimuli (Chollet and Valentin, 2000; Hollins et al., 1993; Lawless, 1989; Lawless et al., 1995). Within Givaudan, the sensory scientist usually chooses free sorting for the study of 10–30 products. What can be perceived as a difficult task for the flavourists is to base their grouping task on their overall flavour perception instead of deconstructing it and focusing on various details. This difficulty is actually also valid for sensory expert panellists who are trained on QFP methodology. They are indeed more used to accurately and individually describe each of the products.

18.4.2.2 Benefits to flavourists

The first benefit is a better understanding of sensory methodologies. Some flavourists collaborate more often with sensory experts in their day-to-day work than others, depending on the type of projects they work on, but also their location. Givaudan sensory resources are indeed allocated to support the most important customer projects and/or the most strategic markets. Whenever flavourists, who do not often collaborate with sensory experts, work in a project in which a flash profile or free sorting will be conducted, they are invited to come and perform the test themselves in parallel with the sensory expert panellists. The objective is to make them understand the methodology and learn about the product set in order for them to better understand the insights. Their raw data are consequently removed from the whole data set before the data analysis.

When the consensual product positioning and clustering is available, it is compared to that performed by the flavourist. In case of critical differences, a discussion between the sensory scientist and the flavourist occurs to determine their origin. For instance, it happened that a flavourist invited to perform a free sorting on poultry flavours came up with a slightly different product clustering from the one obtained with the sensory expert panellists. Two flavours were put in two very different clusters by the sensory expert panellists, whereas they were grouped in the same cluster by the flavourist. After some discussion, it appeared that the flavourist had recognised these two flavours and remembered that they were labelled as turkey flavours, which explained why he put them in the same cluster, even though they were perceived to be very different from each other. This finding enabled the flavourist to better understand the importance of being unbiased, and to rely on objective descriptive information to complement his own evaluation.

The second benefit is a better understanding of the market. As already explained, Givaudan flavourists are not necessarily native to, or living in, the country for which their flavours are designated. Some regions of the world are very fragmented, with very different cultures and food habits. It is very hard for the flavourists to know each market, and consequently understand well what the customers and consumers are used to in terms of flavour profile. Therefore, sensory experts can provide flavourists with support, by guiding them in their proactive market product evaluations. Thanks to regular free sorting tests performed on product sets varying in applications and markets, flavourists can regularly update their market knowledge and ensure a good overview of flavour directions existing on various markets. The information gained via these free sorting sessions is purely internal flavourists' knowledge and is as a result not communicated as sensory insights.

The third benefit is a better understanding of Sense It™ language. Some proactive sensory initiatives are conducted in Givaudan to try to align Sense It™ and flavourists' languages and thereby enable a better communication between both types of experts. In general, several product sets encompassing varying product applications and countries are selected to study important flavour icons (e.g. orange).

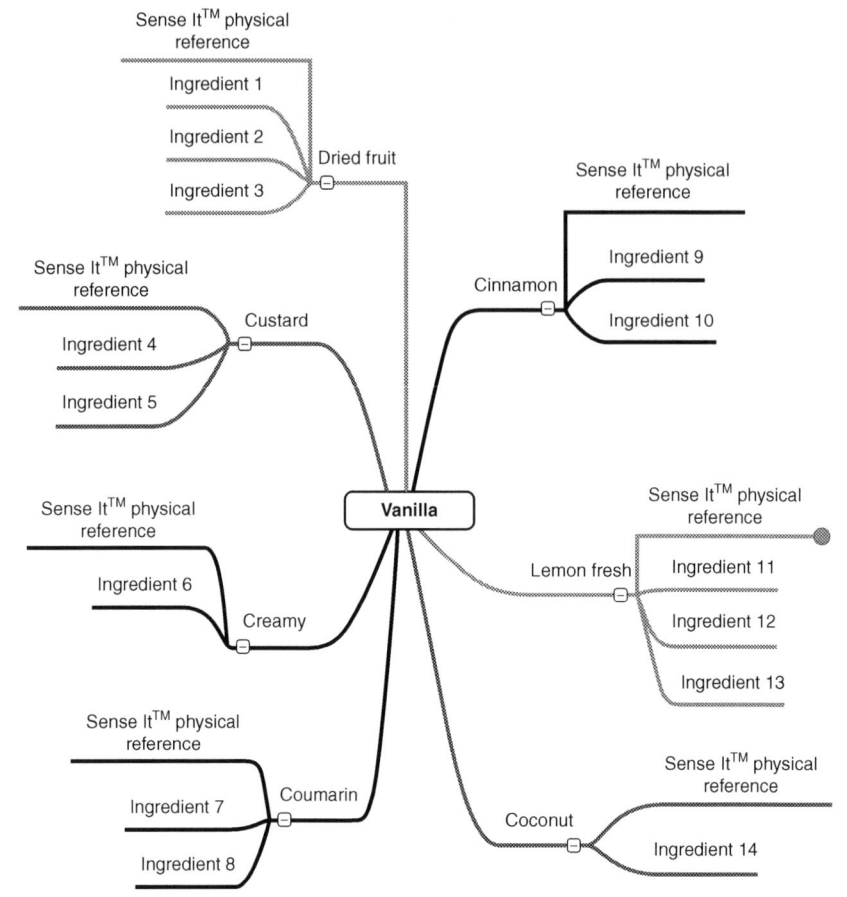

Figure 18.1 Example of links made between Sense It™ references and flavourists' ingredients to illustrate descriptors of the vanilla flavour language.

A group of flavourists and a team of sensory expert panellists perform in-parallel sensory evaluations. The generated data are analysed and compared by sensory scientists.

A typical outcome is a link between one Sense It™ descriptor and one or several flavourist' word(s), or between one flavourists' word and one or several Sense It™ descriptor(s) (Fig. 18.1). Thanks to these proactive initiatives, sensory experts are enabled to adapt their speech when they communicate their insights to flavourists, who will better understand them and therefore successfully use them in their creation processes.

Nevertheless, a drawback exists in this strategy. It is often difficult to extract insightful language information out of this type of comparison. The mental process used by flavourists when they perform sensory evaluations is very different from that used by sensory expert panellists. As explained previously, flavourists use an analytical process to deconstruct their flavour perception, in the sense that they try to find what molecules or ingredients were combined to generate each flavour direction. This will lead them to focus on very specific flavour aspects of the products, which might not be a focus for sensory expert panellists.

The positioning and clustering of the product set obtained by both types of experts are therefore usually very different from each other. Consequently, the link between the two types of language is very difficult to build.

18.5 Future trends

As explained in this chapter, flavourists are chemists with artistic minds. They are experts in deconstructing their flavour perceptions. This enables them to create new flavours, which are either copies or modified versions of what already exists or completely new flavour directions. Nowadays, they need additional sensory inputs to better understand the consumers' perceptions of flavours. They need to understand what consumers from each market are used to consume, appreciate, or even expect in the future. Indeed, these insights are key in providing Givaudan customers with successful flavour solutions which differentiate their food products and make them appealing to consumers. The aim of sensory experts in Givaudan is to put flavour insights at the fingertips of the flavourists to inspire, guide and validate their efforts. This chapter has listed the challenges of communication between sensory experts and flavourists that are due to their different backgrounds, fields of expertise and ways of working. It then explained which strategies Givaudan has implemented to improve their communication and benefit from the combination of their two types of expertise. As a consequence, sensory experts understand the needs of the flavourists better, and therefore provide them with actionable insights which are translatable into flavour solutions.

The communication between sensory experts and flavourists has certainly been improved, but it is still not fully optimised. New strategies to improve it are constantly being investigated. These vary in feasibility of implementation and in short-, mid- or long-term time frames. An example of a new strategy that could be implemented in Givaudan is to regularly propose short "internships" of a few days to their new employees. On the one hand, every new sensory scientist and panel leader would follow a short "flavour creation internship" in order to follow the day-to-day work of flavourists and better understand their ways of working and challenges. On the other hand, every new flavourist would participate in a short "sensory internship" to be better aware of the kinds of sensory supports they can be provided with, their added value, and the way they should be understood and translated into flavour creations.

A similar strategy could be extrapolated to the studies of sensory scientists and flavourists. Thanks to extra university modules respectively of flavour creation and sensory, these two types of experts would be better prepared to start their career in the flavour industry and would quickly optimise their collaboration.

In conclusion, the goal of Givaudan is to have all flavourists and sensory experts well aware of each other's fields of expertise and completely fluent in each other's languages and ways of thinking, in order to better understand and translate consumer and customer wants into finished flavours.

Coming back to the central theme of this book, rapid profiling techniques and related methods, I would like to highlight the fact that rapid descriptive methods are an integral part of the work of sensory scientists in Givaudan. Indeed, in today's challenging economic context, the flavour industry, like every other industry, must optimise as much as possible the time and resources allocated to sensory studies.

However, "rapid" does not mean any compromise of the quality and relevance of the sensory information. As previously stated, before selecting any rapid descriptive method, the sensory scientist must check that it fits the objectives of the study, and the expected type and precision of the resulting information. Flash profile and free sorting are rapid techniques that are often used in Givaudan. But as every study is different, Givaudan constantly needs to increase its flexibility by diversifying its pool of rapid descriptive methods. Some rapid methods have already been tested but not kept for various reasons. One of them is the projective mapping, which is a comparative sensory descriptive method in which the subjects are asked to use their own criteria to position objects directly on a map according to the rule that the closer two objects are placed on the map, the closer their perceived characteristics (Goldstone, 1994). In Givaudan, it was at first tested with consumers (Veinand et al., 2011), and difficulties in using the spatial positioning were observed. Actually, careful observation revealed that many consumers in fact performed a simple free sorting instead of a real projective mapping. The discrimination ability of this method was also lower than the other tested methods. Later, similar problems were encountered with sensory expert panellists. No obvious added value was finally seen in comparison to free sorting, which is the reason why projective mapping was not selected as a usual Givaudan method.

Sensory symposiums are good opportunities for Givaudan sensory scientists to discover new rapid profiling techniques and to test their suitability for the flavour industry. One of the hottest topics is, for instance, to find a rapid technique which facilitates the evaluation of large product sets with tough temperature constraints (such as warm soups and icecreams), as it is well known that any change of temperature dramatically modifies the flavour profile of food products.

References

Amerine, M. A., Pangborn, R. M. and Roessler, E. B. (1965) *Principles of Sensory Evaluation of Food*, New York, N.Y., Academic Press.

Chollet, S. and Valentin, D. (2000) Le degré d'expertise a-t-il une influence sur la perception olfactive? Quelques éléments de réponse dans le domaine du vin, *L'année psychologique*, **100**, 11–36.

Danzart, M., Sieffermann, J.-M. and Delarue, J. (2004) New developments in preference mapping techniques: Finding out a consumer optimal product, its sensory profile and the key sensory attributes. *6th Sensometric Meeting*, Davis, CA USA.

Delwiche, J. F., Buletic, Z. and Breslin, P. A. S. (2001) Covariation in individuals' sensitivities to bitter compounds: Evidence supporting multiple receptor/transduction mechanisms, *Perception Psychophysics*, **63**, 761–776.

Faye, P., Brémaud, D., Durand Daubin, M., Courcoux, P., Giboreau, A. and Nicod, H. (2004) Perceptive free sorting and verbalization tasks with naive subjects: An alternative to descriptive mappings, *Food Quality and Preference*, **15**(7–8), 781–791.

Gilbert, A. N., Martin, R. and Kemp, S. E. (1996) Cross-modal correspondence between vision and olfaction: The color of smells, *American Journal of Psychology*, **109**, 335–351.

Goldstone, R. (1994) Methods and designs. An efficient method for obtaining similarity data, *Behaviour Research Methods, Instruments, and Computers*, **26**, 381–386.

Hollins, M., Faldowski, R. and Rao, S. (1993) Perceptive dimension of tactile surface texture: A multidimensional scaling analysis, *Perception & Psychophysics*, **54**(6), 697–705.

Kemp, S. E. and Gilbert, A. N. (1997) Odor intensity and color lightness are correlated sensory dimensions, *American Journal of Psychology*, **110**, 35–46.

Lawless, H. T. (1989) Exploration of fragrance categories and ambiguous odors using multidimensional scaling and cluster analysis, *Chemical Senses*, **14**(3), 349–360.

Lawless, H. T., Sheng, N. and Knoops, S. S. C. P. (1995) Multidimensional scaling of sorting data applied to cheese perception, *Food Quality and Preference*, **6**, 91–98.

Meilgaard, M., Civille, C. V. and Carr, B. T. (2007) *Sensory Evaluation Techniques*, Boca Raton, FL, USA, Fourth Edition, CRC.

Mosciano, G. (2006) *Successful Flavors – From Formulations to QC to Applications and Beyond*, Carol Stream, IL, USA, Allured Publishing Corp.

Stampanoni, C. R. (1993) The "Quantitative Flavour Profiling" Technique, *Perfumer and Flavourist*, **18**, 19–24.

Stillman, J. (1993) Color influences flavor identification in fruit-flavored beverages, *Journal of Food Science*, **58**, 810–812.

Veinand, B., Godefroy, C., Adam, C. and Delarue, J. (2011) Highlight of important product characteristics for consumers. Comparison of three sensory descriptive methods performed by consumers, *Food Quality and Preference*, **22**, 474–485.

Werber, B. (1993) *Le livre secret des fourmis*, France, J'ai Lu.

Zellner, D. A., Bartoli, A. M. and Eckard, R. (1991) Influence of color on odor identification and liking ratings, *American Journal of Psychology*, **104**, 547–561.

Zellner, D. A. and Kautz, M. A. (1990) Color affects perceived odor intensity, *Journal of Experimental Psychology: Human Perception and Performance*, **16**, 391–397.

Zellner, D. A. and Whitten, L. A. (1999) The effect of color intensity and appropriateness on color-induced odor enhancement, *American Journal of Psychology*, **112**, 585–604.

Projective Flash Profile from experts to consumers: a way to reveal fragrance language

S. Ballay, E. Loescher, G. Gazano
LVMH Parfums & Cosmetiques, Paris, France

19.1 Introduction: an industrial approach to the assessment of fragrances

Fine fragrances are a crucial spearhead for the beauty industry, representing not only one of the most profitable categories but also contributing to the brands' images. Nevertheless, perfumes are not easy to describe and their evaluation usually requires a high level of expertise. Perfume creators are the best placed to describe fine fragrance, due to their acknowledged expertise. However their technical descriptions can be difficult to communicate to consumers. Indeed, sensory perception of a perfume is complex (Lawless, 1999), requires many attributes and cannot easily be summarized in one or two words.

On the other hand, from the consumer's point of view, it is difficult to understand the olfactory characteristics of a perfume solely on the basis of its description by the beauty consultants. Most of the vocabulary is indeed common to very different fragrances (i.e. "*Floral*" or "*Fresh*"), and descriptions are often not so discriminating.

The question thus arises whether it is possible to describe the perfume in a discriminating and consumer-relevant way in order to communicate more specifically, whatever the country and the culture.

19.2 Flash Profile of fragrances: perfumers vs consumers

The Flash Profile is a quick sensory descriptive method derived from Free Choice Profiling (Williams and Langron, 1984; Williams and Arnold, 1985). It was initially developed as a way to rapidly position products according to their major sensory characteristics (Dairou and Sieffermann, 2002; Sieffermann, 2002). The method allows semantic flexibility, since it does not rely on a consensual description. As a result, Flash Profile can be used to rapidly assess a whole set of stimuli, from which to extrapolate insights into the way subjects perceive factors under investigation (Ballay *et al*. 2003; Delarue *et al*., 2004). The method allows quick access to the overall sensory structure of a sample set, which is important in understanding the main sensory differences, or similarities, between products associated with a semantic description (Delarue and Sieffermann, 2004).

In the typical Flash Profile methodology, each subject is asked to compare and describe the differences between the products by using a ranking protocol based on their own personal attributes. We thus imagined that Flash Profile would be a good method to be used with professional experts who have a strong individual vocabulary and the capacity to describe fragrances in fine detail. They are used to analysing different olfactory proposals and should therefore not be overwhelmed by the comparison task. Professional perfumers also represent the reference in terms of fragrance evaluation, and they are the decision makers during fragrance developments. However, their technical language can sometimes be difficult to communicate to consumers.

Would consumers be able to perform the same comparison tasks as the perfumers? Would they arrive at the same product positioning, and would they be able to describe the fragrances with relevant attributes?

Two Flash Profiles were performed in 2005 with a similar methodology to compare perfumers' and consumers' perceptions and descriptions of the same fragrance set (Gazano et al., 2005). Experimental conditions, results and limitations are described in the following section.

19.2.1 Materials and methods

19.2.1.1 Subjects

In the first experiment, subjects were six highly experienced French professional experts (nez), dedicated to selective fragrance creation. It is not usual to include several such experienced subjects in sensory experiments, except maybe in studies with oenologists. These experts were two men and four women, aged from 26 to 55 years old. One subject was a fragrance analytical development expert for LVMH brands, and the others were perfumers dedicated to perfume creation from different companies.

In the second experiment, a consumer panel was composed of 89 naïve French consumers. They were women aged from 27 to 55 years old. They all were heavy users of perfumes from the selective market, but none of them had prior experience in the sensory descriptive profiling of perfumes.

19.2.1.2 Products

Twelve perfumes (one "Parfum," eight "Eaux de Parfum" and three "Eaux de Toilette") (Table 19.1) were evaluated during the study. Perfume selection was based on the best-selling brands on the European market in 2005, when the study was carried out.

The 12 perfumes were sampled in 15 mL glass spray bottles, ensuring that all samples of a specific perfume belonged to the same manufacturing batch. They were then randomly coded with letters from A to L.

19.2.1.3 Methodology

Perfumers. Flash Profile with perfumers consisted of one session per perfumer. All sessions were individual and lasted from 1 h to 2 h 30 min depending on the subject.

Table 19.1 Codes, short and full names of the 12 perfumes

Code	Perfumes short names	Perfumes full names
A	AngelEP	Angel Thierry Mugler Eau de Parfum
B	CinemaEP	Cinema Yves Saint Laurent Eau de Parfum
C	PleasuresEP	Pleasures Estée Lauder Eau de Parfum
D	AromaticElixir	Aromatic Elixir Clinique
E	LolitaLempickaEP	Lolita Lempicka Eau de Parfum
F	N5ET	N°5 Chanel Eau de Toilette
G	InstantEP	Instant Eau de Parfum
H	JadoreET	J'Adore Eau de Toilette
I	JadoreEP	J'Adore Eau de Parfum
J	PurePoisonEP	Pure Poison Eau de Parfum
K	ShalimarET	Shalimar Eau de Toilette
L	CocoMademoiselleEP	Coco Mademoiselle Eau de Parfum

All sessions were carried out individually, on separate days and separate locations but within a quiet environment with neutral ambient odour.

Perfumers were familiar neither with sensory analysis nor with the Flash Profile procedure. They were therefore first given a brief outline of the methodology and procedure. They were then introduced to the 12 samples of perfumes simultaneously. Paper strips were provided for evaluation.

Perfumers were asked to individually generate attributes, which should allow enough discrimination to rank the samples. Secondly, perfumers proceeded to the evaluation on a ranking mode on each of their own generated attributes. Ties were allowed and subjects were free to evaluate the samples as many times as they liked. They were free to have breaks whenever they needed.

The data were collected on Microsoft Excel® spreadsheets. Generalized Procrustes analysis (GPA) (Gower, 1975) was applied to the data from Flash Profile to assess the consensus between perfumers' sensory maps.

Score plots of the perfumes and loadings of the attributes on the circle of correlations were calculated to describe inter-product differences.

In addition to this, hierarchical cluster analyses (Euclidean distances, Ward's criterion) were performed, both on attributes and product coordinates, following GPA to help understand product differences and semantic interpretations.

All multidimensional data treatments were performed using the XLSTAT® add-in for Microsoft Excel, with the exception of GPA and (RV-cluster) that were performed using Matlab® routines.

Consumers. Flash Profile with consumers consisted of one session. All sessions were individual and lasted from 45 min to 1 h 30 min, which was shorter than perfumers' Flash Profile sessions. Methodology was similar to the perfumers' Flash Profile, except that for consumers all sessions were carried out individually in a partitioned sensory booth.

Sessions took place on separate days and separate locations. The consumers' data were treated in a similar way to the perfumers' data presented in the previous paragraph.

19.2.2 Flash Profile with perfumers: results

All perfumers were very dedicated, and complied with the Flash Profile task. They were interested by this new approach and found it amusing even though they took it very seriously. They did not find the task too difficult, but were eager to get feedback regarding their positioning compared to the other experts.

19.2.2.1 Semantic description

Perfumers used from 6 to 12 attributes, with an average of 7.8 attributes, to describe the set of 12 samples. Thirty-one semantic different words or expressions were collected (Fig. 19.1).

An analysis of the frequency of quotation of each attribute can be attempted, even though at this stage it is difficult to tell if a given word used by different perfumers has

Figure 19.1 Wordcloud of the 31 different attributes generated by the six perfumers. The higher the quotation frequency, the bigger the word.

the same meaning. Most attributes relate to notions that are very classically used to describe fragrances: "Fruity" and "Green" were used by five out of the six perfumers. They were followed by "Woody" (four out of six), "Floral" and "Green" (three out of six each) and "Aldehyde" and "Flowery," "Musc" and "Gourmand", quoted by two out of the six perfumers. The other 31 attributes were mentioned only by one of the perfumers, showing a large semantic diversity with very specific attributes or more individuals' object evocations, such as "from classic musk to powdery musk found in violet" or "fresh rose like in sweet rose water."

19.2.2.2 Products positioning

The GPA allowed positioning the 12 perfumes within two main dimensions (representing 67% variance). A hierarchical clustering analysis (HCA) was also performed to obtain clusters based on the product similarity (Fig. 19.2).

The 12 perfumes were split into five distinct clusters from the most "woody, leather, oriental and patchouli" such as Aromatic Elixir and Shalimar ET to the most "floral, citrusy, green and musk" perfumes, such as Pleasure EP, J'Adore EP and ET. The second axis (24% variance) compared the "aldehydic" perfumes, such as N°5 ET, to the most "fruity" perfumes, such as Angel EP.

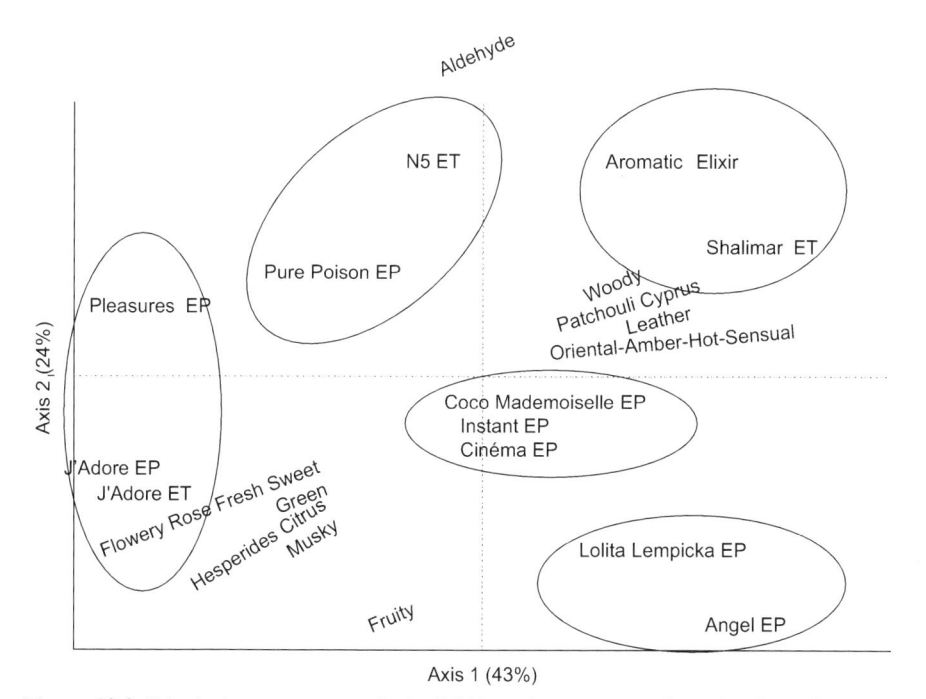

Figure 19.2 Principal component analysis (PCA) on the consensus from the six perfumers obtained by GPA. Interpretation of the individual attributes led to nine semantic items represented on the map. An item was obtained when there were several similar individual attributes highly correlated. Ellipses represent product clusters after HCA.

19.2.3 Flash Profile with consumers: results

All consumers managed to complete the task. No consumer gave up the study despite the large number of products. They all fully understood the ranking procedure. However, there was more interaction with the experimenter, because some consumers asked for clarification or a check during the session. Moreover, they took globally less time than perfumers to complete the task.

19.2.3.1 Semantic description

The consumers used from 3 to 11 attributes to describe the 12 perfumes. The number of attributes used by the experts and by the consumers was globally similar but experts used on average slightly more terms than consumers (7.8 terms compared to 6.3 on average). One hundred and ninety six semantic different words or expressions were collected (Fig. 19.3).

The main attributes used by the two panels are summed up in Table 19.2. A large number of terms (20) were used by both the consumers and the expert perfumers. However, an even larger number of terms (27) were specific to the consumers, some of them being used by many consumers. Also, 11 attributes from the expert subjects were not used by the consumers.

Regarding the attributes, as could be expected, perfumers used more technical and precise terms, sometimes in reference to ingredients or raw materials, whereas consumers used a different semantic register, including more evocations such as "provocative" or "bewitching." More consumers' terms were related to fragrance usage or sensations.

19.2.3.2 Comparison of sensory maps and product clusters

The sensory maps obtained after GPA of the data from the two panels show a similar basic structure (Figs 19.2 and 19.4). In both cases, the two main sensory dimensions account for about 65% of the total variance (67% for the experts' panel, 65% for the consumers' panel).

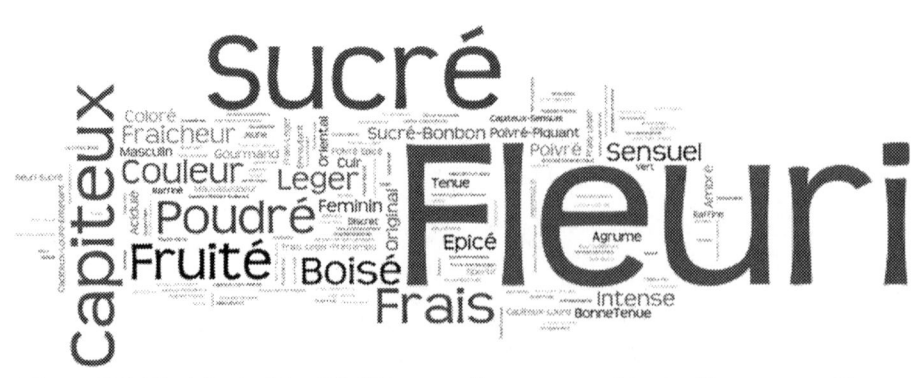

Figure 19.3 Wordcloud of the 196 different attributes generated by the 89 consumers. The higher the quotation frequency, the bigger the word.

Table 19.2 Most frequently used attributes from the Flash Profile conducted either with experts or with consumers

Descriptive terms used both by consumers and experts	Descriptive terms only used by consumers	Descriptive terms only used by experts
Flower–Floral–Flowery 78	Fresh–Freshness–Cold 47	Aldehyde 2
Sugary 54	Heady 46	Rural–Lavender–
Fruit–Fruity 31	Light 34	Hesperide 1
Sensual–Sensuality 29	Feminine 17	Autumn 1
Powder–Powdery 26	Intensity 17	Food 1
Woody 24	Colour–Colourful 15	Light 1
Spicy 21	Peppery 15	Raw 1
Green 18	Strong–Strength 14	Grass 1
Hot–Heat 14	Heavy 11	Hesperides 1
Gourmand 13	Long-lasting	Orange tree 1
Oriental 13	Tenacious 11	Ostentatious 1
Vanilla 11	Discrete 10	Dark 1
Musk–Musky 10	Young–Youth–	
Citrus 7	Juvenile 10	
Amber 6	Masculine 9	
Cyprus 4	Natural 9	
Leather 4	Spicy 8	
Nauseated 4	Sophisticated 8	
Animal 3	Acid–Tart 7	
Vegetal 3	Lemon–Citrus 7	
	Overwhelming 7	
	Candy 6	
	Sweetness–Mildness 6	
	Bewitching 6	
	Original 6	
	Spring 6	
	Elegant 5	
	Enveloping 5	
	Powerful 5	

The attributes are translated from French and sorted according to the number of times they were used in each case.

In both cases, the first and main axis separates products [PleasuresEP, JadoreET, JadoreEP] from products [AromaticElixir, ShalimarET] with products [CinemaEP, N5ET, InstantEP, PurePoisonEP] in between.

In both cases, the second axis separates products [AngelEP, LolitaLempickaEP] from the others.

However, some differences between the consumers and experts are also clearly visible. One of these differences is the relative position of the product [CocoMademoiselleEP]; another difference lies in the greater apparent discrimination between the perfumes as seen by the experts.

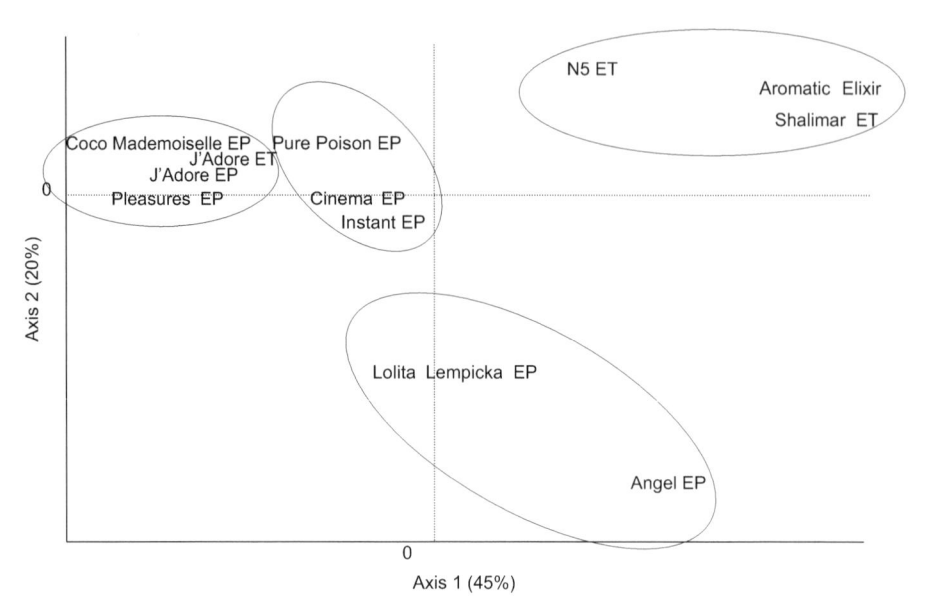

Figure 19.4 PCA on the consensus from the 89 consumers obtained by GPA.

In order to better assess the link between the two datasets, we applied a cluster analysis respectively to the two Flash Profile datasets after GPA. This way, all the sensory information is taken into account in the interpretation.

The two cluster analyses confirm the results observed using the score plot from the GPA; that is to say, consumers differentiated less the perfumes and grouped them in a simpler way (Fig. 19.5).

Even if similarities exist between perfumers' and consumers' perceptions of the products positioning, perfumers seemed more precise because they were able to distinguish more clusters on the 12 perfumes.

19.2.3.3 Summary and opportunities

Four main results were found.

1. All consumers were able to use the Flash Profile methodology. No consumer gave up the study despite the large number of products. They all fully understood the ranking procedure and managed to complete the task. This was also the case for perfumers; even if they were not familiar with sensory analysis protocols, they completed the Flash Profile session even quicker than the consumers, probably because of their ability to smell several perfumes. It is also possible that perfumers recognized some perfumes.
2. Results showed a great variability among the descriptive approaches. One hundred and ninety-six terms were generated, each consumer contributing between 3 and 11 attributes. In accordance with what we expected, the semantic approaches were both hedonic and descriptive; however, very few terms were strictly hedonic, all others incorporating some kind of analytical semantic significance. This use of descriptive words proved the ability

(a) Experts   (b) Consumers

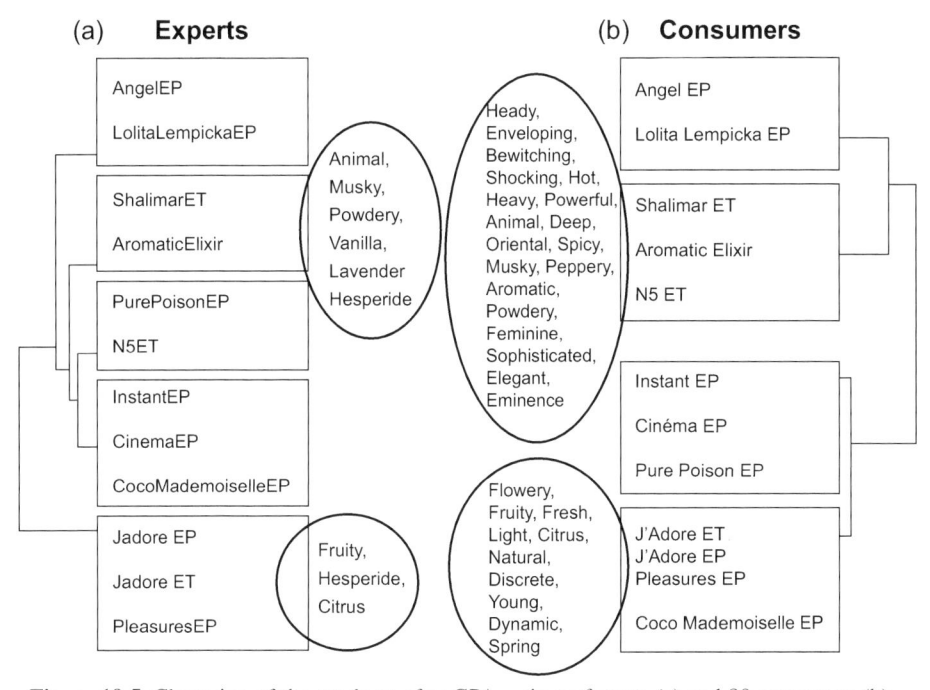

Figure 19.5 Clustering of the products after GPA – six perfumers (a) and 89 consumers (b).

of most of the so-called naïve consumers to provide some sort of analytical information to complement their hedonic judgement.

3. The perfumers' descriptions were more detailed than the consumers', but there was a strong similarity between the global structure of the experts' multidimensional sensory map and the consumers' one. This major result strengthens both the consumer and the expert studies as complementary, with two connected fragrances descriptions: a more detailed and technical vision from perfume creators and more global and related to evocations from the consumers.

4. Concerning the sensory attributes (whose interpretation should always be done very carefully when using Free Choice Profiling-based methods), both experts and consumers led to very different corpuses of semantic attributes. We were then able to identify and quantify the consumers' terms and, in some cases, to correlate them with the more precise experts' attributes. This was particularly helpful to translate some technical terms into consumer words to communicate to our customers.

Using consumers with the Flash Profile methodology proved to be an effective way to understand specific consumer sensory perceptions (Ballay *et al.*, 2004). It allowed us not only to correlate specific perfumers' attributes with major consumers' impressions, but also to identify consumers' semantic attributes which still need to be investigated and fully understood. It also allowed us to reduce the time limit of the studies and therefore to be more reactive and operational.

However, consumers seemed more likely to use associations rather than technical attributes. It might be helpful to guide them in this direction with different categories of attributes that would not be strictly limited to sensory perception but also included evocations and emotions. In the following section, we present a study conducted in a cross-cultural context in order to test this possibility.

19.3 An extension to Flash Profile of fragrances with consumers: beyond sensory description

As shown in the first study presented above, it is possible to obtain relevant descriptive information from consumers using Flash Profile. Our results, however, have shown that consumers tend to naturally use evocations or mental associations, or even specific usage, to describe fragrances. One step to further describing fragrances with consumers would thus be to be able to measure these evocations which are elicited by the smell of fragrances.

Moreover, previous studies have shown that consumers are quite capable of using imagery to consciously express their fragrance experience (Ingersoll, 1997; Ingersoll and Winter, 2003).

Consequently, critiquing the subliminal may be quite challenging and require some fresh and innovative thinking. Typically, more indirect approaches using for instance free associations, symbols, metaphors and imagery would seem more pertinent.

We thus decided to extend the Flash Profile method to the measurement of emotions, mental images and lifestyles associated with the smell of fragrances. Our goal was to tap into this imagery process on the emotive side of the consumer in a manner that is meaningful to both the perfumer and marketing researchers (Ballay *et al.*, 2006).

For an international company such as LVMH, cultural diversity is a key issue that needs to be addressed. Naturally, we would expect this diversity to be even larger when measuring emotions, images and lifestyle compared to the strict sensory perception. In the fragrance industry, there is usually a distinction to be made between the European market and its expectations and those of the American and Asian markets, even if obviously there are intra-continental differences. Therefore, we decided to conduct this study of an extended version of Flash Profile with consumers in three countries: China, France and the United States.

19.3.1 Materials and methods

19.3.1.1 Subjects and countries

The study was conducted in three different countries: China, France and United States. In each country, the panel was composed of 100 consumers: 70 were women and 30 were men, and between the ages of 18 and 55. All participants were heavy users of perfumes from the selective market, but had no previous experience of sensory descriptive analysis of perfumes.

Table 19.3 Generic perfumes classification

Code	Generic perfumes classification
A	Floral
B	Floral, fruity, sensual
C	Oriental
D	Fresh, fruity
E	Green chypre
F	Sweet chypre
G	Aromatic fresh
H	Floral woody
I	Aromatic, woody, powdery

19.3.1.2 Products

Nine LVMH perfumes, six women's perfumes and three men's perfumes were evaluated during the study. According to their generic classification they are quite different, and two of them were common to the first study above (Table 19.3).

One day preceding each evaluation, blotters were dipped into bottles of fragrance. The blotters were then air-dried for 1 min by a fan and placed in their corresponding test bottles with caps. Assessors had to open and squeeze the test bottle in order to make a sample evaluation. Doing so allowed them to evaluate the same olfactory note from the beginning through the duration of the test. Indeed, in the first study, fragrances were sprayed on paper strips. Consequently, as a Flash Profile evaluation is free, the evaluation of the perfumes could be done at different times, which, for a perfume, is susceptible to match very different notes (Top note, Heart note or Base note).

All samples were presented blind with an alphabetic code (instead of a three-digit code) in order to avoid mistakes when reporting data.

19.3.1.3 Methodology

Projective Flash Profile methodology. All consumer panels (France, China and United States) evaluated exactly the same products. Consumers were given no indication of the perfume brands and names. The sessions were conducted on separate days and in locations that were specific to each country. The evaluation took place in a sensory booth according to detailed guidelines. Participants were allowed to take breaks whenever they wished.

Regarding the methodology, *The Projective Flash Profile* was carried out in two stages:

First, a qualitative free association stage, followed by a quantitative evaluation stage. During the first stage, a briefing on the methodology was held before the products were distributed to the participants. Perfumes were evaluated in monadic sequential fashion.

Each assessor was asked to individually evaluate the perfumes in accordance with four predefined themes and to describe, in his/her own words, the sensations he/she perceived for each perfume against the following categories of perceptions or associations (each category was addressed using a specific question):

- *Olfactory Description* (What does this perfume smell like?)
- *Associated Emotions* (What do you feel? What is your mood when you smell this perfume?)
- *Mental Images* (What images immediately cross your mind when you smell this perfume?)
- *Lifestyles* (What lifestyle, woman, man or brand would fit with this perfume?)

The order in which they described the categories was the same for all subjects. Thus, at first, they were asked to describe the perfume's *Olfactory* characteristics. Second, they were asked to describe the *Associated Emotions*. Third, we requested they come up with some associated *Mental Images*. Finally, they were asked to associate *Lifestyles* with the perfume samples.

In the second stage, the subjects kept the list of the attributes they had generated but we shifted the focus to a transversal product comparison, as in the original Flash Profile procedure. The whole product set was therefore presented simultaneously.

Each subject was asked to focus on the major differences he perceived among the products in accordance with each of the four previously described themes. Each selected transversal attributes from the ones he previously came up with. Items were chosen based on the assessor's ability to discriminate the nine perfume samples. Differentiation had to be sufficiently distinct to be able to rank the samples.

Then, the subjects ranked the perfumes against the chosen attributes. For instance, for the *Olfactory* theme, if a subject had chosen the items *Flower* and *Fresh,* he then had to rank all the perfumes, from the least to the most intense, for each of these two attributes successively. Ties were allowed, and assessors could re-evaluate the samples as much as they liked.

The entire evaluation took between 2 and 4 h for each participant.

Statistical methodology. The statistical analysis mainly consisted of GPA and cluster analysis. Cluster analysis on the panel means and consensus were performed, in order to compare the product clusters obtained by the consumers' procedures. Cluster analysis was performed on both terms and products following GPA, to help in understanding product discrepancies and to assist in semantic interpretation.

19.3.2 Results and contribution

19.3.2.1 Semantic results

Regarding the semantic quantitative comparison, as shown in Fig. 19.6, it can be seen that in all categories, French consumers generated more items than either the US- or China-based assessors. Moreover, US and China generated about the same number of items for each category, except for *Lifestyles.*

Additionally, for China and the US, the *Olfactory* and *Emotional* results were quite similar. However, the latter did not hold true for French assessors, who used more *Olfactory* than *Emotional* wordings.

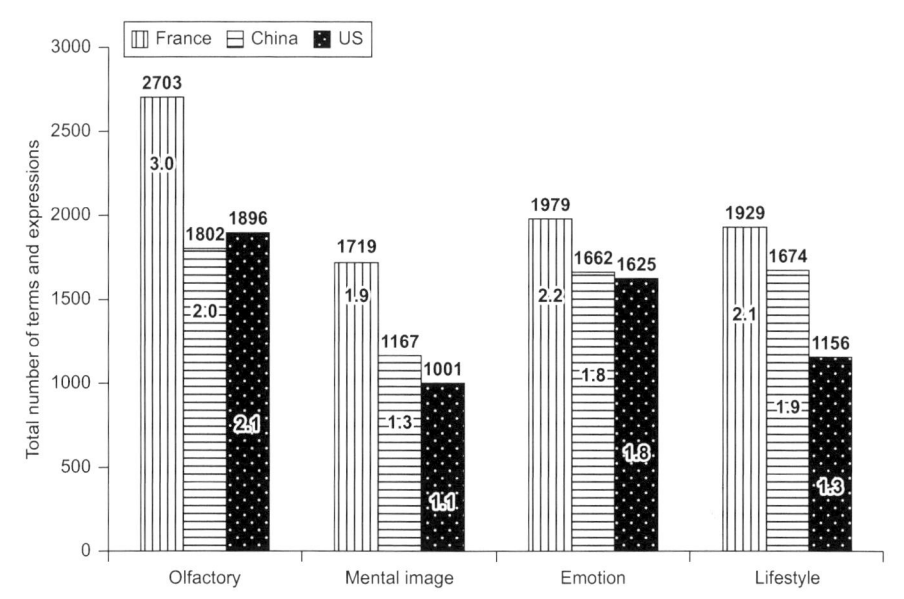

Projective Flash Profile from experts to consumers 413

Figure 19.6 The total number of terms and expressions freely generated by consumers in each category by country and the mean of terms and expressions generated per consumer in each category by country.

Finally, for all countries, fewer descriptions of *Mental Images* were generated compared to those of other categories.

A closer look at the different items which have been generated shows that French and US consumers use more terms related to *Emotions* while the Chinese assessors generated more *Mental Images* and *LifeStyles* items. Descriptions of *Olfactory* were comparable (Fig. 19.7).

As regards the semantic qualitative comparison, whatever the country, great variability was demonstrated in idioms and expressions, which ranged from simple words to complicatedly detailed phrases (Tables 19.4 and 19.5).

This individual variability has not prevented the generation of recurrent items in the three countries.

- Some items were common to most of the perfumes and to the three countries.
- For instance, as shown in Figs 19.8a, 19.8b and 19.8c, "Sweet" and "Floral" are the most often used items in China (Fig. 19.8a) for *Olfactory Description* and they are also the most quoted items in the US and France (Figs 19.8b and 19.8c).

Depending on the fragrances, the frequency of the item "Floral" is between 42 and 13 for China, 44 and 8 for the US and between 52 and 16 for France. The frequency of the item "Sweet" was between 40 and 5 for China, 47 and 11 for the US and 33 and 10 for France. Both of these items seem to be commonly used for describing perfumes, whatever the country studied and the fragrance.

Some items were common to the three countries (Fig. 19.9).

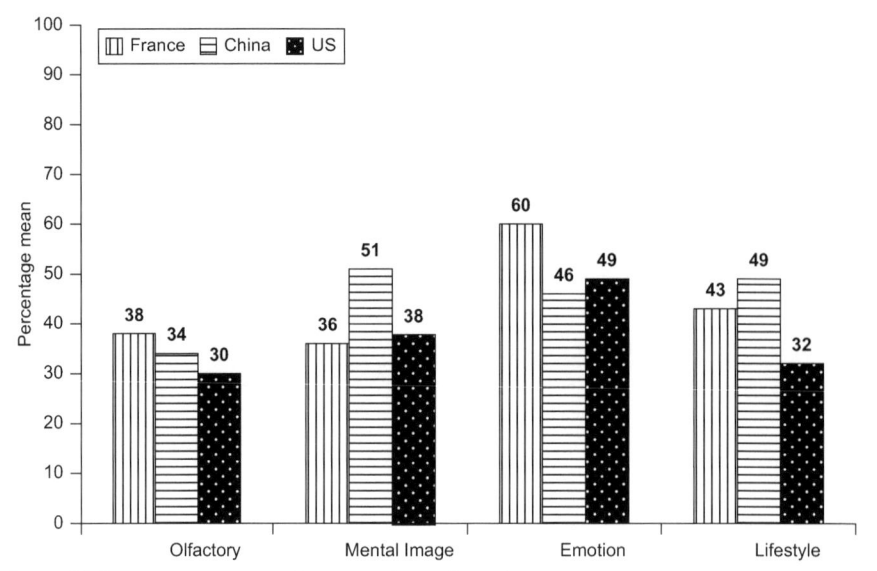

Figure 19.7 Percentage mean of separate items in each category.

Table 19.4 **Examples of simple idioms and phrases used by consumers for *Mental Images* in China**

Examples of simple idioms
Happy
Relaxed
Energetic
Sexy
Messy
Outdoor
Paris
Shopping
Flowers
For discotheque
Shrewd woman
Lawn

Concerning *Olfactory Description*, "Fruity," "Fresh" and "Powdery" were used in the three countries. In the same way, "Relaxed," "Happy" and "Comfortable" were cross-cutting items for *Emotion*, "Bath.Shower," "Garden.Park," "Flowers" and "Love.Romance" for *Mental Images* and "Young," "Working person" and "Sport" for *Lifestyles*.

However, there was also some variability between countries, with specific associations between countries and themes (Fig. 19.9).

Projective Flash Profile from experts to consumers 415

Table 19.5 Examples of detailed idioms and phrases used by consumers for *Mental Images* in China

Examples of more detailed phrases
A garden is full of blooming and sweet-scented flowers. The bees are busy gathering nectar
On the way to work, repeat the same thing every day. There is no difference
A lovelorn person is drinking in a bar to lighten pain
Shopping on a pedestrian street, such as Huaihai Road
Litchi garden with rich fruits
A group of schoolgirls are playing on the street
An attractive and natural beauty spot with an orchard, mountains and rivers
In a spacious and top-grade office
In a luxurious party, releasing energy and passion
Watching sitcoms in a room in cold colour while thinking about other things
Middle-aged men are having cigars and drinking wine with enchantment
A lady is answering questions from the general manager confidently in the assembly room

(a)

The 10 most frequently generated

C
Floral/flowers	30
Sweet	30
Heavy	23
Creamy	11
Light	11
Elegant	7
Fresh	7
Sexy	7
Fruity	6
Harsh	6

B
Fresh/refreshing	27
Fruity	27
Sweety	26
Light	19
Floral	18
Elegant	9
Natural	9
Sexy	7
Cantaloup	4
Comforting	4

D
Floral	31
Fresh	20
Light	19
Sweet	12
Natural	11
Elegant	9
Fruity	9
Heavy	8
Harsh	6
Green/grass/herbal	5

E
Harsh	43
Heavy	30
Floral	13
Toilet-water/fresh	10
Medicine smell	9
Chemical	8
Alcohol	6
Light	5
Spicy	4
Exciting	3

A
Fresh/flowers	42
Light	23
Heavy	16
Fresh	11
Elegant	10
Sweet	10
Harsh	8
Comforting	4
Soap/lux soap	4
Clean/cleany	3

F
Sweet	40
Floral	20
Fruity	20
Heavy	13
Light/not heavy	11
Fresh	8
Elegant	6
Comforting	4
Sexy	4
Acid	3

G
Harsh	26
Heavy	23
Floral	17
Fresh	10
Fruity	6
Herbal	6
Light	6
Elegant	5
Sweet	5
Alcohol smell	4

H
Heavy	23
Light	18
Floral	15
Harsh	12
Fresh	11
Herbal	9
Elegant	4
Fruity	4
Nicotian	4
Musky	3

I
Heavy	22
Harsh	21
Light	16
Floral	14
Fresh	8
Alcohol	6
Woody	5
Cool	4
Elegant	4
Mint smell	4

Figure 19.8 (a) Common items to most of the perfumes in China. (b) Common items to most of the perfumes in the US. (c) Common items to most of the perfumes in France.

Continued

(b)

The 10 most frequently generated

C	
Sweet	39
Powdery	28
Floral	22
Musky	10
Fruity/Fruit	8
Light	8
Vanilla	8
Fresh/Refreshing	7
Clean	6
Soft	5

B	
Sweet	42
Floral	40
Fruity	19
Clean	16
Fresh/Refreshing	15
Soap	9
Citrusy	7
Light	6
Powdery	4
Shampoo	4

D	
Floral	35
Sweet	34
Citrus/Citrusy	14
Clean	13
Fruity	11
Soapy	8
Fresh	7
Herbal	5
Light	4
Pine	4

E	
Floral.Flower	24
Strong	20
Powder/Powdery	18
Harsh	11
Sweet	11
Clean/Cleaning	10
Musky	9
Chemical	8
Spicy/Spice	8
Woodsy	7

A	
Floral/flowers	34
Sweet	34
Powder	25
Strong	11
Clean	10
Fresh/Refreshing	10
Perfumy/Perfumey	9
Light	6
Musky	5
Soap/Baby Soap	5

F	
Sweet	47
Floral/Flowery	44
Fruity/Fruit	15
Powdery/Powder	10
Fresh	8
Strong	8
Citrus/Citrusy	7
Clean	7
Perfumy	7
Soapy/Soap	6

G	
Sweet	19
Clean/Cleaners	18
Strong	12
Spicy	11
Fresh/Refreshing	10
Chemical	9
Floral/Flower	8
Citrus	5
Harsh	5
Powdery	5

H	
Sweet	23
Floral	13
Spicy	13
Woody	10
Strong	9
Musky	8
Clean	7
Herbal	7
Masculine	6
Medicinal	6

I	
Sweety	25
Powdery	18
Clean	16
Floral	15
Fresh	8
Spicy	8
Soapy	6
Medicinal	5
Herbal	4
Light	4

(c)

The 10 most frequently generated

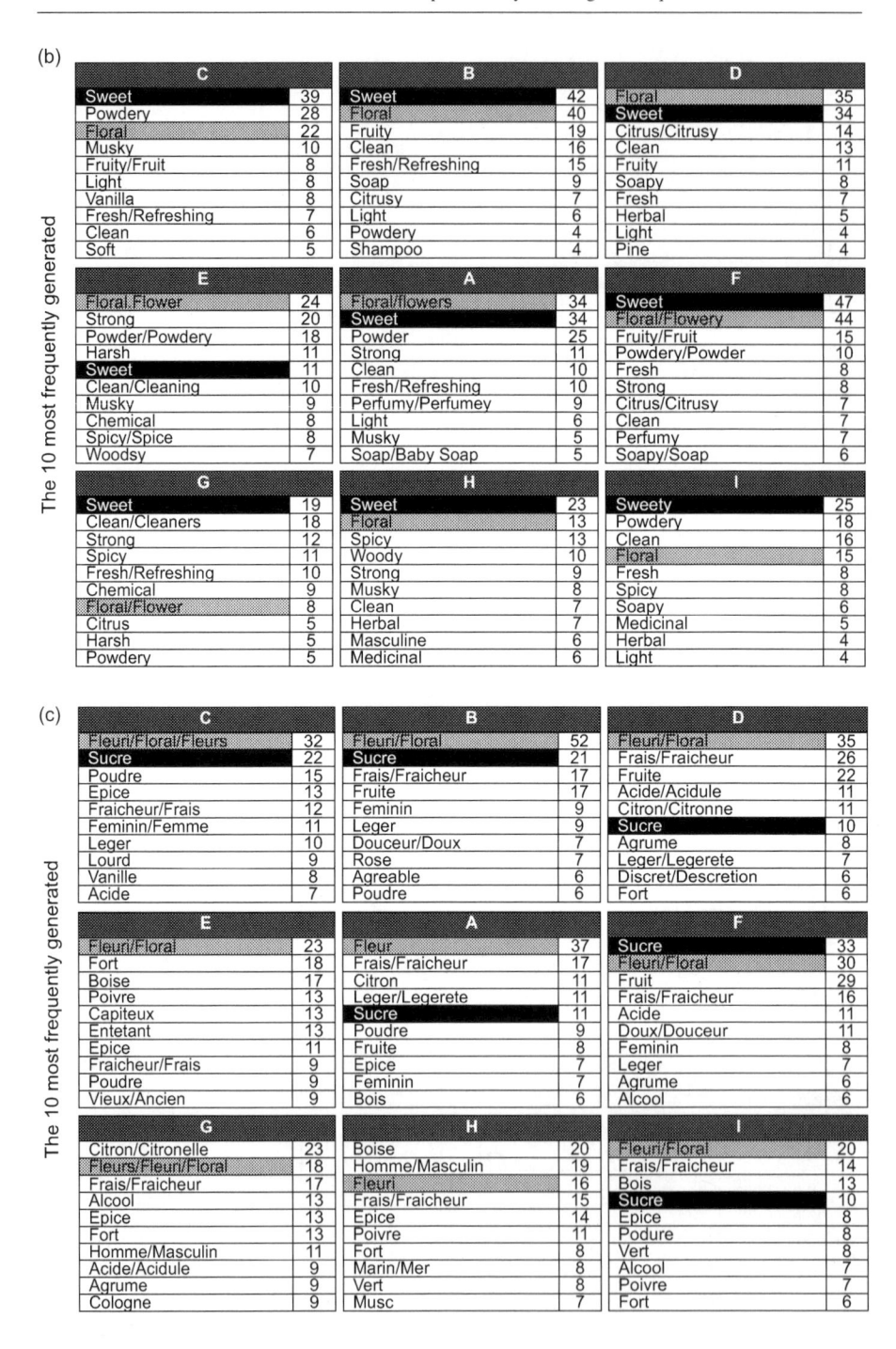

C	
Fleuri/Floral/Fleurs	32
Sucre	22
Poudre	15
Epice	13
Fraicheur/Frais	12
Feminin/Femme	11
Leger	10
Lourd	9
Vanille	8
Acide	7

B	
Fleuri/Floral	52
Sucre	21
Frais/Fraicheur	17
Fruite	17
Feminin	9
Leger	9
Douceur/Doux	7
Rose	7
Agreable	6
Poudre	6

D	
Fleuri/Floral	35
Frais/Fraicheur	26
Fruite	22
Acide/Acidule	11
Citron/Citronne	11
Sucre	10
Agrume	8
Leger/Legerete	7
Discret/Descretion	6
Fort	6

E	
Fleuri/Floral	23
Fort	18
Boise	17
Poivre	13
Capiteux	13
Entetant	13
Epice	11
Fraicheur/Frais	9
Poudre	9
Vieux/Ancien	9

A	
Fleur	37
Frais/Fraicheur	17
Citron	11
Leger/Legerete	11
Sucre	11
Poudre	9
Fruite	8
Epice	7
Feminin	7
Bois	6

F	
Sucre	33
Fleuri/Floral	30
Fruit	29
Frais/Fraicheur	16
Acide	11
Doux/Douceur	11
Feminin	8
Leger	7
Agrume	6
Alcool	6

G	
Citron/Citronelle	23
Fleurs/Fleuri/Floral	18
Frais/Fraicheur	17
Alcool	13
Epice	13
Fort	13
Homme/Masculin	11
Acide/Acidule	9
Agrume	9
Cologne	9

H	
Boise	20
Homme/Masculin	19
Fleuri	16
Frais/Fraicheur	15
Epice	14
Poivre	11
Fort	8
Marin/Mer	8
Vert	8
Musc	7

I	
Fleuri/Floral	20
Frais/Fraicheur	14
Bois	13
Sucre	10
Epice	8
Podure	8
Vert	8
Alcool	7
Poivre	7
Fort	6

Figure 19.8 Continued

		China	France	USA
Common items	Olfactory	Flowery/sweet/fruity/fresh/powdery		
	Emotion	Relaxed/happy/comfortable		
	Mental Image	Bath.shower/garden.park/flowers/love.romance		
	Lifestyle	Young/working person/ sport		
Specific items	Olfactory	Elegant	Spicy	Clean
	Emotion	Delighted	Smooth	Clean
	Mental Image	Party/dancing	Holiday/woman	Clean/disinfect
	Lifestyle	White collar	Elegance/class	Housewife/mother

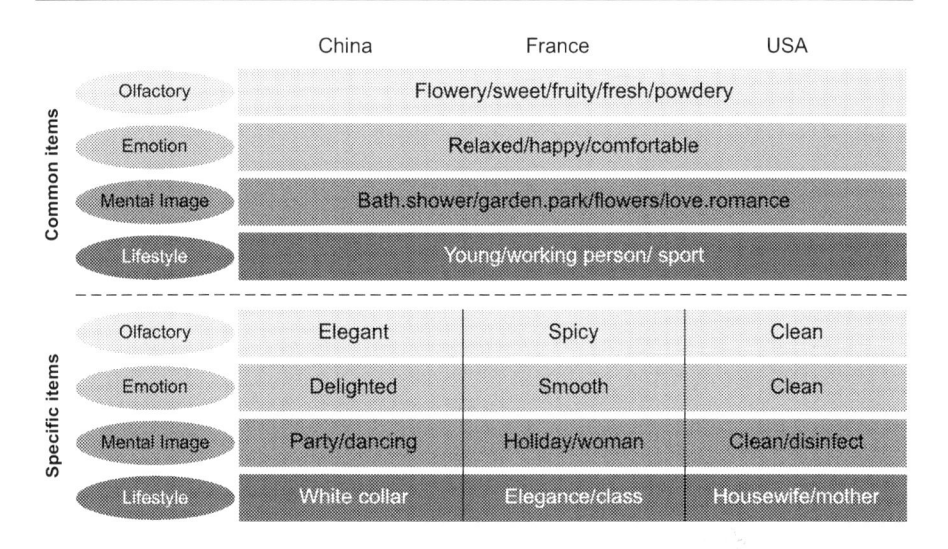

Figure 19.9 Common and specific items for the three countries.

For instance, relating to respectively *Olfactory Description* and *Emotions* "Elegant" and *"Delighted"* were only used by Chinese, "Spicy" and "Smooth" by French and "Clean" for both by Americans.

Moreover, regarding the semantic result in each category, the generated items do not fit exactly with the expected items. Indeed, "Elegant" is not a strictly Olfactory item, and "Clean" is not an Emotion.

19.3.2.2 Results of the quantitative description

GPA and cluster analysis carried out on Flash Profile results for the three countries provided rich information regarding the consumers' perceptions and product clustering. Product sensory and descriptive maps (Fig. 19.10), and product clusters (Fig. 19.11) were obtained for each country and category. They were compared and analysed between countries.

Results of the "Olfactory Description" profile. Figure 19.10 shows the olfactory maps that were calculated from the *Olfactory* profile data for all the consumers in each of the three different countries. It can be seen that, whatever the country, perfumes G, H and I are on the right side of the map. This result is not surprising, since they are the three men's fragrances. They are separated from women's fragrances because of their woody and aromatic characteristics. Interestingly, one women's perfume is positioned on the same side as men's fragrances. It is perfume E, and this is due to its chypre characteristic which relates to woody notes. Regarding other women's perfumes, it can be noticed that B and D are always very close: they both have a fruity hint.

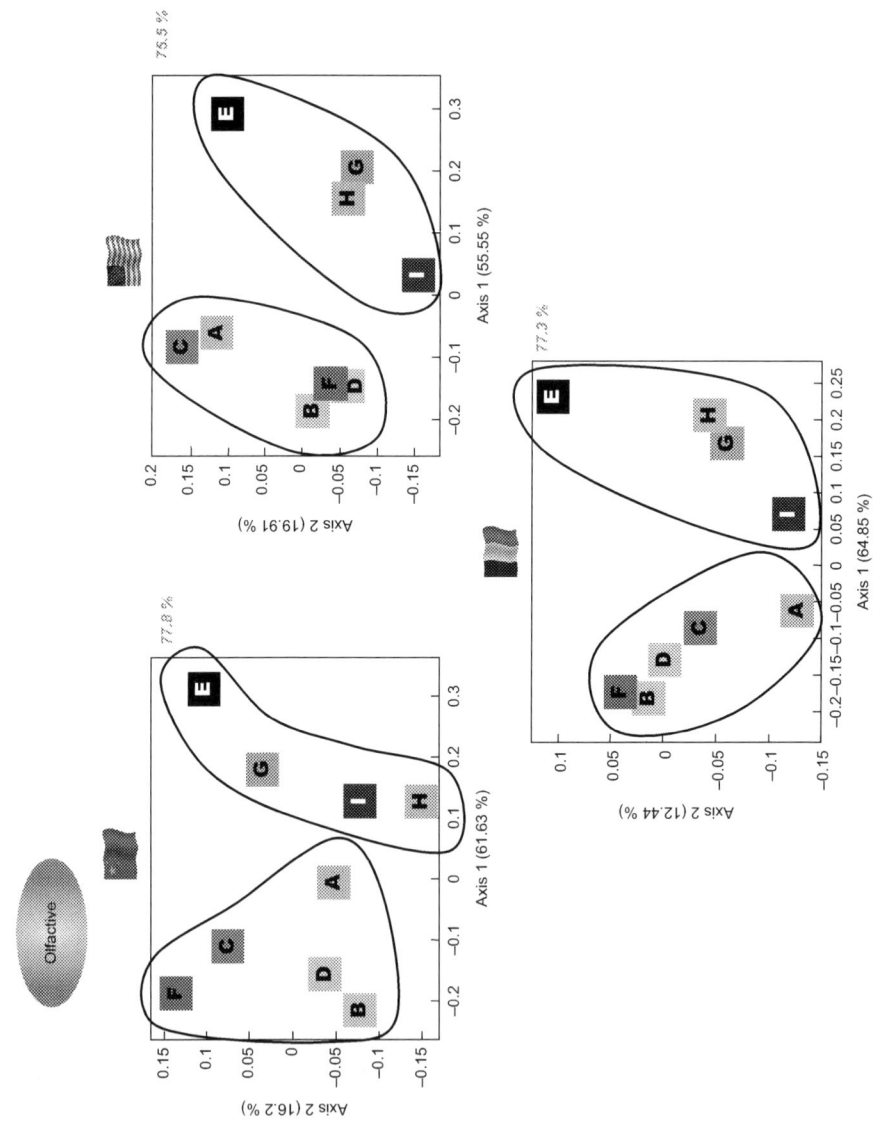

Figure 19.10 Olfactory sensory data: products consensual sensory maps after GPA.

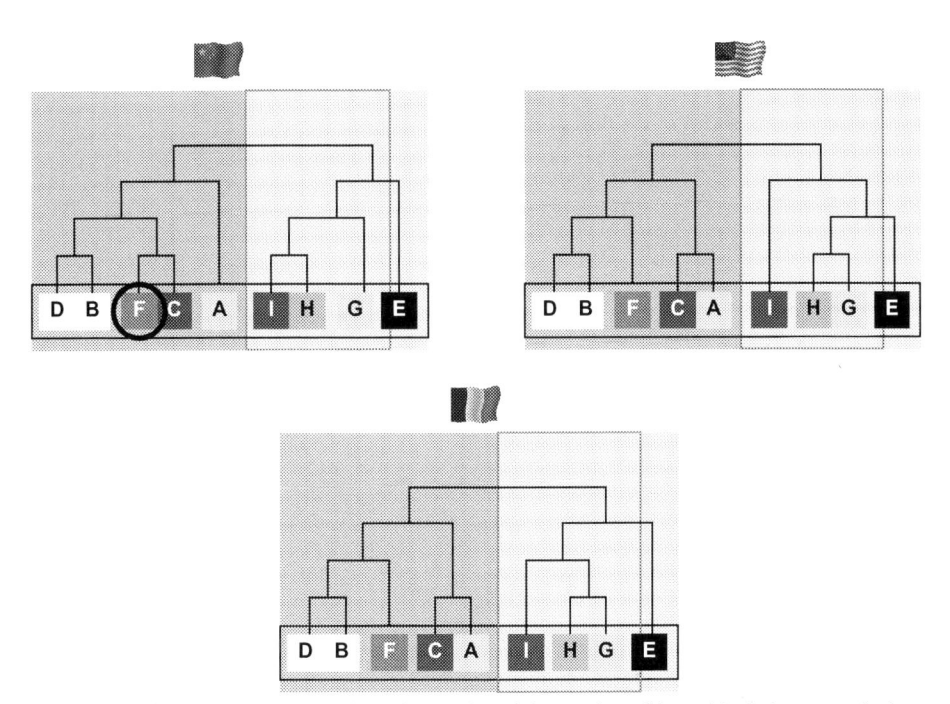

Figure 19.11 Olfactory sensory data: the results of the products hierarchical cluster analysis carried out on products after GPA.

However, there are also some differences: according to the French and the Americans, perfume F is very close to perfumes B and D, whereas the Chinese position is closer to C.

As follows, the cluster analysis associated with those maps illustrates that the product groups are very similar across countries (Fig. 19.11).

Actually, very few fragrances differ in the way they are related to the other fragrances, from one country to another.

Results of the "Associated Emotions" profile. As regards the *Emotions*, G is apart from H and I for France and China, unlike the *Olfactory* positioning. And for France, C is apart from women's fragrances and E from men's fragrances. Moreover, clusters slightly break up for France. (Fig. 19.12). The perception of these perfumes through emotions has evolved versus the strictly olfactory description, into an even finer discrimination for French consumers.

Results of the "Mental Images" profile. Concerning the *Mental Images* category, the positioning of women's fragrances is quite similar in France and in the US, but

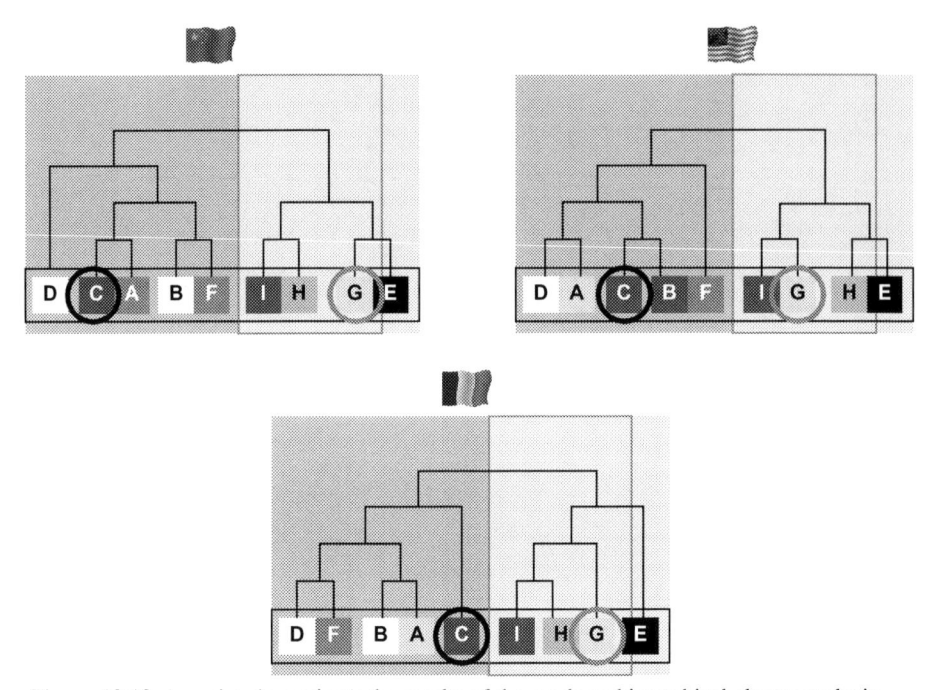

Figure 19.12 Associated emotions: the results of the products hierarchical cluster analysis carried out on products after GPA.

the cluster breaks up for the Chinese (Fig. 19.13). Globally, with *Mental Images*, perfumes stand out more precisely.

Results of the "Lifestyle" category. Finally, regarding the *Lifestyles* category, the men's fragrance I is close to the women's fragrances in the US and the E women's fragrance is apart from the others in China and in the US, whereas the French position is close to men's fragrances (Fig. 19.14).

Thus, when we study the four categories, we can observe a striking result: going from the *Olfactory* perception to the associated *Emotions*, *Mental Images*, and finally regarding the resulting *Lifestyles*, fragrances are better discriminated and differences between countries are more and more highlighted (Fig. 19.15).

For instance, for China, the number of groups is different and the fragrances in each group are not the same. For the *Mental Images* and *LifeStyles,* the way the consumers perceive and associate the fragrances may vary from the perception they had when they were asked to describe the *Olfactory* perception.

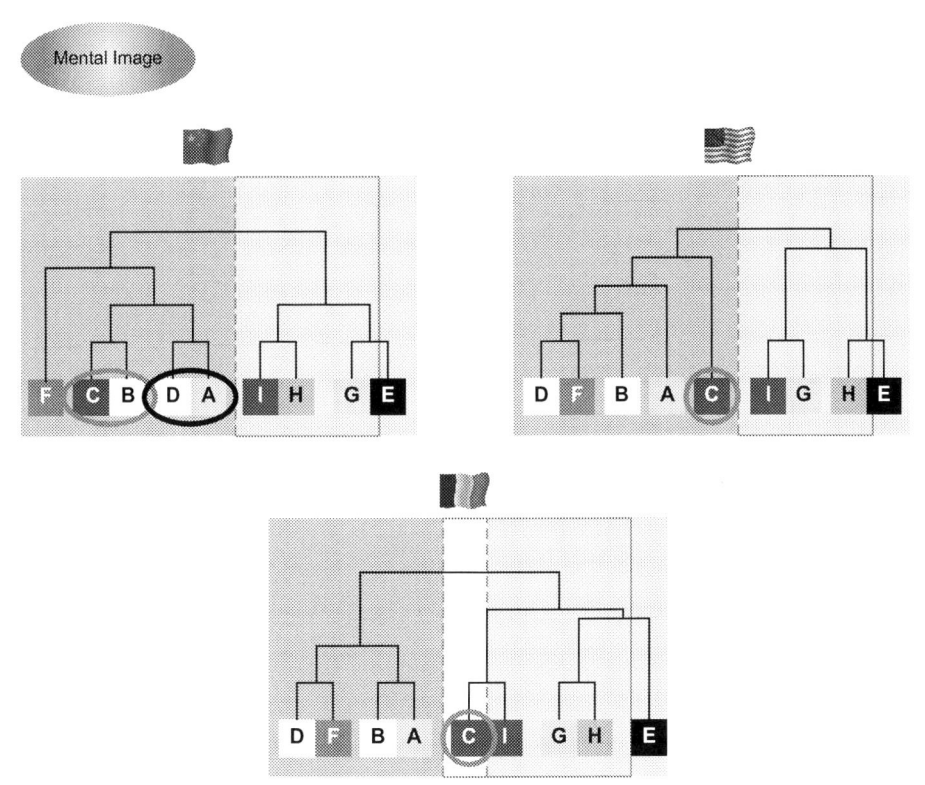

Figure 19.13 Mental Images: the results of the products hierarchical cluster analysis carried out on products after GPA.

It is also true in the US, as well as in France. This clearly shows the influence of the cultural background.

This is for us the confirmation that if we ask the consumer to think about the four categories, he will come up with strong differentiations. Thus, more relevant information is obtained and it is much more culturally dependent.

We shall not, therefore, depend only on the *Olfactory* orientated perception when we want to understand and communicate on the way the consumer uses and associates the fragrances he or she chooses.

19.3.2.3 Advantages of extensions to Flash Profile

Similar to what was observed with the previous *Flash Profile* consumer studies, all consumers were able to use the *Projective Flash Profile* methodology. Moreover, no consumers had trouble finding and imagining mental images, emotions or associations with the different perfumes. This is a confirmation of the highly imaginative and very evocative potential of fragrances.

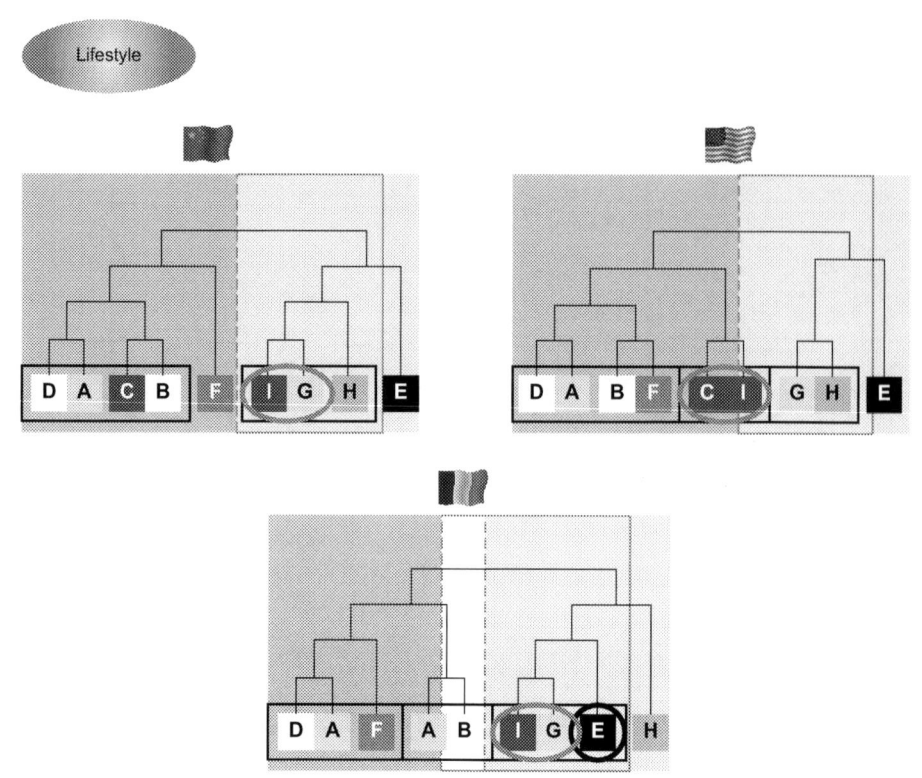

Figure 19.14 Lifestyles: the results of the products hierarchical cluster analysis carried out on products after GPA.

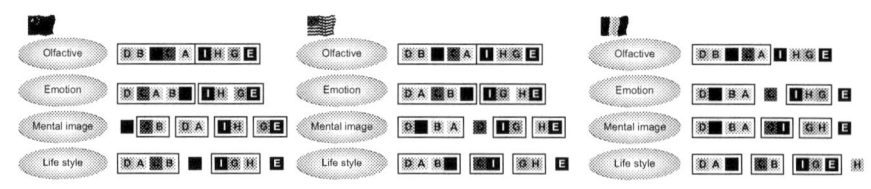

Figure 19.15 The results of the products hierarchical cluster analysis carried out on products after GPA per country and category.

Regarding the global *Olfactory* description, there is a quite good consensus among the three countries.

For example the citrus characteristic of the feminine perfume E brings it, for the consumers, towards the masculine fragrances rather than the feminine ones. This is true, whatever the country.

But as soon as you mention *Emotion* or want to talk about *Mental Images*, culture takes over.

Actually, descriptive *Olfactory* consensus does not imply common meaning and associations.

Second, the method we used and adapted from *Flash Profile* provides us with a more relevant way to understand the consumers.

Third, it gives us a means to better communicate with the consumers on a more emotional level based on his/her individuality and culture.

19.3.2.4 Intercultural knowledge

Concerning the quantitative results of the Projective Flash Profile (mental perception), whatever the country, *Emotions*, *Mental Images* and *Lifestyles* lead to better discrimination of the fragrances. If this differentiation is transversal, the fragrances which have been differentiated are not the same. The semantic results allow us to understand these differences, which are cultural. Actually, depending on their culture, people identify some characteristics with more acuity.

For instance, French people are the only one who use *Spicy* for *Olfactory Description*, *Trip/Holidays/Sun/Women* for *Mental Images* and *Elegance/Class* for *Lifestyles*.

Chinese People are more interested in the social approach; as a result they use *Elegant* for *Olfactory Description*, *Party/Dancing* for *Mental Images* and *White collar/Blue Collar* for *Lifestyles*.

Finally, US people are very concerned about family and cleanness. They use specifically *Clean* for *Olfactory Description* and *Emotions*, *Clean/Disinfect* for *Mental Images* and *Housewife/Mother* for *Lifestyles*.

19.4 Discussion and conclusion

To conclude, a fragrance induces a perception, and not necessarily solely an olfactory conclusion.

These studies show how it is possible to bring to the perfume universe a specificity on the basis of a "consumer language," with descriptive terms but also emotions and images, and not only on a predefined "coded" language.

The quick and highly reactive methodology we used has allowed us to get a huge amount of information in a relatively short time. The main, and very original results, are highly informative consumers' maps of the relative proximities of the products taking into account olfactory perceptions, emotions, images and lifestyles for the three countries.

In addition to that, the semantic flexibility of the methodology has allowed an international exploration. The cultural differences we spotted led to specific terms generated for each country, and have revealed some common items (for example "fruity" or "happy") being directed towards different significations for French, Chinese and American consumers.

The cultural differences were, of course, more pronounced regarding the evoked emotions and images.

For the first time, we were able to compare, on the same underlying product maps, the main specificities regarding French, Chinese and American consumers at four different interpretative levels.

Consequently, this methodology shows:

- The influence of emotions and evocations on consumers.
  For the global olfactory description, there is quite a consensus among the three countries; Nevertheless, once you mention emotion or mental images, culture takes over.
  Indeed, descriptive olfactory consensus does not imply common meaning and associations.
- A way to better understand the consumers
  The Projective Flash Profile, adapted from Flash Profile, provides us with a more relevant way to understand the consumers.
- A way to better communicate to the consumers
  The Projective Flash Profile gives us a means to better communicate with the consumer. The information we have about each perfume may be used to illustrate the perfumers and creative team efforts in order to create a profile sheet for each perfume.

This will allow us to do it in a way that will directly talk to the consumer on a more emotional level based on his/her individuality and culture.

Finally, the Projective Flash Profile used with products having an evocative power gives us several added values:

- Better understanding the consumer mind and specificities, whatever the country, culture, language and experiences.
- Creating a link between the experts' world and the consumers' views.
- Introducing a new communication link between the creative and the development teams, and between the brand and its consumers.

Thereby, Projective Flash Profile has become a key LVMH methodology.

Actually this methodology is regularly applied to fragrances; however, it was further deployed on such other axes as skincare for instance. From our experience, the more evocative the products are, the more relevant this projective methodology tends to be.

References

Ballay S, Sieffermann JM, Danzart M and Gazano G (2006) A new Fragrance language: Intercultural knowledge and emotions, *IFSCC 24th Congress*, 16–19 October, Osaka.

Ballay S, Sieffermann JM and Gazano G (2003) Sensory evaluation of cosmetic essences using two different profiling techniques: *The 5th Pangborn Sensory Science Symposium*, 20–24 July, Boston.

Ballay S, Sieffermann J-M and Gazano G (2004) Flash profile with consumers: A new method to understand specific Japanese moisturizing expectations on cosmetic products, *IFSCC 23rd Congress*, 24–27 October, Orlando.

Dairou V and Sieffermann JM (2002) A comparision of 14 jams characterized by conventional profile and a quick original method, the Flash Profile, *Journal of Food Science*, **67**, 826–834.

Delarue J, Danzart M and Sieffermann JM (2004) Flash profile gives insights into human sensory perception, *5th Meeting of the International Multisensory Research Forum*, 2–5 June, Sitges (Barcelona), Spain.

Delarue J and Sieffermann JM (2004) Sensory mapping using Flash profile. Comparison with a conventional descriptive method for the evaluation of the flavour of fruit dairy products, *Food Quality and Preference,* **15**, 383–392.

Gazano G, Ballay S, Eladan N and Sieffermann JM (2005) Flash Profile and Fragrance Research – The world of perfume in the consumer's words, *ESOMAR*, 15–17 May, New York.

Gower JC (1975) Generalized procrustes analysis, *Psychometrika*, **40**, 33–51.

Ingersoll DW (1997) The challenge of using the "Inarticulate" consumer as an R&D partner in cosmetic product development. *Surfactants in Cosmetics* (2nd Edn.), **68**, M.M Rieger and L.Rhein (Ed.), Marcel Dekker, NY, 533–556.

Ingersoll DW and Winter F (2003) Building cross-cultural consumer insights of women's colognes with interactive multivariate statistical tools. *ESOMAR*, 27 February–March 1ST, Lausanne.

Lawless HT (1999) Descriptive analysis of complex odors: Reality, model or illusion? *Food Quality and Preference*, **10**, 325–332.

Sieffermann JM (2000) Le profil Flash – un outil rapide et innovant d'évaluation sensorielle descriptive. *Proceedings of AGORAL 2000*, XIIèmes rencontres "L'innovation: de l'idée au succès," Montpellier, France, 335–340.

Sieffermann JM (2002) Flash profiling. A new method of sensory descriptive analysis. In *AIFST 35th Convention*, 21–24 July, Sidney, Australia.

Williams AA and Langron SP (1984) The use of free-choice profiling for the evaluation of commercial ports, *Journal of the Science of Food and Agriculture*, **35**, 558–568.

Williams AA and Arnold GM (1985) A comparison of the aromas of 6 coffees characterized by conventional profiling, free-choice profiling, and similarity methods, *Journal of the Science of Food and Agriculture*, **36**, 204–214.

Use of rapid sensory methods in the automotive industry

D. Blumenthal[1], N. Herbeth[2]

[1]UMR1145 Ingénierie Procédés Aliments, AgroParisTech, INRA, Cnam, Massy, France; [2]Renault SAS, Guyancourt, France

20.1 Introduction

This chapter is the result of more than 15 years of experience of sensory evaluation in the car industry. During this time period, a great number of sensory and consumer studies have been conducted by the sensory science team of the Renault R&D department. Such studies have rarely been published in the sensory science literature, although there have been a few communications in conferences (Astruc and Blumenthal, 2004; Astruc *et al.*, 2005, 2006, 2007, 2008; Blumenthal, 2004, 2008; Blumenthal and Bouillot, 2007, 2010; Blumenthal and Dairou, 2007; Blumenthal *et al.*, 1998, 2000, 2001, 2003, 2004; Boivin and Blumenthal, 2009; Dairou *et al.*, 2003; Herbeth, 2007; Herbeth and Blumenthal, 2007, 2010; Herbeth *et al.*, 2007; Petiot *et al.*, 2010). This chapter is thus an opportunity to step back and share our experience on the use of sensory analysis, through three selected examples. We believe that this experience will be of profit to the reader in other domains of application.

After explaining the context of automotive sensory evaluation and the need for sensory science in the car industry, we will develop three examples of studies, chosen to highlight the interest of rapid sensory methods.

20.1.1 Why the automotive industry needs sensory science

Like other industries, car companies do care about their consumers. Apart from obvious price considerations, consumers do have needs and expectations. These needs and expectations concern not only technical specifications linked to the "performance" of the car (top speed, fuel consumption, safety, etc.) but also subjective aspects, such as the aesthetics of the exterior design. With both of these, there is room for sensory science, because many aspects involve the sensory perception of the car.

Thus, sensory science is used in the car industry to:

- To better understand the perceptions of the consumers,
- To develop new products matching consumers' expectations,
- To help communication on sensations inside the company,
- To improve the training of driving experts,
- To establish links between the perceptions of consumers and the technical measures, which are the only measures that can be communicated to suppliers in their requests for proposal and specifications.

Beyond the evaluations of materials and odours that have been well-publicized, sensory science has a wide range of applications in the automotive industry. Some of them deal with the systems used during driving: the gearbox, the steering system, the braking system, the roll and lateral support, the sound of motor when idle, the overall size. Others deal with the cockpit: the perception of space in the car, the shape and materials of the dashboard, the seat and the human–machine interfaces. Additionally, some studies focus on very specific aspects of the car: the key card, the text-to-speech voices, etc.

20.1.2 The automotive context and its specificities

20.1.2.1 The products

The primary specificities of sensory science in the car industry deal with the product itself. Automotive vehicles are very complex products. A car can be divided into a large number of sub-products, including the seat, the dashboard, the steering wheel, each of these sub-parts being a potential object of sensory and consumer study.

For example, many components in front of the driver can be appraised: the dashboard, the steering wheel, the instrument panel, the windscreen. The dashboard has its own sensory characteristics (shape, materials, acoustics), but it is also made up of several elements that can be considered separately with their own sensory properties: the radio buttons, the ventilation, the headlights, the vent slots, the speakers, etc.

Finally, the variety of elements that can be tested with sensory science methodologies make it impossible for the company to manage separate expert panels.

20.1.2.2 Static or dynamic conditions

Depending on the objectives, a study can be conducted in static condition or in dynamic conditions. The static condition is quite similar as the usual sensory science condition: the subject has to focus on his/her sensations and (almost) nothing else. In driving conditions, the situation is very different, and more difficult to deal with, in several ways. Driving imposes a high mental load, even though it can become an automatic action with driving experience. Whatever is at stake in the study, the main instruction given to the participants is to follow traffic rules and to focus on driving safety. So, while driving safely, participants in sensory tests under dynamic conditions have to isolate their sensations from a lot of other information, which can be quite challenging.

20.1.2.3 The experts

Due to the complexity of the product and the experimental conditions, designing and assessing a vehicle is usually a matter for technical experts. These specialists have not only technical skills, but also a wide knowledge of the consumers. They design parts of the car, taking into account consumers' expectations and technical constraints.

They have a huge influence on car design, because they are not only responsible for the sensory and hedonic aspects, but also for the safety of the car. Therefore, their judgements are decisive. These experts use their own words to define the sensations, with their own test procedures, and score their perceptions according to a scale that has been developed inside the company for that specific purpose. Sensory scientists have to take into account their expertise and skills.

20.1.2.4 Price, availability and anonymous evaluation of cars

For sensory studies as well as for consumer studies, the price and availability of the tested cars are always an issue. Renting a specific car is expensive, which is why each study is generally limited in the total number of days needed. In addition to this, only one sample of each product is available.

To avoid bias, products are usually evaluated anonymously in most sensory studies. However, in the automotive industry, it may be difficult or even impossible to make products anonymous. In effect, subjects can easily recognize cars, even without brand identification. This is quite a problem, because subjects might not disregard the brand or anything else that could influence their evaluations.

20.1.3 The purposes for which rapid sensory methods can be used

The specificities of the automotive industry make the use of conventional profiling difficult. There are two main issues: first, conventional profiling is time consuming and when cars must be rented for that purpose, the cost of a study can be astronomical; second, due to the availability and the various experiences of the experts of the company, it is almost impossible to reach a consensus on sensory aspects in less than 6 months with one 3 h session a week. As the automotive company cannot manage specific sensory panels for each element of the car, the training and the assessments make the conventional profiling very time consuming.

Therefore, rapid sensory methods are used to substitute for conventional profiling. They are used as a first step to select a subset of products to study. In terms of sensory description, we use them to develop a first base sensory lexicon, as described in the example on idle noise in diesel engines. They are also used to compare experts' and consumers' perceptions. Indeed, the experts' point of view is sometimes so accurate that it can be too precise, and way above the perceptions of the consumers. Rapid sensory methods are then used to "calibrate" the experts according to the sensations of consumers in terms of intensities and vocabulary: this aspect will be illustrated by the example on "Gearboxes sensations and comfort."

Finally, for several aspects of the usage of the car, the variability in consumers' perception due to anthropometric cues must be taken into account. For this reason, rapid descriptive methods are used to raise the number of individuals that can be taken into account. This will be illustrated by the example on "Roll and lateral support perception."

20.2 Example 1: gearbox sensations and comfort

Transmission test drivers are specialists in the assessment of the sensations conveyed by transmissions: they use specific technical terms to describe gearboxes, and tune them accordingly by anticipating consumers' expectations. However, they sometimes have difficulty in explaining consumers' complaints solely on the basis of this technical vocabulary.

20.2.1 Objective: to compare experts' and consumers' perceptions

The objective of our study is to compare consumers' and test drivers' gearshift sensory terms. In particular, we want to verify that the gearshift lexicon is exhaustive and adapted to the analysis of consumers' complaints. Differences between consumers' and test drivers' sensory descriptions will be evaluated.

20.2.1.1 Specific constraints of gearshift sensory evaluation

- The "product" being described is actually a function of the vehicle. It is not an object, as in most sensory studies. Therefore, the sensory description must occur during driving. There is a high chance that "non-expert" drivers may not be able to describe their sensations while driving with this particular "product," especially since in normal driving situations it is used in a barely conscious, automatic way. As a first step for consumer description while driving, we wanted the task to be as simple as possible and to reflect the ability of these "non-experts" to perform it.
- It should be noted that gearshift sensations are influenced by the temperature of the gearbox, the engine speed, the vehicle speed and the way the driver shifts up or down. The vehicles must thus be driven before the evaluation to be tested under heat.
- Evaluation procedures require careful attention. Some phenomena which are considered as defects appear in an unpredictable fashion: the gearbox, for example, may offer once strong resistance to shifting at one point, then not do it again in the session.

To be able to compare consumers' and technical vocabulary, participants must not influence each other, whether they be consumers or test drivers. In fact, very experienced experts have a strong influence on each other. Therefore, a free choice profiling method that does not require discussion with the rest of the panel is preferable. Besides, test drivers are available for less than 10 h a month, so a fast methodology is needed. Due to the complexity of the products and the constraints of the evaluations, we choose to use Flash Profile (FP) (Dairou and Sieffermann, 2002; Delarue and Sieffermann, 2004; Taréa et al., 2003).

20.2.2 Materials and methods
20.2.2.1 Method

FP is derived from free choice profiling: sessions are always individual. Every assessor uses his own list of sensory terms. Consequently, FP potentially allows comparing test drivers' technical terms and consumers' attributes. FP is based on a comparative evaluation: ranking is easier than scoring for untrained assessors. Here, we decided to use FP in its three-session version, as described by Dairou and Sieffermann (2002):

- First session: generation of descriptive terms by testing the whole set of products.
  - The objective of this session is to generate the maximum number of terms to describe gear shifting sensations exhaustively. To be exhaustive, the evaluation must take place in different driving situations and at different vehicle speeds. We thus designed a 10 km route: City (start, rising and falling passages 1–2 and 2–3), Road (upshifts and down 3–4, 4–5, 5–6) and Parking Lot (start, reverse, handling static). Each participant drove each car on this route. The session lasted about 3 h for each participant.
- Second session: setting up of the final lists of terms, with definitions and evaluation procedures.
  - We pooled all of the terms used by participants during the first session (experts and novices together), and submitted this list to the participants so they could check they had not forgotten important sensory information to describe their sensations exhaustively.
  - As gearshift sensations are complex to evaluate, we wanted the subjects to be accurate for each term defining and establishing an evaluation protocol. These precisions are given by the assessors on an individual basis.
  - These definitions have also other interest. First, they help to interpret the results. Second, they help the test drivers to develop the protocols for further use after the study, as we want them to have a methodology reflecting the sensations of consumers.
- Third session: products are ranked on an ordinal scale for each attribute.
  - For each term, subjects rated the vehicles in order of increasing (or decreasing) intensity of the sensation perceived.
  - Unlike evaluation of food, a direct comparative evaluation of the products is not possible here because of the 10 km route. As we think that assessors know what strategy fits them the best, they are given great latitude in the way they evaluate the products. As a result, driving experts chose a monadic evaluation, scoring one vehicle at a time, for all terms. Consumers chose to operate by subgroups of sensory terms (e.g. they ranked the vehicles for all static operations, then in reverse gear and finally when driving).

20.2.2.2 Products

To limit bias in ratings due to the driving position and the brand image, we choose six Renault vehicles equipped with different manual transmissions. They were all sedan cars. No minivans were chosen because the driving position is very different from sedans. The description is focused on gearshift sensations; there is no description of the pommel, or of the place of the lever gearshift, or of the characteristics of the clutch.

20.2.2.3 Assessors

Our panel is composed of ten assessors: five test drivers (T1–T5) and five "consumers" (C1–C5). "Consumers" means people who are not trained to evaluate their gearshift sensations, although they have had previous experience in sensory evaluation while driving; they could also be considered as "sensory panellists."

20.2.3 Results

20.2.3.1 Generation of descriptions of assessor's gearshift sensations

During the *first session*, each assessor generated his own list of attributes to describe his gearshift sensations. Test drivers chose their terms from their usual gearshift sensory lexicon (although the lexicon was not presented during the test). The five "consumers" initially generated a total of *210 terms (193 of which were unique)*, and the five test drivers generated *196 terms (160 of which were unique)*.

During the *second session*, the "consumers" retained 65 *terms* to evaluate the six manual transmissions. Among these terms, 46 were not linearly correlated (non-significant Pearson coefficient at $\alpha = 5\%$), 71% of the evaluated attributes. The test drivers retained *103 terms*. Only 47 attributes were not linearly correlated for the test drivers (46% of the evaluated attributes). Figure 20.1 shows the results assessor by assessor. A first result is that "consumers" obviously managed to generate and to evaluate sensory terms while driving.

Thanks to the definitions and evaluation procedures provided by the assessors, we could distinguish between eight types of attributes depending on the aspects they referred to (Force, Vibrations, Precision, Travel, Noise, Pattern, Reverse gear and Others). We observe that consumers and test drivers did not give the same importance to each modality, as they did not generate the same numbers of terms (Fig. 20.2). However, even if test drivers seemed to speak more about force, vibrations and

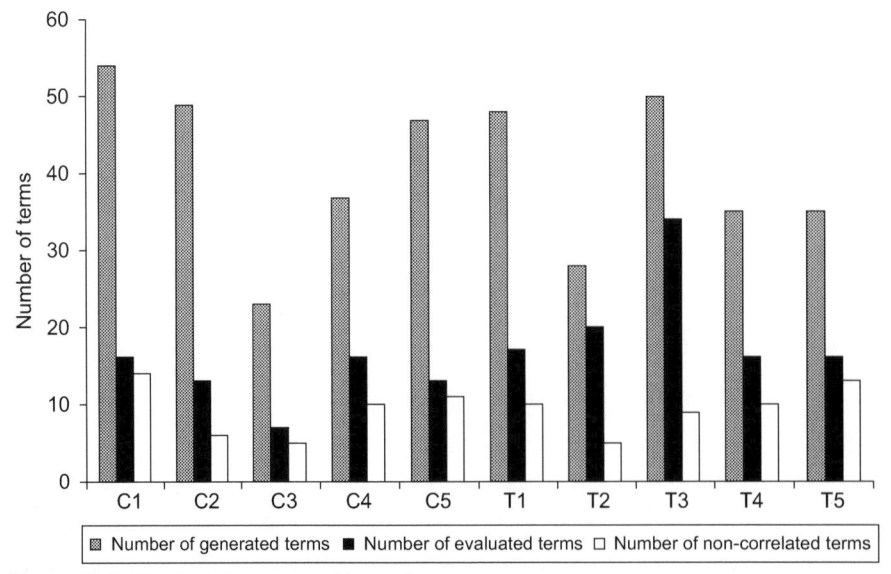

Figure 20.1 Number of terms used by each assessor.

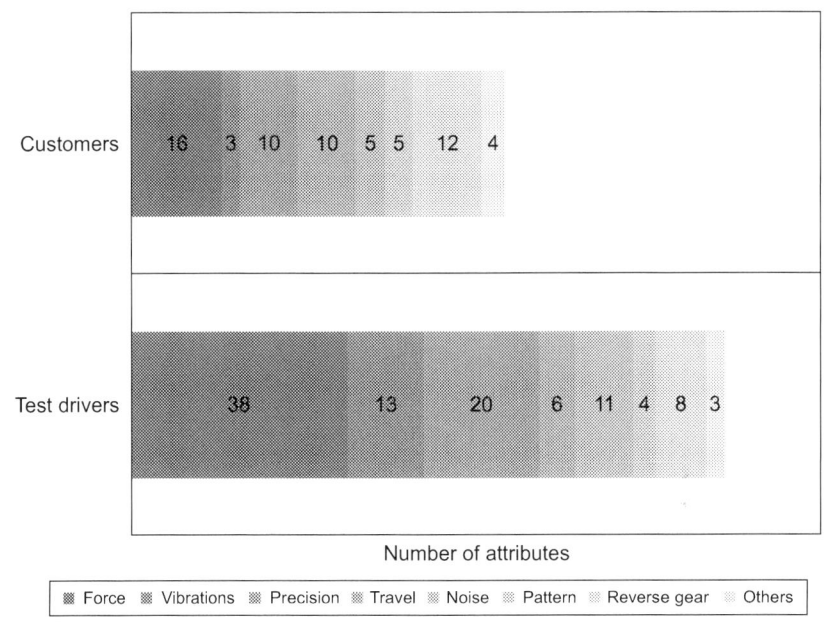

Use of rapid sensory methods in the automotive industry 433

Figure 20.2 Distribution of the terms in eight modalities.

Customers: 16, 3, 10, 10, 5, 5, 12, 4
Test drivers: 38, 13, 20, 6, 11, 4, 8, 3

Legend: Force, Vibrations, Precision, Travel, Noise, Pattern, Reverse gear, Others

precision, there is no link between the numbers of terms by modality and the type of participants (χ^2 test showed no significant difference at $\alpha = 5\%$).

There are two main results here. All assessors, "consumers" and test drivers seem to have managed to describe and define their sensations when gear shifting. Moreover, the glossary of gearshift comfort used by test drivers is exhaustive, in the sense that it covers all the terms generated by the consumers.

20.2.3.2 Consistency between consumers and test drivers descriptions

To compare the ten sensory maps provided by the ten assessors and to search for a consensus between them, we used generalized Procrustean analysis (GPA) (Gower, 1975). Taking into account the criteria of final procrustean distance and the confusion between products on the maps, our results show that there is no consensus between the ten assessors. In fact, two subgroups of assessors had to be composed to find a consensus in the perception of the manual transmissions:

- Group 1 is composed of C2, C4 and T5. The corresponding GPA sensory map is shown on Fig. 20.3.
- Group 2 is composed of T1, T2 and T3. The corresponding GPA sensory map is shown on Fig. 20.4.
- The rest of the assessors (C1, C3, C5 and T4) show no consensus, each assessor having his own view of the manual transmission. Perceptual spaces must thus be analysed by individual principal component analysis (PCA).

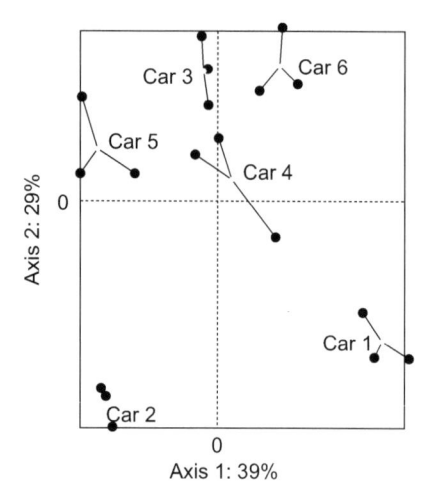

Figure 20.3 Sensory map from GPA of Group 1.

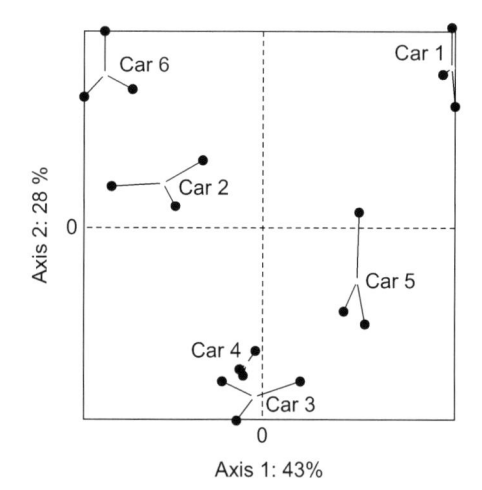

Figure 20.4 Sensory map from GPA of Group 2.

On Figs 20.3 and 20.4, the first two axes explain, respectively, 68% and 71% of the total variance. The consensual position of the vehicles is represented by the names of the cars, and assessors' evaluations by the connecting black dots.

On Fig. 20.3, all the cars are discriminated (Car 4 is well represented on Axis 3, and Cars 2 and 6 are completely opposed on Axis 2). In this group, T5 allowed us to compare "consumer" and technical vocabulary. Group 2 (Fig. 20.4) is exclusively composed of test drivers. They group together Cars 3 and 4. Cars 2 and 6 are close. It is striking that only one of our test drivers (T5) showed consistency with two of the "consumers." Three other test drivers provided similar sensory descriptions, but they did not fit to any description from our five "consumers."

20.2.4 Discussion and conclusion

Despite the complexity of gear shifting sensations, "consumers" managed to describe their sensations. They used 65 attributes to compare the six gearboxes. The terms were precisely defined. Technical terms were used by the test drivers (103 terms kept for the description of the gearboxes). However, some attributes are not understood in the same way: T4 and T5 did not agree with the other test drivers on the description of the gearboxes. Finally, except for T5, there is no consistency between consumers' and test drivers' sensory descriptions. Thanks to the expert T5 and GPA, we can establish links between the "consumers" description and a more technical vocabulary. Unfortunately, we cannot supply illustrative detail for reasons of confidentiality.

It is important to note that there is a possibility that the other experts would have shown consistency with "consumers" if a greater number of "consumers" had participated.

The rapid description study allows us to verify that the test protocols of experts are close to the solicitations of consumers about gear shifting.

20.3 Example 2: role and lateral support perception

20.3.1 Objective: to understand a complex perception

The perception of driving is quite different between standard cars and minivans. Minivans are often perceived as having poor road-holding properties. This is believed to be linked to the fact that the driver is seated in a higher position, which induces quite different sensations from those in standard cars.

To improve the driving sensations provided by the minivans, we need to better understand this phenomenon of road-holding.

The road-holding sensation seems to be associated with two of the vehicle's functions: the roll of the car, and the lateral support of the seat. To understand the impact of these two functions on the road-holding sensation, we carried out a sensory description study and a consumer study on seven minivans under driving situations.

Although the focus of this chapter is on descriptive analysis, the consumer study will be presented because the results will emphasize the fact that even rapid descriptive methods can be used to understand consumers' perceptions, through External Preference Mapping for instance.

20.3.2 Materials and methods

20.3.2.1 Products

Seven minivans were evaluated in both a sensory profile and a consumer study (Table 20.1): three cars from the European market and four prototypes that were combinations of several rolling chassis and seats from different cars.

Table 20.1 **The seven minivans evaluated in both sensory profile and consumer study**

	Seat 1	Seat 2	Seat 3	Prototype seats	
Rolling chassis 1	RC1-S1	RC1-S2			
Rolling chassis 2	RC2-S1	RC2-S2			
Rolling chassis 3			RC3-S3	RC3-P1	RC3-P2

We wanted to understand the impact of the rolling chassis and the seat on these perceptions. We thus needed to combine several sorts of chassis and seats. We bought two samples of the cars: RC1-S1 and RC2-S2 are the cars as they are sold. For RC1-S2, we used the chassis RC1 of Car 1 but we took out its seat and put in the seat S2 of Car 2.

Using these combinations, we could compare :
- the seats on the same chassis (for example, the products RC1-S1 and RC1-S2 allow us to compare the seats S1 and S2 on the same chassis, RC1),
- the chassis with the same seat (for example, the products RC1-S1 and RC2-S1 allow us to compare the chassis RC1 and RC2 with the same seat S1).

20.3.2.2 Sensory description

Our objective is to obtain a sensory description of the cars in a situation where roll and lateral support will be highly perceptible. As this description should be used for preference mapping, we initially aimed at conducting a conventional profile and we designed the first tests accordingly.

However, the first session allowed us to understand the difficulty of the task.

Firstly, to generate the type of sensations that we needed in order to study road-holding, the drivers had to follow a particular sequence involving simulation of a roundabout at 60 km/h (Fig. 20.5). The task was so exhausting for them that we had to use a quicker approach.

Secondly, we also quickly realized that consensus among assessors would be difficult to obtain, as the sensations they mentioned were quite different. We therefore chose to adapt the FP methodology.

On one hand, as we still have the objective of preference mapping, a minimum set of consensual sensations is required. On the other hand, we wanted to preserve the inter-individual differences of perception between the participants. Consequently, our methodological approach has been based on both individual and group sessions. The final list of sensations is composed of terms used by all participants, and terms used by just part of the group.

Even though we developed a quicker procedure, 11 sessions of 2 h were still needed because of the difficulty to focus on the road-holding sensations. Nine sessions were individual, and two were group sessions (Fig. 20.6). Figure 20.7 shows an example of an evaluation sheet for a term chosen and defined by all the panellists.

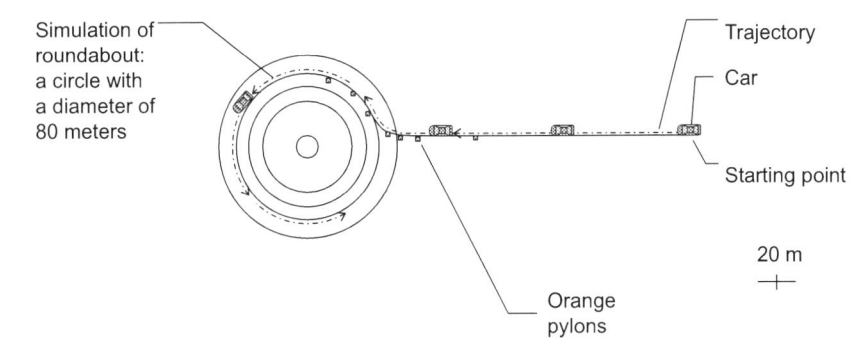

Use of rapid sensory methods in the automotive industry　437

Figure 20.5 Driving context of the sensory study, a roundabout.

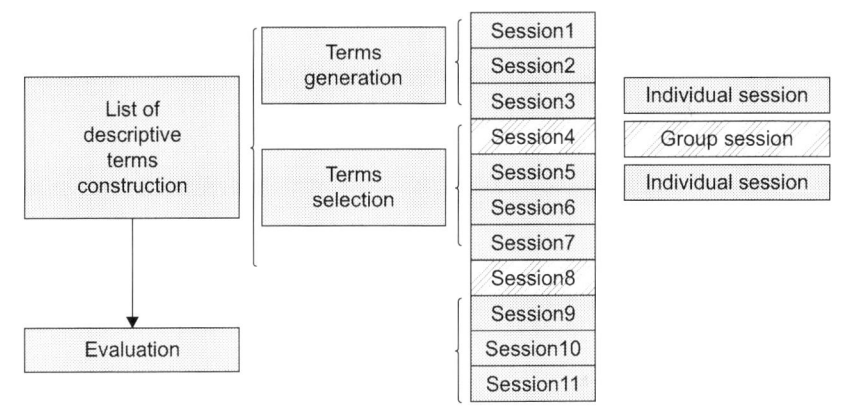

Figure 20.6 Sensory profiling sessions.

Pressure intensity of the back on the lateral supports

Protocol: In the left curve, having your speed and trajectory stabilized you have to evaluate the intensity of the pressure of your back on the side-support backrest

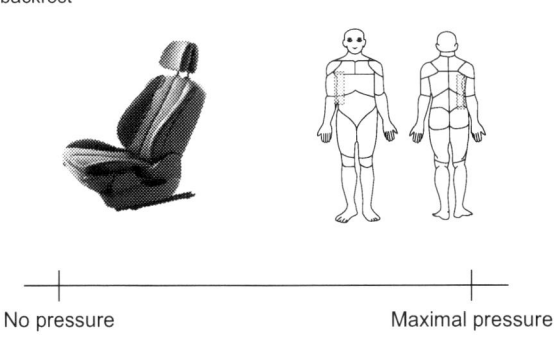

No pressure　　　　　　　　　　　　Maximal pressure

Figure 20.7 Example of evaluation sheet.

Most of the assessors chose to evaluate a set of terms within the same ride on the roundabout (Fig. 20.5). As they had to concentrate on the vehicle trajectory, they needed a visual reminder of the area of the seat and of their body for each term. A quick look at the sheets was a great help in focusing on the sensation before starting the simulation of the roundabout, as there was no one to take notes during the driving. These sheets were also a way to communicate within the group.

Road-holding perceptions are also influenced by the driver's anthropometry. In fact, anthropometry has an impact on the driving position, which in turn has an impact on the road-holding sensations. Thus, in addition to inter-individual differences in sensory sensitivity, it is important also to take into account subjects' sizes and weights in such studies.

Ten Renault employees experienced either in sensory science or in vehicle behaviour (technical experts) took part in the study. They were chosen to be representative of the sizes and weights of the consumers.

20.3.2.3 Consumer study

One hundred and fifty-one drivers took part in the study. We recruited regular minivan drivers because the rolling sensations between a minivan and a sedan are very different and a sedan driver would be disappointed by any minivans he/she would drive. As the driving occurs on Renault private tracks that are only open for employees (for both insurance and confidentiality reasons), the drivers were Renault employees.

Each consumer took part in a unique 2-h session for the assessment of the seven cars. The instructions were vague in terms of purpose: to evaluate the behaviour of the cars in curves, and the pleasantness of sensations while driving. The drivers rated each car from 0 (this car is unpleasant) to 10 (this car is very pleasant).

The driving procedure was closer to reality for consumers than for assessors: motorways with curves (100 km/h) and traffic circles (35 km/h) on a private track in order to induce road-holding sensations (Fig. 20.8).

20.3.3 Results

20.3.3.1 Sensory description

Even though we had used a quicker approach than conventional profiling, each assessor drove a total of more than 25 h and 600 km.

About 500 terms were generated by the assessors. From these 500 terms, 81 were considered by the assessors to be important. Finally, each assessor retained from 18 to 26 terms. The final list of 29 descriptive terms covered three themes: five terms describing the movement of the car, nine the movement of the body of the driver in the car and 15 more particularly on the body of the driver (Table 20.2).

The ten assessors did not use the same terms: only seven terms were used in common. There are several explanations to these differences in selected terms.

First, differences in drivers' anthropometric data, such as height and weight, explain why they do not have the same sensations. For instance, four drivers are not

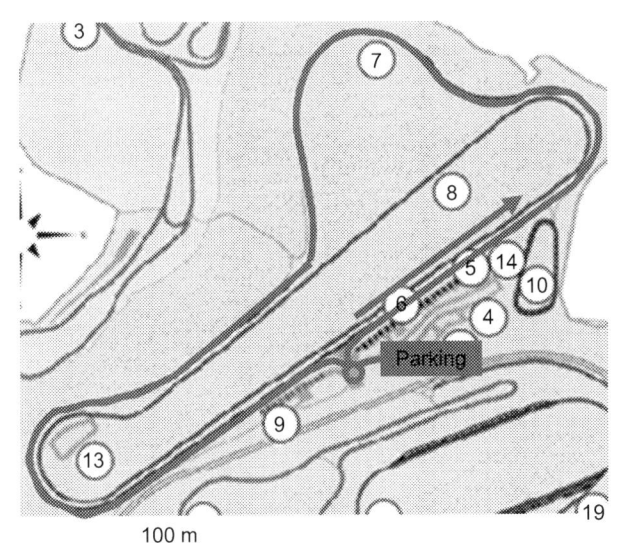

Figure 20.8 Test drive track for consumers.

in contact with the central console and consequently do not use the corresponding term. The torso of three of them is so broad that it is completely wedged between the lateral supports and does not move. The term "Wedging of the torso on the seat" has no relevance for them.

Second, several assessors have more difficulty than others in focusing on subtle sensations during driving. Given our objective, we considered that this was not really a problem, as it would reflect the consumers' perception and limited ability to notice these subtle sensations.

According to their individual PCA and GPA, we can distinguish two groups of drivers who do not differentiate the cars in the same way:

- A first group of eight drivers who discriminated the minivans with the rolling chassis (Fig. 20.9). The first axis represents 70% of the total variance. On this axis, there are three groups of cars: on the left, the cars with the chassis RC1; in the middle, the cars with the chassis RC2 and, on the right, the cars with RC3. The axis concerns sensations of the three types: body, movements of the body and movements of the car.
- A second group of two drivers who discriminated the cars with the seat and then the rolling chassis (Fig. 20.10). On the first axis, the cars with the seat S2 are on the right, the seat S1 in the middle and the last ones on the left. The differentiations between the chassis come second and are based on the movements of the car.

20.3.3.2 Consumer study

Thanks to proper ANOVA, we checked that the presence of co-pilots in the car, the weather conditions and the order of presentation did not bias mean hedonic scores. The 151 drivers significantly discriminated the seven minivans. They preferred the

Table 20.2 The terms used to describe roll and lateral support

The sensation concerns	Terms	Used by all the assessors
The body of the driver	Pressure intensity of the left foot on footrest	
	Pressure intensity of the right leg on the central console	
	Pressure homogeneity of the right leg on the central console	
	Pressure intensity of the right thigh and buttock pressure on the seat	X
	Pressure homogeneity of the right thigh and buttock pressure on the seat	
	Pressure intensity of the back on the seat	X
	Pressure homogeneity of the back on the seat	
	Muscular stress intensity of the back	
	Pressure intensity of the back on the lateral supports	X
	Pressure homogeneity of the back on the lateral supports	
	Lateral flexion of the torso	X
	Rotation of the torso	
	Transverse torsion of the torso	
	Pressure intensity of the belt on the collarbone	
	Muscular stress intensity of the neck	
The movements of the body of the driver	Amplitude of the lifting of the left buttock and left thigh from the seat	
	Amplitude of the sliding of buttocks and thighs on the seat	
	Wedging of the right buttock on the seat	
	Amplitude of the lifting of the back from the seat	X
	Amplitude of the sliding of the back on the seat	X
	Amplitude of the exit of the back from the seat	
	Wedging of the torso on the seat	
	Amplitude of the exit of the head from the headrest	
	Wedging of the head on the headrest	
The movements of the car	Contribution of the wheel for wedging on the seat	
	Steering wheel linearity	
	Car roll	X
	Car tilt	
	Time to achieve balance in rolling	

rolling chassis RC3 and the preferred car is RC3-S3, one of the cars available on the market (Fig. 20.11).

According to the results of analysis of variance, we can estimate that the drivers differentiated the rolling chassis RC3 from the two others, and discriminated the three seats S1, S2 and S3.

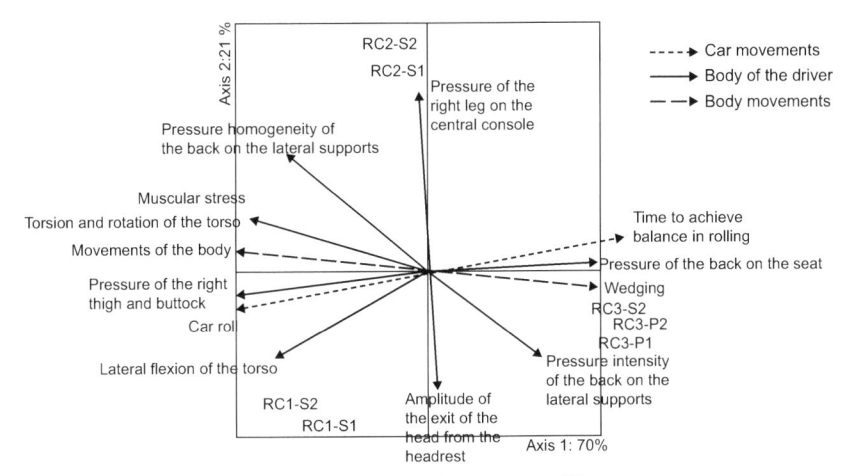

Use of rapid sensory methods in the automotive industry   441

Figure 20.9 Sensory map from GPA of eight of the ten panellists.

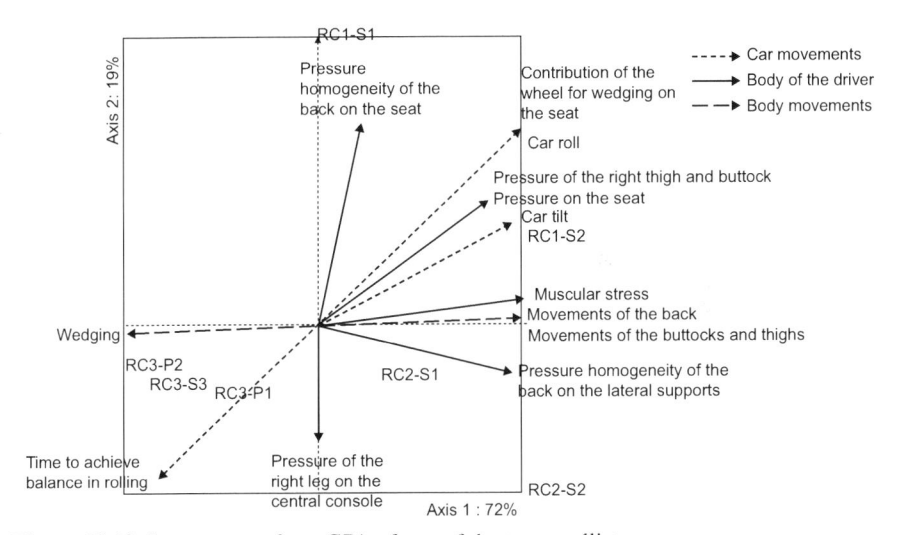

Figure 20.10 Sensory map from GPA of two of the ten panellists.

A Hierarchical Ascendant Classification (Euclidean distance, Ward's criterion) was performed on the hedonic scores of the entire population: four clusters can be distinguished (Fig. 20.12). Even if the number of consumers was low for Clusters 1 and 2, it seemed of interest to consider these drivers separately because they have separate opinions on RC3-P2. Cluster 1 differentiated the cars according to their seat, regardless of the chassis. Cluster 2 liked RC3-S3 and rejected RC3-P2. Cluster 3 differentiated the cars according to their chassis and not their seat. Finally, Cluster 4 differentiated the cars according to their chassis and then to their seat.

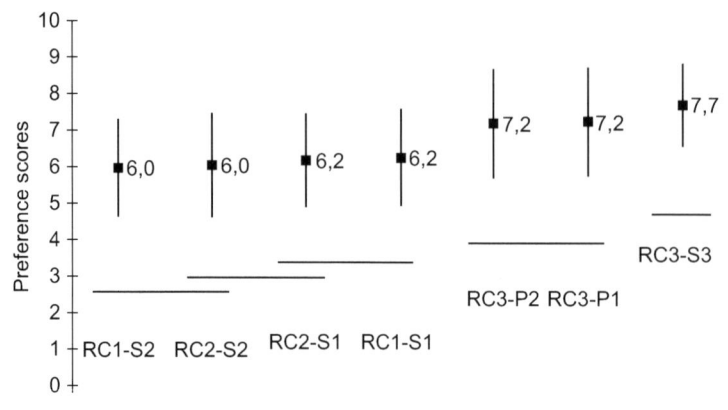

Figure 20.11 Hedonic scores of the consumers study (ANOVA: $F(6,1050) = 40.452$, $p < 0.001$ – Turkey's honest significant difference (HSD) 5%).

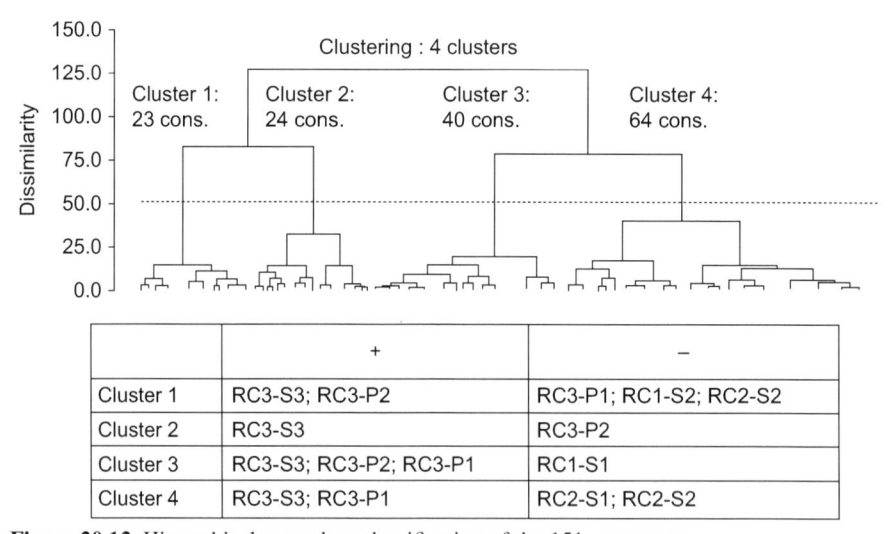

	+	−
Cluster 1	RC3-S3; RC3-P2	RC3-P1; RC1-S2; RC2-S2
Cluster 2	RC3-S3	RC3-P2
Cluster 3	RC3-S3; RC3-P2; RC3-P1	RC1-S1
Cluster 4	RC3-S3; RC3-P1	RC2-S1; RC2-S2

Figure 20.12 Hierarchical ascendant classification of the 151 consumers.

20.3.3.3 External Preference Mapping

To pursue our objective to understand the impact of roll and lateral support on the road-holding sensation, we used External Preference Mapping with quadratic model. The individual models of the 151 drivers with the consensus sensory map are quite good: the average R^2 coefficient is 0.87 (Fig. 20.13). As anticipated with the raw results of the consumer study, the best area in terms of preference is near RC3-S3 but this is not the main point we would like to raise here.

External Preference Mapping with quadratic model on 151 consumers

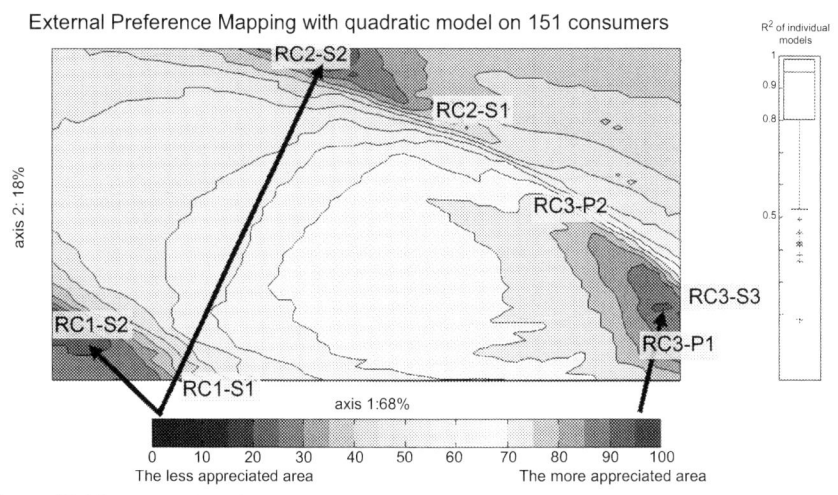

Figure 20.13 External Preference Mapping on the consensus map.

Indeed, in order to better interpret consumers' preferences, we performed two additional External Preference Mappings using the description provided either by the first or by the second group of assessors. Interestingly, some consumers are "better" modelled with the description of the first group, while others are "better" modelled with the second one. In fact, the 151 drivers can be clustered in two groups, depending on which of the two groups of assessors best fits their preferences:

- The hedonic scores of 90 consumers are better modelled by the sensory profile provided by the eight assessors who differentiate the cars with the rolling chassis;
- The hedonic scores of 61 consumers are better modelled by the sensory profile provided by the two assessors who differentiate the cars with both the seat and the chassis.

20.3.4 Discussion and conclusion

First, the sensory description of the cars that we obtained would not have been possible with conventional profiling because of the time frame acceptable for assessors and the fatigue generated by the test driving. Thus, a "quicker" approach was the only option and, even with this approach, the amount of time needed to describe the cars is quite impressive.

Second, not forcing assessors to use the same descriptive terms, as in free choice profiling, was finally a good idea because their perceptions are indeed quite different.

The perceptions of roll and lateral support are decomposed into 29 elementary sensations. Mostly due to anthropometric differences, the assessors do not have the same perceptions while driving. Furthermore, according to the sensory description, the assessors do not discriminate the minivans with the same sensations, and can be separated into two groups. Eight of the ten participants are more sensitive to the rolling chassis and pay more attention to visual sensations such as car rolling. The two other drivers are more sensitive to the seats and to sensations such as the pressure under the thighs.

What is particularly interesting in this case is that these differences of points of view help us to understand the consumers and their complaints about road-holding of their car. Indeed, even if the road-holding of a car is satisfying in terms of security, thanks to this study, we know that there are more than visual sensations of roll which can be disturbing for consumers and can cause worries. From a methodological point of view, this improvement of External Preference Mapping results a posteriori validates our choice to have a free choice profiling approach for measuring the sensations and a clustering approach of the consumers. We are convinced that it is a real added value for the understanding of the consumers' behaviour.

20.4 Example 3: idle noises of diesel engines

20.4.1 Objective: to test the feasibility of a rapid sensory profiling technique on products recognized as requiring high expertise

Usually, the studies of noises in the car are reserved for the expertise of acousticians. This exclusivity can raise some issues concerning the development of the vehicle. Among these issues, the very specific vocabulary of these experts can complicate the exchange of information between the different R&D departments involved.

Of course, we do not question the expertise of the acousticians, but we think it would be of great value to verify that their description of idle sounds is close to what consumers perceive and to share a more common lexicon within the company. In order to address this question, we conducted a study that had three underlying objectives:

- To estimate the feasibility of the FP methodology to describe engine sounds,
- To estimate the common points and the differences between the descriptions of these sounds by acoustic experts, employees of the marketing department and consumers,
- To obtain a sensory map of 11 idle noises.

20.4.2 Materials and methods

20.4.2.1 Products

Eleven idle noises of diesel engines were evaluated anonymously. They are coded in this paper by their short names (Table 20.3). These cars were chosen to represent different types of engine (3-, 4-, 6- and 8-cylinder).

The sounds were evaluated in a listening room and played at 104 dB (A) with two speakers and a subwoofer (2.1 Trans aural system). For this study, we wanted to facilitate the description of the noises rather than to be representative of actual driving conditions. Therefore, the chosen level of listening is higher than the actual level in a car cockpit.

Table 20.3 **The 11 cars used for idle noise**

Code	Car	Type
Hyu	Hyundai Matrix	3-cylinder engine
Polo	VW Polo	3-cylinder engine
407	Peugeot 407 hdi	4-cylinder engine
A3	Audi A3 1.9 TDI	4-cylinder engine
Corsa	Opel Corsa	4-cylinder engine
Focus	Ford Focus 2.0 TCDI	4-cylinder engine
Golf	VW Golf	4-cylinder engine
X73	X73 G9T 2.2 dCi	4-cylinder engine
BMW	BMW 730d	6-cylinder engine
A6	Audi A 62.5 TDI	V6 engine
S400	Mercedes S400	V8 engine

20.4.2.2 Assessors

Ten Renault employees participated in the study: four acousticians (coded E1–E4), two employees of the marketing department (P1 and P2) and four persons considered as novices on acoustics, and hence closer to consumers (C1–C4).

20.4.2.3 Method

Flash Profiling was conducted in two stages, each lasting 1 h. The first stage was dedicated to finding the words to describe the sounds, and the second stage was dedicated to evaluations. Assessors were asked to follow the instructions below.

The objective of this first session is to describe 11 sounds. At the end of this session, you will have developed a list of terms of your own.

You can place yourself in the following context: you will have to describe eleven sounds in a meeting to any person, without being able to make him/her listen. You will need to find terms to describe the sensations generated by listening to the sounds.

- *To ensure that your party can well represent the sounds with your description, it must be precise. This accuracy implies that you do not use terms like "beautiful sound," "unpleasant noise." Indeed, these hedonic terms are directly related to a level of pleasure, and depend on the individual who uses them. In this case, it is particularly interesting to force oneself to think about the "why": "why is this sound more beautiful than another?" The sensations are often in this "why" and therefore the terms necessary for the description.*
- *For this description to be as effective as possible, it is not necessary to use redundant terms as synonyms or antonyms.*
- *In addition to this, it is important that the terms are used to differentiate the sounds. For example, if these sounds have the same loudness level, this term is not useful.*

Using the interface, our recommendation is to create groups of sounds that you find close. Once these groups are formed, you can try to identify the sensations that make the sounds of the same group close and that make two groups sound different. You can repeat these groupings as many times as you wish.

At the second session, you will rank the sounds for each of these terms on a line segment.

To facilitate the ranking, you can give a definition for each term and a synonym or antonym if you deem it necessary. For each term, you will rank the intensity of the sensation perceived when listening. It is therefore important that you identify the lower bound and the upper bound of sensory intensity.

The ranking scores for each attribute were recorded and analysed using GPA.

20.4.3 Results

Sixty-eight words were generated by the ten participants in the first session, and all of them were used in the second session. Experts used slightly more terms to describe the sounds (between 6 and 11 words) than marketers and novices (between 4 and 8 terms) (Fig. 20.14).

The first plane of the GPA accounts for 79% of sensory information provided by all terms (Fig. 20.15). Each participant is represented by a vector whose length represents the distance from the consensus. For a given product, the shorter the vector, the closer is the perception of the participant to the consensus. Products for which the consensus is good, that is to say where the participants agreed, are shown in bold. The first axis opposes S400 and BMW vehicles, respectively 8-cylinder and 6-cylinder with Polo and Hyundai vehicles, 3-cylinder. Four-cylinder engines, positioned in the middle of the first axis, are not differentiated on this axis. The second axis differentiates 4-cylinder engines, and opposes Focus with Golf and Corsa.

To summarize the information provided by the terms, we have identified five types of sensations, represented by vectors in Fig. 20.16. These vectors represent the

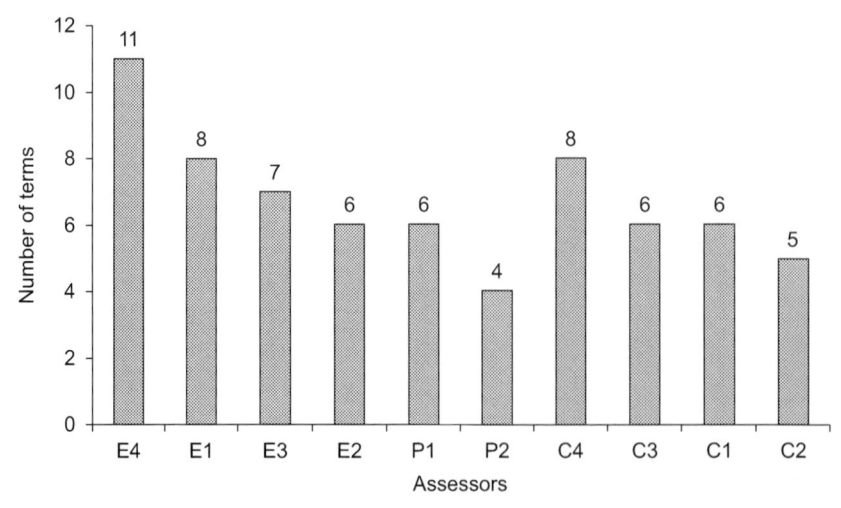

Figure 20.14 Number of terms used by each assessor to describe idle sounds.

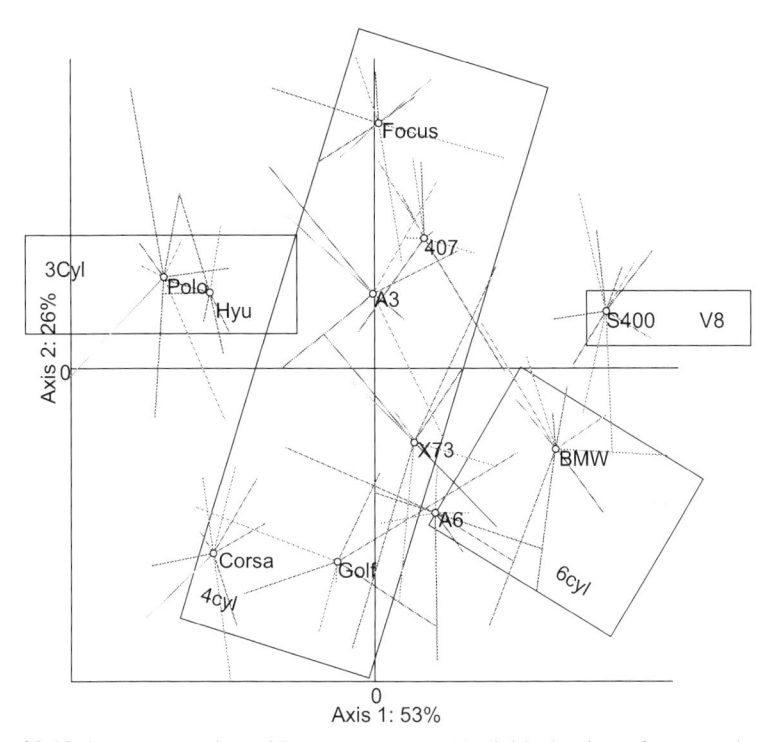

Use of rapid sensory methods in the automotive industry 447

Figure 20.15 Sensory map from GPA (consensus and individual points of assessors).

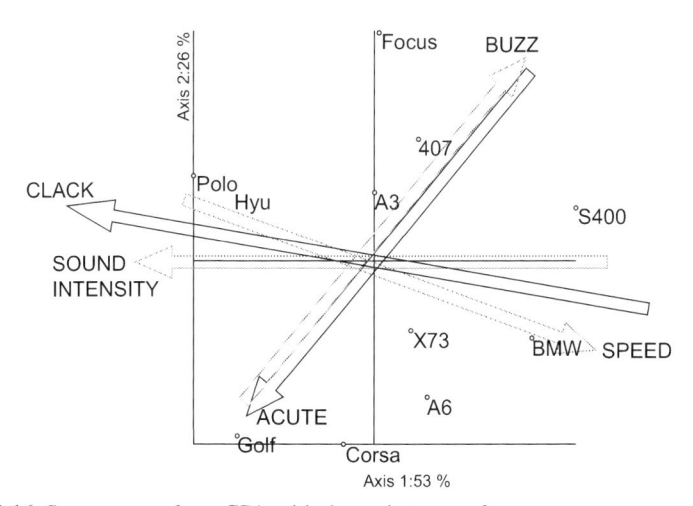

Figure 20.16 Sensory map from GPA with the main types of terms.

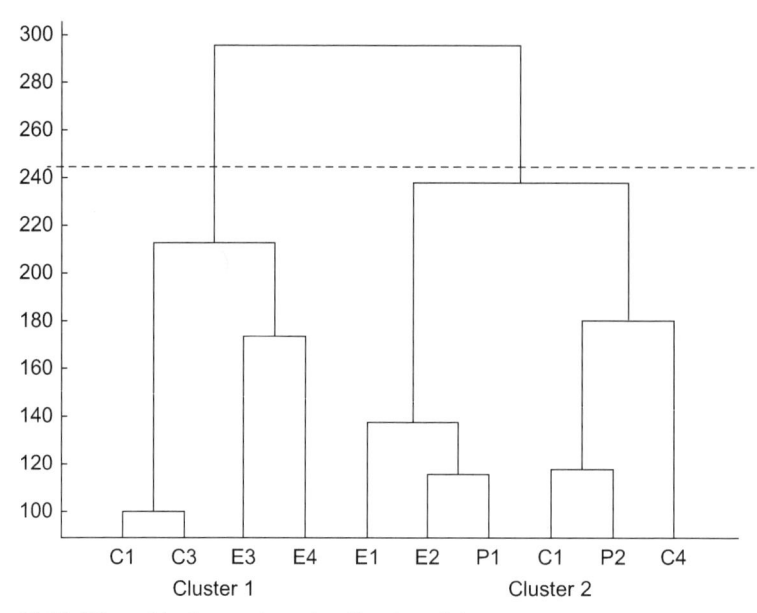

Figure 20.17 Hierarchical ascendant classification of the assessors.

terms that are most correlated with each other and which cover the same sensation perceived as defined by the participants. They include all the terms for which sounds are ranked in the same way.

- CLACK: includes terms "clack intensity," "claquance" (this term was used by an acoustician), three terms "farm tractor noise" and one "jackhammer noise."
- SOUND INTENSITY: includes terms about the intensity or the overall sound level, but also "noise," "loudness" and "intensity level."
- SPEED: includes terms describing the pace of a typical diesel sound: "speed fluctuation," "noise burst" and "speed of the pattern."
- ACUTE: the sound tone from low to high pitch. It includes terms "tone," "bass/treble," "tone of the engine" and "low pitch sound."
- BUZZ: includes two terms "buzz" and "hum level" cited by experts and "deafening."

The consensus is good for Huyndai, 407, S400, Golf, Polo and BMW. On the contrary, there is more important dispersion for X73, A6, Corsa, A3 and Focus. Therefore, it is of interest to cluster assessors according to their perceptions of the products. Cluster 1 is composed of four people: C1, C3, E2 and E4. Cluster 2 is composed of six people: E1, E2, P1, C1, P2, C4 (Fig. 20.17). These clusters are composed of both experts and novices. We can thus conclude that there is no expert and novice distinction at the sensory description of these particular sounds.

The dimension "clack" opposes BMW and S400, which have a low level of clack, to Huyndai and Polo, which have a high level of clack. This is the same for Cluster 1 as for Cluster 2.

The assessors from Cluster 1 differentiate the sounds on the "acute" dimension. Corsa and Golf are perceived as very acute, A6, Polo, Huyndai, A3 and BMW X73 as averagely acute, S400 and 407 as low pitch sound, and finally Focus seems to have the lowest pitch. The assessors of Cluster 2 differentiate three groups of sounds with this dimension: Golf, Corsa, X73 and A6 are very acute, Huyndai, Polo, A3 and BMW are moderately acute, and 407 and S400 Focus have low pitch sounds.

The main difference between Clusters 1 and 2 concerns the perceptual dimension "Buzz." It opposes Corsa, Golf and Polo S400 and BMW for Cluster 1. It opposes Golf, Corsa, X73 and A6 and 407, Focus for Cluster 2. Furthermore, the dimension "buzz" is anti-correlated with the "acute" dimension to Cluster 2, which is not the case for Cluster 1.

20.4.4 Discussion and conclusion

The first conclusion of this study is the suitability of a quick profiling method such as Flash Profiling, even for novices, to describe products recognized as requiring high expertise: idle engine noises.

As can be seen from Fig. 20.18, the sensory information is not distributed in the same manner for all assessors. For example, 86% of sensory information from the expert E1 is shown with Dimensions 1, 2 and 3 on the sensory map, while the first two dimensions of the PCA are sufficient to represent 88% of the information for the novice C1. Sensory information E1 is richer than that of C1. For six subjects (the four experts, one marketer and one novice), three dimensions are needed to summarize 80% of the information provided by their evaluations. For the second marketing employee and three novices, two dimensions are sufficient to synthesize 80% of the information. According to the participants' sensory descriptions, sensory information can be summarized in two or three dimensions. The dimensionality of the information provided by novices is thus similar to that of experts, suggesting that their description is not poorer than that of experts.

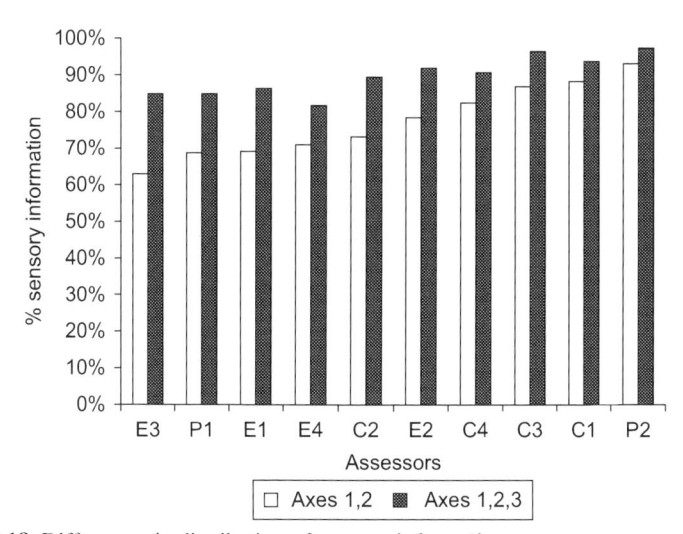

Figure 20.18 Differences in distribution of sensory information among assessors.

Another conclusion concerns the relationships between the terms used by novices, marketing department employees and experts. A Hierarchical Ascendant Classification of the terms allows us to group terms used by each type of assessors. For instance, the term "Claquance" used by E2 is quite technical, but it is highly correlated with more understandable terms such as: "farm tractor noise" or "jackhammer noise." The terms "Central frequencies" and "Middle High Frequencies" are technical, but knowing that they are anti-correlated to "Engine Power," used by several novices, allows us to understand how consumers can perceive this particular aspect of the idle noises.

20.5 Conclusion: pros and cons of rapid sensory methods in the automotive context

There are several significant advantages to the use of rapid sensory methods in our particular context.

The first is obvious: time is of the essence. A majority of sensory studies would not have occurred if they had to use conventional profiling. No manager wants to involve his/her team in a sensory study that could take more than 15 2-h sessions, which would be the average number of sessions for conventional profiling in the studies previously mentioned. Moreover, for cost reasons, the assessors are internal employees. Participating in a sensory profiling is time consuming in their day-to-day work. It comes as an addition to it, and very often urgent daily tasks quickly take over. That is why, usually, the motivation and the participation of the panellists decrease over time.

Therefore, as an alternative to the use of conventional profiling, the duration of Flash Profiling can be seen as a relief. Assessors are usually involved for 4–8 h. In the portfolio of sensory studies at Renault, an FP study can thus be considered as a preliminary study. If the results are not completely satisfactory, but promising, assessors and their management usually agree to continue the study. If the results are not satisfactory at all, the study can be ended rapidly without much harm done. FP could be a way to get "a foot in the door."

Moreover, this methodology, if correctly understood, is quite flexible. Therefore we do not hesitate to adapt the protocol according to the specific objectives of the studies. This is the case for the roll and lateral support perception study, where we needed a common minimum set of sensations but wanted to preserve the inter-individual differences of perception between the participants.

The second advantage of FP here is linked to the free choice aspect of the description and could concern a wider range of products, beyond cars.

For an industry where expertise is so important, a methodology involving both experts and consumers in the same test without having the first ones strongly influencing the second ones is very helpful. Even with great skills in panel leadership, it can be difficult to preserve the opinions of consumers or novices in group sessions involving experts, the ones who, at the end, will put the product on the market, engaging their responsibility. Therefore, using free choice profiling can preserve both the precise description of an expert and the naivety and diversities of consumers. Ultimately, it allows us to compare side by side the two types of description and find a "consumer" match for almost all the technical terms used by the experts. These correspondences between the vocabularies of experts and novices are essential for the marketing department.

Beyond the semantic aspects, the use of a free choice method is quite interesting when one knows that assessors cannot perceive the same sensations. In the automotive industry, this situation occurs when anthropometry plays a role in the perception. How do we ask an 80 kg-man to have the same sensations as a 55 kg-woman in an automotive seat? The example on roll and lateral support demonstrates that this can be a huge asset for External Preference Mapping, when there are several clusters of consumers with different perceptions.

Unfortunately, there are also several drawbacks for rapid sensory methods such as Flash Profiling.

The simplest to understand is the difficulty obtaining a useful consensus in the sensory description. As assessors use their own vocabulary, this consensus is mainly obtained mathematically by such techniques as GPA and Hierarchical Ascendant Classification. Therefore, the correlation of two terms after GPA is the only clue that they can mean the same thing. Sadly, it is not sufficient. Two terms with different meanings can be highly correlated because of the products evaluated.

This also has consequences for product conception when such techniques as External Preference Mapping are required. We can find an optimal product, but there are as many sensory explanations as assessors. Consequently, the difficulty of interpretation makes the results less usable.

A more subtle drawback concerns the sensory description when consumers are the only assessors of the study. We do not illustrate this particular point in this chapter, but we have met with the case several times. In effect, during the FP preliminary session, the consumers tend to generate a huge number of terms. Then, for the second session, when the time comes to rank the products for all the terms they select, they change their minds. In fact, they select the easiest terms to score, and get rid of the others. Usually, the more subtle sensations (i.e. the harder to score) disappear. The main consequence is that the description becomes poorer. We thus advise the reader to pay much attention to such biases, as they might concern other industrial domains as well.

As a final conclusion, a rapid sensory method such as FP is useful because it can lead to "quick wins" for the industry. But, even if the method is quite simple to understand, it is our opinion that the experimenter should have a solid experience in conventional profiling to obtain the best of the methodology.

References

Astruc, C. and Blumenthal, D. (2004). L'analyse sensorielle appliquée à l'automobile: caractérisation sensorielle des perceptions de maintien latéral et de roulis en dynamique. In: *3ème Congrès SIA/CTTM*, 12–13 October 2004, Le Mans, France.

Astruc, C., Blumenthal, D., Delarue, J. and Danzart, M. (2005). How do drivers evaluate a new car? A study under driving situation. In: *6th Pangborn Sensory Science Symposium*, 7–11 August 2005, Harrogate, UK.

Astruc, C., Blumenthal, D., Delarue, J., Danzart, M. and Sieffermann, J.M. (2006). How to construct a global driving procedure for dynamic hedonic tests? In: *8th Sensometrics Conference*, 2–4 August 2006, ÅS, Norway.

Astruc, C., Sieffermann, J.M., Delarue, J., Danzart, M. and Blumenthal, D. (2007). The influence of assessment context on the consumer responses in a driving situation. In: *7th Pangborn Sensory Science Symposium*, 12–16 August 2007, Minneapolis, MN, USA.

Astruc, C., Blumenthal, D., Delarue, J., Danzart, M. and Sieffermann, J.M. (2008). An original use of Pearson's correlation to construct a unique assessment procedure from individual ones for dynamic hedonic tests of cars. In: *9th Sensometrics Conference*, 20–23 July 2008, St. Catharines, ON, Canada.

Blumenthal, D. and Bouillot, S. (2007a). The comparison of instrumental and sensory data to understand the customers: a case study on static seat comfort. In: *7th Pangborn Sensory Science Symposium*, 12–16 August 2007, Minneapolis, MN, USA.

Blumenthal, D. and Bouillot, S. (2010). CARTOPTI: a Tool for Automotive Seat Conception Using Regression Models and Customers Studies. In: *International Conference on Kansei Engineering and Emotion Research (Keer)*, 2–4 March 2010, Paris, France. pp. 2188–2197.

Blumenthal, D. and Dairou, V. (2007b). The use of sensory profiling to develop an evaluation grid on automotive cockpit Human-Machine Interfaces. In: *7th Pangborn Sensory Science Symposium*, 12–16 August 2007, Minneapolis, MN, USA.

Blumenthal, D. (2004). How to obtain the sensory scores of the optimal product according to preference mapping with quadratic model? In: *7th Sensometrics Conference*, 27–30 July 2004, Davis, CA, USA.

Blumenthal, D. (2008). Exploring leads to make external preference mapping more operational in an automotive context. In: *9th Sensometrics Conference*, 20–23 July 2008, St. Catharines, ON, Canada.

Blumenthal, D., Dairou, V. and Sieffermann, J.M. (2000a). How improve the sensory information provided by Free Choice Profiling in Preference Mapping using individual maps? In:*5th Sensometrics Conference*, 9–11 July 2000, Columbia, MO, USA.

Blumenthal, D., Danzart, M. and Sieffermann, J.M. (2000b). "Application de la méthode du profil libre en automobile, application sur le confort de sièges." *Revue Française de Marketing*, 179/180: pp.143–156.

Blumenthal, D., Lino, F., Danzart, M. and Sieffermann, J.M. (1998). Free choice profiling and preference mapping on non-food products. In: *3rd Pangborn Sensory Science Symposium*, 9–13 August 1998, Alesund, Norvège.

Blumenthal, D., Priez, A., Sieffermann, J.M. and Danzart, M. (2001). Sensory profiling of complex products: the front passenger compartment of cars. In: *4th Pangborn Sensory Science Symposium*, 22–26 July 2001, Dijon, France.

Blumenthal, D., Priez, A., Sieffermann, J.M. and Danzart, M. (2003). Relative impact of sensory and brand image attributes on consumers' preferences: a case study on cars. In:*5th Pangborn Sensory Science Symposium*, 20–24 July 2003, Boston, MA, USA.

Blumenthal, D., Priez, A., Sieffermann, J.M. and Danzart, M. (2004). How to use stepwise regression to measure the relative impact of sensory and brand image attributes on consumer's preferences. In: *7th Sensometrics Conference*, 27–30 July 2004, Davis, CA, USA.

Boivin, L. and Blumenthal, D. (2009). Identifying the perception differences according to the way of representation. In: *8th Pangborn Sensory Science Symposium*, 26–30 July 2009, Florence, Italie.

Dairou, V. and Sieffermann, J.M. (2002). A comparison of fourteen jams characterized by conventional profiling and a quick original method, the flash profile. In *Journal of Food Science*. Dallas: Institute of Food Technologists, pp. 826–834.

Dairou, V. Priez, A., Sieffermann, J.M. and Danzart, M. (2003). An original method to predict brake feel: a combination of design of experiments and sensory science. In: SAE Technical Paper 2003-01-0598.

Delarue, J. and Sieffermann, J.M. (2004). Sensory mapping using Flash profile. Comparison with a conventional descriptive method for the evaluation of the flavour of fruit dairy products. *Food Quality and Preference*, **15**(4), pp. 383–392.

Gower, J.C. (1975). Generalized procrustes analysis. *Psychometrika*, **40**(1), pp. 33–51.

Herbeth, N. and Blumenthal, D. (2007). Etude de l'acceptabilité du conducteur de lois de direction innovantes (Steer-By-Wire). In: *14th conference on Vehicles Dynamics – SIA*, 20–21 June 2007, Lyon, France.

Herbeth, N. and Blumenthal, D. (2010). Investigating not only sensations but also emotions to increase visual comfort of car seats. In: *International Conference on Kansei Engineering and Emotion Research (Keer)*, 2–4 March 2010, Paris, France. pp. 88–97.

Herbeth, N. (2007). Comparing customers and test drivers' vocabulary: the case of gearshift sensations. In: *7th Pangborn Sensory Science Symposium*, 12–16 August 2007, Minneapolis, MN, USA.

Herbeth, N., Meillon, S. and Blumenthal, D. (2007). Customer test on innovative products: the steering system. In: *7th Pangborn Sensory Science Symposium*, 12–16 August 2007, Minneapolis, MN, USA.

Petiot, J.F., Poirson, E., Aliouat, E., Boivin, L. and Blumenthal, D. (2010). Interactive user tests to enhance innovation. Application to car dashboard design. In: *International Conference on Kansei Engineering and Emotion Research (Keer)*, 2–4 March 2010, Paris, France. pp. 2021–2030.

Taréa, S., Sieffermann, J.M. and Cuvelier, G. (2003). Use of Flash Profile to build a product set for more advanced sensory study. Application to the study of particle suspensions. In *IUFoST XIIth World Congress of Food Science and Technology*. Chicago, IL.

Testing consumer insight using mobile devices: a case study of a sensory consumer journey conducted with the help of mobile research

21

D. Lutsch, R. Möslein, M. Strack, S. Kunze
isi GmbH & Co. KG, Niedersachsen, Germany

21.1 Mobile research: status quo

Opinion and market research had started with in-street polls about two centuries ago, and a century later continued with advertising testing and consumer focus groups (e.g. Dichter, 1947). Also, diary studies appeared in the 1940s. Today, these older techniques are still in use, such as in-street recruitment for lab tastings. It seems that old and modern techniques coexist and coevolve.

In the last five decades, advances in technology have allowed computer-assisted telephone interviews, as well as computer-assisted web interviews, that have enabled more complex interview flows and a more economic data collection. Today the "CA" in acronyms as CATI or CAWI is taken for granted; the current discussions focus more on the participant's data entry devices: the personal computer, the tablet or even smaller smartphones. Although most research is still web based (92%) and only a minor part is done on mobile devices (7%), there is a hype about mobile assessment tools in market research that has just emerged in the past few years. Smartphones progressively substitute personal computers in market research because they are smaller and thus mobile, always at hand, yet nevertheless almost as capable as a PC in accessing the Internet, which is the main platform for data collection (Macer and Wilson, 2011).

According to Microsoft, since 2013 more tablets than PCs will have been sold. Almost every fourth mobile phone that is used is a smartphone. The comScore MobiLens study from October 2012 shows that 48% of German, 51% of French, 63% of Spanish and 62% of British inhabitants (13+ years) use smartphones. The OurMobilePlanet study (Google, 2013) shows that more than 60% of smartphone users daily go online with it. Smartphone users even tend to access the Internet only via mobile devices, e.g. in the UK already 25% of the users use only their mobile devices when browsing the Internet. The trend is still growing (Quinn, 2013). It is estimated that in 2015 in Western Europe, 95% of all mobile phones sold will be smartphones (Schöttelndreier and Helferich, 2012).

Multiple conferences are held around the globe that focus on mobile research, e.g. the "Market Research in the Mobile World" conference took place in Amsterdam in 2012, 2013 in Kuala Lumpur (Malaysia), as well as in Minneapolis (USA), and in October in London (UK). Public opinion research follows with a session "Facing the Challenges of Data Collection via Mobile Internet" at the World Congress of Sociology in Yokohama, Japan, 2014.

Scholars discuss research technologies for multi-device surveys, research practices that take advantage of the new opportunities ("ecological momentary assessment", Kuntsche and Labhart, 2013) and potential changes to the global market research space. According to the Market Research Software Survey Report (2011), self-completion questionnaires via mobile devices were the fastest developing method. This is probably due to the fact that personal or telephone interviews are getting more and more obsolescent, whereas the method "online surveys via PC" is more or less saturated. As the proportion of standard PCs in panels will continuously decline while the number of tablets and smartphones increases, standard in-home-use-tests (IHUT) become more difficult. The question whether to do mobile research or not is already answered by the consumers, since in the US and Europe between 10% and 15% of supposedly "online" surveys are already accessed via mobile devices. In the future, studies suitable only for standard PCs might be addressed solely to older target groups who still use PCs; surveys focusing more at a younger target group need to adapt in such a way that they can be accessed from smaller-sized mobile devices (Macon and Wilson, 2011).

21.2 Mobile sensory research: a new mobile research method

The current mobile research market is reduced to typical "online" studies and "new surveys" such as "micro surveys," "gamification," "in-the-moment surveys" and "passive measurement" in which the respondents allow their smartphone to monitor them (e.g. by Global Positioning System (GPS) tracking), and digital qualitative, as well as ethnography, studies (Johnson, 2012). People involved in sensory analysis might now ask themselves whether they should add sensory research as a new segment to the mobile research market. To answer this question, it is necessary to understand the advantages of mobile sensory research in comparison to standard sensory testing, as well as its limitations.

21.2.1 Advantages of mobile sensory research

Sensory evaluation is defined as "a scientific discipline used to evoke, measure, analyse and interpret those responses to products that are perceived by the senses of sight, smell, touch, taste and hearing" (Stone and Sidel, 1993). It applies principles of experimental design and statistical analysis to evaluate consumer products. Its methods are divided into two sub-sections (Scharf, 2000): analytical methods (descriptive and discriminative testing) and consumer testing (preference tests, acceptance tests and evaluation tests).

Nowadays, standard consumer testing involves testing of products preferably in sensory labs or in standardized central locations (CLT). In these test set-ups, consumers test products and answer questions, either on paper/pencil questionnaires or by filling in computer supported online questionnaires. Since test facilities have a limited number of test booths, large sample sizes require testing over a longer period of time. Due to their high level of standardization, CLTs have a high internal validity. However, their external validity is relatively low, because the products are not tested in their natural environment and consumers of few locations are interviewed (limited representativeness). Consequently, the results cannot be generalized to the total population offhand (Scharf, 2000). Using mobile devices for CLTs does not really deliver any additional benefit, especially not in comparison to data collection via PC. However, they might be a good alternative for surveys in test kitchens (e.g. evaluating the process of preparation of a meal) or wet rooms, i.e. for locations where it is difficult or impossible to use fixed installed PCs. Apart from that, the usage of mobile devices in standard CLTs does not deliver real new benefit, which is why in that context we would not speak of a new research method.

The story is quite different, however, in the context of IHUTs, which are getting more and more popular since producers are increasingly also interested in high external validity. In IHUTs, consumers test products as they normally use them, and answer a paper/pencil questionnaire afterwards or go to their PC and fill in an online questionnaire. For this type of research, switching to mobile devices does offer clear advantages compared to data collection via paper/pencil or PCs that are linked to different problems. Paper/pencil questionnaires need to be posted back to the research agency and coded in order to be analysed. Online questionnaires that are answered on the respondent's PC often go hand in hand with a place and time break (i.e. products are e.g. tested in the kitchen and the questionnaire is answered in the workroom or living room). This results in such problems as biased data due to distorted memory, especially for sensory and affective experiences.

Here is where mobile research comes into play and significantly improves the research method: IHUTs supported by online surveys run on mobile devices do not only enable a much higher external validity, which is of high interest to better estimate a product's market potential, but also enable instant results. Especially when it is crucial to access "in-the-moment" experiences and emotions, mobile devices provide a window into people's lives and generate responses in the specific context and moment, sharing instantaneous impressions, due to the fact that smartphones are always at hand. Mobile users naturally like to comment and share opinions. Hence, evaluating product experiences via mobile devices can be easily integrated into their daily routine without much effort. Furthermore, smartphones are often used for entertainment purposes, such as listening to music, watching TV, surfing the Internet, playing games, or reading a book/newspaper. Hence, using a smartphone is more often associated with fun than is the case for PCs, which are cognitively related to work and task orientation. This could be one reason why answering questionnaires on a smartphone is still more entertaining than answering questions while sitting in front of a PC, and thus better meets the wish for gamification[1]. In addition to "in-the-moment" data collection, mobile sensory research also offers the opportunity to assess products in different situations. For example, it enables the evaluation of product usages/applications, especially of products that are used in different occasions or rooms where normally no PC is available.

Moreover, survey apps offer various opportunities regarding monitoring, data collection, etc.: alerts that remind to access the study, geo-tracking via GPS to activate surveys only in specified regions (e.g. in a supermarket), dynamical questionnaire flow, answer categories adopted to the participant's answer behaviour and media input (e.g. taking a picture/video of the consumption situation, using a barcode scanner). This makes participating in such surveys less fatiguing and increases the participant's commitment, resulting in better data quality. In addition, survey apps offer great potential for better field control, data monitoring and efficient interaction. Since those apps automatically synchronize with the company's online server, all data are securely stored, thus do not require manual data entry (saves time and money) and even enable interim results before the fieldwork is over (Schöttelndreier and Helferich, 2012).

The best way to recruit participants for mobile research is to involve a pre-recruited panel. Online panels can provide a deep profiling of the panel members (e.g. socio demographics, usage behaviour), good responsiveness and large sample sizes. A big advantage of mobile research involving an online panel is that it is not limited to cities offering sensory facilities, but respondents can be recruited from all over the country – even, if desired, worldwide. Moreover, mobile research is not dependent on office-hours. Hence, reaching the working population becomes much easier. Experts even assume that smartphones will be the only way in the future of reaching the difficult target group of young male professionals who are always on the move and thus have no time or readiness to participate in sensory studies. Since their smartphone is always with them, not only on the job but also at home, at work, at restaurants, when commuting or waiting, taking part in a survey is easier, less time consuming and with less effort involved. Also incentives for online recruited respondents are lower than for respondents, who need to come to a specific facility for a specific time, which affects costs.

21.2.2 Concerns and limitations of mobile sensory research

But as for any other trend, there are not only pros but also cons to mobile sensory research. The development of an own-research app is first of all connected to high costs. Like all innovations in their early adoption phases, apps too require new competencies. Since the knowledge on how to implement new technologies is mostly non-existent in the early phase, services need to be bought in either as individual or off-the-shelf solutions. This saves costs but reduces freedom and independence; in-home solutions secure independence but are mostly rather time and cost intensive. Mobile research has further critical limitations, such as limitations on the number of

---

[1] Gamification is the use of game-thinking and game mechanics in non-game contexts in order to engage users and solve problems. Gamification is used in applications and processes such as surveys done in market research to improve e.g. user engagement and data quality.

questions. To avoid overstressing participants, surveys have to be shorter than standard questionnaires, because only one or two questions per page can be displayed. Other limitations arise because of the smaller screens of smartphones: pictures can be difficult to detect, and text can be too small to be deciphered. Hence, it is not enough to just use a standard layout of an online survey for mobile research on smartphones – individual solutions have to be developed. In general, tablets are less affected by these problems. Therefore, intelligent mobile research solutions need to identify from which device (smartphone, tablet, laptop, etc.) a survey is being accessed and display the questionnaire in an appropriate layout. Participants can also face technical issues e.g. low battery power, bad Internet connections (if research is done via online questionnaire and not app) and difficulties when answering open-ended questions.

Other concerns among researchers are regarding privacy, especially when using passive data collection. Therefore, all big market research networks e.g. ESOMAR worldwide and BVM in Germany, are publishing privacy guidelines for mobile research. Also, some researchers criticize mobile research for not being representative, because mostly a younger target group can be reached. But in comparison to standard CLTs, respondents are recruited all across the country, and also the working population can be more easily reached via mobile research, which make it comparatively more representative. The increasing market penetration of smart phones further improves the representativeness of "mobile online panels" each year. Generally speaking, we can summarize by saying that the mobile target group is representative of the segment with highest spending power, and for early technology adopters.

Taking all previously gathered information into consideration, we conclude that mobile research delivers benefits additional to those of "traditional sensory research," such as high external and ecological validity as well as rapid data delivery (see Table 21.1). That is why we believe it to be worth being considered as new approach in the sensory community.

21.3 Case study: a sensory consumer journey conducted with the help of mobile research

21.3.1 Reasons for considering a consumer journey

Consuming a product often involves an enduring consumer journey. For example, promotion of a ready meal in a supermarket evokes a desire to consume. Hence, the meal is bought, and afterwards cooked at home – which in some instances might be quite troublesome. Eating the food might require special dishes and cutlery. The dining process itself may be accompanied by drinks and conversation. Thereafter, leftovers and packaging are to be disposed of, and dishes cleaned. Since all those elements can have an important effect on how the product is evaluated overall, they become an essential part of its consumer journey.

Consumer journeys are event sequences, which psychologists conceptualize in the form of scripts (Schank and Abelson, 1977). These scripts drive consumer

Table 21.1 IHUT via mobile devices vs CLT

IHUT via mobile devices		CLT	
+	High external *validity*, "in-the-moment" surveys	+	High *internal validity*, standardized conditions
+/–	High geographical *representativeness/* (mostly rather representative for target groups with high income/spending power)	–	Limited geographical *representativeness* (CLT often in one or few cities)
+	*Easy to reach target groups*: younger/working population	+	*Easy to reach target groups:* older target groups
–	*Small screen size:* • limited numbers of questions • high complexity of stimuli not possible	+	*Large screen size:* • large numbers of questions possible • high complexity of stimuli possible
+	*Data entry:* scales via touch-screen; data entry also via taking pictures/videos, voice recording, scanning barcodes, GPS tracking; text entry rather difficult	+	*Data entry:* keyboard allows easy answering of open-ended answers; usage of scales
+/–	*Costs:* • low costs for recruitment and incentives, less manpower intensive, no facility costs • more test products needed for over-recruitment	+/–	*Costs:* • less costs for test set-up (online questionnaire, no product shipment) • higher recruitment costs/incentives • facility costs
+	*Time:* less time intensive, more rapid data delivery (depending on sample size)	–	*Time:* often longer field work; less instant results (depending on sample size)

expectancies and are the comparison level for the individual product experiences. They therefore determine the overall product liking. For example, although the ready meal tasted good, if its packaging causes inconvenience customers become angry and might look for an alternative.

Since, especially for consumer journeys, it is important to measure product liking at the right place and time, we have chosen the consumer journey of a body care product as case study because it nicely offers the opportunity to assess the strengths and weaknesses of mobile sensory research.

21.3.2 A consumer journey that focuses on the perception and evaluation of sensorial attributes

Generally, consumer journeys describe successive phases of a product experience from its first product contact, its purchase, usage or consumption, to its disposal. In all the different phases, expectancies arise that have a major impact on product perception and evaluation in the succeeding phases. Currently, there are various models restricted to the consumer purchase decision process, such as the attention-interest-desire-action (AIDA) model but omitting the consumption phase which is crucial for the estimation of consumer product potential.

Another perspective offers psychology with models stressing the phase specificity of drivers of behaviour: one is the "Rubicon – model of motivation" (Heckhausen and Gollwitzer, 1987). It involves four phases: the first two involve the pre-decisional phase, in which different options are assessed, and the implementation of the chosen option is planned. The third phase is referred to as the behavioural phase, in which the action plan is carried out. In the final phase, the actual target achievement is assessed, which might result in a new assessment phase. Generally, this model is helpful to understand the basic psychological processes of goal formation and action execution, but leaves out the important aspects of product usage and its sensorial experience.

The first "complete" consumer journey models were developed for service engineering (e.g. the service blueprint, Shostack, 1984) to assess and optimize all consumer touch points to consequently increase customer satisfaction. Although already being very close to the idea of a consumer journey, it still does not take physical products and their sensorial characteristics into account. In different phases different sensorial modes may predominate. For instance, at the point of sale, a product is predominantly optically perceived, mediated by its package design. When consumers pick up a product, haptical cues are perceived, initiated by weight, consistency and surface of the product. That this "pre-consuming" information is important was described by Piqueras-Fiszman and Spence (2012), who showed a transference of the product's packaging on the oral-somatosensory texture perception of a foodstuff. Unpacked products allow a smell or a taste sample at the point of sale and create expectations that later on influence the product experience. Chemoreception senses often only become involved when the package is opened, and the final preparation of the consumption starts. The chemical information sent by the product at this point of contact in the journey can still be different to that in the later phase of actual consumption. During the actual product consumption, taste and smell might predominate, but the eye (and some mechanoreceptors) must be pleased as well to result in a high overall liking. In the post-consumption phases, satiety and after-taste play a role, but optical and olfactory messages sent by the leftovers can become critical drivers of the overall liking, too.

So, taking all this discussed information into consideration, how would then a sensory consumer journey look?

21.3.3 How to assess time-critical experiences during a consumer journey

A typical consumer journey is related to a physical journey through time and space: the inspiration to buy a specific product may occur in front of a TV at home, or during a talk with friends in the streets. The purchase is conducted at a point of sale in a shop, or online via a PC or a mobile device at home. The preparation and the consumption phases may be spread between private, work and diverse leisure places ("mobile"). Assessment authenticity is maximized if the time span of memorization between experience and report is minimized. To follow such a journey in space and time, techniques of diary studies were developed, which usually demand a self-administering procedure, ease of wording and a reminding function. However, conveniently transportable paper/pencil solutions afford discipline (e.g., a funny p&p drink-tracker card http://rethinkingdrinking.niaaa.nih.gov/toolsresources/DrinkingTracker Cards.asp). Consequently, they suffer from low response rates due to the intention–behaviour gap. Participants generally consent to participate, but forget their commitment at the time of action (e.g. Bolger *et al.*, 2003 called it "honest forgetfulness"). A convenient solution to this problem is offered by modern mobile devices, which include mostly acoustical push technologies ("beeping") to invite and remind the participants at the appropriate time. So-called electronic diary studies often equipped the participants with special tools, mostly mere signallers or pagers but, since the 1990s, also handheld computers with special software (e.g. Party *et al.*, 1992). With the ongoing prevalence of mobile phones they have taken over the reminder function, either by receiving an SMS message or giving acoustic signals.

21.3.4 Material and methods

The set-up of a consumer journey usually divides into *two steps*: one qualitative preparatory step, and one quantitative and confirmatory consumer journey analysis.

In order to nicely illustrate the effect of different product touch points during a consumer journey involving different sensorial stimulations, it was decided to have a hair wax of *AXE as research object*. AXE launched four hair wax variants in Germany in 2012. Hair wax offers diverse sensorial experiences: first, consumers feel the plastic box when opening it, then smell the fragrance of the hair wax. But the most interesting characteristic of hair wax is the constant change of its visual, olfactory and haptic character between the in-box, in-hand and in-hair phases. Within the box, the wax looks quite opaque and feels quite viscous. Also, its odour seems rather intense. When putting it on the fingertip, it becomes transparent, and when spread in the hair, invisible, and the hair itself surprisingly flexible. Also, the odour tends to fade more and more when applied to the hair.

The *participants* of each research step were screened and recruited from a German online panel (http://www.myonlinepanel.de/) according to product usage and Internet access, including the operating system of their smartphone. To avoid differences across operating systems, ownership and daily usage of an android smartphone were the screening criteria for the quantitative study. Furthermore, only men in the age range of 18–35 (typical adult AXE target group) were recruited (Franke, 2013).

For the *first qualitative study*, which aimed at the phase segmentation of a behavioural consumer journey of hair wax, we recruited five young men aged 18–25 years (Freter, 2013) using wax or gel to style their hair (heavy users). In a conventional half-standardized 20 min interview we explored their behavioural contact points with the product. As a result, several contact points segmenting the consumer journey could be derived.

For the *quantitative study*, 30 young men aged 18–35 (mean 26 years) were recruited, all being heavy users of hair wax. Overall, a good brand distribution was strived for: users of Swiss O Par, got2b, L'Oréal and private label brands (e.g. REWE, Rossmann). Each respondent evaluated one AXE product over a longer period of time: AXE Spiked Up ($n = 16$) and AXE Smart Look ($n = 13$). Out of the 30 respondents, there was one drop-out, who could not be motivated to carry on with the survey. Fieldwork took place in April 2013. Each participant evaluated one test product at six different contact points during one day, and answered various questions on their mobile phone.

21.3.5 Results

21.3.5.1 Results of the consumer journey

The qualitative preparatory step provided a segmentation of the consumer journey in different touch points: (1) the *Inspiration Phase* sums up the information funnel from first contacts with the product category and the brand till the building of at least weak purchase intention; (2) the *Purchase Phase* sums up experiences at the point of sale purchase intention; (3) the *Application Phase* is the most interesting, concerning sensory product experiences – it is divided into four sub-phases (see Fig. 21.1); (4) in the *Wearing Phase* the main product purpose has to be fulfilled; (5) A *Restyling Phase* is facultative; (6) a shower to clean the hair is obligatory. Nevertheless, nearly half of the participants go to bed with the wax still in their hair.

Based on Fig. 21.1, we decided to quantitatively assess the sensory consumer journey at seven different touch points:

T0: Concept (with closed can) → *T1: Wax in the open can* → *T2. Wax at one finger* → *T3: Rubbed at hands* → *T4: Just applied to the hair* → *T5: Wearing until midday* → *T6: Wearing up to night-time.*

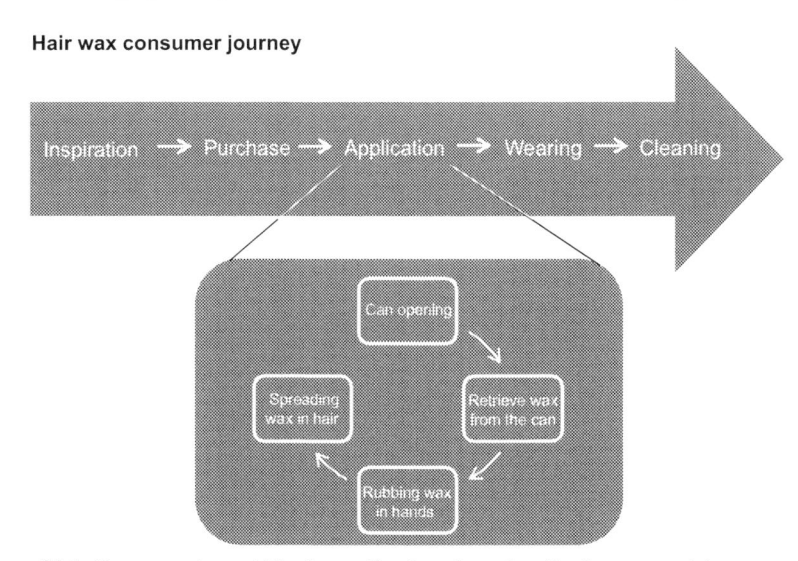

Figure 21.1 Contact points within the application phase (qualitative pre-study)

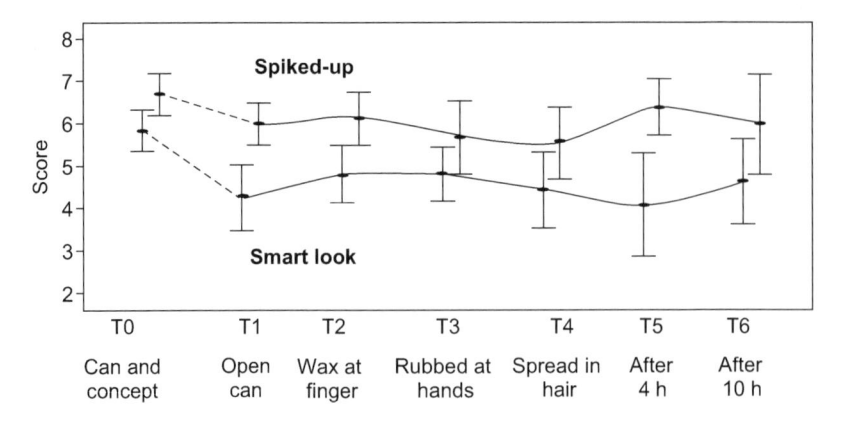

Figure 21.2 Overall liking across the sensory consumer journey of two hair wax samples.

Whereas the first half of the usage chain could also be studied in a lab (CLT), the second half, the wearing journey, could only be studied with the help of a mobile monitoring device.

The results of the quantitative phase show the overall liking and intensity ratings of specific sensory dimensions at different touch points. Figure 21.2 shows the overall linking of the two AXE variants aggregated over the 29 participants.

Evidently the variant Smart Look dissatisfied some of the users, specifically at the concept perception phase T0, as well as at the first sensory contact. Smart Look displays an unexpected blue glittering wax in the can. After four hours of wearing, the liking of both variants diverged again.

To explain the overall liking by the liking of sensorial attributes, a moderated regression was conducted including a linear predictor indicating the phase of the journey (beginning with T1). Table 21.2 shows the numerical and Fig. 21.3 the graphical results: modelling the phases linearly, eight significant effects on overall liking emerged; two of them are phase moderations of drivers.

The strongest driver of overall liking was the appearance liking (Table 21.2); its driver strength was moderated by the samples. The second driver was the liking of consistency, which was moderated both by the samples and by the phase. As Table 21.2 shows, the later the phase the less important becomes the driver of consistency liking. Therefore, the likings of the wax consistency in the can and on the fingers are more important drivers than the liking of the consistency in the hair during the day. The third driver, the odour liking, was moderated in a three-way interaction: odour turned up to be a stronger driver for Spiked Up in later phases, but for Smart Look in early phases of the journey.

To sum up, the benefit of the Consumer Journey approach is given by the phase specificity of the importance of consistency liking in early phases, and of odour liking in early phases for Smart Look but in late phases for Spiked Up. Therefore, the product optimizations cannot only be derived for specific sensorial dimensions in

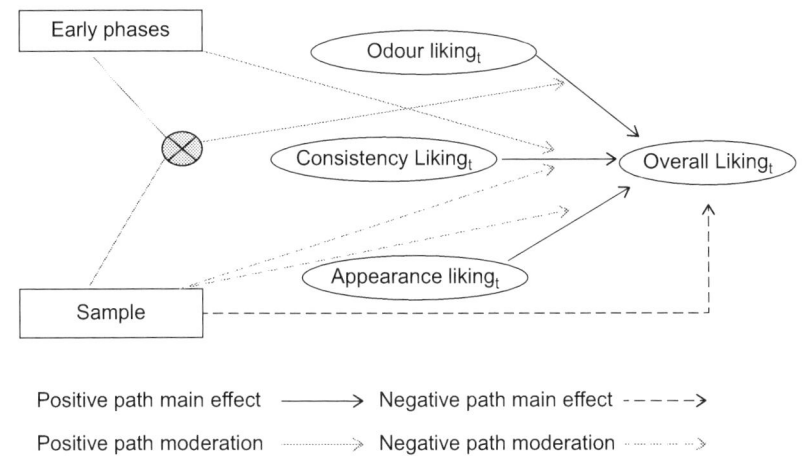

Figure 21.3 Phase specificity of sensorial drivers (moderated regression per GLM).

Table 21.2 **The effect of sensorial attributes on overall liking over product touch points**

General Linear Model (GLM) parameters ($R^2 = 0.78$)	Raw coefficient
Constant	5.445
Odour liking	0.277**
Odour liking × Sample × *Phase*	0.180*
Consistency liking	0.646**
Consistency liking × Sample	−0.338**
Consistency liking × *Phase*	−0.150*
Appearance liking	0.742**
Appearance liking × Sample	0.284**
Sample (coded as −1, +1)	0.159*

*$p < 0.05$.
**$p < 0.01$.

general, but also for specific phases and contact points, when the dimensions are of highest importance.

21.3.5.2 Mobile sensory research: method assessment

The previously presented and discussed case study of a consumer journey conducted with the help of mobile research was not only carried out to better understand consumer journeys of a sensorial perceivable product, but also to gather insights on the methodology of "mobile research." In the course of the survey we as researchers, but also our respondents, gained experiences that were positive but also, to some degree, negative: the recruitment of the respondents proved to be more challenging because we did not only need to consider the "standard recruitment" criteria but also had to make sure that our respondents were users of the mobile Internet possessing a relevant device. We also experienced a so-called media cut because the respondents were recruited via an online questionnaire and had then to download the app to their mobile phone, which caused a drop-out of some already recruited respondents. Also, the allowance of access rights to their mobile phone (e.g. for alert systems) might have caused some further drop-outs. However, now that our "mobile panel" is growing, the recruitment of respondents for successive surveys is even easier because the respondents only need to be contacted via their app and can be directly invited to surveys, which makes the recruitment process in the long run more rapid.

Further problems were reported from our respondents during fieldwork. During the phases T2–T4, in which the respondents had direct contact with the hair wax, the participants were reluctant to use their touch pad because of wet and sticky fingers. Also, some participants were annoyed by the alert system that reminded them to fill in their questionnaire.

Furthermore, the questionnaire had to be designed so as to meet the requirements of a small screen. Hence, we decided to do without long statement batteries.

Nevertheless, taking everything into consideration, the method proved to be more beneficial than disadvantageous: since the product evaluation happened in exactly the moment of product contact, we were able to gather more "honest" and less distorted emotions and opinions. Since the product experience was still present in the participant's mind, it was easier for the respondents to express their opinions, and minor details could also be gathered.

The participants did not only report that the survey was "more fun" than standard surveys, which resulted in a high commitment, but also the survey itself – in comparison to an online questionnaire on a PC – was more rapid and flexible because the PC did not have to be switched on and the questionnaire could even be answered while out and about. This aspect was especially helpful for the evaluation of lastingness at different points of time and place. We also had the feeling that we could gather insights that could not have been gained in a standard test: e.g. most participants went to bed with the wax still in their hair; hence, some of them complained about residuals on their pillow when they woke up the next day.

Overall, we can conclude that next to recruitment and questionnaire answering, the data collection was also more rapid than paper/pencil questionnaires because all data were automatically stored on our server.

21.4 Summary and discussion

A journey assessment is especially relevant for products that are consumed over a longer period of time and for which a change of their sensorial characteristics occurs and assessments (answering a short questionnaire) that need to be done not only at different points of time but also at different locations. Since mobile research offers the opportunity of a flexible product assessment with regard to time and place, a journey assessment was chosen as case study to better understand the *pros and cons of mobile research*.

By implementing mobile research (e.g. with the help of mobile Internet or a sensory app) during such a consumer journey, *ecologic validity* is increased because the information is gathered in the product's natural environment at the right time and place and offers thus the possibility to catch the moment. This advantage, of course, goes hand in hand with the disadvantage of a *lower internal validity*, which can be better guaranteed in a CLT in which products are tested under standardized conditions.

Furthermore, for some *products* it only makes sense to be tested at the *consumer's home* (e.g. shampoo, lotion, home appliances) or a *longer testing phase* (e.g. anti-ageing lotion, baby soothers) is required. For the assessment of those products, the flexibility of mobile research offers a clear advantage.

Also, the *data collection* offers vast opportunities and advantages: nowadays mobile phones/smartphones are always at hand and allow thus a *more flexible, more effortless and more rapid data collection* because the respondents do not need to go to their PC, switch it on, enter the Internet and answer an online questionnaire, but can access their mobile questionnaire with a single click. Another disadvantage of completing a questionnaire e.g. in the evening via a PC is the so-called *time and place break*: i.e. respondents have to call up their product experience from their memory, which can lead to a distortion of the memory contents and "less honest" emotions. Also paper/pencil questionnaires are no better alternative: of course, here we do not run the risk of time and place breaks, but the questionnaires do need to be posted back to the researcher, and *coded and entered* into a statistics program, whereas the data entered into a mobile device are *instantly securely stored* on an online server. Since nowadays all smartphones are equipped with a camera, *information* can be *quickly captured* merely by taking a picture, instead of laborious descriptions via text entry. Also, GPS tracking offers easy and rapid information gathering. Overall, data entry via a smartphone/mobile device is often perceived as more *entertaining* and thus less *fatiguing*, which can result in a *higher commitment* of the respondent and thus more valid data. Nevertheless, one has to keep in mind that standard questionnaires that are normally used in CLTs need to be adjusted to the mobile phones' *small screens* and also be *reduced in length*.

Moreover, the longer a survey is, the more *drop-outs* must be expected. Hence, a sufficiently high over-recruitment should be considered. Also, a specially programmed *smartphone app* that reminds respondents to participate via push messages can reduce drop-outs.

Also, for *difficult to reach target groups*, mobile research can be a solution: e.g. those respondents that cannot be easily invited to come to a test facility, as is the case for pets (e.g. dogs and cats), babies, or professionals with long working hours. Additionally, in contrast to a CLT, a higher geographical *representativeness* can be reached. However, the other side of the medal is that at the moment it is mostly younger consumers with high income/spending power who can be reached via mobile research, whereas older consumers can hardly be recruited.

To sum up whether mobile research offers more pros or cons, a simple answer cannot be given. In fact, it highly depends on the research objective (e.g. high external validity vs high internal validity), the test products themselves and the target group.

21.5 Conclusion

To draw a final conclusion, we would like to come back to the title of this book – rapid sensory profiling techniques. So is mobile research a rapid method?

To answer this question, we first need to understand that mobile research is not a sensory method as such, but rather an auxiliary that can be applied to facilitate existing sensory methods. Mobile devices can be used in standard CLTs and IHUTs – two possible ways of gathering information about products. Since CLTs – as the name already implies – take place at one or few location(s), mobile devices are more related to difficulties (e.g. small screens, more difficult data entry) than advantages and can thus not really be recommended.

However, when used in the context of surveys that are either carried out at the consumer's home, or even on the move, mobile research can offer vast possibilities as already described (e.g. data entry at the right time and place, possibility to catch the moment, GPS tracking, taking pictures, quick access to the questionnaire, etc.) and can thus be seen as an innovative approach to gathering information with high external validity in a rapid way.

So under which circumstances is mobile research rapid?

- *Recruitment*: as previously described, building up a panel of regular smartphone users has proved to be more difficult and time consuming than standard recruitment. However, when such a panel finally exists, *recruitment* of participants for further surveys is very quick, because the panel members only need to be invited by their app.
- *Data collection from a participant's point of view*: also from a participant's point of view, mobile research compared to the participation in a CLT or an IHUT with a standard online questionnaire is more rapid. When comparing mobile research with a standard CLT, mobile research is more rapid because the participants do not need to go to the test facility but can answer the questions more or less everywhere they are using/consuming the test product. When comparing an IHUT supported by a mobile app with a standard online questionnaire, mobile research proves to be more rapid and convenient because the app-based questionnaire can be answered everywhere, at home or even outdoors, and a long switching-on phase of the PC can be avoided.
- *Data collection from a researcher's point of view*: the data collection via mobile research, compared with standard paper/pencil questionnaires, is more rapid because tedious data coding and entry afterwards is no longer necessary – the data are already safely stored on an online server.
- *Analysis and reporting*: regarding data analysis and reporting, mobile research is not more rapid compared to other data collection methods. However, real-time data make it possible to rapidly get intermediate results.

To sum up, mobile research can under certain circumstances offer more rapid results (recruitment and data collection) and enrich standard consumer research but is not per se the right method for any research question.

References

Bolger, N., Davis, A. and Rafaeli, E. (2003). Diary methods: capture live as it is lived. *Annual Review of Psychology,* **54**, 579–616.

Dichter, E. (1947). Psychology in market research. *Harvard Business Review,* **25**, 432–443.

Franke, A. (2013). Eine sensorische Consumer Journey. Eine quantitative Studie von phasenspezifisch ausgelösten Motiven [A sensory consumer journey. A quantitative study of phase specific triggered motives]. Unpublished Bachelor Thesis. Georg-August-University of Göttingen, Germany.

Freter, S. (2013). Momente der Wahrheit: Eine qualitative Analyse der Consumer Journey am Produktbeispiel Haarwachs [Moments of truth – a qualitative examination of the consumer journey with hair wax]. Unpublished Master Thesis. University of Applied Science Nordhausen, Germany.

Google (2013). OurMobilePlanete study. http://services.google.com/fh/files/misc/omp-2013-de-en.pdf; http://services.google.com/fh/files/misc/omp-2013-uk-en.pdf, and other country reports.

Heckhausen, H. and Gollwitzer, P.M. (1987): Thought contents and cognitive functioning in motivational versus volitional states of mind. *Motivation and Emotion,* **11**, 101–120.

Johnson, A.J. (2012). Putting mobile research in context. Mobile technology is changing consumers' lives and how we conduct market research. *Research World,* **34**, 50–51.

Kuntsche, E. and Labhart, F. (2013). ICAT: development of an internet-based data collection method for ecological momentary assessment using personal cell phones. *European Journal of Psychological Assessment,* **29**, 140–148.

Macer, T. and Wilson, S. (2011). *Globalpark Annual Market Research Software Survey 2010.* http://www.meaning.uk.com/resources/reports/2010-Globalpark-MR-software-survey.pdf.

Paty, J.A., Kassel, J.D. and Shiffman, S. (1992). The importance of assessing base rates for clinical studies: an example of stimulus control of smoking. In: M. DeVries (Ed.): *The Experience of Psychopathology.* Cambridge University Press: Cambridge, pp. 347–352.

Piqueras-Fiszman, B. and Spence, C. (2012). The influence of the feel of product packaging on the perception of the oral-somatosensory texture of food. *Food Quality and Preference,* **26**, 67–73.

Quinn, S. (2013). Optimizing mobile research: How we can realise its true potential? *OnDevice Research Newsletter,* 3 June 2013, http://ondeviceresearch.com/blog/optimising-mobile-research:-how-we-can-realise-its-true-potential.

Schank, R.C. and Abelson, R.P. (1977). Scripts. In: R.C. Schank and R.P. Abelson (Eds.): *Scripts, Plans, Goals and Understanding. An Inquiry into Human Knowledge Structures.* New Jersey: Lawrence Erlbaum Associates, Inc, Chapter 3, pp. 207–233.

Schötteldreier and Helferich, (2012). Mobile Research: Konkurrent oder Ergänzung? Bereicherung gängiger Erhebungsmethoden durch neue Techniken. *Planung & Analyse,* 1/2012, 40–43.

Shostack, G.L. (1984). Designing services that deliver. *Harvard Business Review,* **62**, 133–139.

Stone, H. and Sidel, J.L. (1993). *Sensory Evaluation Practices.* 2nd edn. Academic Press: San Diego.

Part Four

Applications in sensory testing with specific populations and methodological consequences

Sensory testing in new product development: working with children

S. Nicklaus

Centre des Sciences du Gout et de l'Alimentation CNRS, INRA and Université de Bourgogne, Dijon, France

22.1 Introduction

Sensory evaluation typically measures several aspects of products: quality, intensity, temporality and hedonic value. Studying all these aspects with children may prove difficult, due to their incomplete cognitive development, which may have practical implications. We will consider here verbal children, from ~2 years until early teenage, because older children have sufficiently developed cognitive abilities to behave as adults in sensory tests.

Children have a different approach to products from adults. Even teenagers have a different perception of product quality from their parents (Bech-Larsen and Jensen, 2011). Children's basic sensory perceptions are likely to differ from those of adults (Ganchrow and Mennella, 2003; Nicklaus et al., 2005), which results in different perceptions of food products and, ultimately, in different hedonic evaluations and different optimal product formulation (Hough et al., 1997). Diminished sensitivity may be specifically true for boys when simple food stimuli are concerned (James et al., 1997), but not in more complex matrices (James et al., 1999).

Children's cognitive abilities are not as developed as those of adults and they are in constant evolution, as has been well summarized elsewhere (ASTM, 2013; Guinard, 2001). This makes any definite conclusions about what type of testing they are able to conduct difficult to draw. Moreover, the lack of systematic comparisons with different methodologies conducted with children from various age groups also makes it difficult to delineate their actual possibilities. Briefly, it is possible to describe the main cognitive development stages of children as the following (Piaget and Inhelder, 2003):

- Pre-operational stage (from 1.5–2 to 6–7 years): beginning of the symbolic thinking (language, mental image, drawing, symbols), but the reasoning is still limited,
- Concrete operation stage (from 6–7 to 11–12 years): the reasoning concerning objects in terms of categories and relations develops, thoughts become logical, but operations are still limited to concrete objects,
- Symbolic operation stage (from 11–12 to 14–15 years): stage of hypothetico-deductive thinking; combinatory thinking develops and judgements can be reversed.

The limited reasoning abilities during the pre-operational stage may involve difficulties for children in performing the analytical tasks required during sensory evaluation.

Moreover, it is possible that the abstract nature of chemosensory stimuli (odours, tastes in aqueous solutions) may account for children's difficulties in performing sensory tasks, even those that are generally considered as "easy", such as discrimination tasks (Nicklaus and Monnery-Patris, 2003). It is conceivable that children's abilities to process chemosensory stimuli may be trained by sensory education, as was shown by some researchers, especially in children aged 7–8 years (Mustonen et al., 2009; Reverdy et al., 2010), although the impact of this training is not always obvious (Sune et al., 2002). It can certainly be improved by training prior to sensory testing for specific sensory tasks, as is usually done with adults.

22.2 Reasons for studying sensory aspects in children

It has been shown quite clearly that children's sensory preferences may differ from those of adults (e.g., Zandstra and de Graaf, 1998), and that growing from childhood into adulthood may alter preferences, in particular for sweet tastes (Desor and Beauchamp, 1987). Children's sensory perceptions may also differ from those of adults, especially when intensity scaling is used (Zandstra and de Graaf, 1998), and to a lesser extent when pair comparisons are used (Mennella et al., 2003). However, the extent of differences in terms of perceptions between children and adults, or between children from various age groups, is often inferred from hedonic judgements, making a "true" comparison of children's and adults' sensory abilities difficult. One cannot rule out the possibility that children differ more from adults in terms of affective responses than of perceptual abilities.

For these reasons, for products directly targeted at children, it seems to make sense to involve children in the product development process (Chambers, 2005), which may be achieved through sensory evaluation testing (Laing, 2003). There is no doubt that hedonic ratings may differ between children and adults, as mentioned previously, and thus that conducting hedonic evaluation with children is relevant; but whether children should be involved in descriptive sensory analysis of products is an open question, and some argue that such types of evaluation are more easily performed by adults (Guinard, 2001). There are not much data to speculate on the potential advantage(s) of conducting descriptive analysis with children; but the cost involved in organizing such tests may certainly be a limitation in the product development process. Perception or discrimination abilities of children have not been characterized as a whole, and when some differences exist (see below), they tend to show a lower discrimination ability between products in children than in adults, and a lower analytical ability, probably in relation to a lower attentional capacity. For these reasons, performing descriptive analysis with children may present a real challenge, and should be considered cautiously.

As underlined before, most of the published researches in the sensory field conducted with children aimed at evaluating preferences. This has increased knowledge about the impact of age and culture on taste preference (Lanfer et al., 2013). Some studies help in understanding the possibilities of running different types of hedonic ratings with children by conducting systematic comparisons of methods (Chen *et al.*, 1996; Knof *et al.*, 2011; Léon *et al.*, 1999; Liem *et al.*, 2004). Such studies are also useful in underlining the abilities of children to discriminate products in terms of their sensory properties (Cordelle *et al.*, 2004): in some instance, children have been shown to be able to discriminate apple juices differing in terms of sourness and sweetness (Brueckner *et al.*, 2007), or acidified milks different in terms of sourness (Kildegaard *et al.*, 2011a, b). On the other hand, some studies have reported a lower ability in children to discriminate hedonic differences among a set of products compared to adults (Cliff *et al.*, 1997; Guinard and Marty, 1997). Such studies tend to show that applying the same method to study children's and adult's perceptions and preferences may result in a different formulation of optimally liked food products for these two targets (Hough *et al.*, 1997). Some studies show that preference data in teenagers (11–16 years) may predict real choice of snack bars (Mielby *et al.*, 2012).

When preference mapping is applied to hedonic data from children, it is noteworthy that the sensory profile is generated by an adult panel (Arditti, 1997; Hough and Sánchez, 1998; Kühn and Thybo, 2001; Thybo *et al.*, 2004), except in one study where the sensory profile was generated by children (Baxter *et al.*, 2000).

22.3 How to organize sensory evaluation testing with children

This section will present how to organize sensory evaluation testing with children, by considering legal issues and practical recommendations.

22.3.1 Organizing sensory testing with children: legal issues

Children may demonstrate enthusiasm for participating in sensory research; however, they are not responsible for themselves and it is the responsibility of the researcher/test organizer to make sure that all legal steps have been taken in order to protect subjects participating in research, according to the principles established in the Declaration of Helsinki. Children are considered a vulnerable population, so any research conducted with children should have been approved by a regulatory institution such as Internal Review Board, Committee to Protect People, or Ethical Committee, according to specific national regulations. It may happen that in some countries, some ethical committees consider research in the sensory area as non-invasive, because usual stimuli are used (e.g. foods, toys, packaging), and as such, they therefore consider that this type of research does not necessitate formal approval. However, such committees should always be consulted before conducting any study with children. Moreover, parents should clearly be informed about the objectives and constraints of the research, and they should provide the researcher with a signed consent form. In addition, when children are old enough to express a personal judgement (generally after the age of 7, but it may be before if the task is easy to describe), they should be informed personally about the objectives and conduct of the research, and they should have the possibility to refuse to participate in a given study, even if their parents have approved their participation. When the products are to be ingested, as for testing with adults, children presenting allergy in particular for the products under scrutiny should not be included in the study.

22.3.2 Organizing sensory testing with children: practical recommendations

Several authors have formulated recommendations for conducting sensory evaluation with children (ASTM, 2013; Laing, 2003), especially concerning the organization of hedonic evaluation (Kimmel *et al.*, 1994; Léon and Ivent, 2001).

The youngest children may be impressed by unknown settings, so a moment devoted to context familiarization is necessary. The instructions describing the task to be performed should be simple, and the vocabulary should be adapted to children. The duration of the testing should be short, and aligned with children's concentration ability. Sensory sessions should not last more than 10–15 min with children under 6. The environment may also be modified from the usual neutral environment of sensory evaluation rooms towards more warmth, provided it is not too distracting. Making sure that each child has understood the study instructions requires time, and the younger the children, the more staff should be devoted to the organization of the testing. Moreover, children may be impressed by unknown persons. For instance, children aged 3–4 years have provided more discriminant hedonic evaluation of orange beverages varying in sweetness when they were interviewed by their mothers than when they were interviewed by the study experimenter (Popper and Kroll, 2005). Enjoyment of the sensory session may constitute a motivation for children; moreover, instructions aiming at encouraging children to compete with one another may increase their discriminating abilities (Liem and Zandstra, 2010).

22.4 Application of different sensory evaluation techniques to children of different ages

As previously noted, there are no systematic published evaluations or comparisons of which sensory evaluation techniques may be applied at each age. Table 22.1 presents an overview of the published studies that have reported results from sensory studies with children in different age groups. The following section comments this table, and highlights specific aspects from these studies, in terms of methodology or of results.

22.4.1 Discrimination tests

Discrimination tests, such as pair comparison or triangular tests, have been widely used in order to evaluate the sensory abilities of children. Discrimination tests in the olfactory domain have indicated a lower ability to discriminate odours in 2- to 4-year-old children compared to children over 5 years (Richman *et al.*, 1995). Discrimination

Table 22.1 **Summary of the principal analytical sensory evaluation tasks used in children from different age ranges (in years)**

Age range	2–4	5–6	7–12	12–15
Discrimination tests (pair comparisons, triangular test, tetrad tests)	Kimmel *et al.*, 1994; Liem *et al.*, 2004; Richman *et al.*, 1995	Kimmel *et al.*, 1994; Le *et al.*, 2007; Liem *et al.*, 2004; Richman *et al.*, 1995	Garcia *et al.*, 2012; James *et al.*, 1997; Kimmel *et al.*, 1994; Mennella *et al.*, 2003; Richman *et al.*, 1995	Alexy *et al.*, 2011; Richman *et al.*, 1995
Intensity ranking	Kimmel *et al.*, 1994	Kimmel *et al.*, 1994; Liem *et al.*, 2004	Bouhlal *et al.*, 2013; Kildegaard *et al.*, 2011; Kimmel *et al.*, 199)	
Intensity rating		Zandstra and de Graaf, 199)	Bouhlal *et al.*, 2013; Feeney *et al.*, 2014; James *et al.*, 2003; Lavin and Lawless, 1998; Mustonen *et al.*, 2009; Zandstra and de Graaf, 19)	Lavin and Lawless, 1998; Zandstra and de Graaf, 199)
Identification/ naming	Lumeng *et al.*, 2005; Monnery-Patris *et al.*, 2009	Lumeng *et al.*, 2005; Monnery-Patris *et al.*, 2009	de Wijk and Cain, 1994; Monnery-Patris *et al.*, 2009	de Wijk and Cain, 1994
Description/profiling		Rose *et al.*, 2004a	Baxter *et al.*, 1998; Baxter *et al.*, 2000; Mustonen *et al.*, 2009; Rose *et al.*, 2004a; Sune *et al.*, 2002	

tests have also been applied to the study of the discrimination of sweetness in solutions: 4-year-old children had trouble in differentiating the sweetness, whereas 5-year-old children succeeded better, although not as well as adults (Liem et al., 2004). In this study, even the youngest children were able to clearly express their preferences, so the lack of discrimination may be related not to the lack of sensitivity to the solutions, but rather to the type of questioning in relation to their limited cognitive abilities, and/or limited abilities to process abstract stimuli such as taste solutions. The preference task is better understood than the analytical task, perhaps because it is more intuitive and more global than the analytical task, and more motivating. The reproducibility of the hedonic pair comparison was compared in children from 4 to 10 years; it was very low at 4 or 5 years, and slightly better in older children (Léon et al., 1999). A comparison of triangular and tetrad tests with 6-to-11-year-old children showed that the tetrad test may be more powerful than the triangular test in identifying differences among products, even in children (Garcia et al., 2012). Such studies based on discrimination tests help to show, for instance, that 6-to-8-year-old children, as adults, may detect iron fortification in common foods (Le et al., 2007).

In sum, the lack of consistency in responses of children below the age of 5, even for simple discrimination tests such as pair comparisons when they are orientated in an analytical way, limits the use of discrimination tests for children below the age of 5 (Liem et al., 2004). The application of pair comparison to evaluate preference is less problematic, even with children aged 2–3 years (Kimmel et al., 1994).

22.4.2 Ranking

Ranking has been used a lot for hedonic evaluations with children, and probably less for describing characteristics of the products. Intensity ranking is problematic with children below the age of 6 (Kimmel et al., 1994). As for pair comparison, it was shown that 4-year-old children could not rank sweet solutions according to their perceived sweetness intensity, but were able to rank them according to their preferences (Liem et al., 2004). However, intensity ranking was successfully applied to the study of salt perception in usual foods (green beans, pasta) with children aged 8–11 years (Bouhlal et al., 2013), or to the study of the impact of sourness in fruit beverages with children aged 9–14 years (Kildegaard et al., 2011b). Children aged 9–10 performed less well than teenagers (14–16 years) (de Graaf and Zandstra, 1999).

22.4.3 Intensity rating

Children aged 8–9 were shown to be able to estimate the intensity of sweetness in a suitable way compared to adults, by comparing category scale, magnitude estimation and hand separation (James et al., 2003). A nine-point scale was successfully applied to the evaluation of the intensity of sweetness and creaminess in children aged 8–10 and 11–14 years, but the researchers used pair comparisons to evaluate perceptions in children aged 5–7 years (Lavin and Lawless, 1998). A five-point scale was used to describe perception of sweetness intensity throughout the age span from 6 years:

children were able to use the scale but those from the 6–12 year-group provided different concentration–intensity patterns from the older groups (Zandstra and de Graaf, 1998). More recently, generalized Labelled Magnitude Scales were applied to the evaluation of bitterness and sweetness with 7–13-year-old children, and only a few children had to be discarded from data analysis because of their lack of understanding of this type of scale (Feeney *et al.*, 2014).

22.4.4 Descriptive analysis

Because children below the age of 8 years have difficulties to use sensory tasks in an analytical way (whereas they may perform well with the same tasks with a hedonic orientation), it appears quite challenging to apply more classic sensory descriptive techniques to the study of children below the age of 8; and in fact there are very few reports of such studies in the literature. Above 8 years, children can be expected to know how to read and write fluently, which is often necessary for descriptive sensory analysis, although computerization may help to deliver only oral or visual (not written) instructions.

Conducting sensory descriptive analysis involves generating terms or attributes to describe products. When foods are concerned, these attributes may be related to the odours of the products. It has been shown that the ability to identify odour develops with age between 3 and 6 years (Lumeng *et al.*, 2005), and that in 8-year-old children, it is below that of adults (de Wijk and Cain, 1994). Moreover, within children aged 4–12 years, odour identification is impacted by verbal ability, which is higher in girls, and may account for the reported higher ability to identify odours in girls (Monnery-Patris *et al.*, 2009). For this reason, it may be useful to characterize verbal ability of children involved in sensory tests which require the use of verbal skills, such as identification or description.

There are very few published studies with children as trained assessors. In one study, children (7–11 years) have been asked to describe freely the sensory aspects of different types of bread (Mustonen *et al.*, 2009). This highlights the higher verbal fluency of older children, who individually named on average 14 terms against eight terms for the younger ones. More formally, the generation of attributes was generally conducted using the repertory grid method: it was applied to the description of vegetables with children aged 8–11 years (Baxter *et al.*, 1998, 2000), and to the description of chocolate bars with children aged 9–11 years (Sune *et al.*, 2002). In this latter study, naive children (who were only trained 2 h in sensory evaluation principles and vocabulary) generated 110 terms; however, adult panellists trained to evaluate chocolate generated 94 terms with the QDA method (Stone *et al.*, 1974). The results obtained by children and adults' panel reveal some similarities for the visual and textural dimensions, but not for the flavour dimension (Sune *et al.*, 2002). Untrained children provided about the same product descriptive maps as children trained for 10 h with a general training method focused on sensory aspects of foods, but not specific to sensory analysis (Vivien and Sune, 2009). In another attempt to develop descriptive analysis, two groups of children were compared, 6–7 vs 10–11 years (Rose *et al.*,

2004a). Each group of children was trained individually by one interviewer for a 5 h session, and then asked to describe four types of meat. Both groups of children generated between 32 and 38 descriptors per type of meat, mostly related to visual appearance, but the descriptors that best predicted liking were slightly different among age groups: they were related to texture for the younger ones and to smell/taste for the older ones. The same researchers also studied the children's abilities to evaluate how much they liked different sensory characteristics of three types of meat, with similar age groups (6–7 vs 10–11 years) (Rose *et al.*, 2004b). Although this is not the typical application of sensory profile, which generally aims at evaluating intensity of the sensory characteristics, this confirmed that texture (mouthfeel and afterfeel) was the main driver of liking in 6–7-year-old children, whereas taste was more important for 10–11-year-old children.

22.4.5 Temporal measurements

Temporal measurements were not often conducted with children (Temple *et al.*, 2002). This work compared time–intensity measurement of 8–9-year-old children and adults, and it shows that the intensity of the sweet taste is higher in children than in adults, and that the sweetness perception decreases more rapidly in children than in adults. Even if temporal perceptions may differ between children and adults, it appears unlikely that sophisticated, training-intensive techniques such as multi-attribute time–intensity (Kuesten *et al.*, 2013) or temporal dominance of sensations (TDS) (Labbe *et al.*, 2009) may be applied to routine studies with children.

22.5 Conclusion

Applying descriptive sensory techniques to children presents a real challenge for children under the age of 8. Even when possible, sensory testing with children is especially time-consuming, because children often require a one-by-one approach to make sure that instructions are understood properly, and to monitor the test. Nevertheless, some profiling methods have been applied successfully with children and help to highlight that the sensory characteristics that impact the hedonic responses to a product may evolve across development.

22.6 Future trends

Even if today's children evolve rapidly, the pace of cognitive development is relatively fixed and limits the potential application of sensory techniques with children. However, one may imagine that the application of the most recent ways of collecting sensory data (i.e. using internet and tactile pads), which are appealing to children, may open new ways to motivate them to take part in sensory evaluation.

As described earlier in this chapter, conducting descriptive sensory evaluation with children represents a challenge and should not be attempted with children below the age of 8 years. Because they involve less extensive training and a quicker evaluation, rapid sensory profiling techniques, such as Flash profiling in particular, may be well suited to the exploration of children's responses, although, to the best of our knowledge, such applications have not been reported so far. Sorting of pictures has been applied with children aged 7–10 years in the frame of a statistical validation (Cadoret *et al.*, 2011). More applications of sorting with other product types (i.e. foods) may be useful to understand the potential of applying this technique to children, although it does not provide extended description of the product *per se*.

Applying simplified profiling techniques to children (Giacalone *et al.*, 2013; Jaeger *et al.*, 2013), such as CATA techniques developed to explore consumers' sensory perceptions of products, may prove an interesting direction when evaluation of a product's sensory characteristics through the eyes of a child is considered necessary.

22.7 Sources of further information

- *Food Quality and Preference*
  http://www.journals.elsevier.com/food-quality-and-preference/
- *Journal of Sensory Studies*
  http://onlinelibrary.wiley.com/journal/10.1111/(ISSN)1745-459X
- Declaration of Helsinki
  http://www.wma.net/en/30publications/10policies/b3/index.html
- Pangborn Sensory Science Symposium
- EuroSense Conference (European Conference on Sensory and Consumer Research)

References

Alexy, U. T. E., Schaefer, A., Sailer, O., Busch-Stockfisch, M., Huthmacher, S., Kunert, J. and Kersting, M. (2011) "Sensory preferences and discrimination ability of children in relation to their body weight status," *Journal of Sensory Studies*, **26**, 409–412.

Arditti, S. (1997) "Preference mapping: A case study," *Food Quality and Preference*, **8**, 323–327.

ASTM. (2013) *Standard Guide for Sensory Evaluation of Products by Children*. West Conshohocken, PA: ASTM.

Baxter, I. A., Jack, F. R. and Schröder, M. J. A. (1998) "The use of repertory grid method to elicit perceptual data from primary school children," *Food Quality and Preference*, **9**, 73–80.

Baxter, I. A., Schröder, M. J. A. and Bower, J. A. (2000) "Children's perceptions of and preferences for vegetables in the west of Scotland: the role of demographic factors," *Journal of Sensory Studies*, **15**, 361–381.

Bech-Larsen, T. and Jensen, B. B. (2011) "Food quality assessment in parent–child dyads – A hall-test of healthier in-between meals for adolescents," *Food Quality and Preference,* **22**, 614–619.

Bouhlal, S., Chabanet, C., Issanchou, S. and Nicklaus, S. (2013) "Salt content impacts food preferences and intake among children," *PLoS ONE,* **8**, e53971.

Brueckner, B., Schonhof, I., Schroedter, R. and Kornelson, C. (2007) "Improved flavour acceptability of cherry tomatoes. Target group: Children," *Food Quality and Preference,* **18**, 152–160.

Cadoret, M., Lê, S. and Pagès, J. (2011) "Statistical analysis of hierarchical sorting data," *Journal of Sensory Studies,* **26**, 96–105.

Chambers, E. (2005) "Commentary: Conducting sensory research with children," *Journal of Sensory Studies,* **20**, 90–92.

Chen, A. W., Resurreccion, A. V. A. and Paguio, L. P. (1996) "Age appropriate hedonic scales to measure food preferences of young children," *Journal of Sensory Studies,* **11**, 141–163.

Cliff, M. A., King, M. C., Scaman, C. and Edwards, B. J. (1997) "Evaluation of R-indices for preference testing of apple juices," *Food Quality and Preference,* **8**, 241–246.

Cordelle, S., Lange, C. and Schlich, P. (2004) "On the consistency of liking scores: insights from a study including 917 consumers from 10 to 80 years old," *Food Quality and Preference,* **15**, 831–841.

de Graaf, C. and Zandstra, E. H. (1999) "Sweetness intensity and pleasantness in children, adolescents, and adults," *Physiology and Behavior,* **67**, 513–520.

de Wijk, R. A. and Cain, W. S. (1994) "Odor identification by name and by edibility: Life-span development and safety," *Human Factors,* **36**, 182–187.

Desor, J. A. and Beauchamp, G. K. (1987) "Longitudinal changes in sweet preferences in humans," *Physiology and Behavior,* **39**, 639–641.

Feeney, E. L., O'Brien, S. A., Scannell, A. G. M., Markey, A. and Gibney, E. R. (2014) "Genetic and environmental influences on liking and reported intakes of vegetables in Irish children," *Food Quality and Preference,* **32**, Part C, 253–263.

Ganchrow, J. R. and Mennella, J. A. (2003), "The ontogeny of human flavor perception." In R. L. Doty, *Handbook of Olfaction and Gustation,* New York, Dekker, M., 823–846.

Garcia, K., Ennis, J. M. and Prinyawiwatkul, W. (2012) "A large-scale experimental comparison of the tetrad and triangle tests in children," *Journal of Sensory Studies,* **27**, 217–222.

Giacalone, D., Bredie, W. L. P. and Frøst, M. B. (2013) "All-In-One Test" (AI1): A rapid and easily applicable approach to consumer product testing,' *Food Quality and Preference,* **27**, 108–119.

Guinard, J.-X. (2001) "Sensory and consumer testing with children," *Trends in Food Science and Technology,* **11**, 273–283.

Guinard, J.-X. and Marty, C. (1997) "Acceptability of fat-modified foods to children, adolescents and their parents: Effect of sensory properties, nutritional information and price," *Food Quality and Preference,* **8**, 223–231.

Hough, G. and Sánchez, R. (1998) "Descriptive analysis and external preference mapping of powdered chocolate milk," *Food Quality and Preference,* **9**, 197–204.

Hough, G., Sánchez, R., Barbieri, T. and Martínez, E. (1997) "Sensory optimization of a powdered chocolate milk formula," *Food Quality and Preference,* **8**, 213–221.

Jaeger, S. R., Chheang, S. L., Yin, J., Bava, C. M., Gimenez, A., Vidal, L. and Ares, G. (2013) "Check-all-that-apply (CATA) responses elicited by consumers: Within-assessor reproducibility and stability of sensory product characterizations," *Food Quality and Preference,* **30**, 56–67.

James, C. E., Laing, D. G., Jinks, A. L., Oram, N. and Hutchinson, I. (2003) "Taste response functions of adults and children using different rating scales," *Food Quality and Preference*, **15**, 77–82.

James, C. E., Laing, D. G. and Oram, N. (1997) "A comparison of the ability of 8–9 year-old children and adults to detect taste stimuli," *Physiology and Behavior*, **62**, 193–197.

James, C. E., Laing, D. G., Oram, N. and Hutchinson, I. (1999) "Perception of sweetness in simple and complex taste stimuli by adults and children," *Chemical Senses*, **24**, 281–287.

Kildegaard, H., Løkke, M. M. and Thybo, A. K. (2011a) "Effect of increased fruit and fat content in an acidified milk product on preference, liking and wanting in children," *Journal of Sensory Studies*, **26**, 226–236.

Kildegaard, H., Tønning, E. and Thybo, A. K. (2011b) "Preference, liking and wanting for beverages in children aged 9–14 years: Role of sourness perception, chemical composition and background variables," *Food Quality and Preference*, **22**, 620–627.

Kimmel, S. A., Sigman-Grant, M. and Guinard, J.-X. (1994) "Sensory testing with young children," *Food Technology*, **48**, 92–99.

Knof, K., Lanfer, A., Bildstein, M. O., Buchecker, K., Hilz, H. and Consortium, I. (2011) "Development of a method to measure sensory perception in children at the European level," *International Journal of Obesity*, **35**, S131-S136.

Kuesten, C., Bi, J. and Feng, Y. (2013) "Exploring taffy product consumption experiences using a multi-attribute time–intensity (MATI) method," *Food Quality and Preference*, **30**, 260–273.

Kühn, B. F. and Thybo, A. K. (2001) "The influence of sensory and physiochemical quality on Danish children's preferences for apples," *Food Quality and Preference*, **12**, 543–550.

Labbe, D., Schlich, P., Pineau, N., Gilbert, F. and Martin, N. (2009) "Temporal dominance of sensations and sensory profiling: A comparative study," *Food Quality and Preference*, **20**, 216–221.

Laing, D. G. (2003) "Sensory analysis – what children can do," *Food Australia*, **55**, 47–51.

Lanfer, A., Bammann, K., Knof, K., Buchecker, K., Russo, P., Veidebaum, T., Kourides, Y., de Henauw, S., Molnar, D., Bel-Serrat, S., Lissner, L. and Ahrens, W. (2013) "Predictors and correlates of taste preferences in European children: The IDEFICS study," *Food Quality and Preference*, **27**, 128–136.

Lavin, J. G. and Lawless, H. T. (1998) "Effects of color and odor on judgments of sweetness among children and adults," *Food Quality and Preference*, **9**, 283–289.

Le, H. T., Joosten, M., van der Bijl, J., Brouwer, I. D., de Graaf, C. and Kok, F. J. (2007) "The effect of NaFe EDTA on sensory perception and long term acceptance of instant noodles by Vietnamese school children," *Food Quality and Preference*, **18**, 619–626.

Léon, F., Couronne, T., Marcuz, M. C. and Köster, E. P. (1999) "Measuring food liking in children: a comparison of non verbal methods," *Food Quality and Preference*, **10**, 93–100.

Léon, F. and Ivent, S. (2001), "Adaptation des méthodes hédoniques classiques aux enfants." In I. Urdapilleta, C. Ton Nu, C. Saint Denis and F. Huon de Kermadec, *Traité d'évaluation Sensorielle. Aspect Cognitifs et Métrologiques Des Perceptions*, Paris, Dunod, 169–180.

Liem, D. G., Mars, M. and de Graaf, C. (2004) "Consistency of sensory testing with 4- and 5-year-old children," *Food Quality and Preference*, **15**, 541–548.

Liem, D. G. and Zandstra, E. H. (2010) "Motivating instructions increases children's sensory sensitivity," *Food Quality and Preference*, **21**, 531–538.

Lumeng, J. C., Zuckerman, M. D., Cardinal, T. and Kaciroti, N. (2005) "The association between flavor labeling and flavor recall ability in children," *Chemical Senses*, **30**, 565–574.

Mennella, J. A., Pepino, M. Y. and Beauchamp, G. K. (2003) "Modification of bitter taste in children," *Developmental Psychobiology*, **43**, 120–127.

Mielby, L. H., Edelenbos, M. and Thybo, A. K. (2012) "Comparison of rating, best–worst scaling, and adolescents' real choices of snacks," *Food Quality and Preference*, **25**, 140–147.

Monnery-Patris, S., Rouby, C., Nicklaus, S. and Issanchou, S. (2009) "Development of olfactory ability in children: sensitivity and identification," *Developmental Psychobiology*, **51**, 268–276.

Mustonen, S., Rantanen, R. and Tuorila, H. (2009) "Effect of sensory education on school children's food perception: A 2-year follow-up study," *Food Quality and Preference*, **20**, 230–240.

Nicklaus, S., Boggio, V. and Issanchou, S. (2005) "Les perceptions gustatives chez lenfant," *Archives de Pédiatrie*, **12**, 579–584.

Nicklaus, S. and Monnery-Patris, S. (2003) "Poids de la prime enfance dans la formation des préférences alimentaires: présentation des méthodes d'étude et enjeu de l"approche écologique," *Psychologie Française*, **48**, 23–38.

Piaget, J. and Inhelder, B. (2003) *La psychologie de l"enfant*. Paris: PUF.

Popper, R. and Kroll, J. J. (2005) "Conducting sensory research with children," *Journal of Sensory Studies*, **20**, 75–87.

Reverdy, C., Schlich, P., Köster, E. P., Ginon, E. and Lange, C. (2010) "Effect of sensory education on food preferences in children," *Food Quality and Preference*, **21**, 794–804.

Richman, R. A., Sheehe, P. R., Wallace, K., Hyde, J. M. and Coplan, J. (1995) "Olfactory performance during childhood. II. Developing a discrimination task for children," *The Journal of Pediatrics*, **127**, 421–428.

Rose, G., Laing, D. G., Oram, N. and Hutchinson, I. (2004a) "Sensory profiling by children aged 6–7 and 10–11 years. Part 1: a descriptor approach," *Food Quality and Preference*, **15**, 585–596.

Rose, G., Laing, D. G., Oram, N. and Hutchinson, I. (2004b) "Sensory profiling by children aged 6–7 and 10–11 years. Part 2: a modality approach," *Food Quality and Preference*, **15**, 597–606.

Stone, H., Sidel, J., Oliver, S., Woolsey, A. and Singleton, R. C. (1974) "Sensory evaluation by quantitative descriptive analysis," *Food Technology*, November, 24–34.

Sune, F., Lacroix, P. and Huon de Kermadec, F. (2002) "A comparison of sensory attribute use by children and experts to evaluate chocolate," *Food Quality and Preference*, **13**, 545–553.

Temple, E. C., Laing, D. G., Hutchinson, I. and Jinks, A. L. (2002) "Temporal perception of sweetness by adults and children using computerized time-intensity measures," *Chemical Senses*, **27**, 729–737.

Thybo, A. K., Kühn, B. F. and Martens, H. (2004) "Explaining Danish children's preferences for apples using instrumental, sensory and demographic/behavioural data," *Food Quality and Preference*, **15**, 53–63.

Vivien, M. and Sune, F. (2009) "Two four-way multiblock methods used for comparing two consumer panels of children," *Food Quality and Preference*, **20**, 472–481.

Zandstra, E. H. and de Graaf, C. (1998) "Sensory perception and pleasantness of orange beverages from childhood to old age," *Food Quality and Preference*, **9**, 5–12.

Sensory testing in new product development: working with older people

I. Maitre[1], R. Symoneaux[1], C. Sulmont-Rossé[2]

[1]UPSP GRAPPE, Groupe ESA, SFR QUASAV 4207, Angers, France; [2]INRA, UMR 1324, CNRS, UMR 6265, Université de Bourgogne, Centre des Sciences du Goût et de l'Alimentation, Dijon, France

23.1 Introduction

In the context of an ageing population, the development of products meeting the elderly's needs and expectations becomes a major challenge for the food industry as well as for society. As a result, it is highly important to use sensory assessment tools with elderly subjects "as their perceptions may not be interchangeable with those of adults who are under 60 years of age" (Murray *et al.*, 2001). However, few sensory descriptive studies exist which include a panel of older subjects, and very few publications have compared the performances of young and elderly subjects in sensory tests.

After a brief emphasis on the social and economic challenge that represents the development of products aimed at the elderly, we will describe first the specificity of this population with a main focus on its heterogeneity and its consequences on sensory tasks. Second, we will present a study which aims at comparing the capacities of young and elderly subjects to use a discrete scale. This trial, based on a hedonic measure, has allowed the assessment of the discriminatory power and repeatability in elderly subjects living at home or in nursing homes. This trial has also enabled us to gain knowledge about the practical organization of a sensory test for an elderly population (number of samples, size of scale, etc.). In the last part, we will present the precautions in undertaking a sensory test among elderly subjects. However, it is important to keep in mind that there exist very few studies today that validate the use of sensory tools among elderly subjects. We are, nonetheless, hopeful that these first recommendations may be a good start for future studies. Furthermore, this chapter will deal with applied sensory assessment in the food sector, in alignment with the authors' skills, but we also hope that future studies will further expand our conclusions to other fields.

23.2 The elderly market: a challenge between needs and pleasure

In most developed countries, the proportion of elderly people in the population is growing quickly and the elderly market represents a key issue. The global population aged 60 years and above has doubled since 1980, and should reach two billion by 2050. Europe has 17% of 65 years old and above, and 4.7% 80 years old and above (Eurostat, 2011). The population of elderly people in Europe is increasing fast and, in 2060, there will be nearly 30% over 65 years old, and 12% over 80 (Eurostat, 2011). Developing age-friendly products and services for this target is one of WHO's recommendations to face the specific health challenges for the twenty-first century, caused by the population ageing (WHO, 2012).

Ageing is a universal, natural and physiological mechanism which is imposed upon somebody without choice. During this period, a healthy human being is affected by impairments which can weaken him. Cognitive capacities, functional deficiencies and energy storage progressively decrease during ageing, inducing a vulnerability which increases the risk of disease, aggravating dependence and so on to the end of life. The question of the needs of this population in nutritional and sensory terms in order to allow a longer life and a better ageing without deficiencies is one that the food industry and scientists have to address.

Having a "good" diet is considered an important factor to assist the elderly in maintaining optimal levels of health and preventing the onset of disease (HAS, 2007; PNNS, 2011)[1]. But a "good" diet does not only meet the nutritional needs of the elderly population. It also maintains "eating pleasure," an essential component in the regulation of food intake (Grunert *et al.*, 2007). This becomes even more crucial when elderly people have to delegate all or part of their meal shopping and/or preparation to others (home help, meals-on-wheels, nursing homes, catering service) because of the onset of disabilities (lack of skills, physical disability, cognitive impairment).

Recently, we conducted a large survey of French people over 65 years old (AUPALESENS survey). Different categories of the elderly were interviewed: elderly people living independently at home ($n = 289$), elderly people living at home with help unrelated to food activity (housekeeping, gardening, personal care) ($n = 74$), elderly people living at home with help, including help related to food activity (food purchasing, cooking; home meal delivery) ($n = 101$), and elderly people living in a nursing home ($n = 95$). Results showed that "pleasure to eat" remained the same whatever the level of dependence ("Is eating a source of pleasure?"), whereas satisfaction with the consumed meals decreased by 22% with culinary dependence at home or in nursing homes ("Do you enjoy your meals these days?") (Sulmont-Rossé *et al.*, 2012). In parallel, it was observed that 46% of the elderly dependent for their meals, whether living at home or in an institution, suffered from malnutrition or were at risk of malnutrition, as against 8% among autonomous elderly persons (Maitre *et al.*, 2014).

In view of these results, it is essential to develop foods for the elderly that meet both their nutritional needs and their sensory preferences, to make eating still

---

[1] Several French political organisations and national institutions (Ministère Français de la Santé et de la Solidarité, Institut de la Veille Sanitaire, Agence Française de Sécurité Sanitaire des Aliments – AFSSA, Assurance Maladie, Institut National de Prévention et d'Education pour la Santé – INPES) promoted a plan for 2006 to 2010 entitled "National Health Nutrition Plan" (Plan National Nutrition Santé – PNNS) to prevent malnutrition in the elderly population.

enjoyable in this population. Consumer testing with this population is therefore a major requirement, in order to develop nutritious food that meets the preferences of older adults. However, frailty and the appearance of cognitive alteration question the relevance of sensory tests, in particular with very old or dependent people.

23.3 The heterogeneity of the elderly

At what age does one become a senior? Even the notion of a senior is variable. For the professionals in marketing, the consumer becomes a senior at the age of 50 years old. For the public authorities, the person is considered a senior between 60 and 65 years old (the access age to certain social financial supports). For health professionals the age is over 70 because it is at the age of 73 on average that the first health accident occurs. In the survey AUPALESENS previously described, the age of 80 represents a decisive moment when nutritional frailty seems to appear (Maitre et al., 2013). In this book, we chose to consider as senior as everyone over 65.

Beyond the physiological age, a high heterogeneity is seen in the elderly population in terms of health and physical condition. Comparing a young retiree in good health with an elderly person aged over 90 would be considered a serious error of method. Likewise, one 78-year person can do long walks, whereas another person of 67 will struggle gardening. Two persons of the same age may show dissimilar functional, sensory and cognitive abilities (Ferry, 2009).

Conducting techniques of sensory analysis in a specific population requires know-how in order to establish the most appropriate criteria for recruitment, and also to forecast difficulties the assessors may be confronted with during sensory tests. The following chapters will review the factors likely to affect the recruitment of a panel of elderly tasters and/or the response of the elderly during a sensory test.

23.3.1 Physical and psychological health and dependency

According to a Health and Social Protection survey conducted in France by CREDES in 2002, 43% of the elderly aged between 60 and 65, and half of the elderly over 70, had at least one chronic disease (against 23% in 30–50 year olds) (Auvray et al., 2003). In addition to eye and dental pathologies, the diseases most frequently encountered are cardiovascular diseases, endocrine and metabolic diseases (diabetes) and osteoarticular diseases. The high prevalence of these pathologies goes along with a frequent use of drugs which are likely to have harmful side effects on chemosensory abilities and/or the oral and dental status of the elderly. On average, persons over 60 take on a regular basis three different drugs per day (four in the elderly aged over 80) against one in 30–50 year olds.

In addition to clinical pathologies, anxiety and depressive states are disorders more frequently encountered in the elderly than in younger subjects. The prevalence of depression is assessed at 8–15% in the elderly (Alaphilippe and Bailly, 2013). Depression could be associated with olfactory anhedonia or/and olfactory negative alliesthesia (Atanasova *et al.*, 2008). Although depressed people are reluctant to participate in a sensory study, an under-representation of this segment of seniors can be a problem if the aim is to assess acceptability of products among an aged population, particularly with frail and dependent people.

Finally, the life of an elderly person can undergo changes (such as illness of a spouse, or widowhood) leading to the appearance of a physical or cognitive inabilities, and hence the necessity to delegate a part or all the daily activities (cleaning, shopping, grooming, meal preparation).The last stage of dependence is the admission to an institution. The survey "Disabilities-Dependence-Incapacities" carried out between 1998–2001 in homes and institutions in France counted 10% of dependent people in 80 year olds, 25% of dependent males and over 35% of dependent females in 90 year olds (Duée and Rebillard, 2006). Moving to nursing homes, whether it is a personal decision or not, disrupts the elderly person's life: the habits, and the physical and social environment, are modified. Apart from the adaptation to community, the elderly have to accept the rules imposed by a third party, in particular in terms of eating habits. In most nursing homes, the time and the location where food is served, as well as the composition of the menu, are imposed on the residents. Obviously, the elderly who are more dependent and, a fortiori, those who live in nursing homes are less fit, more depressive and show more cognitive disorders compared to the autonomous elderly living at home. For these reasons, conducting a sensory assessment study of dependent elderly people would be costly and difficult.

23.3.2 Chemosensory capacities

Several authors have shown that ageing is accompanied by a deterioration in taste and smell, which is reflected by an increase in detection thresholds, a decrease in perceived intensity of supra threshold concentration, and a decrease in the ability to distinguish between odours (for a review, see Methven *et al.*, 2012; Murphy, 1986; Schiffman, 1993; Sulmont-Rossé *et al.*, 2010, 2012). Compared to young adults, the elderly have also shown stronger olfactory adaptation and slower recovery (Stevens *et al.*, 1989). Concerning taste, more than 60% of studies cited in Mojet's review of the literature (2001) reported an increase in the detection threshold of sapid molecules with age (also see Murphy, 1986; Stevens *et al.*, 1991). In their own experiments, Mojet and collaborators (2001, 2003) observed an overall decrease in taste detection and taste intensity perception with ageing, but a larger difference was observed between the young and the elderly for salt detection than for sweet and sour detection. Beyond this "mean" effect of age on chemosensory abilities, ageing is also accompanied by inter-individual variability in olfactory performance scores and, to a lesser degree, in taste performance scores (increased dispersion of scores) (Laureati *et al.*, 2008; Stevens and Cain, 1987; Stevens and Dadarwala, 1993; Thomas-Danguin *et al.*, 2003). Thanks to the AUPALESENS survey previously described, we evaluated the chemosensory capabilities of respondents through olfactory and gustatory tests. Four profiles were observed among our French elderly sample. A first group (43% of the respondents) showed preserved chemosensory capabilities, while 21% of the respondents faced an olfactory and gustatory decline. A few people (3%) were not able to perceive odours (close to anosmia), while they were still able to taste salt. Finally, 33% of the respondents showed preserved olfactory capabilities but a strong alteration of salt perception. The number of people with preserved sensory capabilities is higher among the group of those living independently, and makes up to 54% of this group. On the other hand, the cluster with low olfactory and gustatory abilities rises to 39% of people living in nursing homes.

Several factors have been reported to account for age-related changes in olfaction, such as the drying out of the mucous layer, and reduction of the production of new sensory cells (Boyce and Shone, 2006; Schiffman and Graham, 2000). In addition to these neurophysiological changes associated with the normal ageing process, extrinsic factors may also impact on the chemosensory decline, such as the number of diseases, medication, prior upper respiratory tract infection, prior exposure to certain air-bone chemicals known to affect odour sensitivity adversely (Murphy *et al.*, 2002). These factors, closely related to the life story of each individual, probably account for a large part of the variability observed in the elderly population.

23.3.3 Oral health status

The cumulative effects of physiological ageing, diseases and drugs, impact on different aspects of oral physiology, which plays a key role in eating behaviour (for a review, see Mioche *et al.* (2004)). With regard to the masticatory function, ageing goes along with a reduced strength in jaw muscles of mastication and an alteration of dental status (Mioche *et al.*, 2002). While the elderly are able to cope relatively well with muscular weakness by extending the time cycle of mastication, losing teeth on the other hand highly diminishes masticatory efficiency. According to Steele *et al.* (1997), the conservation of at least 21 well-distributed teeth is necessary to maintain a good masticatory function. However, the average number of lost teeth increases with age. A US survey showed that, in the age range 65–69 years, individuals have, on average, 18 remaining teeth (Carlos and Wolfe, 1989). In France, according to a survey conducted by the CREDES in 2002, 35% of 65–79 year olds and 56% of over 80 year olds claim they have lost all or roughly all their teeth, as against 11% in 40–64 year olds (Auvray *et al.*, 2003). Wearing a prosthesis may restore the masticatory function; however, this is less efficient compared to natural dentition (Fucile *et al.*, 1998; Veyrune and Mioche, 2000).

With age there may also appear swallowing disorders and dysphagia. The prevalence of dysphagia could reach 13–35% of elderly people living independently (Forster *et al.*, 2011). However, this prevalence is probably underestimated, as these disorders are not always detected in their early stages. Ekberg and Feinberg (1991) conducted a radiological study among 56 seniors without dysphagia (average age: 83 years old). The findings showed that only 16% of the participants had normal swallowing function. The oral function was abnormal in 63% of the participants showing inaccurate initial insertion and foodstuff control, drooling and rapid movements of the tongue, longer chewing and delayed swallowing response. Swallowing disorders, when aggravated, may even be harmful when foods or liquids enter the airway. Finally, ageing may also induce symptoms such as dry mouth, or xerostomia, likely to make food intake painful.

As for chemosensory capacities, there is a large variability in elderly people regarding oral health. While some alterations in the oral functions can be linked to an aged-related physiological state, others depend highly on the life story of each subject: the appearance of pathologies, the use of drugs, good oral hygiene practices, access to dental care. The alteration of oral health in elderly adults may induce eating difficulties (avoidance of hard and fibrous foods, loss of appetite when food intake becomes painful). This may in turn lead to an alteration in the elderly's nutritional status (Bailey et al., 2004; Maitre et al., 2013). For a sensory manager, the alteration of oral health may impact on food perception by changing texture perception (Mioche et al., 2002; Veyrune and Mioche, 2000) and the release of flavour components (Duffy et al., 1999). However, very few studies on this topic exist. The research studies on meat texture produced by Mioche and collaborators showed no age-related effect in terms of texture perception when comparing young subjects and old adults with good dental health (Mioche et al., 2002). On the other hand, Veyrune and Mioche (2000) noticed that subjects with complete dentures were more sensitive to changes in juiciness of meat samples compared to dentate subjects. Furthermore, it has been shown that, for elderly subjects, retronasal flavour perception can be more impaired by age than orthonasal odour perception, which could be associated with a change in oral food manipulation (Duffy et al., 1999).

23.3.4 Cognitive abilities

In the case of normal (non-pathological) ageing, age can induce an alteration in cognitive functions, namely an alteration of the functions allowing one to acquire, use and store information. In their review, Alaphilippe and Bailly (2013) report that age does not affect the different cognitive functions in the same way. Age has very little effect on the semantic memory, which stores factual knowledge about the world (knowing that an apple is fruit), or the procedural memory, dealing with skills (knowing how to peel an apple). However, age can induce an alteration of the episodic memory (Did I buy some apples yesterday?) and the working memory. Working memory is the system that stores and manipulates transitory information in the mind, where it can be manipulated. This processing fluency involves reasoning and problem solving, and makes them available for further information-processing. This system of transitory memory is particularly requested for classic tests of difference used in sensory assessment (Is the apple I am eating right now identical to or different from the one I have just had?). Finally, ageing can also lead to a decline in attentional capacities, namely in capacity to select relevant information in a flow of information (selected attention) and capacity to analyse several items of information in parallel (divided attention) (Alaphilippe and Bailly, 2013). For instance, an older adult will have more difficulties than a younger subject in following a tasting protocol requiring attention both to oral movements and sensory properties ("Carry the product to the mouth, chew it for ten seconds, rate sugar intensity, chew until swallowing, rate sugar intensity before swallowing, swallow, etc.").

With regard to the cognitive treatment of chemosensory inputs, many studies have shown a decrease in the capacity to identify odours with age (see for instance, (Larsson and Backman, 1997; Lehrner et al., 1999). Hedner et al. (2010) have recently tried to dissociate the impact of perceptive versus cognitive processes on the capacity of older adults to detect, discriminate and identify odours. In this study, participants completed olfactory tests (the use of Sniffin' Sticks test developed by Hummel et al. (1997)) as well as a cognitive test covering executive functioning (memory span), semantic memory (e.g., verbal fluency) and episodic memory (word recall and recognition). Results showed that proficiency in executive functioning and semantic memory contributed significantly to odour discrimination and identification performance, whereas all of the cognitive factors proved unrelated to performance in the odour detection test.

In addition to aged-related physiological effects on cognitive capacities, ageing can also go along with more serious cognitive disorders ("senile dementia"), generally induced by neurodegenerative diseases (Alzheimer's disease, Parkinson's disease), strokes, or sometimes just a simple decline in overall health status. The prevalence of dementia has been estimated to be 5–10% in over 65 year olds and up to 16% in octogenarians in Belgium (De Deyn et al., 2011). The prevalence of cognitive disorders in nursing homes shows that Alzheimer's disease represents 72% of diagnosed dementias with a prevalence of 0.5% before 65 years old, 2–4% over 65, reaching 15% in 80 year olds (INSERM, 2007). The aforementioned disorders in the elderly may be an obstacle for the application of surveys and classic tests used in sensory assessment. Besides, it is important to notice that neuro-degenerative disorders such as Alzheimer's disease are often associated with a severe decline in odour identification, even at the beginning stage of the disease (Doty et al., 1987, 1991; Wong et al., 2010).

In conclusion, all the factors that have been described in this section emphasize the heterogeneity of the elderly population in terms of their ability to perceive sensory stimuli, their difficulties during food intake, their capacity to understand and perform a task and their wish to participate in sensory tests.

23.4 Impact of age and dependence on performance at a sensory task: key findings on scale use in a monadic sequential presentation

Several authors used scales with elderly panels in sensory tests, to get both hedonic and intensity scores. However, these studies were usually carried out with healthy elderly people, who were able to come into the lab, rather than with frail and dependent elderly people (Kremer et al., 2007; Methven et al., 2012; Mojet et al., 2001; Withers et al., 2013). Furthermore, these authors were interested in the evaluations obtained thanks to the scales rather than in the *way* elderly people use scales. The capacity of the elderly to execute sensory tasks effectively is still an outstanding issue.

Table 23.1 **Experimental design**

	Consumers	Warm-up	Block 1	Block 2
Session 1	1 to 15	E	ABC	ADE
	16 to 30	E	ABD	ACE
	31 to 45	E	ADE	ABC
	46 to 60	E	ACE	ABD
Session 2	1 to 15	E	ADE	ABC
	16 to 30	E	ACE	ABD
	31 to 45	E	ABC	ADE
	46 to 60	E	ABD	ACE

I totally dislike			I can't say if I like or dislike the product			I like a lot
☹☹☹	☹☹	☹	😐	☺	☺☺	☺☺☺

Figure 23.1 Hedonic scale developed in the AUPALESENS programme.

23.4.1 Description of a case study

Thanks to the AUPALESENS programme, we set up a study aiming at assessing whether elderly people (including frail and dependent people) are able to carry out a classic sensory methodology. Specifically, we assessed the ability of the elderly to carry out discriminative and repeatable judgement on hedonic scales and the suitability of a monadic sequential presentation for this population. This study was also an opportunity to gain insights on practical considerations that should be considered when one is willing to run a sensory test with elderly people.

For this study, we recruited three groups of participants: young participants (n = 64; 18–49 years), elderly participants living at home (n = 55; 65–80 years) and elderly participants living in a nursing home (n = 22; 74–93 years). To be recruited, elderly candidates had to score at least 20 in the Mini Mental Scale Examination (MMSE) (Folstein et al., 1975). The MMSE screens for cognitive impairment: scores greater than or equal to 26 points (out of 30) indicate normal cognition. Below this, scores can indicate severe (≤ 15 points), moderate (16–19 points) or slight (20–25 points) cognitive impairment (Derouesné et al., 1999). Five dairy dessert creams from the French market including three oral nutritional supplements (A,B,C,D,E) were tested. Consumers attended two sessions with a one-week interval. During each session, participants received a monadic series of seven samples, which included a warm-up sample and two samples corresponding to the same product (A1 and A2). The presentation order was designed to allow testing for intra- and inter-session repeatability (Table 23.1). For each sample, participants were asked to taste it and to give a liking score on a 7-point categorical scale combining labels and pictograms, which has been adapted thanks to in-depth interviews (Fig. 23.1). Sessions were conducted in a sensory lab for the young and the independent elderly. For the dependent elderly, sessions took place in a meeting-room at the nursing home.

23.4.2 Running a sensory test in nursing home: practical considerations

The course of the sessions carried on with the autonomous elderly people was quite similar to that with young respondents. For both groups, sessions took place in a sensory room through collective sessions (12–16 respondents came at the same time), with respondents seated in individual booths. However, in the nursing home we had to adapt ourselves to the constraints of the environment and to respondents' frailty. With regard to the environment, it was not always possible to seat respondents at individual tables (the room allocated to the test was too small, not enough tables). On several occasions, two or three participants were seated at the same table. In this case, we insisted throughout all the sessions that the presentation order of the products was different between respondents to reassure our volunteers in their capacities to perform the task ("you may have a different opinion on this product from your neighbour as you did not receive the same product"). With regard to respondents' disabilities, the instructions were read loudly by the investigator. The scale was presented in a large size (60–80 cm) on the table. Respondents were asked to put the sample on the scale level corresponding to their personal liking; the answer was written on a questionnaire sheet by the investigator (respondents were not asked to write anything). For several participants, instructions had to be repeated by the investigator for each product. One investigator was able to take care of two to four subjects (only one in the case of severe disability such as blindness or deafness).

23.4.3 Impact of age and dependence on performance

23.4.3.1 Preserved discrimination and inter-session repeatability

Liking scores were submitted to ANOVA with sample, session and subject as factors for each group in order to test discrimination and inter-session repeatability. Results showed a decrease in the discrimination level with age and dependency, but significant differences between the products were still observed for the elderly group (Young people: $F_{sample} = 100.71$; $p < 0.001$; $F_{sample \times session} = 0.82$; $p = 0.54$; Independent elderly people: $F_{sample} = 36.19$; $p < 0.001$; $F_{sample \times session} = 1.34$; $p = 0.25$; Dependent elderly people: $F_{sample} = 3.75$; $p < 0.01$; $F_{sample \times session} = 0.25$; $p = 0.94$). Figure 23.2 shows with box plots how liking scores are distributed for each age group (product names on the left, and the homogenous means in small letters on the right). However, product ranking was similar for the three groups of individuals. Furthermore, no significant sample×session interaction was observed: repeatability between sessions seems to be preserved with ageing and dependence, even if it is easier to be repeatable for dependent elderly people as they are less discriminant.

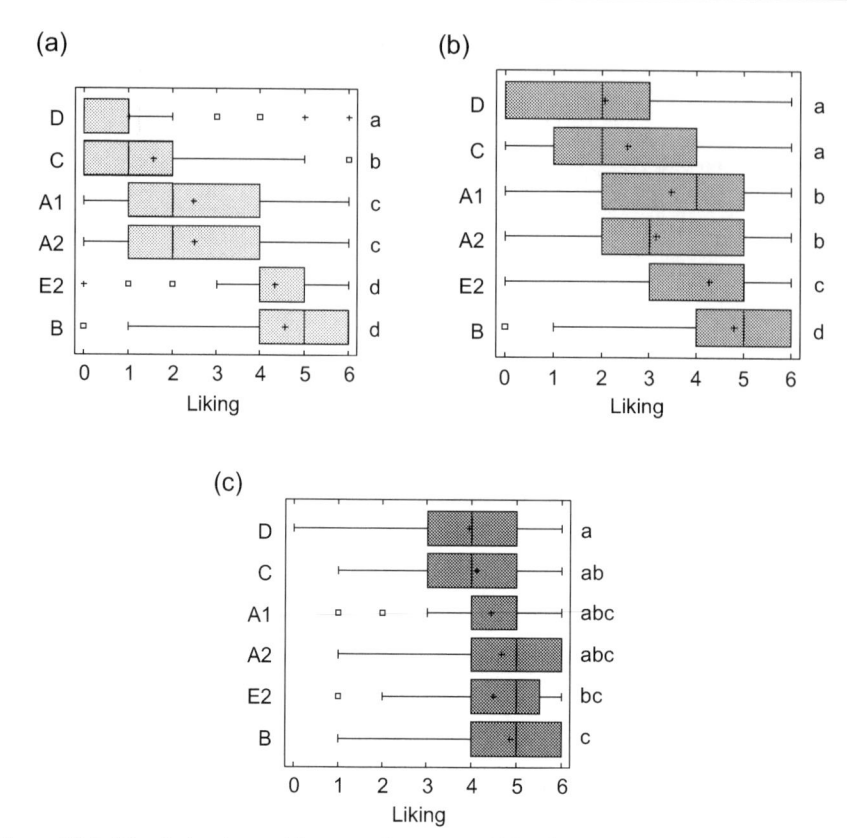

Figure 23.2 Discrimination and inter-session repeatability for (a) young people, (b) independent elderly people and (c) dependent elderly people.

23.4.3.2 No evidence for a fatigue effect through the monadic series

To check intra-session repeatability, we submitted the liking scores of the samples A to ANOVA with block and session as factors plus interaction. Results showed no significant difference between the liking score of A in the first block and the liking score of A in the second block, whatever the group (Young people: $F_{block} = 0.57$; $p = 0.45$; Independent elderly people: $F_{block} = 2.97$; $p = 0.08$; Dependent elderly people: $F_{block} = 0.13$; $p = 0.72$). In fact, the distribution of the difference between the liking score of A in the first and in the second block revealed no significant difference between the three groups (Fig. 23.3).

Liking scores were submitted to ANOVA per group with position in the series (rather than product) and subject as factors. Results showed no significant effect of position ($F_{position} = 0.96$; $p = 0.44$). In other words, when asked to rate seven samples, elderly people, and even dependent elderly people, did not exhibit any boredom or fatigue effects (Fig. 23.4).

Sensory testing with the elderly

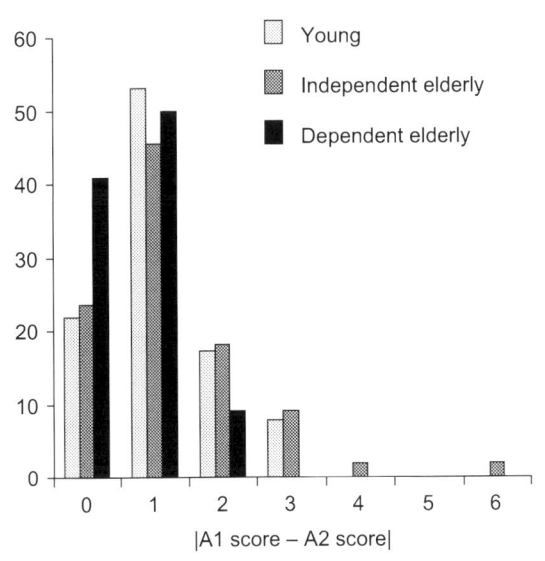

Figure 23.3 Distribution of differences between A1 and A2 liking score for each group (session 1).

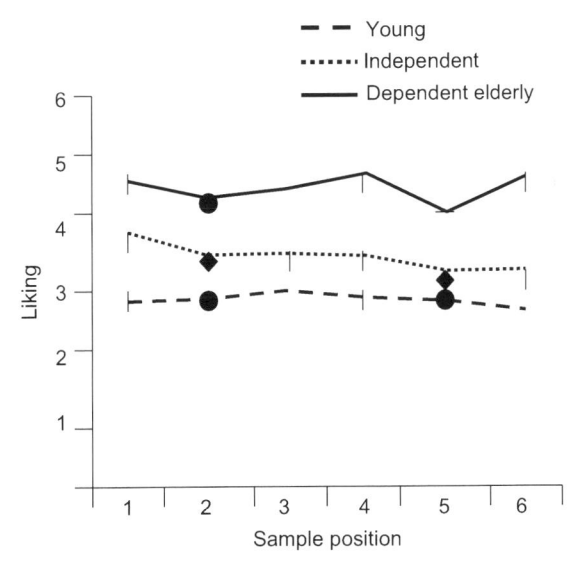

Figure 23.4 Scale use and boredom.

23.4.3.3 Bias toward the positive side of the scale

The previous ANOVA revealed a strong group effect (F_{group} = 112.10; $p < 0.001$): on average, dependent elderly people gave higher liking scores (used the right side of the scale more) than autonomous elderly people, who themselves gave higher liking scores than young people (Fig. 23.4). A similar result had already been observed in several works (Cordelle et al., 2004; d'Hauteville et al., 1997; Kälviäinen et al., 2003; Tuorila et al., 1998). It is possible that this result reflects a true difference in acceptability between young and elderly participants, especially since the elderly could perceive at a lower intensity off-notes of oral nutritional complement. However, another plausible explanation is that the elderly's scoring is tainted by a positive bias. This could be due to the fact that the present generation of elderly people have been educated not to complain about food (Doty, 1991). This could also be due to the fact that elderly people want to please the investigators, who are "taking care" of them during the sensory tests.

23.4.3.4 Impact of cognitive status on performance

In order to evaluate the effects of cognitive status on rating performances, we compared the elderly who had a score over or equal to 26 at MMSE (i.e. 50 elderly living at home and 13 living in a nursing home) with those who had scored between 20 and 25 (five elderly people living at home and nine living in a nursing home). Although the results should be considered with caution, as the number of testers in the two groups was not equal, we observed that participants who had a weaker MMSE score had a tendency to be less repeatable, and gave rating scores on average higher than participants who had an MMSE score over 25.

23.4.4 Key findings

These results clearly demonstrate that the elderly, even those living in a nursing home, were able to use a liking scale in a monadic sequential way, were repeatable, and were fairly discriminant when they did not suffer from any cognitive alteration (e.g., score MMSE > 25). However, if the ranking of the products was identical among the different age groups, elderly participants, and the dependent ones even more so, made fewer differences between the products than the young participants.

From a practical point of view, we noticed no weariness or tiredness effects among the aged population, provided that the number of products was not too high (seven in the present study, which could be considered as a maximum). This validates the use of a monadic sequential design with elderly people. However, older participants were more likely to give higher liking scores than younger ones. Whatever the reason, and on the basis of current knowledge, this result pleads for not using rating scales that are too small.

Finally, this study also emphasized the difficulty in working with frail and dependant elderly people. Investigators have to adapt themselves to the constraints of the environment and to the difficulties of the respondents (face-to-face interviews, repetition of the instructions, collection of respondent's answers, etc.). Obviously, such a close relationship between an investigator and a respondent will frequently lead to a positive emotional context, which may in turn bias the elderly's answers during the test (this positive emotional context is even stronger in a nursing home, where sensory tests represent a disruption in the daily routine and an opportunity to "chat" with a "visitor"). This constitutes an important challenge when running sensory tests with frail and dependent elderly people: investigators are requested to make elderly respondents comfortable with the test procedure, while not influencing tests results (being nice but neutral). In other words, running sensory tests with elderly people, and specifically frail elderly people, requires being trained and well aware of possible biases.

23.5 Running sensory descriptive analysis with an elderly panel: recommendations

Based on the limits and results mentioned in the previous sections, is it relevant to conduct a descriptive sensory analysis with a panel of elderly people? The descriptive sensory tool has a major role in understanding preferences and translating what the consumer expects into sensory indicators, which R&D can integrate to orientate product development. Very often, even though the product aims at a particular target (children, elderly, a certain category of patients or disabled people), it is the results from "young" (20–50 years old) and healthy panellists which are used. In this case, the descriptive sensory tool is considered an analytical tool and the experimenter assumes that this measure is not an accurate reflection of the target's perception. There are two risks associated with this way of studying: either the sensory analytical tool is not sensitive enough, and some preference key driver descriptors are not detected and measured; or the analytical tool is more sensitive than the target, and there is a risk of uselessly modifying the products with descriptors that are not perceived by the target. For example, some elderly people will be less sensitive to tastes and flavours than younger adults, whereas others may be more sensitive to some texture characteristics, depending on their oral health status.

Based on our experience with elderly people, we believe that using a descriptive sensory tool with a panel of elderly people is not indispensable. Indeed, if the product targets healthy and autonomous elderly people, then a panel of younger tasters is enough to describe this product. My analytical tool (my panel) may be more sensitive than the target population (identification of differences which an elderly panel would not perceive), but the risk of taking a wrong decision is low (decision of carrying on a development project if, according to the sensory results, the sensory objective has not been reached). However, if the product is dedicated to a specific segment of the elderly population (for example, people with dentures or people suffering from dry mouth), then we believe it is particularly important to have this product tested by the target population. In the hearing aid industry, it is already normal practice to work with panels of people with hearing difficulties (Legarth *et al.*, 2013). According to the same logic, it would be of interest to recruit panels of testers with specific difficulties to have them test food products or food packaging.

23.5.1 Recruitment of an elderly panel

23.5.1.1 A long and costly recruitment

Whether you recruit a panel of healthy and autonomous elderly people, or a panel of elderly people with a specific disability (for example elderly people with masticatory problems), the key issue is to recruit a homogenous panel among a very heterogeneous population. Indeed, as for any well-conducted descriptive sensory analysis, panel homogeneity, its consistency with the objective and its ability to carry out the sensory tests should be sought. In the first case (panel of healthy and autonomous elderly), the challenge will be to select subjects presenting no pathology, taking no drugs and with good sensory, masticatory and cognitive aptitudes. In the second case (panel of elderly people with a specific disability), the challenge will be to recruit subjects presenting the difficulty which is targeted by the study but, regarding their other capacities, relative homogeneity. In both cases though, recruiting a panel of elderly testers will be longer and more costly than recruiting younger testers. More precisely, it will be more difficult to recruit people with specific problems, as they are usually frail and dependent and therefore less willing to take part in this type of study and/or cognitively not capable of carrying out sensory tests.

23.5.1.2 Selection tools

From a practical point of view, a number of tools have already been validated with a population of elderly people, which can help the experimenter to characterize his sample. The most relevant for this population include:

The Mini Mental State Examination (MMSE) (Folstein *et al.*, 1975). The MMSE screens for cognitive impairment: scores greater than or equal to 26 points (out of 30) indicate normal cognition. Below this, scores can indicate severe ($\leq$15 points), moderate (16–19 points) or slight (20–25 points) cognitive impairment.

The European Test of Olfactory Capabilities (ETOC) (Thomas-Danguin *et al.*, 2003) or Sniffin' Sticks (Hummel *et al.*, 1997). These tests screen for olfactory impairment. The ETOC test consists of 16 blocks of four vials presented one by one to the participants. Only one vial of the four in the block contains an odorant. For each block, participants are asked to point out the odorous vial (forced-choice detection task) and then to identify the odour by choosing a name for the odour out of four proposals (forced-choice identification task). Sniffin' Sticks comprises three tests of olfactory function: odour threshold (n-butanol), odour discrimination (16 pairs of odorants) and odour forced-choice identification (16 common odorants).

The General Oral Health Assessment Index (GOHAI) (Atchison, 1997; Tubert-Jeannin *et al.*, 2003). This 12-item questionnaire measures self-report oral functional problems together with the psychosocial impacts associated with oral disease. The maximum score is 60. A score of 57–60 corresponds to a good oral quality of life, while a score of 50 or less reflects a poor oral quality of life.

The Geriatric Depression Scale (GDS) (Sheikh, 1986; Yesavage *et al.*, 1983, 2000). This questionnaire of 30 questions screens for depression. Scores from 0–10 are considered as normal; more than 11 probability of depression and more than 21 is very depressed.

The Instrumental Activities of Daily Living (IADL) scale (Lawton and Brody, 1969). This scale screens for functional decline by assessing a person's ability to perform such tasks as using a telephone, doing laundry and handling finances.

23.5.1.3 Ethics committee agreement

As elderly people are perceived to be a frail population, European regulation requires presentation of the research protocol to an ethics committee. Directive 2001/20/EC regulates and harmonizes ethical rules governing clinical trials in Europe. The ethics committee is an independent body made up of healthcare professionals and non-medical members. Its responsibility is to protect the rights, safety and well-being of human subjects involved in a clinical trial, and to control the protocol, the suitability of investigators and the documents used to inform the subjects. Consequently, it is the responsibility of each investigator to ensure compliance of the sensory studies with the ethics laws of his country.

23.5.1.4 Societal commitment as a source of motivation

Depending on the place, the elderly play a more or less important role in society. Young active retirees often take responsibilities within their family, in their community or in associations. The older they are, the lower these responsibilities are. They have to leave aside their commitments and it is hard to keep some in nursing homes. Our developed countries have difficulties finding a valued role for our wise persons. Our experience showed that the main motivation source for our volunteers was to be involved in a research project, as a vital link in the research process, and to have a critical role to play. This involvement gives them the societal responsibility they have mostly lost, and they appreciate being considered as full consumers. As usual, panellists should be rewarded with a gift, and, in our case, they perceived it as proof of the importance of their role in the research.

23.5.2 Conducting sensory analysis trials among the elderly

23.5.2.1 The discriminative tasks

Previous research on the perception threshold has shown that the elderly were able to perform 2-AFC or 3-AFC type tasks (Mojet *et al.*, 2001); for a review see Methven *et al.* (2012). As a consequence, it may be relevant to conduct discriminative tasks with elderly people so as to, for example, check if any improvement brought to the product has actually been perceived by this population or not. Yet, as attentional capacities decline with age, we recommend choosing duo–trio tests, or paired comparison tests, rather than a triangle test or tetrad. Indeed, the latter require comparing simultaneously more samples than the former, and are therefore more costly from a cognitive point of view.

23.5.2.2 Classification and sorting tasks

Tasks involving "ranking" or "sorting" products according to their sensory proximity are often proposed as fast and costless methods to describe the structure of a product space or even to describe the sensory properties of products (napping task, free sorting task). To our knowledge, no study has until now been carried out to assess the relevance of such tasks with a panel of elderly people. Yet, in the light of current knowledge, we do not recommend using this type of task with an elderly population, especially if dependent, for two reasons. First, this type of task is very costly from a cognitive point of view (large number of products to compare, need to remember the already formed groups, etc.). Yet, as stated in previous sections, age goes with impairment of working memory and falling attentional capacities, even in the normal ageing process. Second, elderly people have also shown stronger olfactory adaptation and slower recovery, which may be problematic for such a task where allowing a break between two stimuli is not always easy to monitor.

23.5.2.3 The sensory profile

Although we are not aware of any study that has formally compared performances of young and older subjects as part of a sensory profile, we believe it is possible to carry out a sensory profile with elderly people provided that they are in very good shape, both physically (not too sensitive to tiredness) and cognitively (e.g., score MMSE > 25). Indeed, on the one hand, research made as part of the AUPALESENS programme has shown that elderly people are still capable of verbalizing their sensory feeling with respect to a product in a meaningful way (see case study Section 23.5.2.4); on the other hand, several studies have shown that elderly people were very capable of assessing the intensity of a sensory attribute by using a rating scale (Mioche *et al.*, 2002; Veyrune and Mioche, 2000), even they were not interested by the *way* elderly people use scales and perform the task. For example, Kremer *et al.* (2007) have asked young testers (aged 18–35) and older testers (aged 60–85) to evaluate the intensity of different texture and flavour attributes (fattiness, elasticity, airiness, dry, swallowing effort, after feel, vanilla or cheese flavour, sweetness and saltiness) in waffles. Results showed that the elderly differed from the young in their perception of texture and flavour but, apparently, the elderly did not have a lower performance than the younger ones on this task (unfortunately, the authors did not comment on the ability of elderly people to use the scales, but examination of the results does not show higher standard deviations for the elderly than for the young).

Even if the profile is done with elderly people in good physical and mental shape, adapting the organization of the profile will be necessary so as to take into account the effect of age on learning and work capacities. Indeed, it will probably be necessary to organize less "intensive" sessions (e.g., fewer products to taste, fewer descriptors per product) and a longer training time than for younger participants. It will also be necessary to carefully monitor the performances of the elderly throughout the training, so as to rapidly detect any difficulty that may arise. In other words, if the sensory profile is long and costly with a young panel, it will be even more so with an older panel.

Finally, as regards the choice of scale, Doty (1991) argued that the simplicity and robustness of category scales make them very appropriate to elderly studies. In fact, in the study described in Section 23.4, it was observed that older subjects were using a seven-point hedonic scale in a repeatable and discriminative way. In Kremer *et al.*'s study (2007), evaluations were made from visual analogic scales. All the previous works on scales (Borg Griep *et al.* 1998) support the recommendation to use category scales to score intensities.

23.5.2.4 Free description in hedonic tests: a rapid methodology alternative to descriptive analysis

One of the objectives of sensory profiling is to correlate descriptive data with liking data collected on the same products, in order to determine the sensory key drivers of preferences (preference mapping). When targeting the elderly, a first option is to carry out hedonic measures with elderly people and the sensory profile with a panel of younger testers (see for example Koskinen *et al.*, 2003). As mentioned earlier, the limitation of such a method is to overestimate or underestimate the importance of some sensory attributes with regard to elderly people's abilities. The second option is to carry out hedonic measures *and* descriptive measures with an elderly panel. However, as seen previously, implementing a sensory profile with an elderly panel may be longer and more costly than with a younger panel. In addition, such a method can be conducted only with people in good shape, from a physical and cognitive point of view, which, again, does not avoid the risk of underestimating or overestimating the importance of some sensory attributes with regards to the abilities of the more fragile and dependent elderly people.

As part of the AUPALESENS programme, we tested the capacity of autonomous and dependent elderly people to verbalize the sensory defects and qualities of some products. Two methods were tested: the focus group and the use of open-ended questions as part of a hedonic test.

A first option is to carrying on a focus group with elderly people. When dealing with broader issues, such as asking people to react to food issues, to the quality of food products, or to sensory determinants, use of qualitative methods is possible. Focus groups enable the testers to obtain a great amount of information in a short time. However, it is necessary to check that the participants do not find it difficult to speak in a group of ten people or so. That is why pre-interviewing participants by phone enables the investigator to evaluate their speaking capacities. If physical disabilities (moving, visual, etc.) do not constitute a limit as such, it is important to take into account the socio-demographic characteristics of the people attending the meetings so as to recruit a variety of profiles, especially considering their professional activity before retirement, their age, their eating habits, etc. As part of the AUPALESENS programme, we organized focus groups with independent elderly people who reacted to this method positively, expressing themselves easily, and accepting to project themselves in the situations proposed while respecting the operating rules, to suggest improvements in the food products that were being tested. Organizing focus groups in nursing homes is possible but quickly shows limits with regards to people's autonomy and their ability to speak in a group. In that case, it is recommended to do individual interviews. This is the option we chose for the AUPALESENS project. There is no specific methodology during such interviews with the elderly, but there are key principles: taking the time to do things, listening, allowing digressions, encouraging association of ideas.

A second option is to use open-ended questions in hedonic tests. As part of the AUPALESENS programme, 103 elderly people living at home (44 men/59 women; age range 65–82 yrs) and 63 elderly people living in a nursing home (8 men/55 women; age range: 67–98 yrs) tested various recipes of the same meat dish (e.g., blanquette de veau), of cheese, of bread and dairy desserts. Five varieties were tested per session. The sessions took place in a sensory evaluation laboratory for the independent elderly, and in the common area of the nursing home for the dependent ones. For each variety, participants were asked to give an evaluation mark on the seven-point scale presented in Fig. 23.1. At the end, we asked them to choose two samples, the one they preferred and the one they liked the less, and to specify "what they like in the product" and "what they do not like in the product." Two separate questions were asked, so as to clearly identify, in consumers' answers, what was related to the qualities of the product and what was related to its defects, as one general question could have led to ambiguous answers (Symoneaux et al., 2012). Depending on the participants' state of fatigue, this question can be asked for all products, the majority of products, or for the least and most appreciated products. Ideally, the qualities and defects of each product should be collected so as to compile a more reliable contingency table following the text analysis. While these evaluations were being made, the sensory profile of each variation of the same product was established by a panel of young participants.

Results showed that participants were using the descriptors of taste, flavour intensity and texture more than those of aromas to describe the qualities and defects of the products tested. Also, the people living in nursing homes generated a smaller quantity and a less rich vocabulary than autonomous elderly people (on average, 3.9 descriptors for dependent elderly respondents and 4.9 descriptors for autonomous elderly respondents). Nevertheless, the two groups of subjects agreed globally on the qualities and defects they found in the samples, which argues for using such a methodology in laboratory as much as in nursing homes.

23.5.3 Practical recommendations

Whatever the task carried out with a panel of seniors, a number of precautions should be taken for them to feel comfortable in their taster role:

Giving easy-to-read instructions and questionnaires (airy layout, large print, large scales) so as not to put seniors suffering from visual impairment in an awkward position.

Giving instructions both orally *and* in writing. As many elderly people suffer from hearing impairment, it is recommended to give the instructions collectively, then to individually check that each participant has understood well.

It may be worth including a few training or warming sequences at the beginning of each session, with a product different from the target product, so as to make sure that the participants have understood the task they have to perform.

Elderly people often have digestion problems, so it is recommended to check that the total amount of ingested food does not exceed the size of a standard portion, and that the product is not likely to be painful.

Elderly people show stronger olfactory adaptation and slower recovery. So a strict and appropriately long rinsing protocol must be respected between each tasting.

The use of computer tools with an old population must be carefully considered as many "young" seniors master computers perfectly and will be totally capable of performing the sensory tests with this type of tool (probably they will be even more so in the future). Yet this is not the case for all aged people, and a paper questionnaire should be systematically available. In the future, tools specifically developed for tablets may get around some difficulties, as far as they are developed for people with visual impairment. Finally, some elderly people may not be able to write (which frequently happens in nursing homes). In that case, an experimenter should take notes of the participant's answers without influencing him.

Unfortunately, resilience (dropping out of the study) is more frequent in this population than in a younger one. It is therefore recommended to recruit more participants than the final number needed.

Finally, it should be expected that sessions will be longer with elderly people than with younger subjects (when a session lasts 45 min with a student, it may last 1–1 h 15 with a senior). Also, the latter will take a longer time to read the instructions and perform the task, often with a view to "doing things well." Furthermore, seniors chat with the experimenter more than young adults at the end of a session.

As a relationship is developed between subjects and investigators, it is essential to train the latter to be as neutral as they can, to explain patiently the instructions, to pay attention to the way the elderly perform the task and to interfere as little as possible in the elderly's evaluation.

23.5.4 Rapid sensory descriptive analysis: perspectives

As a main conclusion, "rapid" and "elderly" do not seem to be compatible concepts. As already discussed, as soon as one intends to target the elderly population in a sensory task, the recruitment takes longer and the elderly also take more time to complete the tests than young people do. In addition, the amount of work you can ask from an elderly is less than from a younger adult, due to elderly people being more easily tired. Considering all these limits, and thanks to our previous work and expertise in conducting sensory tests with the elderly, we confirm that: (1) conducting sensory descriptive tasks may be possible with elderly people with a good cognitive health; (2) rapid sensory techniques which let elderly people using their own vocabulary to describe the products may be possible. However, with these rapid sensory methodologies, the elderly should probably be more assisted than young people to obtain a sufficient number of verbal descriptions. This is how we proceeded for the hedonic test (Section 23.5.2.4). Further work is needed on the elderly's abilities to perform rapid sensory analysis.

23.6 Conclusion and future trends

From our point of view, the main limiting factor in using sensory tools with a panel of elderly people is tiredness and their cognitive status. Indeed, a senior in good physical shape and presenting good cognitive capacities will be perfectly capable of performing sensory evaluation tests. However, implementation of these protocols will be longer and more costly than with younger panellists. Also, the senior population is a *very* heterogeneous population. As a consequence, recruiting aged tasters will require a more thorough selection process (evaluation of the cognitive status, of the sensory capacities, of the oral dental health, etc.) than with younger tasters. In addition, even within the framework of normal ageing (with no pathology), age goes with slower cognitive and motor functions. In this case, tests should be simplified and the workload during tasting sessions should be reduced (more time for instructions, for the training phase, fewer products to taste, etc.). Anyway, working with the elderly should be less rapid than working with younger people.

Due to these difficulties, using a descriptive panel of elderly people should only be considered if it brings additional information to that which would have been provided by a younger panel. In most cases, a good compromise is to select a panel of young tasters to establish the sensory profile of the most interesting products, while, at the same time, carrying out a hedonic test with elderly people and collecting free comments. As these comments are often descriptive ones (appreciation of the texture, flavour), they can help to determine the defects and qualities of the products. Yet, we believe it is particularly relevant to develop panels of elderly people presenting specific difficulties (e.g., mastication difficulties, impaired odour perception, protein deficiencies) so as to develop and/or improve the products aimed at this population. Within the context of an ageing population, developing products which will meet the specific needs of seniors while satisfying their sensory and hedonic expectations already constitutes a major societal and economic challenge.

To date, too few sensorialists have been interested in targeting the elderly. Thanks to our experience with this population, we have tried to come up with a number of recommendations for anyone wishing to use sensory evaluation tools with the elderly. However, as we have mentioned several times, there are almost no *methodological* studies enabling evaluation of the impact of age on performance during sensory tests. In the future, many sensory studies may hopefully be carried out with elderly populations, so as to improve the sensory tools, more particularly by integrating people with impaired health and/or cognitive state. It is also hoped that sensorialists will share their experiences when running sensory evaluation with the elderly, to overcome difficulties that may be encountered and propose new solutions.

Acknowledgements

The programme AUPALESENS – Improving pleasure of elderly people for better ageing and for fighting against malnutrition – was funded by the French National Research Agency (ANR – ALID call). The authors thank Pauline De Facq, Françoise Durey, Valérie Feyen, Christophe Martin, Corinne Patron, Isabelle Saillard and Jérémy Tavarès for their help during the experiments; Sylvie Issanchou, Nathalie Bailly and

Virginie Van Wymelbeke for providing advices; Anca Bioteau and Véronique Hébrard for providing language help.

References

Alaphilippe, D. and Bailly, N. (2013). *Psychologie de l'adulte âgé*, De Boeck.
Atanasova, B., Graux, J., El Hage, W., Hommet, C., Camus, V. and Belzung, C. (2008). Olfaction: a potential cognitive marker of psychiatric disorders. *Neuroscience and Biobehavioral Reviews*, 32, 1315–1325.
Atchison, K. (1997). The general oral health assessment index. *Measuring Oral Health and Quality of Life*. Chapel Hill: University of North Carolina, 71–80.
Auvray, L., Doussin, A. and Le Fur, P. (2003). Santé, soins et protection sociale en 2002. CREDES.
Bailey, R. L., Ledikwe, J. H., Smiciklas-Wright, H., Mitchell, D. C. and Jensen, G. L. (2004). Persistent oral health problems associated with comorbidity and impaired diet quality in older adults. *Journal of the American Dietetic Association*, 104, 1273–6.
Borg, G. (1982). A category scale with ratio properties for intermodal and interindividual comparisons. . In: Petzold, H. G. G. A. P. (ed.) *Psychophysical Judgment and the Process of Perception*. Berlin: Deutscher Verlag der Wissenschaften.
Boyce, J. M. and Shone, G. R. (2006). Effects of ageing on smell and taste. *Postgraduate Medical Journal*, 82, 239–241.
Carlos, J. P. and Wolfe, M. D. (1989). Methodological and nutritional issues in assessing the oral health of aged subjects. *The American Journal of Clinical Nutrition*, 50, 1210–1218.
COMMISSION_DIRECTIVE (2001)/20/EC of 4 April 2001 on the approximation of the Laws, regulations and administrative provisions of the Member States relating to the implementation of good clinical practice in the conduct of clinical trials on medicinal products for human use. *Official Journal of the European Communities*, L 121/34.
Cordelle, S., Lange, C. and Schlich, P. (2004). On the consistency of liking scores: insights from a study including 917 consumers from 10 to 80 years old. *Food Quality and Preference*, 15, 831–841.
D'hauteville, F., Aurier, P. and Sirieix, L. (1997). A sensory approach to consumers preferences for rice. First results of a European survey (France, Greece, Netherlands, Spain). *International Symposium on rice Quality*, Nottingham, UK.
De Deyn, P. P., Goeman, J., Vervaet, A., Dourcy-Belle-Rose, B., Van Dam, D. and Geerts, E. (2011). Prevalence and incidence of dementia among 75–80-year-old community-dwelling elderly in different districts of Antwerp, Belgium: The Antwerp Cognition (ANCOG) Study. *Clinical Neurology and Neurosurgery*, 113, 736–745.
Derouesné, C., Poitreneau, J., Hugonot, L., Kalafat, M., Dubois, B. and Laurent, B. (1999). Le Mini-Mental State Examination (MMSE): un outil pratique pour l'évaluation de l'état cognitif des patients par le clinicien. Version française consensuelle. Groupe de Recherche sur les Évaluations Cognitives (GRECO). *La Presse Médicale* 28, 1141–1148.
Doty, R. L. (1991). Olfactory system. In: M. L. Getchell, R. L. D., L. M. Bartoshuk and J. B. Snow (ed.) *Smell and Taste in Health and Disease*. New York: Raven Press.
Doty, R. L., Perl, D. P., Steele, J. C., Chen, K. M., Pierce, J. D., JR., Reyes, P. and Kurland, L. T. (1991). Olfactory dysfunction in three neurodegenerative diseases. *Geriatrics*, 46 47–51.
Doty, R. L., Reyes, P. F. and Gregor, T. (1987). Presence of both odor identification and detection deficits in alzheimer's disease. *Brain Research Bulletin*, 18, 597–600.

Duée, M. and Rebillard, C. (2006). La dépendance des personnes âgées: une projection en 2040. *Données sociales – La société française.*
Duffy, V. B., Cain, W. S. and Ferris, A. M. (1999). Measurement of sensitivity to olfactory flavor: application in a study of aging and dentures. *Chemical Senses,* **24,** 671–677.
Ekberg, O. and Feinberg, M. J. (1991). Altered swallowing function in elderly patients without dysphagia: radiologic findings in 56 cases. *AJR American Journal of Roentgenology,* **156,** 1181–1184.
Eurostat (2011). "Structure et vieillissement de la population" – Statistics Explained http://epp.eurostat.ec.europa.eu/statistics_Explained/index.php/Population_structure_and_ageing/Fr, 2012/8/0.
Ferry, M. (2009). *Prévention des problèmes liés au vieillissement: rôle de la nutrition et de l'activité physique* [Online]. Available: www.inserm.fr/content/download/4898/40212/file/ferry.pdf (Accessed).
Folstein, M., Folstein, S. and Mchugh, P. (1975). Mini-Mental-State: a practical method for grading cognitive state of patients for the clinician. *Journal of Psychiatric Research,* **12,** 189–198.
Forster, A., Samaras, N., Gold, G. and Samaras, D. (2011). Oropharyngeal dysphagia in older adults: a review. *European Geriatric Medicine,* **2,** 356–362.
Fucile, S., Wright, P. M., Chan, I., Yee, S., Langlais, M. E. and Gisel, E. G. (1998). Functional oral-motor skills: do they change with age? *Dysphagia,* **13,** 195–201.
Griep, M. I., Borg, E., Collys, K. and Massart, D. L. (1998). Category ratio scale as an alternative to magnitude matching for age-related taste and odour perception. *Food Quality and Preference,* **9,** 67–72.
Grunert, K. G., Dean, M., Raats, M. M., Nielsen, N. A. and Lumbers, M. (2007). A measure of satisfaction with food-related life. *Appetite,* **49,** 486–493.
HAS. (2007). Stratégie de prise en charge en cas de dénutrition protéino-énergétique chez la personne âgée. recommandations professionnelles. Available: http://www.has-sante.fr/portail/upload/docs/application/pdf/denutrition_personne_agee_2007_-_recommandations.pdf.
Hedner, M., Larsson, M., Arnold, N., Zucco, G. M. and Hummel, T. (2010). Cognitive factors in odor Detection, odor Discrimination, and odor identification tasks. *Journal of Clinical and Experimental Neuropsychology,* **32,** 1062–1067.
Hummel, T., Sekinger, B., Wolf, S. R., Pauli, E. and Kobal, G. (1997). 'Sniffin' sticks': olfactory performance assessed by the combined testing of odor Identification, odor discrimination and olfactory threshold. *Chemical Senses,* **22**(1), 39–52.
Kälviäinen, N., Roininen, K. and Tuorila, H. (2003). The relative importance of Texture, taste and aroma on a yogurt-type snack food preference in the young and the elderly. *Food Quality and Preference,* **14,** 177–186.
Koskinen, S., Kälviäinen, N. and Tuorila, H. (2003). Perception of chemosensory stimuli and related responses to flavored yogurts in the young and elderly. *Food Quality and Preference,* **14,** 623–635.
Kremer, S., Mojet, J. and Kroeze, J. H. A. (2007). Differences in perception of sweet and savoury waffles between elderly and young subjects. *Food Quality and Preference,* **18,** 106–116.
Larsson, M. and Backman, L. (1997). Age-related differences in episodic odour recognition: the role of access to specific odour names. *Memory,* **5,** 361–78.
Laureati, M., Pagliarini, E. and Calcinoni, O. (2008). Does the enhancement of chemosensory stimuli improve the enjoyment of food in institutionalized elderly people? *Journal of Sensory Studies,* **23,** 234–250.

Lawton, M. P. and Brody, E. M. (1969). Assessment of older people: self-maintaining and instrumental activities of daily living. *Gerontologist*, **9**, 179–186.

Legarth, S.V., Simonsen, C.S., Dyrlund, O., Bramsløw, L., Jespersen, C.T., Le Ray G. and Zacharov, N. (2013). A hearing impaired panel for the hearing aid industry. *Pangborn*. Rio.

Lehrner, J. P., Gluck, J. and Laska, M. (1999). Odor identification, consistency of label use, olfactory threshold and their relationships to odor memory over the human lifespan. *Chemical Senses*, **24**, 337–46.

Maitre, I., Van Wymelbeke, V., Amand, M., Vigneau, E., Issanchou, S., and Sulmont-Rossé, C. (2014). Food pickiness in the elderly: Relationship with dependency and malnutrition. *Food Quality and Preference*, **32**, Part B(0), 145-151.

Maitre, I., Van Wymelbeke, V., Sulmont-Rossé, C., Cariou, V., Bailly, N., Ferrandi, J. M., Salle, A. and Vigneau, E. (2013). Food behaviour and health patterns in the French elderly population. *Pangborn*. Rio.

Methven, L., Allen, V. J., Withers, C. A. and Gosney, M. A. (2012). *Ageing and Taste*, Cambridge, Royaume-Uni, Cambridge University Press.

Mioche, L., Bourdiol, P. and Monier, S. (2002). Changes in chewing behavior induced by aging and consequences in texture perception during meat consumption. In: Degraaf (ed.) *Tenth Food Choice Conference*. Wageningen, The Netherlands.

Mioche, L., Bourdiol, P. and Peyron, M. A. (2004). Influence of age on mastication: effects on eating behaviour. *Nutrition Research Reviews*, **17**, 43–54.

Mojet, J., Christ-Hazelhof, E. and Heidema, J. (2001). Taste perception with age: Generic or specific losses in threshold sensitivity to the five basic tastes? *Chemical Senses*, **26**, 845–860.

Mojet, J., Heidema, J. and Christ-Hazelhof, E. (2003). Taste perception with age: generic or specific losses in supra-threshold intensities of five taste qualities? *Chemical Senses*, **28**, 397–413.

Murphy, C. (1986). Taste and smell in the elderly. *Clinical Measurement of Taste and Smell*. New York: Macmillan.

Murphy, C., Schubert, C. R., Cruickshanks, K. J., Klein, B. E. K., Klein, R. and Nondahl, D. M. (2002). Prevalence of olfactory impairment in older adults. *JAMA: The Journal of the American Medical Association*, **288**, 2307–2312.

Murray, J. M., Delahunty, C. M. and Baxter, I. A. (2001). Descriptive sensory analysis: past, present and future. *Food Research International*, **34**, 461–471.

PNNS (2011). Programme National Nutrition Santé 2011 2015. Ministère du Travail de l'Emploi et de la Santé, DICOM S-11-047.

Schiffman, S. S. (1993). Perception of taste and smell in elderly persons. *Critical Reviews in Food Science and Nutrition*, **33**(1), 17–26.

Schiffman, S. S. and Graham, B. G. (2000). Taste and smell perception affect appetite and immunity in the elderly. *European Journal of Clinical Nutrition*, **54** 3, 54–63.

Sheikh, J. I. (1986). Geriatric depression scale (GDS) recent evidence and development of a shorter version. *Clin Gerontol*, **5**, 165–173.

Steele, J. G., Ayatollahi, S. M. T., Walls, A. W. G. and Murray, J. J. (1997). Clinical factors related to reported satisfaction with oral function amongst dentate older adults in England. *Community Dentistry and Oral Epidemiology*, **25**, 143–149.

Stevens, J. C. and Cain, W. S. (1987). Old-age deficits in the sense of smell as gauged by Thresholds, magnitude Matching, and odor identification. *Psychology and Aging*, **2**, 36–42.

Stevens, J. C., Cain, W. S., Demarque, A. and Ruthruff, A. M. (1991). On the discrimination of missing ingredients: aging and salt flavor. *Appetite*, **16**, 129–140.

Stevens, J. C., Cain, W. S., Schiet, F. T. and Oatley, M. W. (1989). Olfactory adaptation and recovery in old age. *Perception,* **18,** 265–76.

Stevens, J. C. and Dadarwala, A. (1993). Variability of olfactory threshold and its role in assessment of aging. *Perception and Psychophysics,* **54,** 296–302.

Sulmont-Rossé, C., Maître, I. and Issanchou, S. (2010). Age, perception chimiosensorielle et préférences alimentaires. *Gérontologie et société,* **3** (n°134), 87:106.

Sulmont-Rossé, C., Maître, I. and Van Wymelbeke, V. (2012). Aupalesens: improving pleasure of elderly people for fighting against malnutrition In: Vitagora (ed.) *7ème Congrès international goût nutrition santé.* Dijon.

Symoneaux, R., Galmarini, M. V. and Mehinagic, E. (2012). Comment analysis of consumer's likes and dislikes as an alternative tool to preference mapping. A case study on apples. *Food Quality and Preference,* **24,** 59–66.

Thomas-Danguin, T., Rouby, C., Sicard, G., Vigouroux, M., Farget, V., Johanson, A., Bengtzon, A., Hall, G., Ormel, W., DE Graaf, C., Rousseau, F. and Dumont, J.-P. (2003). Development of the ETOC: a European test of olfactory capabilities. *Rhinology,* **41,** 142–151.

Tubert-Jeannin, S., Riordan, P. J., Morel-Papernot, A., Porcheray, S. and Saby-Collet, S. (2003). Validation of an oral health quality of life index (GOHAI) in France. *Community Dentistry and Oral Epidemiology,* **31,** 275–284.

Tuorila, H., Anderson, Å., Martikainen, A. and Salovaara, H. (1998). Effect of product Formula, information and consumer characteristics on the acceptance of a new snack food. *Food Quality and Preference,* **9,** 313–320.

Veyrune, J.-L. and Mioche, L. (2000). Complete denture wearers: electromyography of mastication and texture perception whilst eating meat. *European Journal of Oral Sciences,* **108,** 83–92.

WHO. (2012). *10 facts on ageing and the life course* [Online]. Available: http://www.who.int/features/factfiles/ageing/fr/index.html (Accessed 2013/08/28 2013).

Withers, C., Gosney, M. A. and Methven, L. (2013). Perception of thickness, mouth coating and mouth drying of dairy beverages by younger and older volunteers. *Journal of Sensory Studies,* **28,** 230–237.

Wong, K. K., Muller, M. L. T. M., Kuwabara, H., Studenski, S. A. and Bohnen, N. I. (2010). Olfactory loss and nigrostriatal dopaminergic denervation in the elderly. *Neuroscience Letters,* **484,** 163–167.

Yesavage, J., Brink, T. and Rose, T. (2000). Geriatric depression scale (GDS). *Handbook of Psychiatric Measures.* Washington DC: American Psychiatric Association, 544–546.

Yesavage, J. A., Brink, T. L., Rose, T. L., Lum, O., Huang, V., Adey, M. and Leirer, V. O. (1983). Development and validation of a geriatric depression screening scale: a preliminary report. *Journal of Psychiatric Research,* **17,** 37–49.

Empathy and Experiment™: dealing with the algebra of the mind to understand and change food habits

24

H.R. Moskowitz[1], M. Reisner[2], L. Ettinger Lieberman[2], B. Batalvi[3], M. Beg[3]
[1]iNovum LLC, White Plains, NY, USA; [2]Moskowitz Jacobs Inc., White Plains, NY, USA; [3]SB & B Marketing Research, Pty, Toronto, Canada and Lahore, Pakistan

24.1 Introduction

Over time, researchers have become increasingly involved in the study of food habits. Sensory researchers have a scientific interest in foods, whereas studies by consumer researchers lead naturally to research on how and why foods are consumed. A cursory inspection of the sensory and consumer research literature reveals that in the early days (the 1960s–1980s), food habits were not of interest. It was the product that was the focus of attention. We see this in reviews, but more concretely in books and bibliographies dealing with food (Amerine *et al.*, 1965; Pangborn and Trabue, 1967). When food habits were studied, they would be studied by sociologists or anthropologists. One or another university department of food science would offer the occasional course on food habits, often perceived to be soft, non-scientific, and certainly not helpful to anyone applying for his or her first job. As a science matures, its scope widens. That truism applies to the world of foods, especially sensory analysis, as well as to the world of consumer research. With maturity comes widening focus, beyond stuff, beyond the thing, beyond the easy-to-control, eventually leading to the existential leap into the unknown. In our case, that leap is a leap beyond the product itself, to the mind of the consumer, specifically the Pakistani consumer who is encouraged to eat rice which contains vitamin A precursors. Although one could focus this chapter on the Golden Grain product itself, we use it as the focus of attention in order to learn about dealing with human food habits through application of a behavior research technique more typically used to understand preferences between different items in everyday commerce. We show how to CREATE a rapid deep understanding of the consumer mind (empathy) and, equally, a rapid discovery of what to say. What took years now becomes a short streamlined process taking weeks.

24.2 The origins of the study

A lack of vitamin A creates the conditions for blindness (Rahi *et al.*, 1995). Adding vitamin A to food through fortification, or making it exist in its precursor form in the food, are strategies that might well go a long way toward eradicating the blindness associated with vitamin A deficiency. To many, the promise of preventing blindness would seem to be no problem whatsoever.

There are problems, however, and it is those involved with new groups of consumers that constitute the focus of this chapter. The recurring problem is food habits and the negative response to foods outside of our daily repertoire. Many of us limit our diet to foods with which we are familiar, rejecting new foods. This rejection (neophobia, or fear of the new) is common with food, especially with those who are new to the culture (Pliner and Hobden, 1992). Indeed, there is a generalized fear of the strange and the unusual. Speakers of a language who invent new words to describe fears are tapping into deep psychological processes.

We often learn best from case histories. In that spirit, this chapter presents work on responses to the notion of introducing a new rice product into a third-world population of poor consumers, accustomed to eating wheat rather than rice. Working with these respondents on a specific task, understanding their mind-set with respect to this rice, tells us a lot about how to deal with new populations. We present the approach, faced with distinctive issues, grounded in both qualitative and quantitative research, showing how this method might be a template for further studies across different populations.

We designed this study after consulting with the Bill and Melinda Gates Foundation, based in Seattle, Washington. The issue was how to better understand the prospective customer for the grain known as Golden Rice, a product well known in the world of agriculture, which turns out to have clear benefits, but also negatives:

1. It is rice, not wheat. How does one convince people accustomed to wheat products to buy and accept a new variety of rice?
2. Golden Rice is genetically modified. How does one convince people who could live in far better health to accept a grain which is genetically modified? This is a general worldwide problem, not a particular one limited to this study.
3. Golden Rice has a color resulting from the vitamin A precursors. How does one convince a poor population to accept a new sensory characteristic in a grain? We know that people begin to eat "with their eyes." The gold color itself, which many might not consider an issue, could well turn out to be a major problem if the color were to signify quality, psychological or cultural/ethnic issues. According to Jill Morton, who served as a visiting professor in the School of Visual Arts at Beaconhouse National University in Lahore, Pakistan, during February–March, 2009, although there are no absolutes, there are logical sources for the range of complex and sometimes contradictory meanings of colors, which include: cultural, political and historical, religious and mythical, linguistic associations, as well as contemporary usage and fads. Among the people, landscape, architecture, food, she found timeless examples of gold and yellow reflecting fruit, flowers, kite and basant (yellow) festival, ox harnesses which represented happiness, luminosity, hope, warmth and nourishment (Morton, 2011). That being the case, color might not be an issue in the introduction of this new grain.

24.3 Background: Golden Rice – the positives

Dietary micronutrient deficiencies, such as the lack of vitamin A, iodine, iron or zinc, are a major source of morbidity (increased susceptibility to disease) and mortality worldwide. For the most part, these deficiencies adversely affect children, impairing their immune systems and normal development, causing disease and ultimately death. The best way to avoid micronutrient deficiencies is by way of a varied diet, rich in vegetables, fruits, and animal products. The second best approach, especially for those who cannot afford a balanced diet, is by way of nutrient-dense staple crops. Sweet potatoes, for example, are available as varieties that are either rich or poor in provitamin A. Those producing and accumulating provitamin A (orange-fleshed sweet potatoes) are called biofortified, as opposed to the white-fleshed sweet potatoes, which do not accumulate provitamin A. In this case, what needs to be done is to introduce the biofortified varieties to people used to the white-fleshed varieties. Unfortunately, there are no natural provitamin A-containing rice varieties (see www.goldenrice.org).

Biofortification is the development of micronutrient-dense staple crops using the best traditional breeding practices and modern biotechnology. According to Nestel and other researchers affiliated with HarvestPlus, International Food Policy Research Institute, Washington, DC and HarvestPlus, Centro International de Agricultura Tropical, Cali, Colombia, this approach has many advantages: (1) it capitalizes on the regular daily intake of a consistent and large amount of food staples by all family members; (2) after the one-time investment to develop seeds that fortify themselves, recurrent costs are low, and germplasm can be shared internationally; (3) the biofortified crop system is highly sustainable, and continues to be grown and consumed year after year, even if government attention and international funding for micronutrient issues fade; (4) biofortification provides a feasible means of reaching undernourished populations in relatively remote rural areas, delivering naturally fortified foods to people with limited access to commercially marketed fortified foods that are more readily available in urban areas. Biofortification and commercial fortification, therefore, are highly complementary; and (5) breeding for higher trace mineral density in seeds will not incur a yield penalty (Nestel *et al.*, 2006).

Biofortified rice has been recognized as contributing to the alleviation of life-threatening micronutrient deficiencies in developing countries. Rice plants produce β-carotene (provitamin A) in green tissues but not in the endosperm (the edible part of the seed). The outer coat of the dehusked grains – the so-called aleurone layer – contains a number of valuable nutrients, e.g., vitamin B and nutritious fats, but no provitamin A. These nutrients are lost with the bran fraction in the process of milling and polishing. While it would be desirable to keep those nutrients, the fatty components are affected by oxidative processes that make the grain turn rancid. Thus, unprocessed rice – also known as brown rice – is not suited to long-term storage. Even though all required genes to produce provitamin A are present in the grain, some of them are turned off during development. In rice-based societies, the absence of β-carotene in rice grains manifests itself in a marked incidence of blindness and susceptibility to disease, leading to an increased incidence of premature death of small children, the weakest link in the chain (www.goldenrice.org).

24.4 Background: Golden Rice – the negatives

How would you feel if you were told that a group of scientists came to Canada and fed a group of 24 children between the ages of 6 and 8 with a potentially dangerous product that had yet to be fully tested or understood? What if you learned that federal authorities came out publicly against this very experiment, and yet the experiment continued? You would be pretty outraged, right (see www.greenpeace.org/canada/en/Blog/golden-rice-the-true-story/blog/42223/)?

Not Margaret Wente, a columnist at *The Globe and Mail*. She preferred to criticize Greenpeace for issuing an alarming press release on that very issue. In September, 2012, she wrote a column ("Greenpeace's Golden Rice stand should appall us all") that addressed the genetically engineered (GE) Golden Rice trial in China. She ended her article by asserting that Greenpeace and its allies would rather have millions of children go blind than be given this "safe" solution (Wente, 2012).

Such hyperbole is not worthy of Canada's national newspaper and the reporter could not be further from the truth. Ms. Wente is feeding the illusion that Golden Rice is a safe solution. The truth is that the safety of GE food for humans and feed for animals is still unknown. There are no available independent studies that prove otherwise. On the other hand, there are studies that prove that GE crops certainly have the potential to cause allergic reactions.

24.5 Empathy and Experiment™: the two halves of the approach

One way to work with new populations of respondents is to listen to them, understand their feelings, and then select a solution which works, ideally the optimum solution. The two halves are the approach we call Empathy (getting to the respondent's heart) and Experiment™ (understanding what works).

24.5.1 Getting to empathy: origins in the psychology of personality

For more than a century, consumer researchers have recognized a need for a deeper psychological understanding of consumers. While many researchers have ventured down that path, most have shied away from completing the journey and halfway through they either lose track or find themselves traversing landscapes they do not particularly want to explore. Consequently, most marketing and communication efforts tend to remain peripheral and/or superficial, often missing the mark, unable to impact thought, appeal to emotions or change behavior.

Enter 5KEYS™, a holistic qualitative research paradigm emerging from the business practice of one of the authors (Batool Batalvi). The approach attempts to explore, in an efficacious, pragmatic way, the psychic apparatus of consumers in its various dimensions. In a typical 5KEYS™ study, the researcher uses any one, or even a hybrid mix, of qualitative techniques such as observation, ethnographies, in-depth interviews, focus groups, storytelling, online bulletin boards, and so forth. These methods, often used as stratagems for interviewing, engage consumers, pull out responses, and generate insights. It should be noted here that the 5KEYS™ method is actually one of many alternative, disciplined, qualitative processes used by consumer researchers to pull out information from the respondent (Moskowitz et al., 2014).

In business practice, these insights are formalized by assigning them to one of five groups/key response dimensions: personality, cognition (automatic thoughts), affect (moods), behavior (motor) and physiological responses (e.g., sweating, pupil dilation), respectively. These five dimensions are not necessarily independent at a functional level. A change of behavior in any one dimension may result in the behavior changing on other dimensions, often in ways not easy to predict.

The qualitative data for the 5KEYS™ are collected in a systematic, hypothesis-testing manner. This first, soft, approach relies on methods we often associate with the process of psychotherapy and its intuition-oriented approach to understanding. The knowledge-development process is rigorous, despite the softness often associated with this therapeutic treatment. Typically the research begins with an "Assessment" designed to lead to a "Hypothesis" about the nature of the mechanisms causing or maintaining consumer reactions. The process ends with a customized "Action Plan." As fieldwork proceeds, with respondents being interviewed and observed, the researcher collects data, generally descriptive, with the goal to evaluate the effects of the proposed action plan on the five aforementioned dimensions.

Historically, we trace the formulation of the organization into the five dimensions and the specific investigative methods to Beck's work on cognitive theory and therapy (1976). Protocols for data analysis rely on Kohut's Self-Psychology (specifically Kohut's framework of Empathy, Self-object, Mirroring, Idealizing, Alter-Ego, Tripolar self (1971) and, finally, invoke Carl Jung's typological approach to personality typing (1976)).

At the very basis of 5KEYS™ is the psychological construct called the schema, a non-conscious, latent structure for a person to organize his world and take action. The schema enables the person to interpret his or her experiences. External stimuli, situations, and life events activate latent schemas, triggering specific positive or negative behavioral patterns. It is these external stimuli which are to be studied through experimentation, but first understood through the schema.

24.5.2 Empathy as the source of raw material to understand the respondent

Now that we have the basis for understanding a person and his behavior, how can we use empathy to guide us in the development of raw material that will be used in the EXPERIMENT phase, when we test ideas?

We begin by recognizing that schemas, although latent, organize external behavior (Beck, 1983). That means that Golden Rice can be first understood in the mind of the consumer, and afterwards studied by experiment. Observing the consumer's external behavior in the context of an environment gives one an idea of the underlying schema, or what is going on to organize this behavior. That underlying schema will generate ideas about how the consumer perceives and values the world, especially as the world involving food. Furthermore, it is through observing how an individual talks about and behaves in an environment that the researcher can get at thought, mood, physiological response and behavior, giving a sense of the structure of the person's schema (Beck et al., 1993).

It is the researcher's task to talk with the respondent, observe behavior, develop a sense of the underlying schema and, finally from that schema, intuitively extract ideas, sound bites, thought bites. These extracted ideas are the basis of the stimuli to be tested in the EXPERIMENT phase. The extracted ideas are both objective (referring to the topic), and subjective (referring to the experience). The therapist would move forward to use these ideas to change the person's behavior, e.g., you should spend an extra dollar on this superior brand of rice because it is better for health. The experimenter would move in a different direction, using these insights to create simple descriptive phrases to be used in experimentation, the second part of the research.

If we were to summarize the output of EMPATHY, we might say the following:

> ... those rewarding junctures in the narrative, when a singular moment
> cracks open a door, shining light on vaguely imagined possibilities.
> The crack widens as empathy takes hold...revealing novel, unexpected
> glimpses of people and situations, presenting intriguing new scenarios
> and solutions, overlooked by competition....
> What else do we get from these pivotal moments of transformation?
> Consumer Insight.......! A strong brand idea brings sudden clarity...
> When you get it right it is the awakening of old knowledge...Which
> is introspectively self-evident...A feeling of "aahhh... I already knew that"
> (Moskowitz et al., 2014).

24.5.3 Guidelines about gathering qualitative data

Although it might seem that anyone can execute qualitative studies, simply by talking to the respondent or by observing behavior at the home or when shopping, the reality is a bit more complicated. Qualitative research requires skill in knowing how to interact with respondents, as well as understanding what works and what does not. Table 24.1 presents a set of observations that give a sense of what is involved with qualitative research in general, and with the 5KEYS™ approach in particular.

24.5.4 A sense of the actual experience in Pakistan

Most published research involving qualitative interviews, whether in trade journals or in archived journals, involves respondents in developed countries. The same types of interviews are performed in developing countries, but often are one-on-one, with an interviewer talking directly to the respondent. Figure 24.1 shows a photograph of a typical interview from this project. The interviewer, with a computer, is talking with the respondent. The interviewer and the respondent are of the same gender. The interviewer is recording the respondent's answers, taking notes on a computer. Around the

Empathy and Experiment™ 515

Table 24.1 General rules of thumb for qualitative research

- Face-to-face methods of data collection often work better than others (e.g., diary or self-completion).
- Usage of relevant language or dialect is essential, not just dominant language.
- In some markets, additional security measures should be undertaken.
- All geographies should be included (different climate zones, urban/rural, etc.).
- In countries with low literacy, level of information should be presented bit-by-bit and not all together.
- Access to homes is often facilitated via contact in neighborhood.
- Very simple projective techniques should be used (due to the fact that consumers take very literal interpretations of the tasks) and should be selected as appropriate to the country.
- Respondent may also have limited brand experience (especially in rural areas), so some projective techniques, e.g., brand sorting, should not be used or used with caution.
- Focus groups should be conducted in a place comfortable for respondents (e.g., in their homes).
- In rural areas especially, transportation of the respondents to focus group venue may be needed.
- Focus groups work better (people are more open) if respondents are people from a similar social network or community.
- Analysis of the research should be done by or with involvement of a person who knows the lifestyle of respondents and will be able to contextualize the information received.
- In countries with low level of literacy among respondents, graphic/picture stimuli should be used as description materials.
- Less educated respondents can experience difficulties evaluating the information presented in a conceptual way or evaluating too many concepts.

Source: Courtesy SB&B Research, Pty.

couple, the interviewer and the respondent, may be children are playing or attentively watching this rather novel experience. These individual respondents, not the hundreds of respondents that will be needed when time comes to do the experiment, are needed for some research experiments. The particular study described here involved the following individuals:

1. Five story tellers, i.e., respondents who were interviewed in depth
2. Residents of Lahore slums and rural locales
3. Mothers and fathers
4. Families with at least one child 5 years or below
5. Lower SECs: (C/D) (E1/E2)

24.5.5 Developing the information from EMPATHY

In the study, the researcher uses one or a mix of qualitative techniques to engage consumers, pull out responses, and generate insights. Below are some of the formalized insights assigned to one of the five groups/key response dimensions: personality,

Figure 24.1 One of the five interviews in a slum area of Lahore, Pakistan, for the Golden Rice study reported here. This interview was part of the Empathy portion of the Golden Rice project.

cognition (automatic thoughts), affect (moods), behavior (motor), and physiological responses (e.g., sweating, pupil dilation), as shown in Tables 24.2 through 24.4 below.

The work product of qualitative research, whether interviews, focus groups, or ethnography, is stories, insights, a sense of what is happening, a feeling of the deeper reality of the situation. Thus, in the case of our Pakistani interviews, a variety of latent feelings were brought to the surface, together with reactions to one's life and to the food in that life. In some respects, the insights were already present in the mind; it just took a competent researcher/field observer/analyst to paint the insights in language that could be understood, rather than just read.

What specifically can be done with these insights? How does one apply the rigor to know what to do? The insights themselves are a person's effort to make sense out of the structured, yet inherently chaotic, real-life situation. How can one bring disciplined science to the problem, to create an archival set of data, numbers that live the reality of the consumer, yet numbers that live the reality of science? It is here that experimentation comes in, experimentation informed by the qualitative insights (Thomke, 2003).

24.6 The value of experimentation and implementation of Golden Rice evaluations among Pakistanis

There is a whole industry of insight providers, so-called qualitative researchers, who talk to people, observe their behaviors, and then offer hypotheses of why. These insights are important to ground our knowledge of what people want – their needs,

Table 24.2 **Emotion**

Highest charge	Evidence for	Evidence against	Desired emotion	What must change?
Helplessness	The rich feed kids better	I am confident of my kid's needs	Optimism	Feelings of deprivation

Courtesy SB & B Research, Pty.

Table 24.3 **Behavior**

Optimum behavior	Evidence for	Evidence against	Desired emotion	What must change?
Men given precedence	Men buy rice, eat first	Mothers, mothers-in-law, influencers	Cook Golden Rice – for everyone, anytime	Treating (Golden) rice as a specialty, expensive dish

Courtesy SB & B Research, Pty.

Table 24.4 **Sensory**

Burning sensation	Evidence for	Evidence against	Desired sensate	What must change?
Sweetness	Rice desserts made golden with addition of saffron (Zarda, Kheer)	Can yield unpleasant results	Dessert with Golden Rice	Golden Rice has limited options/ sensates

Courtesy SB & B Research, Pty.

desires, and preferences. Yet often these insights remain simply general wisdom around a category, which one or a few practitioners and experienced professionals have, but which is not incorporated into documents that future generations can read. It is here that experiment becomes important.

For the qualitative interview work on Golden Rice, it was clear that the Pakistani respondents were able to articulate various reasons why they liked or disliked the idea of the product. What was not clear, however, was the strength of valence, the degree to which the idea would resonate with the consumer or repel him. What was also not clear was the degree to which the population was homogeneous. In many studies with consumers in one region, one country, or across countries, the reality keeps emerging that the consumer mind is not homogeneous. Rather, people differ from each other, often in profound ways, which makes sense. A further finding is that "almost primaries" exist in population's basic subgroups. When the same set of related ideas is tested in different countries, the same basic groups keep emerging, albeit in different proportions.

The EMPATHY portion of the project generated many different observations, leading to a set of insights. Some of these were clear factoids, such as the feeling that Golden Rice might somehow be contaminated. Others were simple statements, unremarkable in themselves, but taking on more meaning when put into the schema offered by the 5KEYS.™

In order to proceed, it was important to formalize the insights in a structure that could be used for experimental design. The underlying design would mix and match the elements from the EMPATHY portion of the project, create small vignettes from the mixtures, present the vignettes to the is respondent as, instructing the respondent to rate the vignettes as if the vignette represented one single idea. Thus, the experimental design forced the respondent to evaluate a combination, rather than single elements, one element at a time. Although it might seem easier to rate single elements, in actuality evaluating combinations prevents the respondent from providing so-called politically correct answers that would please the interviewer, but would not be true. It is possible to be politically correct with one element, assigning a rating which compromises one's own point of view, in an attempt to be agreeable. One can carefully monitor one's own responses and adjust them. When the respondent is confronted with a mélange of elements, however, it is impossible to be politically correct; too much is happening in a single vignette. The happy result is a more honest answer, simply out of being overwhelmed with information leading in many directions (Moskowitz and Gofman, 2007).

The actual mechanics of experimentation are well known (Box *et al.*, 1978). The approach that we will use is founded in experimental design and straightforward statistical analysis using ordinary least-squares regression (OLS) (Anderson, 1977; Systat, 2007). The actual approach used in this book has been explicated in depth by Moskowitz *et al.* (2005), basing their work on the pioneering approaches of the late Professor Paul Green of Wharton School of Business, The University of Pennsylvania (Green and Srinivasan, 1981; Green and Krieger, 1991).

24.7 Summary of the elements and process of the experiment

Table 24.5 presents each of the five silos (A–E) and a sixth silo (F) presenting some ideas about Golden Rice. The rationale for adding a sixth silo is that a so-called 6 × 6 design (six silos, six elements per silo) generates the most information possible in these experiments, yet is efficient in terms of respondent effort. The 6 × 6 design covers 36 elements, sufficient for six elements for each key, and requires a set of 48 vignettes or combinations for each respondent. Every element appears 5 × among these 48 combinations. Finally, each respondent evaluated a unique set of 48 vignettes. So for 200 respondents, in theory, 9600 different vignettes would be

Table 24.5 The elements used in the experimental design were presented in both Urdu and in English

SILO A: COGNITION	
A1	My children do not get sufficient nutrition on a daily basis.
A2	I do not have enough information about the healthy foods for my children.
A3	I know what a balanced diet is ... and its importance for my family.
A4	Sunday bazaar (fair price bazaar) has poorer quality of food compared to regular stores.
A5	Rice is more expensive than wheat ... and wheat is more filling and nutritious.
A6	I must learn how to choose and cook more nutritious foods
SILO B: EMOTION	
B1	Sometimes, I worry that the nutritional quality of meals I give my family is poor.
B2	I'm confident that I feed my children according to their nutritional needs.
B3	I will feel hopeful if I can learn how to improve the nutritional content of my children's daily diet ... without incurring additional expense.
B4	I feel deprived when I see the rich feed their children better quality food.
B5	Parents will be happy to cook a free sample of new Golden Rice ...especially when they are told it is healthier for their children, compared to white rice.
B6	I feel helpless when my children ask for foods I cannot afford.
SILO C: SENSORY	
C1	I like the look of white rice .. not the look of golden color rice.
C2	New, golden color rice if cooked with white rice could make a tempting Biryani dish.
C3	I could make sweet rice dishes (like zarda), with new Golden Rice.
C4	Adding a touch of saffron would enhance the taste of a new golden color rice dish.
C5	A dish made with new golden color rice will be unappetizing.
C6	I expect a new golden color rice dish to have an odd aroma and flavor.
SILO D: BEHAVIOR	
D1	It is mostly the men in our family who buy grains (like rice and wheat).
D2	Rice is only cooked for special occasions in our family.
D3	We do not cook any special meals for children... they eat what we eat.
D4	Women in our family first feed the men and children before eating themselves.
D5	The increasing cost of living made us cut down on our rice purchase.
D6	I would only feed new golden color rice to my children, if my mother/or mother-in-law said it gives them more strength.
SILO E: REACTIONS TO CONCEPT	
E1	New, golden color rice will make people think the crop is rich and precious in nutrients like gold.

Continued

Table 24.5 Continued

E2	I would be willing to eat this new rice if I'm told its golden color is due to vitamin A, which is important for health.
E3	I would be afraid to feed my kids new golden color rice, even if I'm told it contains the goodness of fruits and vegetables.
E4	A new golden colored rice will make people suspect that the crop has come in contact with wastewater from textile factories.
E5	I would definitely try golden color rice if I were told it contains the goodness of chicken/meat (protein).
E6	It is possible to scientifically improve crops to contain healthier nutrients.
SILO F: PERSONALITY	
F1	I mostly make my own decisions, even if it goes against the opinion of others.
F2	I actively seek out information based on modern science and knowledge, because it is good for me and my family.
F3	I mostly enjoy watching Urdu dramas and Indian movies.
F4	I tend to seek the opinion of others (such as elders/religious leaders) before making major decisions.
F5	I believe age-old wisdom is to be trusted rather than modern-day thinking.
F6	I mostly watch religious programs/sermons on TV.

The study investigated six silos (five silos from the 5KEYS™, one additional silo of elements about Golden Rice).

evaluated. These many different vignettes prevent bias due to an incorrect choice of combinations, perhaps a combination of elements which interact with each other, unbeknownst to the researcher (Gofman and Moskowitz, 2010).

The method used to uncover the impact of the different elements follows a sequence of activities shown in Fig. 24.2. The essence of the approach is to mix and match elements from the different silos, using experimental design. Experimental design can be likened to a recipe book, prescribing the different combinations or vignettes. Each vignette comprises a minimum of three elements and a maximum of four elements, with the average number of elements in a vignette exactly equal to 3.75. Each silo contributes at most one element to a vignette, but often a silo is entirely absent from a vignette, a property which prevents the statistical issue of multi-collinearity wherein the elements are not statistically independent of each other.

Although the vignettes are incomplete, from the point of view of someone who knows the underlying set of silos and elements, the reality is that respondents really never notice the incompleteness. The reason for not noticing is simple. People graze for information, rather than look for complete information. We do not think in terms of silos and elements. We think in terms of issues, factoids, impressions, and so forth. Even knowing the underlying set of silos does not make the vignettes seem incomplete at all. In simple terms, we are just not wired that way, to respond only to complete combinations.

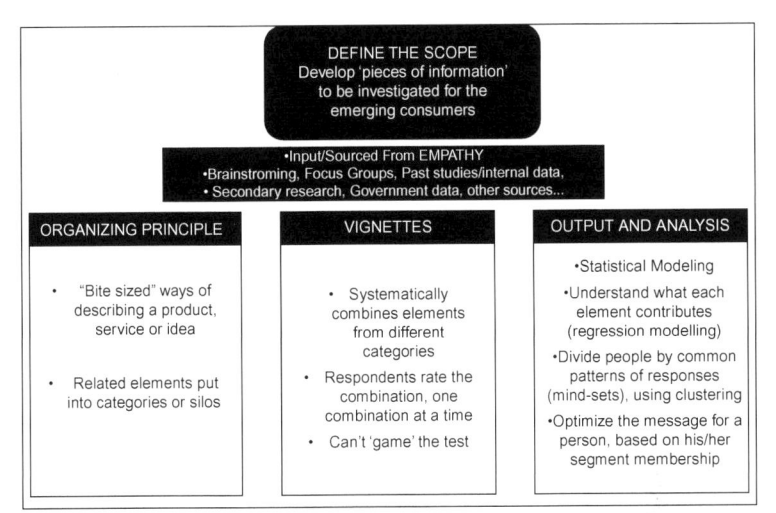

```
                    DEFINE THE SCOPE
                Develop 'pieces of information'
                    to be investigated for the
                       emerging consumers

                   •Input/Sourced From EMPATHY
             •Brainstroming, Focus Groups, Past studies/internal data,
                • Secondary research, Government data, other sources...
```

ORGANIZING PRINCIPLE	VIGNETTES	OUTPUT AND ANALYSIS
• "Bite sized" ways of describing a product, service or idea • Related elements put into categories or silos	• Systematically combines elements from different categories • Respondents rate the combination, one combination at a time • Can't 'game' the test	•Statistical Modeling •Understand what each element contributes (regression modelling) •Divide people by common patterns of responses (mind-sets), using clustering •Optimize the message for a person, based on his/her segment membership

Figure 24.2 Schematic showing how the EXPERIMENT works, using the principles of experimental design, regression modeling, segmentation, and discriminant function analysis. *Source*: Courtesy SB & B Research, Pty.

24.8 The material of the interview and analysis of structured experimental design data

The EXPERIMENT was conducted according to the principles of rule developing experimentation (RDE) (Moskowitz and Gofman, 2007). The study is set up on a computer at a central server. The test stimuli are sent by Internet e-mail invitation to respondents who opt into the study. The actual stimuli are set up on the respondent's own computer. The process is rapid, reducing the amount of waiting time, and ensuring that the respondent remains engaged.

The interview begins with an orientation page. Although in Western countries with today's Internet penetration people feel overly interviewed, the reality is that it is important to have a well-oriented respondent, a respondent who understands what is expected. This requirement holds for Western respondents who are familiar with interviews, and is even more applicable for Asian and African respondents who are unfamiliar with the process of interviewing. Figure 24.3 shows the orientation page, in both Urdu (the language of the respondents in Lahore) and English. Respondents who were unable to read listened to an interviewer who read the orientation page from the computer screen (Fig. 24.4).

The respondent evaluated each vignette, rating the vignette on two five-point scales. Figure 24.5 shows the rating scale. Figure 24.6 shows an example of a vignette as it would appear on the computer screen.

In research it is often tempting to believe that one can inspect a set of data and come up with a hypothesis about what might be occurring. One is inclined to skip the preliminary design work, make the observations, and then search for patterns. As admirable as our intuitive abilities may be, such shortcuts rarely work. When dealing

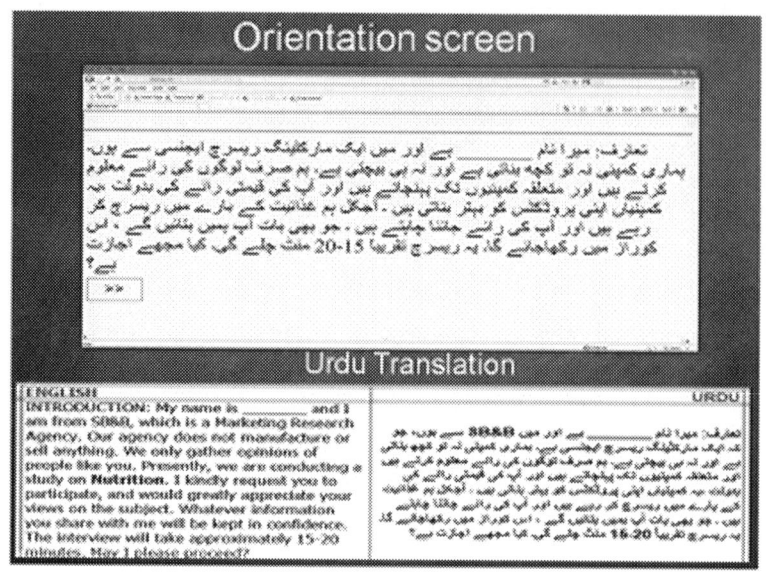

Figure 24.3 The Orientation Page for the study, shown in both English and Urdu.
Source: Courtesy: SB & B Research, Pty.

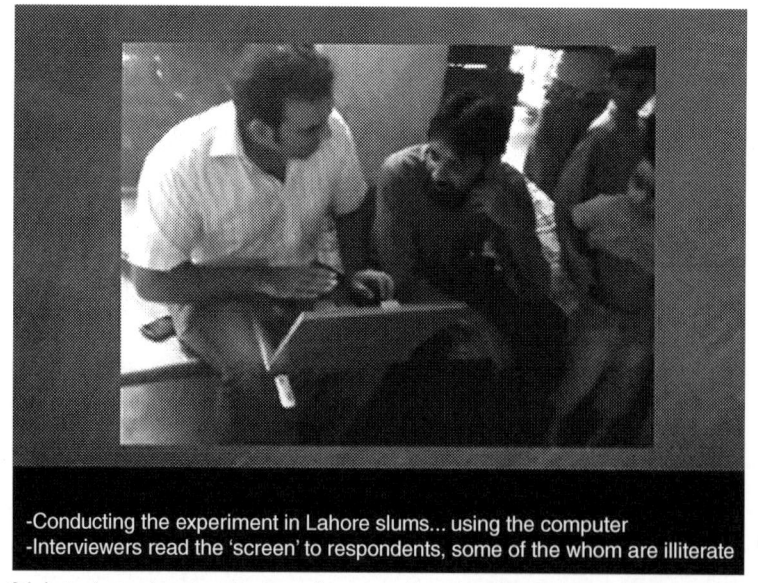

Figure 24.4 A photo showing the interviewer reading the stimulus to an illiterate respondent.
Source: Courtesy: SB & B Research, Pty.

Empathy and Experiment™ 523

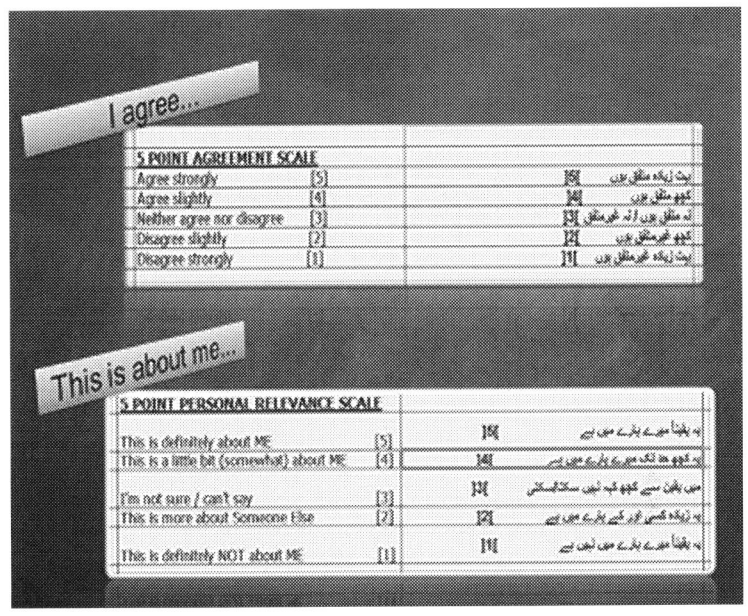

Figure 24.5 The two ratings questions.
Source: Courtesy SB & B Research, Pty.

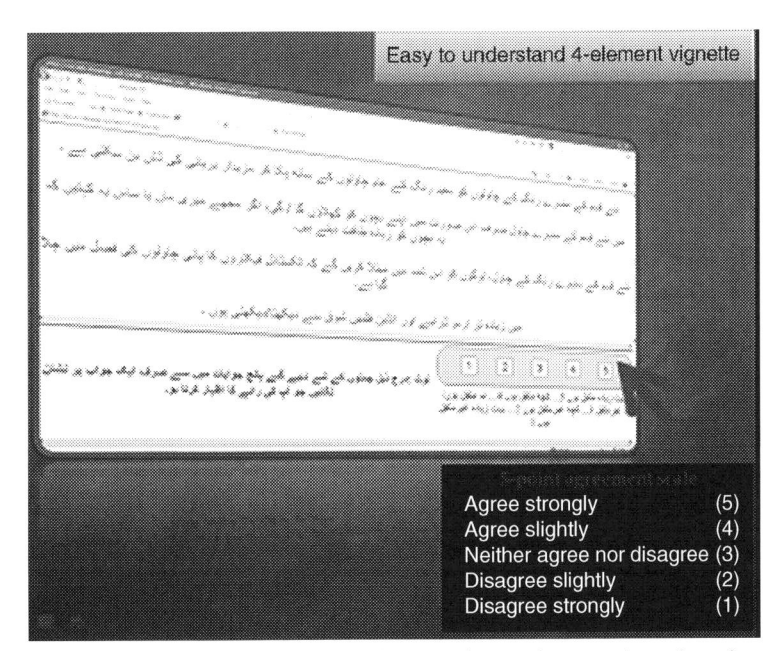

Figure 24.6 Example of a vignette with a rating question as shown on interviewer's computer screen.
Source: Courtesy SB & B Research, Pty.

with new populations without a lot of fundamental knowledge, such shortcuts lead to inevitable disappointment and sometimes to significant failure.

That said, two of the best things about experimental design are its rapidity and ease of analysis. Experimental design forces us to create combinations, systematic variations of the test stimuli. The analysis ends up being simply the recognition of relatively obvious patterns emerging from the different stimuli, rather than the laborious search for patterns lurking in the mass of data, but hard to discern since the stimuli are not connected to each other.

The best way to analyze the statistically designed combinations is with regression analysis, colloquially known as curve fitting. Having set up the combinations so that the elements are truly independent of each other in a statistically valid way (orthogonality), we use regression programs to relate the presence/absence of the elements to the responses as follows:

- Each respondent generates a set of 48 rows, one row for each vignette. The 48 rows have all the information necessary to analyze the results.
- We code the independent variable, the 36 elements, as 0 for being absent, 1 for being present. Each row, therefore, comprises an initial set of 36 columns, one column for each element. Looking across the 36 columns, we will count either three columns containing 1's (for vignettes with three elements), or four columns containing 1's (for vignettes with four elements).
- We code the dependent variables 1–7, corresponding to the ratings assigned to Question #1 (disagree/agree) and Question #2 (fits me).
- For purposes of analysis, we transform the two dependent variables to a binary scale, with ratings 1–5 transformed to the value 0, and ratings 6–7 transformed to the value 100. This transformation follows the conventions of consumer researchers who focus on the membership in a group (not me or 1–5, me or 6–7), rather than on the intensity of feeling.
- We add a small random number to all four dependent variables, i.e., the original two rating scales (seven-point scales, Questions #1 and #2, and the two transformed scales, which are now binary scales). The random numbers, around 10^{-5}, do not affect the subsequent analysis, but they do prevent the regression analysis from crashing due to the respondent choosing the same rating again and again.
- The 48 vignettes evaluated by each respondent constitute an experimental design for that unique respondent, with the array of vignettes different from respondent to respondent. The experimental design allows us to create a simple linear model relating the presence/absence of the 36 elements to the binary rating (0/100) for Question #1, and another simple linear model for the same respondent for Question #2. We express the equation as:

    Binary Rating = k_0 + k_1(Element A1) + k_2(Element A2) ... k_{36}(Element F6).

- The foregoing pair of equations is estimated on a respondent-by-respondent basis, and then aggregated by respondents in a specific subgroup.
- The additive constant tells us the conditional probability that a respondent would assign the value 6–7 to the question in the absence of elements. Of course, all vignettes comprised elements, so that the additive constant is estimated, rather than directly observed. On the other hand, the additive constant is a good baseline for the question and, as we will see in a moment, differs from one group of respondents to another.

- Each coefficient, k_1–k_{36}, shows the part-worth contribution of that particular element to the overall rating assigned to Question #1 or Question #2. We will see dramatic differences across complementary subgroups, giving us a sense of the respondents' minds.

24.9 Explicating the results – the total panel versus gender

The equations themselves constitute a framework in which we believe the respondent may make his decision. We assume that the respondent carries around with him a mental calculator, and in most situations acts intuitively. The respondent may or may not be able to access this mental calculator at a rational level, but behaves as if it were working.

We discover the parameters of this mental calculator, the utility values, through the decomposition and then the summation by key groups. For example, we can look at the results from the total panel by averaging the corresponding parameters of all 200 respondents. We can then divide our respondents into two groups, using gender as the criterion.

Let us focus on Question #2 rating point "describes ME." The results appear in Table 24.6, which shows the subgroups across the top, then base size, the additive constant (basic proclivity to say that the vignette describes the respondent, in the absence of elements), and then the additional part-worth contribution of every element.

One can estimate the total binary score, e.g., the total percent of respondents saying "this vignette describes ME" by summing the additive constant with the contribution of the elements in the vignette designed by the researcher. Each element generates its own coefficient. The elements are statistically independent of each other. The vignette must have at most four elements, with only one or no element from each silo.

Table 24.6 teaches us that:

The additive constant for Question #2 is 45 when we combine all respondents into one set and then average the individual models. The 45 tells us that in the absence of any elements, approximately 45% of the respondents are prepared to say "describes ME," i.e., rate the vignette 6–7.

When we look at gender, we find a radical difference. The males show an additive constant of 55, meaning 55% of the males are likely to say that the vignette describes them. In contrast, the additive constant for the females is 35, meaning only 35% of the female respondents are likely to say that the vignette describes them.

We already get a sense that there are strong gender differences in the patterns of responses, and thus in the structure of the ways males and females respond to these types of stimuli.

Among the total panel, averaging the responses of all groups, no element drives a strong agreement, i.e., no element substantially drives up the rating of "describes ME." The highest positive element has a value of 1, virtually nothing. In contrast, people know what they do not agree with. Here is an element which scores –13.

Table 24.6 How the different elements for Golden Rice fare for the total panel, and for males versus females

		Total	M	F	M–F
	Base size	200	100	100	
	Additive constant	45	55	35	20
F5	I believe age-old wisdom is to be trusted rather than modern-day thinking.	1	4	–2	6
B6	I feel helpless when my children ask for foods I cannot afford.	1	2	–1	3
F4	I tend to seek the opinion of others (such as elders/religious leaders) before making major decisions.	1	2	0	2
E2	I would be willing to eat this new rice if I'm told its golden color is due to vitamin A, which is important for health.	1	–1	3	–4
D4	Adding a touch of saffron would enhance the taste of a new golden color rice dish.	1	–2	4	–6
C1	I like the look of white rice ... not the look of golden color rice.	0	–4	4	–8
A4	Sunday bazaar (fair price bazaar) has poorer quality of food compared to regular stores.	0	–9	9	–18
B3	I will feel hopeful if I can learn how to improve the nutritional content of my children's daily diet without incurring additional expense?	–1	–3	0	–3
B1	Sometimes, I worry that the nutritional quality of meals I give my family is poor.	–1	–3	1	–4
F2	I actively seek out information based on modern science and knowledge, because it is good for me and my family.	–1	–5	3	–8
E6	It is possible to scientifically improve crops to contain healthier nutrients.	–2	–3	–1	–2
A5	Rice is more expensive than wheat... and wheat is more filling and nutritious	–2	–3	0	–3
D1	I like the look of white rice ... not the look of golden color rice.	–2	4	–1	–3
E5	I would definitely try golden color rice if I were told it contains the goodness of chicken/meat (protein).	–2	4	1	–5
B5	Parents will be happy to cook a free sample of new Golden Rice ... especially when they are told it is healthier for their children, compared to white rice.	–3	–1	–5	4
C2	New, golden color rice if cooked with white rice could make a tempting Biryani dish.	–3	–7	1	–8
C6	I expect a new golden color rice dish to have an odd aroma and flavor.	–3	–6	–2	–4
C3	I could make sweet rice dishes (like zarda) with new Golden Rice.	–4	–5	–3	–2

Table 24.6 Continued

		Total	M	F	M–F
D6	I expect a new golden color rice dish to have an odd aroma and flavor.	−4	−6	−2	−4
E4	A new golden colored rice will make people suspect that the crop has come in contact with wastewater from textile factories.	−4	−8	−1	−7
C4	Adding a touch of saffron would enhance the taste of a new golden color rice dish.	4	−8	−1	−7
F6	I mostly watch religious programs/seminars on TV.	−4	−7	0	−7
F1	I mostly make my own decisions, even if it goes against the opinion of others.	−5	−4	−6	2
E1	New golden color rice will make people think that the crop is rich and precious in nutrients like gold.	−5	−7	−2	−5
C5	A dish made with new Golden Rice will be unappetizing.	−5	−8	−1	−7
A3	I know what a balanced diet is… and its importance for my family.	−5	−9	−1	−8
F3	I mostly enjoy watching Urdu dramas and Indian movies.	−6	−5	−7	2
B4	I feel deprived when I see the rich feed their children better quality food.	−6	−8	−4	−4
E3	I would be afraid to feed my kids new golden color rice, even if I'm told it contains the goodness of fruits and vegetables.	−6	−8	−4	−4
A6	I must learn how to choose and cook more nutritious foods.	−6	−13	1	−14
B2	I'm confident that I feed my children according to their nutritional needs.	−7	−7	−6	−1
A2	I do not have enough information about the healthy foods for my children.	−7	−8	−6	−2
D5	A dish made with new golden color rice will be unappetizing.	−7	−11	−4	−7
A1	My children do not get sufficient nutrition on a daily basis.	−8	−11	−5	−6
D2	New golden color rice if cooked with white rice could make a tempting Biryani dish.	13	14	11	3

The data come from responses to question 2 (Fits ME), after the ratings have been converted from the original seven-point Likert scale to a binary scale. The numbers in the body of the table show the additive conditional probability of a response rating the vignette 6–7, either in the absence of elements (additive constant), or when an element is introduced into a vignette.

When this element is introduced into a vignette, 13% fewer respondents feel that it "describes ME": *New, golden color rice if cooked with white rice could make a tempting Biryani dish.*

Males and females agree that this element does not describe them. There are some elements on which the males and females do not agree with each other. For example:

> *Sunday bazaar (fair price bazaar) has poorer quality food compared to regular stores. Females believe that this is the case, while males do not.*
>
> *I must learn how to choose and cook more nutritious foods. Females believe that this is the case, while males do not.*

Table 24.6 provides us with a glimpse into the minds of our Pakistani respondents. Through the experimental design we not only learn how the elements fare, answering the problem of Golden Rice, but also begin to see the patterns of responses across elements, as well as the patterns across different respondents. It is the richness of the information from these designed studies which affords us the insights.

24.10 Culture-mind-set segments

24.10.1 Better understanding of respondents in the culture-mind-set segments

Typically, the consumer researcher divides respondents by conventional subgroups, whether dictated by standard geo-demographics (age, gender, income, education), by life experiences (e.g., losing a child), or by psychographics (the way one thinks about issues, such as food, religion, and so forth). We have already seen what we can learn from gender differences. There is a lot more when we divide the data into other groups as well.

There is an alternative approach to understanding the respondents, which begins not with *who* they are but rather with *how they think* when they respond to a specific topic. By looking at how respondents think about a specific situation, we end up dividing the respondents into a small set of complementary groups called mind-set segments. These individuals may be of different genders, live in different places, even in different countries, share little in common with each other on most topics, but nonetheless think, or at least respond, similarly for a particular topic (Moskowitz et al., 2005).

The notion of mind-set segment is important for understanding the data in this study, and on an on-going basis to understand differences and similarities among people worldwide. Rather than looking at individuals as members of local groups defined by who they are, even defined by what they believe or what they do, we define individuals in terms of responses to a limited set of stimuli. We place individuals into the same groups, for a particular topic, e.g., responses to Golden Rice, when these individuals show similar patterns of coefficients in the models relating the presence/absence of elements to some criterion measure, such as "describes ME."

The mechanics of clustering individuals is not of relevance here, since many statistical textbooks, user manuals for computer programs, and published articles in archival journals deal with clustering. What is important for us here is that we can divide our 200 respondents based upon the patterns of their coefficients in the models relating to the presence/absence of the 36 elements to their rating.

Clustering is a purely statistical procedure, one which separates objects (e.g., people) into a limited set of subgroups based upon some statistical criteria. The criteria for segmentation are that the differences across the centroid or average of the different segments (clusters) must be large, whereas the variation of the objects or people within a segment must be low. In other words, the centroids of the clusters must be far away from each other, whereas the members of a cluster must be close together.

When we follow standard statistical procedures to divide our 200 respondents into groups based upon the aforementioned mathematical criteria (minimum distances within clusters, maximum distance between clusters), we end up with a so-called 1-cluster solution (the total), a 2-cluster solution, a 3-cluster solution, a 4-cluster solution, and a 5-cluster solution. In fact, we could continue the exercise until we reach a 200-cluster solution. How do we know which one is the correct solution?

The notion of correct solution has no absolute answer. Rather, it is defined by one's criteria. The criteria we use here are:

- *Parsimony.* The fewer the number of segments extracted, the better is the solution.
- *Interpretability.* The easier the clusters or segments lend themselves to a story, the more likely the segment is "real," or at least operationally meaningful.

Following the two criteria of parsimony and interpretability we found that we needed four distinct segments, although perhaps five might have been better. We should note that at the outset of the segmentation, we had no expectations of the nature of the segments that might emerge. It might have been that some elements, not seeming to go together reasonably, actually made sense to the local consumer. We were more conservative, using Western interpretability standards. If a Westernized researcher could not tell a meaningful story to an Eastern research respondent, then we assumed the segmentation needed to go to the next level, specifically to add yet another segment to the mix.

Table 24.7 presents the strongest performing elements for each of the four segments. The key things to take away are:

- Whereas no elements for the total panel really described the respondent, at least in the respondent's opinion, and on average, dividing the respondents into mind-set segments reveals elements which respondents feel fits them.
- The pattern of strong performing elements differs by mind-set segment.
- The clustering program does not tell us the meaning of the clusters, but simply allocates respondents to the segments. It is the strong performing elements and, most important, their commonality, which defines the mind-set segment.
- The mind-set segments are not perfectly homogeneous, but they are sufficient to tell four stories, albeit with a few unexpected exceptions.

Table 24.7 Strong performing elements for each of the four mind-set segments (S1–S4) emerging from clustering the 200 respondents based on the pattern of coefficients from question #2

		Total	S1	S2	S3	S4
	Base size	200	54	65	40	41
	Additive constant	45	26	67	30	50
	Segment 1. Responds strongly to messages about health. They believe in science, and will try products which have a health promise. They enjoy traditional tastes. They are prone to be suspicious.
E2	I would be willing to eat this new rice if I'm told its golden color is due to vitamin A, which is important for health.	1	27	–13	1	–10
E6	It is possible to scientifically improve crops to contain healthier nutrients.	–2	22	–14	–6	–10
E3	I would be afraid to feed my kids new Golden Rice, even if I'm told it contains the goodness of fruits and vegetables.	–6	17	–21	–2	–16
D4	Adding a touch of saffron would enhance the taste of a new golden color rice dish.	1	16	–10	–5	5
E5	I would definitely try golden color rice if I were told it contains the goodness of chicken/meat (protein).	–2	16	–12	–6	–5
E4	A new golden colored rice will make people suspect that the crop has come in contact with waste water	–4	13	–11	1	–22
E4	A new golden colored rice will make people suspect that the crop has come in contact with waste-water from textile factories.	–4	13	–11	1	–22
	Segment 2. Almost nothing really turns them on. Seem to be indifferent or negative to messages.
C6	I expect new golden color rice dish to have an odd aroma and flavor.	–3	–11	9	–1	–13
	Segment 3. Wants to make food good and affordable, responds strongly to the perceived authority of others.
F4	I tend to seek the opinion of others (such as elders/religious leaders) before making major decisions.	1	2	–11	18	3
C4	Adding a touch of saffron would enhance the taste of a new golden color rice dish	–4	–9	–5	14	–16
B1	Sometimes, I worry that the nutritional quality of meals I give my family is poor.	–1	3	–20	14	10
B3	I will feel hopeful if I can learn how to improve the nutritional content of my children's daily diet without incurring additional expense.	–1	–6	–14	13	11

Table 24.7 Continued

		Total	S1	S2	S3	S4
A4	Sunday bazaar (fair price bazaar) has quality of food compared to regular stores.	0	5	0	10	−17
Segment 4. Responds strongly to religious influence, and seems to be simply interested in getting help, as long as such activity is permitted. Appears to be the most downtrodden.							
F6	I mostly watch religious programs/sermons on TV.	−4	−7	−10	-6	14
B6	I feel helpless when my children ask for foods I cannot afford.	1	5	−14	6	13
B5	Parents will be happy to cook a free sample of new Golden Rice…especially when they are told it is healthier for their children, compared to white rice.	−3	−3	−14	0	::
B3	I will feel hopeful if I can learn how to improve the nutritional content of my children's daily diet without incurring additional expense.	−1	−6	−14	13	11
B1	Sometimes I worry that the nutritional quality of meals I give my family is poor.	−1	3	−20	14	10

Each segment is defined by the elements to which it responds most strongly.
Source: Courtesy SB & B Research, Pty.

We would have even more compact, coherent stories were we to increase the number of segments to five, six, seven, and so forth. The problem is that as we multiply segments we reduce the usefulness of the mind-set segmentation. By the time we reach four segments, we are stretching the limits of our ability to use the segmentation for practical purposes, such as crafting convincing messages for the respondents to incorporate Golden Rice into their diet.

24.10.2 Difference in response patterns vs similarity in general attitudes

Table 24.7 tells us that our three segments are quite different in the pattern of their responses to the elements about Golden Rice. We know from an analysis of the four segments in terms of geo-demographic composition that the segments are quite similar. That is, if we knew only geo-demographic information about the segments, we would not be able to predict the likelihood that a Pakistani citizen standing in front of us belongs to Segment 1, Segment 2, Segment 3, or Segment 4, respectively. The segments are simply similar in their geo-demographic composition.

What about general attitudes, including those about religion, agriculture, as well as national origins and motives? Do these attitudes co-vary with segment membership?

Our answer emerges from the data in Table 24.8. Without belaboring the point, it is fair to say that the segments are quite similar to each other. That is, the distribution

Table 24.8 **Percent of respondents in the total panel and the four mind-set segments who agree with each statement in a set of eight statements**

	Total	S1	S2	S3	S4
Base size	200	54	65	40	41
Attitudes:					
Boys need to eat more energizing food than girls – Agree	38	33	40	33	44
When every single grain is tamped with the name of its consumer by Allah (dany dany pur mohar hai), the people should not practice family planning – Agree	58 88	61 89	58 85	53 83	56 98
American Government and its policies are detrimental for Pakistani people – Agree					
Modern thinking is not to be trusted as it goes against the interests. of Islam – Agree	69	69	71	70	66
A specially grown crop of golden color rice, introduced by Western scientists is trustworthy, as it will be beneficial for the health of the Pakistani people in the long run – Agree	68	72	65	65	68
Americans have the welfare of Muslim nations at heart – Agree	16	20	17	13	12
It's ok to beat women if they go astray – Agree	20	7	20	28	27
Boys need more education than girls – Agree	24	19	23	23	32

The pattern of agreement is similar across the four mind-set segments.

of agree/disagree responses to questions is the same across the four segments. There are general patterns, but these patterns are held by similar percentages of individuals in each segment. From Table 24.8 we learn that, on average:

- There is a mistrust of the American government and its policies.
- Americans are seen as anti-Muslim.
- There is a conviction that modernity is anti-Islam.
- There is a strong religious streak.
- Fewer than half of the respondents believe in family planning.
- There is conviction that Allah will feed the souls of those who enter this world.

24.10.3 Identifying a person as a member of a segment – the "typing wizard"

We end this chapter with a seemingly simple, but profound problem – assigning a person to a mind-set segment. The data we presented suggests that experimental design of ideas can propel us a fair distance in our journey to understand people, and to find topics and language by which to convince them. In so doing we end up partially customizing the experience, increasing the probability of a desired positive outcome (Rowley, 2002). In this study we deal with a genetically modified product which is good for you, Golden Rice. In another study it might be a question of convincing people to try vegetables, or to comply with a nutritional program.

Over the past three decades it seems clear that Mind Genomics®, the science of experimental design of ideas, works in practice. When deployed in a structured manner, experimental design provides insights about the people whose mental processes lie at the center of interest.

However, there is a nagging problem of assigning a person to a mind-set segment. It is clear that the traditional, what we might call the epidemiological way, does not work particularly well. Knowing who a respondent is, what the respondent does with respect to eating habits, and knowing the respondent's intrinsic moral, social and political values, do not predict membership in a mind-set segment. That is, we cannot use tools which define who a person is, in order to predict what is in the person's mind.

The assignment problem is not impossible, however. We know that the population with which we deal comprises four segments or basic groups. We also know that these groups were developed by RDE, where we presented the respondents with vignettes comprising elements, obtained responses, determined the part-worth contribution of each element to the response of each person, and then clustered these part-worth contributions.

Thinking more deeply, we will use the medical model instead of the epidemiological one. The former model tells us what we think based on who we are, requiring lots of information of who we are, in the attempt to link together the "who" with the "how we respond." The linkage may be quite tenuous. In contrast, the medical model will use the information which creates the segments in the first place, namely response to elements, creating from that information a small test, much like a blood test for enzyme function, or a scratch test for allergy. The response to that test will tell us the segment to which a person belongs.

We will follow a series of steps, beginning with an easy transformation, and move toward the creation of a typing wizard/tool:

1. We use the matrix of data generated by each respondent to help us. We transform the 48 ratings from each respondent for Question #2 (describes ME), so that the original 7-point scale now becomes a 3-point scale. Original ratings 1–3 become 1, original ratings 4–5 become 2, and finally original ratings 6–7 become 3. Step 1 produces a matrix of 48 rows, one row per vignette, with 36 columns corresponding to elements. Each cell is coded either 1 (element present in the vignette), or coded 0 (element absent from the vignette).
2. Add a 37th column, corresponding to the original 7-point rating on Question #1, and then a 38th column, corresponding to the transformed rating (new values = 1, 2, 3).
3. We then use ordinary least-squares (OLS) regression to relate the presence/absence of the 36 elements to the new, transformed rating, now the 3-point scale we created in Step 1. We use OLS on the 48 rows of data from each respondent, respectively, respondent by respondent.
4. Each respondent generates an equation, comprising an additive constant and the coefficient for each element.
5. By summing the additive constant and the coefficient of each element, we can estimate the rating, on a 3-point scale, that each respondent would have assigned to each of the 36

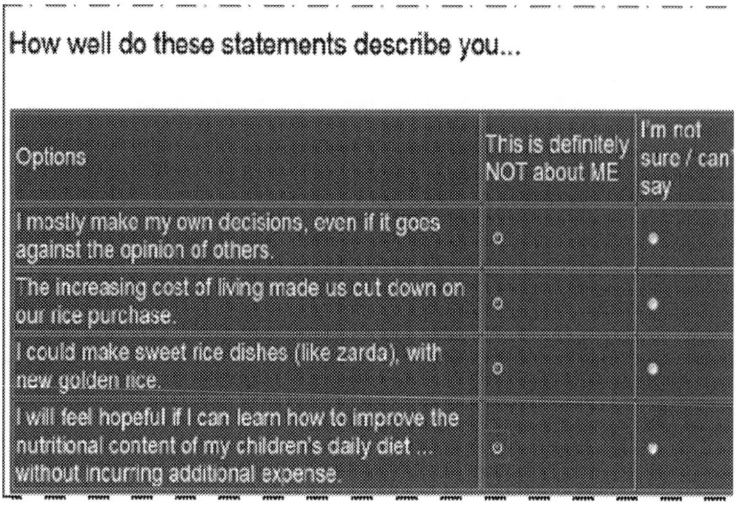

How well do these statements describe you...

Options	This is definitely NOT about ME	I'm not sure / can't say
I mostly make my own decisions, even if it goes against the opinion of others.	○	●
The increasing cost of living made us cut down on our rice purchase.	○	●
I could make sweet rice dishes (like zarda), with new golden rice.	○	●
I will feel hopeful if I can learn how to improve the nutritional content of my children's daily diet ... without incurring additional expense.	○	●

Figure 24.7 Example of the typing wizard in English. The respondent reads each element (row) and rates that element alone on a three-point scale, shown on the top.
Source: Courtesy SB & B Research, Pty.

elements. We now have an estimate of the rating on a 3-point scale each respondent would have assigned to each element.
6. We round the estimated rating for each element, so each respondent has a number 1, 2, or 3, respectively for each element.

Steps 1–6 prepared the data for us. We now know the rating of each element by each respondent, all on a 3-point scale (1, 2, or 3), and the segment membership of that respondent. We know the segment membership of our 200 respondents because we used their data to create the segments in the first place.

Discriminant function analysis (DFA) identifies a small set of elements, with which we will create four equations known as classification functions. The DFA statistical method generates four equations (classification functions) one for each of our four Pakistani segments. Each classification function comprises an additive constant, and four weights or coefficients, one coefficient for each element.

With the four classification functions, we now proceed to instruct an individual to rate the four elements, as questions, using a three-point scale. We see an example of the interview screen (in English) in Fig. 24.7. For the respondent the task is not difficult; simply read the four elements, and rate each on the anchored three-point scale. In actual implementation, the order of the questions would be randomized for each person.

Once the respondent has rated the four questions, generating a pattern of responses, the DFA program calculates the value of each of the four classification functions, using the three-point rating for each element.

The classification function with the highest positive value shows us the mind-set segment to which the respondent belongs. For our worked example the highest value

> **Seg 2**
>
> **It's about ME:**
>
> You are not concerned about what other people are doing.
>
> You would try any rice, irrespective of what is in it or what color it is.
> Nutrition is not much of a concern for you, you just want food on the table.
>
> You don't actively seek out information based on modern science and knowledge.
> Rice is rice, doesn't matter how it looks or if there is some special vitamin in it. You cook rice for any occasion, does not have to a be a special occasion.
>
> >You feel that a new golden color rice dish would probably have an odd aroma and flavor. You don't have any health concerns about yourself or your family. Food is food and everyone gets whatever nutrition they need in what they currently eat.

Figure 24.8 Output from the typing wizard, showing the mind-set segment to which the person belongs and what to say to the person.
Source: Courtesy SB & B Research, Pty.

was achieved by classification function 3, i.e., the classification function corresponding to mind-set Segment 3.

Once we know the mind-set segment to which the respondent belongs, it is an easy matter to print out what to say (positive impact values) and what to avoid. We see an example of the printout in Fig. 24.8.

Finally, the same approach can be followed in Urdu. Only the language changes and, of course, the elements are based upon the original study from which the segmentation was developed.

24.11 Summary and future trends

In summary, our work has led us to think how we might apply high level qualitative and quantitative techniques to convince people about the value of new foods. New foods with healthful properties may improve a person's quality of life, the vitamins and minerals allowing the body to carry on healthfully. Uncovering the mind of the individual through in-depth interviews gives researchers a sense of what worries a person, what relaxes him, and how a person reacts to a strange food in the diet.

But we must move beyond the basic facts, that the product is healthful and acceptable. We must realize that introducing new foods to different populations may require teaching the local population to accept new sources of food, perhaps, even preventing starvation and rioting when consumers run out of foods that they want.

In the future, the methods introduced here will be applied around the world, with this work replicated in many countries and on almost all continents. Changes in climate, weather patterns, war, uprisings and revolutions will continue to remake the planet. The standardized methods of EMPATHY and EXPERIMENT™ using applied computer research tools like Mind Genomics,® IdeaMap®.net will allow social scientists to study populations worldwide and enable them to craft messages that encourage the adoption of healthy and helpful foods and services.

Acknowledgment

The authors wish to thank Dvora Chaiet, Editorial Assistant, Moskowitz Jacobs Inc., for her help in preparing this manuscript for publication. All figures and tables used in this chapter are used with the permission of co-author Batool Batalvi and are © SB & B Marketing Research, Pty.

References

Amerine, M. A., Pangborn, R. M. and Roessler, E. (1965). *Principles of Sensory Evaluation of Food.* New York: Academic Press.

Anderson, N. (1977). Functional measurement and psychological judgment. *Psychological Review,* **77**, 153–170.

Beck, A. T. (1976). *Cognitive Therapy and the Emotional Disorders.* New York: Meridian.

Beck, A. T. (1983). Cognitive therapy of depression: New perspectives. In P. Clayton and J. E. Barrett (Eds.). *Treatment of Depression: Old Controversies and New Approaches* (pp. 265–290). New York: Raven Press.

Beck, A. T., Wright, F. D., Newman, C. F. and Liese, B. S. (1993). *Cognitive Therapy of Substance Abuse,* New York: Guilford.

Box, G. E. P., Hunter, J. and Hunter, S. (1978). *Statistics For Experimenters,* New York: John Wiley.

Gofman, A. and Moskowitz, H. R. (2010). Application of isomorphic permuted experimental designs in conjoint analysis. *Journal of Sensory Studies,* **25** (1), 127–145.

Green, P. E. and Srinivasan, V. (1981). A general approach to product design optimization via conjoint analysis. *Journal of Marketing,* **45**, 17–37.

Green, P. E. and Krieger, A. M. (1991). Segmenting markets with conjoint analysis. *Journal of Marketing,* **55**, 20–31.

Jung, C. (1976). *Psychological Types: A Revision. (The Collected Works of C. G. Jung) (Bollingen Series).* (C. G. Jung, Author), (R. F. C. Hull, Editor), (H. G. Baynes, Translator). Princeton, NJ: Princeton University Press.

Kohut, H. (1971). *The Analysis of the Self.* New York: International University Press.

Morton, J. L. (2011). Color Matters. Retrieved 21 January 2012 from the World Wide Web: http://www.colormatters.com/color-travels/pakistan.

Moskowitz, H., Batalvi, B. and Ettinger Lieberman, L. (2014). Empathy & Experimentation™: Applying Consumer Science to Whole Grains as Foods. Paper presented under title Decoding 'what works' – Science to help consumers embrace the notion of whole grains. In D. Hauge, K. Dammann and L. Marquart (Eds.) Proc. Whole Grains Summit, 2012. St. Paul, MN: AACC International, p. 25.

Moskowitz, H. and Gofman, A. (2007). *Selling Blue Elephants: How to Make Great Products That People Want Before They Even Know They Want Them.* Upper Saddle River, NJ: Wharton School Publishing.

Moskowitz, H. R., German, B. and Saguy, I. S. (2005a). Unveiling health attitudes and creating good-for-you foods: The genomics metaphor and consumer innovative web-based technologies. *CRC Critical Reviews in Nutrition and Food Science,* **45** (3), 191–265.

Moskowitz, H. R., Poretta, S. and Silcher, M. (2005b). *Concept Research in Food Product Design & Development.* Ames, IA: Blackwell Professional.

Nestel, P., Bouis, H. E., Meenakshi, J. V. and Pfeiffer, W. (2006, April) Biofortification of staple food crops. *Journal of Nutrition,* **136** (4), 1064–1067.

Pangborn, R. M. and Trabue, I. M. (1967). Bibliography of the sense of taste. In M.R. Kare and O. Maller (Eds.). *The Chemical Senses and Nutrition* (pp. 45–60). Baltimore, MD: Johns Hopkins Press.

Pliner, P. and Hobden, K. (1992). Development of a scale to measure the trait of food neophobia in humans. *Appetite,* **19** (2), 105–120.

Rahi, J. S., Sripathi, S., Gilbert, C. E. and Foster, A. (1995). Childhood blindness due to vitamin A deficiency in India: regional variations. *Archives of Disease in Childhood,* **72** (4), 330–333.

Rowley, J. E. (2002). Reflections on customer knowledge management in E-business. *Qualitative Market Research,* **5**, 268–280.

SYSTAT (2007). SYSTAT for Windows, Version 11. Chicago, ILL: SYSTAT Software Inc.

Thomke, S. (2003). *Experimentation Matters: Unlocking the Potential of New Technologies for Innovation,* Boston, MA: Harvard Business School Press.

Wente, M. (2012). Greenpeace's Golden Rice stand should appall us all. *The Globe and Mail.* Retrieved 21 January 2013 from the World Wide Web: http://www.theglobeandmail.com/commentary/greenpeaces-golden-rice-stand-should-appall-us-all/article4541042/.

Empathy and Experiment™

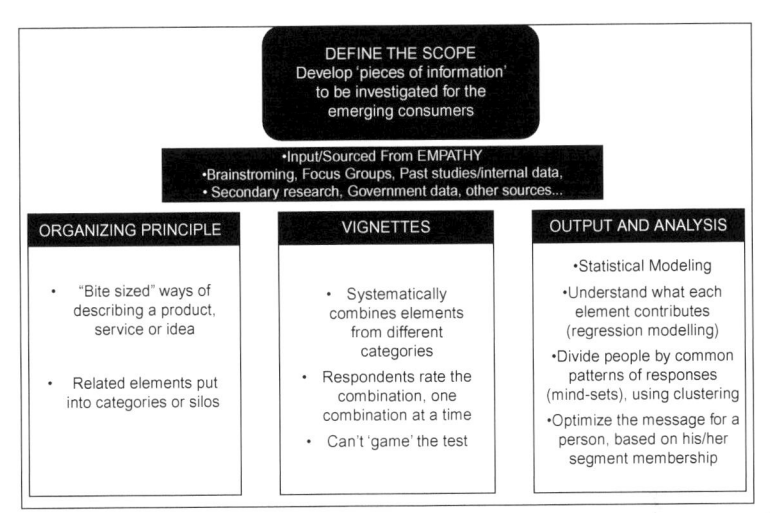

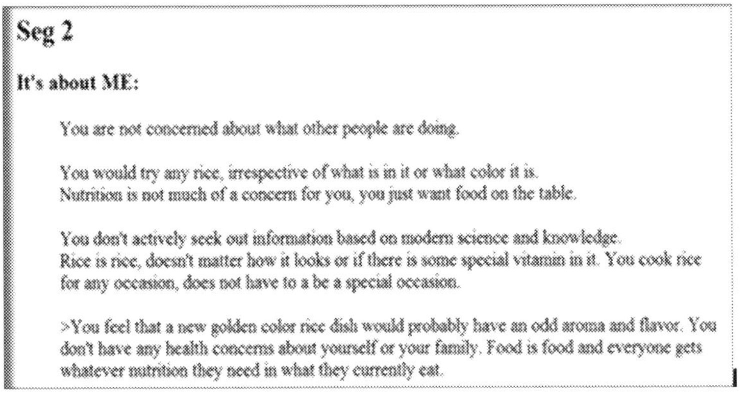

Empathy and Experiment™ 535

Index

A-not-A method, 31
accuracy, 12, 311
   Flash Profile, 133–4
AceK, 58
acute consumption, 61
additive trees, 160, 169–70
adjusted Rand index (ARI), 169
adoption
   Flash Profiling usage in daily new product development and testimonial, 335–43
      Flash Profile as reference methodology, 338–41
      Flash Profile as starting point, 335–8
      limitations and perspective, 341–3
affective methods, 34–7
   9-point hedonic scale, 35
   example of paired preference scorecard, 37
   results from 9-pt hedonic scale listing % responses in each category, 36
ageing, 486, 488, 490
agglomerative hierarchical procedure, 157–8
aleurone layer, 511
alternative methods
   sensory testing advantages and disadvantages, 27–50
      data collection, analysis and reporting, 42–8
      developing descriptive analysis capability, 39–42
      future trends, 49
      important considerations, 37–9
      methods, 30–7
      other descriptive methods, 48–9
      subjects, 28–30
analysis of variance (ANOVA), 35, 219, 440
analytical task, 478

anthropometry, 438
application phase, 463
Application Programming Interface (API), 261
appraisal theory, 73
area under curve (AUC), 285
ascending hierarchical analysis, 168
associated emotions, 412, 419
   results of the products HCA, 420
attention-interest-desire-action (AIDA) model, 461
attribute elicitation, 132
attribute list
   panel training, 272–3
      distributions of attributes selected at least once across all evaluations, 274
AUPALESENS survey, 486, 487, 492, 501
automotive industry
   automotive context and its specificities, 428–9
      experts, 428–9
      price, availability and anonymous evaluation of cars, 429
      products, 428
      rapid sensory methods, 429
      static/dynamic conditions, 428
   case study of understanding the consumer perception of car body style, 170–80
      consensus partition, 175–6
      context and issues, 170–1
      discussion, 180
      materials and methods, 171–2
      proximities between groups, 176–7
      results, 172
      spatial representation of cars, 177–80
   gearbox sensations and comfort, 430–5
   rapid sensory methods, 427–51

automotive industry (*cont.*)
  rationale, 427–8
  samples, 171–2
    car body styles generated by experimental design, 173
    definition of morphological parameters, 172
    description of eight physical parameters and their modalities, 171
  sorting procedures, 172
    number of groups formed by consumers, 174
  subjects, 172
    frequency table of owned car body type, models, age and gender of panel, 174
autonomic nervous system (ANS), 75

balanced incomplete block design (BIBD), 159, 172
balanced rotation order, 232
barycentric, 178, 256
Berridge's animal-based models, 59
best-worst scaling (BWS), 76–7, 91
bias, 230, 495–6
Big Grid napping, 371, 378
binomial test, 282
biofortification, 511
blind testing, 61, 219
bootstrap, 193
  sampling, 221, 232
  technique, 201
brain imaging techniques, 75
Brandphonics process, 104
brew masters, sensory testing, 363–82
brown rice, 511
Bullseye, 91, 103–4
  quantification tool, 109, 113, 116

calculation module, 343
calibration best–worst exercise, 102–3
car body style, 170–80
  proximities between groups, 176–7
    additive tree and representation of consensus partition groups, 177
case studies
  brewers and novices assessing beer, 367–8
    beer samples tested, 368

exploring relationship between derived index of fit-to-brand and new product sales, 113–16
index of fit-to-brand using conceptual profiling with expedited quantification, 108–13
  Glenlivet conceptual profile of product vs brand, 112
  Lagavulin conceptual profile of product vs brand, 112
  summaries of conceptual profiles of Glenlivet and Lagavulin brands, 111
MMR's General Conceptual Lexicon expedited conceptual profiling of three UK High Street pharmacy retailers, 105–8
spice blends and pastes, 371–7
  Big Grid napping, 374
  correlation loading plot from PCA, 375
  results and discussion, 373–7
  score plot from PCA, 375
  spice blends and ingredients, 372–3
categorisation process, 154, 159
category effect, 100
category scale, 55, 500
central locations (CLT), 457
check-all-that-apply (CATA), 3, 16, 100, 252
  data analysis from questions, 233–5
    example of data matrix used for entering data, 233
    penalty analysis, 235
    sample and term configurations, 234
    summary data, 233
  design of questionnaire, 229–31
    consideration of an ideal product, 231
    influence on hedonic scores, 231
    order of terms, 229–30
    type and number of terms, 229
  differences among samples, 233–4
    data matrix for analysing data from one term using Cochran's Q test, 234
  implementation, 229–32
    number of consumers, 232
    number of products, 231–2
  pros, cons and opportunities of application, 240–2
  questionnaires, 364

Index

questions with consumers in practice, 227–42
  case study of application for sensory characterisation of plain yoghurt, 235–40
  examples of sensory and both sensory and non-sensory terms questions, 228
  techniques, 481
chefs, sensory testing, 363–82
chemoreception senses, 461
chemosensory capacities, 57–8, 488–9
chemosensory stimuli, 474
Chi-square, 256
children
  sensory testing in new product development, 473–81
    application to children of different ages, 476–80
    future trends, 480–1
    organisation of sensory evaluation testing, 475–6
    principal analytical sensory evaluation tasks, 477
    sensory aspects, 474–5
classical descriptive methods, 57–8
classification functions, 537
cloud computing, 38
cluster analysis, 100, 167–70, 408, 412, 419
  additive trees, 169–70
    illustration of highlighting the similarities between products P1 to P4, 170
  ascending hierarchical analysis, 168
  consensus partition, 168–9
  general overview, 167–8
clustering, 319, 323, 397, 530
Cochran test, 356
Cochran's Q test, 233–4, 237
cognitive abilities, 490–1
  impact on performance, 496
cognitive disorders, 491
cognitive therapy, 513
cognitive thought processing, 94, 102
Comparative Free Choice Profiling, 336–7
comparative methodology, 215
comparative team tasting, 349–50, 352–3
  description of design, 352
  practical example, 350, 352–3

complex semantics, 194
Compusens, 278
computerised progressive ratio task, 59
concept description
  developing conceptual lexicon, 99–101
  single Malt Scotch whisky category, 101
  use of words and issue of counter-intuitiveness, 99
conceptual associations, 91, 95, 98
conceptual lexicon, 81, 99–101
  category specific, 103, 116
conceptual profiling, 98
  expedited procedures for brands, products and packaging, 91–117
    application and case studies, 105–16
    concepts, conceptualisation and conceptual structure, 95–7
    emotion profiling vs conceptual profiling and theoretical considerations, 97–8
    fundamentals of new product success and failure, 92–3
    measurement using direct scaling, 93–5
  practice, 99–104
    concept description and developing conceptual lexicon, 99–101
    concept description and use of words and issue of counter-intuitiveness, 99
    quantification, 102–3
  rapid methods, 103–4
    Bullseye conceptual profiling interface, 104
    MMR's General Conceptual Lexicon, 103
conceptualisation, 95–7
  concepts and conceptual structure, 95–7
  model linking sensory stimulation to consequent behaviour, 97
concrete operation stage, 473
consensus partition, 168–9, 175–6
consistency, 224, 320–2
  hedonic, 320–2
    assessment of consumers ideals, 322
  ideal data, 317–19
  sensory, 320
    evaluation of ideal products at panel level, 321

constrained sorting, 157
consumer behaviour
  sensory perception measurement, 53–63
    conceptual model and revised model that acknowledge existence of error, 54
    hedonics, 58–60
    linking sensations, liking and intake, 61–3
    product usage and intake, 60–1
    sensation, 54–8
    summary, 61–3
consumer conceptualisations, 81
consumer defined (CD) lexicon CATA approach, 78–9
consumer research, 229
  emotional response measurement in sensory research, 71–87
    application of verbal self-report emotion techniques, 78–84
    approaches, 75–7
    defining emotion, 72–3
    importance, 73–5
    relating sensory properties to consumers, 84–6
    unresolved issues and topics for future research in verbal self-report, 86–7
    verbal self-report emotion lexicon, 77–8
consumer reward, 93
consumer testing, 457
consumers, 11–12, 403–4
  CATA questions, 227–42
    case study of application for sensory characterisation of plain yoghurt, 235–40
    data analysis, 233–5
    implementation, 229–32
    pros, cons and opportunities of application, 240–2
  Flash Profile, 406–10
    clustering of products after GPA, 409
    frequently used attributes, 407
    PCA on the consensus from 89 consumers, 408
    semantic description, 406
    sensory maps and product clusters, 406–8
    summary and opportunities, 408–10
    word cloud of different attributes, 406
  sensory description, 410–23
    common and specific items, for the, 417
    common items to most of the perfumes, 415–16
    detailed idioms and phrases used for mental images, 415
    generic perfumes classification, 411
    materials and methods, 410–12
    percentage mean of separate items different categories, 414
    results and contribution, 412–23
    semantic quantitative comparison, 413
    testing insight using mobile devices, 455–69
consumption emotion set (CES), 77
consumption experience, 84
continuous scale, 217, 218–19
  PSP methodologies, 217
conventional descriptive analysis, 311, 335
conventional sensory profiling, 6, 16, 194
correspondence analysis (CA), 219, 234, 256
cost-effective methods, 248
counter-intuitiveness, 99
cross-cultural validation, 87
culinary professional
  sensory testing, 363–82
culture mind-set segments, 528–36
  better understanding of respondents, 528–31
  difference in response patterns *vs.* similarity in attitudes, 533–6
  percent of respondents in the total panel, 532
  strong performing elements for each of the mind-set segments, 530–1
  typing wizard, 534
curve fitting, 524

daily new product development
  adoption and Flash Profiling usage and testimonial, 335–43
    Flash Profile as reference methodology, 338–41

Flash Profile as starting point, 335–8
  limitations and perspective, 341–3
Danish Microbrew, 367
dark chocolates
  Flash Profile, 124–31
    conclusions, 130–1
    example of transcription of ranking, 126
    instruction and course of session, 125–6
    panel, 124–5
    products, 124
    results, 126–30
    sample preparation, 125
    samples evaluated, 124
data acquisition, 278, 304
  description in different methods, 310
data aggregation, 215
  example, 221
  CA of aggregated data graph, 223
data analysis, 191–3, 218–19
  classical Multiple Factor Analysis output based on FMS study, 193
  comparison between products, 287–90
    analysis of duration of evaluations, 290
    difference testing and randomisation test approach, 288–90
    TDS difference curves, 287–8
  continuous scale, 218–19
    two ways to process PSP data, 218
  projective descriptive methods, 365–7
  representation of product space, 284–7
    individual parameters, 284–5
    parameters from TDS curves, 285–6
    TDS trajectory mapping, 286–7
  representation of sequence, 279–84
    against standardised TDS curves, 279
    individual level, 279
    panel level and TDS curves, 279
    TDS curves and significance limit, 282–4
  triad PSP, 219
data collection, 45–6, 190–1, 467, 468
  analysis and reporting, 42–8, 47–8
    checklist developed from information provided by subjects, 44

compilation of appearance words used by subjects, 43
    example of appearance attributes for scorecard, 46
    example of graphic rating or line scale, 45
    final list of attributes with their definitions, 45
    sensory map or spider plot, 48
data mining algorithms, 255
data pre-treatment, 291
date capture, 38–9
Declaration of Helsinki, 475
dementia, 491
dendrogram, 167
depression, 487–8
descriptive analysis, 271, 474, 479–80
  development capability, 39–42
    language development, 41–2
    recruiting, screening and qualifying subjects, 40–1
  module, 273
  overview, 363–5
  tests, 39–40
descriptive method, 4, 16, 33–4
descriptive team tasting, 349
  example 1, 349
    questionnaire, 351
    result of scale analysis, 352
    results of CATA analysis, 352
    results of scale evaluation and CATA evaluation, 351
  main characteristics, 350
design of experiment (DoE)
  Flash Profile, 139–41
diesel engines, 444–50
  discussion and conclusion, 449–50
  differences in distribution of sensory information, 449
  materials and methods, 444–6
    cars used for idle noise, 445
  objective, 444
  results, 446–9
    hierarchical ascendant classification of the assessors, 448
    sensory map from GPA, 447
    terms used by assessor to describe idle sounds, 446
dietary micronutrient deficiencies, 510–11

direct scaling, 93–5
Directive 2001/20/EC, 499
Disabilities-Dependence-Incapacities, 488
disappearing Bullseye, 103–4
discriminant function analysis, 537
discrimination, 30–3, 493–4
  method, 30–3
    total judgements, correct matches and % correct for statistical significance, 33
  schematic illustration, 494
  tests, 476, 478
discriminative tasks, 499
dissimilarity matrix, 162–4
DISTATIS approach, 165–6
dominance concept, 272
duo-trio difference, 29
duplicates, 134
dynamic condition, 428
dysphagia, 489

ecologic validity, 467
eigenvalue, 202
electroencephalography (EEG), 75
electronic diary studies, 462
elicitation technique, 248–9
emotion
  concept, 98
  definition, 72–3
    four descriptive models of core effect, 73
emotional response measurement
  approaches, 75–7
    PrEmo, 76
  sensory and consumer research, 71–87
    application of verbal self-report emotion techniques, 78–84
    defining emotion, 72–3
    importance, 73–5
    relating sensory properties to consumers, 84–6
    unresolved issues and topics for future research in verbal self-report, 86–7
    verbal self-report emotion lexicon, 77–8
Empathy and Experiment, 512–16
  actual experience in Pakistan, 514–15
  developing information from EMPATHY, 515–16

empathy in understanding respondent, 513–14
experimental design, regression modelling, segmentation and discriminant function analysis, 521
guidelines in gathering qualitative data, 514
psychology of personality, 512–13
episodic memory, 490
EsSense Profile, 78
Euclidean distance, 163
European Test of Olfactory Capabilities (ETOC), 498
evaluation duration
  analysis, 290
    different products averaged across subjects, 290
evaluation task, 4, 8–9, 137
  Flash Profile, 143
Excel, 230
exemplar task, 172
expectation error, 387
expedited procedures
  conceptual profiling of brands, products and packaging, 91–117
    application and case studies, 105–16
    concepts, conceptualisation and conceptual structure, 95–7
    emotion profiling vs conceptual profiling and theoretical considerations, 97–8
    fundamentals of new product success and failure, 92–3
    measurement using direct scaling, 93–5
    practice, 99–104
expedited quantification, 108–13
experimental design, 518, 527
explicative studies
  product benchmarking, 338–40
External Preference Mapping, 232, 257, 339, 442–3
  consensus map, 443
EyeQuestion, 278

facilities, 37–8
FactoMineR, 192, 210, 367
factor analytical techniques, 161–6
  general overview, 161–2

Index 545

MDS on individual dissimilarity
   matrices, 164–6
multidimensional scaling (MDS) on
   overall dissimilarity matrix, 162–4
multiple correspondence analysis, 166
statistical methods of analysing free
   sorting data, 162
fast moving consumer goods (FMCG), 3
fatigue effect, 494–5
   distribution of differences in liking
      score, 495
   scale use, 495
first singular value, 366
fit-to-brand, 93
5KEYS, 512–13, 518
5 point Likert scales, 60–1
fixed sorting, 157
Fizz, 278
Flash Descriptive Analysis, 48
Flash Profile (FP), 7, 394, 398, 430–1, 445,
   450
   evolution, 138–46
      consumer-oriented method, 143–6
      professionals, 141–2
      use with a design of experiment,
         139–41
   fragrance language, 401–24
      industrial approach to fragrances'
         assessment, 401
      perfumers vs consumers, 401–10
      sensory description, 410–23
   industrial context and challenges, 335–6
      diagram showing origin of sensory
         descriptive request within Puratos
         Group, 336
   limitations, 136–8
      control of evaluation conditions,
         136–7
      cross-evaluation comparison and
         build-up of database, 137
      difficulty of the task and number of
         products, 137
      product availability, 136
   limitations and perspectives, 341–3
      current limitations of methodology,
         341–2
      perspectives for further development
         of descriptive sensory analysis,
         342–3

methods and origins, 121–4
   principle, 122
   subjects, 122–4
methological considerations, 132–3
   attribute elicitation, 132
   preliminary session, 132–3
   training a panel, 133
metrological properties, 133–6
   accuracy, 133–4
   level of consensus among subjects,
      135–6
   repeatability, 134–5
   sensitivity and robustness, 135
reference methodology, 338–41
   development and maintenance of
      sensory panel, 341
   explorative studies and product
      benchmarking, 338–40
   sensory description as a tool for
      communication with customers,
      340–1
starting point, 335–8
   advantages in our context, 336–7
   internal development of sensory
      descriptive analysis, 337–8
use in sensory and consumer science,
   121–48
   evaluation of dark chocolates, 124–31
   future trends, 146–8
flash profiling, 16, 41, 249, 481
   usage and adoption in daily new
      product development and
      testimonial, 335–43
   Flash Profile as reference
      methodology, 338–41
   Flash Profile as starting point, 335–8
   limitations and perspective, 341–3
flavour creation, 384
   internship, 397
flavour perception, 385
   deconstruction, 387–8
   description, 388–9
Flavour Profile, 41, 48
flavourists
   different ways of working, 386–90
      deconstruction of flavour perception,
         387–8
      description of flavour perception,
         388–9

flavourists (*cont.*)
  differences between flavourists and sensory experts, 391
  evaluation of products, 387
  individualisation or standardisation of flavour description, 390
  quantification of flavour description, 389–90
  roles and responsibilities, 384–6
    main objectives, 384
    sensory expert panellists' role and responsibilities, 385
    sensory experts' main objectives, 386
    sensory panel leaders' role and responsibilities, 385
    sensory scientists' role and responsibilities, 385–6
    types of sensory experts, 384–5
  sensory testing, 383–98
    future trends, 383–98
  strategies to complement both types of expertise, 390–7
    sensory evaluation performed by flavourists, 394–7
    sensory evaluation sessions observation, 391–4
flexibility, 9
focus groups, 501
food, 74
food frequency questionnaires (FFQ), 61
food habits
  Empathy and Experiment, 509–39
    culture mind-set segments, 528–31
    elements and process of the experiment, 518–20
    future trends, 535
    Golden Rice, 510–12
    interview and structured experimental design data, 521–4
    results, 523–8
    scope of the study, 509–10
food industry
  team tasting improvement, 345–61
    analysis and prospects, 360–1
    analysis of opportunities and constraints linked to project team evaluation, 347–8
    approach adapted to our needs but integrated the limits, 348–57

implementation examples common in R&D field, 357–60
precise analysis of concrete situations, 346–7
food intake, 86
fragrances
  perfumers vs consumers, 401–10
    codes, short and full names of 12 perfumes, 403
    materials and methods, 402–4
    projective flash profile, 401–24
Free Choice Profiling, 16, 41, 48, 121, 249, 364, 430
free comments method, 251
free multiple sorting, 157
  advantages, disadvantages and applications, 194–5
  continuum of sensory assessors, 194
  implementation and data collection, 190–1
  sensory profiling technique, 187–95
    data analysis, 191–3
    future trends and further information, 195
    overview, 187–8
    practical framework and design of experiments, 189–90
    theoretical framework, 188–9
free sorting, 16, 249, 394, 398
  procedure, 154–5
  sensory profiling technique for product development, 153–81
    case study in automotive industry, 170–80
    task, 154–9
  statistical treatment for data, 160–70
    factor analytical techniques, 161–6
    general overview, 160–1
    methods pertaining to cluster analysis, 167–70
    overview of statistical methods of analysis, 162
    transformation of raw data, 161
free sorting task, 154–9, 172, 364
  interests and limitations, 158–60
  overview of usage, 155–6
  procedure, 154–5
  variants and extension, 156–8
free word association, 251–2, 258

frequency threshold, 252
full repetition, 134
functional magnetic resonance imaging (fMRI), 75

gearbox, 430–5
  discussion and conclusion, 435
  experts' vs consumers' perceptions, 430
  materials and methods, 430–1
  results, 432–4
gene association studies, 56–7
General Conceptual Lexicon, 91
  expedited conceptual profiling of UK High Street pharmacy retailers, 105–8
General Oral Health Assessment Index (GOHAI), 498
generalised hedonic scale *see* Global Hedonic Intensity Scales
generalised intensity scales *see* Global Sensory Intensity Scales
generalised Procrustes analysis (GPA), 8, 127, 129, 136, 198, 339, 403, 433, 451
  sensory map, 434
Geneva Emotion and Odour Scale (GEOS) questionnaire, 79–80
geriatric depression scale (GDS), 498
Givaudan, 385
global analysis, 207
Global Hedonic Intensity Scales, 60
Global Sensory Intensity Scales, 55–6
Golden Rice, 510
  elements of total panel and gender, 525–8
  negatives, 512
  positives, 510–11
  value of experimentation and implementation among Pakistanis, 516, 518

habitual intake, 61
HarvestPlus, 511
hedonic consistency, 320–2
hedonic methods, 34
hedonic scale, 35, 492
hedonic scores, 231, 312
hedonic tests, 232, 501–2
hedonics, 58–60
Hellinger distance, 234

hierarchical algorithms, 167
hierarchical ascendant classification, 441, 450, 451
hierarchical clustering analysis (HCA), 127, 405
  GPA per country and category, 422
hierarchical sorting, 157–8
holistic methods, 189
homogeneous groups, 322–4
  consumers clustering, 322–3
    characterisation of two segments of consumers through average liking scores, 324
    dendrogram highlighting two cluster solutions in segmentation procedure, 323
  products, 323–4
    assessment of single or multiple ideals by multivariate analysis, 325
horizontal 15-point intensity scale, 240–1
Hotelling T test, 317

ideal data
  consistency, 317–19
    summary of ideal information provided by consumer, 319
  guideline for analysis, 319
Ideal Mapping (IdMap), 315
  ideal of reference, 324–6
    sensory characteristics of ideal product of reference vs sensory profiles, 326
    solution of IdMap defining ideal reference to match in optimisation procedure, 325
Ideal Profile Analysis (IPA)
  illustration, 319–28
    consistency, 320–2
    homogeneous groups, 322–4
    ideal of reference, 324–6
    optimisation, 326–8
      drivers of liking and disliking obtained by regression on principal component, 327
      product 1 and product 3 using Fishbone method, 327
Ideal Profile Method (IPM)
  additional valuable properties, 316–19

Ideal Profile Method (IPM) (*cont.*)
    consistency of ideal data, 317–19
    guideline for analysis of ideal data, 319
    notion of multiple ideals, 316–17
    in practice, 309–10
    presentation of data obtained from consumer, 310
    principle and properties, 309–12
    tool for product development and product optimisation, 313–16
ideal profiling
    sensory profiling technique, 307–30
    additional valuable properties, 316–19
    illustration of Ideal Profile Analysis (IPA), 319–28
    IPM, tool for product development and product optimisation, 313–16
    principle and properties of Ideal Profile Method (IPM), 309–12
    schematic representation of IPA, 329
IdeaMap, 535
idle noises
    diesel engines, 444–50
IDSORT method, 166
implementation examples
    case 1: characterize eight products in order to select two, 357–8
    case 2: compare two trials to standard, 358–60
    common in R&D field, 357–60
implicit associations, 95
in-home-use-tests (IHUT), 457
    mobile devices vs CLT, 460
indicator variables matrix, 160
individual level
    sequence of data analysis, 279
        handling of TDS data without or with intensity scores, 280
INDSCAL model, 164
informal sensory assessment, 14
Information Age, 261
inkblot test, 197
inspiration phase, 463
instrumental activities of daily living (IADL) scale, 498
intensity ranking, 478–9
inter-session repeatability, 493–4

intercultural knowledge, 423
intermeddle semantics, 194
internal validity, 467
interpretability, 530
introspection, 59
intuitive theory, 96
ISO 8586, 276, 291–2
ISO 8589, 275, 365
iterative algorithm, 163

just-about-right (JAR) procedure, 36–7, 311
just-noticeable-difference, 32–3

knowledge-development process, 513
Kohut's Self-Psychology, 513
Kruskal-Wallis test, 235, 356

Labelled Affective Magnitude scale, 35, 60
labelled hedonic scale (LHS), 60
Labelled Magnitude Scales, 479
language development, 41–2
Latin Square design, 154
lemma, 252
lemmatisation step, 252
level of consensus, 135–6
lexicon, 101
lifestyles, 412
    results of the products HCA, 422
liking–intake relation, 63
liking potential, 317–18
liking score, 313
litmus test, 374
Liverpool Emotion and Odour scale (LEOS), 80

magnitude estimation, 55
Matlab, 403
maximum difference scaling, 102
McNemar test, 287, 356
MDSORT method, 166
mental images, 412
    results of the products HCA, 421
meta-descriptors, 218–19
methodological evolution, 8–14
    diversification of sensory toolkit, 9–10
    flexibility, 9
    rapidity, 8–9
    usage of DA with different types of subjects, 10–11

Index   549

DA with consumers, 11–12
  plant operators, 13
  professional sensory experts, 12–13
  salespersons, 14
  team tasting, 14
metric multidimensional scaling, 163
Mind Genomics, 535
mind-set segments, 530 *see also* culture mind-set segments
Mini Mental Scale Examination (MMSE), 492, 498
mixed Assessor Model (MAM), 219
MMR Research Worldwide, 100–1
mobile devices
  consumer insight, 455–69
    case study, 459–66
    mobile research, 455–6
    new research method, 456–9
mobile research, 455–6
mobile sensory research, 456–9
  advantages, 456–8
  concerns and limitations, 458–9
  method assessment, 465–6
MobiLens, 455
monadic sequence, 232
mood portraits, 75–6
multi-bit temporal dominance sensations, 300–2
  simplified TDS data representations for 12 Nespresso products, 301
multi-sip temporal dominance sensations, 300–2
multidimensional circumflex models, 72–3
multidimensional scaling (MDS), 8, 160, 218
  individual dissimilarity matrices, 164–6
    products spaces, 165
  overall dissimilarity matrix, 162–4
    evolution of stress criterion as function of number MDS axes, 162–4
multinomial logit (MNL), 105
Multiple Affect Adjective Check List (MAACL), 77
multiple correspondence analysis (MCA), 160, 166, 192, 210
multiple factor analysis (MFA), 8, 127, 136, 192, 198, 258, 365–6
multivariate analysis, 210

multivariate analysis on variance (MANOVA), 129
multivariate data analysis, 379

nappe, 199
Napping, 365, 370
  sensory Napping as sensory profile technique, 197–213
    analysis of data using R statistical software, 210–12
    projective tests, 197–206
    sorted Napping, 206–10
  versus project mapping, 382
Natrick 9 point scale, 58
negative emotions, 201
Nestlé Research Centre, 278
neurodegenerative diseases, 491
9-point category scale, 309
no-added-sugar (NAS), 85
non-disappearing Bullseye, 104
non-graphical RV coefficient, 193
non-metric multidimensional scaling, 163
Nordic Food Lab, 371, 378
normalisation, 366
nursing home, 493

older people
  heterogeneity, 487–91
    chemosensory capacities, 488–9
    cognitive abilities, 490–1
    oral health status, 489–90
    physical, psychological health and dependency, 487–8
  sensory testing in new product development, 485–504
    early market, 485–7
    future trends, 503–4
    impact of age and dependence on sensory task, 491–7
    sensory descriptive analysis with elderly panel, 497–503
olfactory description, 412
  sensory data, 418, 419
online panels, 458
open-ended questions, 502
  analysing data of getting valuable outcome from different applications, 255–61

open-ended questions (*cont.*)
  combining spontaneous responses from different modalities, 259
  exploring semantic networks, 261
  finding significant differences between the products, 256
  perceptual maps, 256–7
  reasons for similarities and differences among samples, 259–60
  responses vs different groups of subjects, 258
  general pros and cons, 249–51
    advantages, 249–50
    catches, 250–1
  narrowing spectrum of terms into broader dimensions, 260
  extent to which array of category terms is narrowed down, 260
  processing the answers from raw to clean data, 252–5
    diagram of typical steps usually followed to process responses, 255
    matrix showing occurrences of terms used to describe the products, 254
  sensory testing practice, 247–64
    formats from traditional approach to novel methods, 248–9
    future trends and social media, 261–4
opportunities, 21
oral health status, 489–90
oral pleasantness unpleasantness scale (OPUS), 60
ordinary least-squares regression (OLS), 518, 537
OurMobilePlanet study, 455

paired-comparison method, 31
paired preference method, 36
panel behaviour, 278
panel level
  TDS curves, 279
    data processing, 281
panel performance, 291–5
  performance indexes, 291–5
  protocol and data pre-treatment, 291
  pre-processing for assessment, 292
panel size, 276
  non-exhaustive list of TDS publications, 277
  proposal for minimum number of panellist and replicates, 278
panel training, 272–8, 276
Pangborn Sensory Science Symposium, 16
parsimony, 530
partial least squares (PLS), 235
partial napping
  beer, 368–71
    correlation loading plot from PCA, 369
    score plot from PCA, 369
participant randomisation, 230
partitioning algorithms, 167
penalty analysis, 235
perceptive tasks, 154
perceptual maps
  based on open-ended questions, 256–7
    CA representing which category terms were associated with yoghurt product, 257
performance indexes, 291–5
  example of TDS performance figure, 294
  sequential approach for panel and subject performance diagnostics, 293
perfumers, 402–3
  Flash profile, 404–5
    GPA analysis, 405
    products positioning, 405
    semantic description, 404–5
    word cloud of different attributes, 404
personality typing, 513
plain yoghurt
  application of CATA questions for sensory characterisation, 235–40
    example of evaluation sheet used for evaluation, 236
    mean drop in overall liking as function of percentage of consumers, 239
    number of consumers who used terms of CATA questions, 237
    regression coefficients of PLS model calculated considering overall liking, 239
    sample and term representation in first and second coordinates of CA, 238

plant operators, 13
Polarised Projective Mapping (PSM), 216
polarised sensory positioning (PSP), 4, 137
  birth, 215
  methodologies, 216–17
    continuous scale, 217
    triad, 217
  nature, 216
  rating vs triad, 220–1
    maps obtained with continuous and triad PSP for 88, 60, 40 and 20 consumers, 222
  sensory profiling technique, 215–24
    data analyses, 218–19
    discussion of choice of poles, 223–4
  taste of water, 219–23
    example of data aggregation, 221
    first example, 219–20
polarised triad test, 216
  PSP methodologies, 217
poles, 216
positive emotions, 201
post-coding, 252
pre-operational stage, 473
precise analysis
  concrete situations where this type of evaluation is appropriate, 346–7
preference mapping, 475
preference methods, 34
preference task, 478
principal component analysis (PCA), 126, 192, 199, 234, 256, 284–5, 366, 379, 433
  Scree plots, 127
procedural memory, 490
Procrustes analysis, 38–9
product benchmarking, 338–40
product configuration, 81
product development
  sensory testing in children, 473–81
  sensory testing in older people, 485–504
Product Emotion Measurement Instrument (PrEmo), 75–6
product purchase, 92
products, 39
professional experts, 12–13
  Flash Profile, 141–2
Profiles of Mood States (POMS), 77

project team evaluation
  analysis of opportunities and constraints, 347–8
projective descriptive methods
  data analysis, 365–7
Projective Flash Profile, 411–12, 424
  advantages, 421–3
projective mapping, 198–9, 364–5
  technique, 189
  versus napping, 382
projective tests, 197–206
  analysing Napping data, 199–200
pseudo-barycentric, 256
psychometric rationale, 167
psychophysics, 54–5
purchase intent, 60–1
purchase phase, 463

Q-sort methodology, 156–7
qualitative data
  guidelines, 514
qualitative researchers, 516
qualitative study, 462
quantitative descriptive analysis (QDA), 41, 57, 85, 385, 391
quantitative flavour profiling (QFP), 391–2
quantitative study, 463

R statistical software
  analysing Napping and sorted Napping, 210–12
  fasnt function, 212
Rand index, 168
randomisation test approach
  difference testing, 288–90
ranking, 49, 446, 478
ranking team tasting, 353–5
  description, 354
  practical example, 355
    results of ranking based on Kruskal-Wallis analysis, 356
rapid sensory methods
  automotive industry, 427–51
    gearbox sensations and comfort, 430–5
    idle noises of diesel engines, 444–50
    pros and cons, 450–1
    role and lateral support perception, 435–44

rapid sensory methods (*cont.*)
  usage in R&D and research, 3–22
    aims and needs, 4–5
    consequences on sensory activities, 14–20
    DA in marketing and consumer research, 5–7
    emergence of profiling methods, 4
    evolution in use of descriptive analysis (DA), 3–4
    methodological evolution, 8–14
rapid sensory profiling technique, 194
rapidity, 8–9
rating-polarised sensory positioning (PSP), 216
RebA, 58
recruitment, 468
  elderly panel, 497–9
regression analysis, 527
regression model, 238
repeatability, 291–2
  Flash Profile, 134–5
repertory grid, 16, 121
  method, 479
restyling phase, 463
rheometry, 15
road holding, 435, 438
robustness, 356–7
  Flash Profile, 135
roll and lateral support perception, 435–44
  complex perception, 435
Rubicon model, 461
rule developing experimentation (RDE), 521, 536

sales persons, 14
SAS algorithm, 270
scale type, 273
scale values, 102
schema, 513
self-report measures, 75–6
semantic interpretation, 129–30
semantic memory, 490
semantic networks, 261, 262
sensation, 54–8
  implication for classical descriptive methods, 57–8
  individual differences, 56–7
  group means by genotype, 57
  measurement accuracy, 54–6
Sense It, 389, 392
sensitivity, 30, 32, 63
  Flash Profile, 135
SensoMineR package, 210
sensory activities
  consequences, 14–20
  knowledge and dissemination, 16
    outcomes from 2011 online survey among Pangborn Sensory Symposium, 17
sensory blind tasting sessions, 387
sensory characteristics, 229
sensory consistency, 317–18, 320
sensory consumer journey
  case study, 459–66
sensory data analysis, 210
sensory description, 145–6, 436–8
  adapted FP procedure for consumers' perceptions, 147
  example of evaluation sheet, 437
  sensory profiling sessions, 437
  tool for communication with customers, 340–1
  what's your Texture tool for soft breads, 340
sensory descriptive analysis
  internal development, 337–8
    internal human resources within Puratos Group, 338
    number and type of internal sensory tests for product development, 338
  with elderly panel, 497–503
sensory evaluation, 456
sensory expert panellists', 385, 387
sensory hedonic model, 34
sensory internship, 397
sensory map, 127–9
  dendogram from hierarchical cluster analysis, 128
  FP sensory map, 128
sensory panel leaders, 385
sensory perception
  children, 474–5
  consumer behaviour, 53–63
    hedonics, 58–60
    product usage and intake, 60–1

Index

sensation, 54–8
  summary, 61–3
  linking sensations, liking and intake, 61–3
    exploring the causal chain within group of individuals, 62–3
    influence of sensations on liking, 61–2
    intake prediction, 62
sensory profile, 500–1
  Napping and sensory Napping, 197–213
    analysis of data using R statistical software, 210–12
    projective tests, 197–206
    sorted Napping, 206–10
sensory profiling, 3
  free multiple sorting, 187–95
    advantages, disadvantages and applications, 194–5
    data analysis, 191–3
    implementation and data collection, 190–1
    overview, 187–8
    practical framework and design of experiments, 189–90
    theoretical framework, 188–9
  free sorting for product development, 153–81
    case study in automotive industry, 170–80
    statistical treatment, 160–70
    task, 154–9
  ideal profiling, 307–30
    additional valuable properties, 316–19
    illustration of Ideal Profile Analysis (IPA), 319–28
    IPM, tool for product development and product optimisation, 313–16
    principle and properties of Ideal Profile Method (IPM), 309–12
  polarised sensory positioning (PSP), 215–24
    data analyses, 218–19
    discussion of choice of poles, 223–4
    methodologies, 216–17
    taste of water, 219–23
  temporal dominance sensations (TDS), 269–304
    applications, 295–9

  data analysis and comparison between products, 287–90
  data analysis and representation of product space, 284–7
  data analysis and representation of sequence, 279–84
  experiment and panel training, 272–8
  future trends, 299–302
  overview, 270–1
  panel performance, 291–5
sensory research
  emotional response measurement in consumer research, 71–87
    application of verbal self-report emotion techniques, 78–84
    approaches, 75–7
    defining emotion, 72–3
    importance, 73–5
    relating sensory properties to consumers, 84–6
    unresolved issues and topics for future research in verbal self-report, 86–7
    verbal self-report emotion lexicon, 77–8
sensory scientists, 385–6
sensory-specific effect, 100
sensory task
  impact of age and dependence, 491–7
sensory testing
  alternative methods advantages and disadvantages, 27–50
    data collection, analysis and reporting, 42–8
    developing descriptive analysis capability, 39–42
    future trends, 49
    other descriptive methods, 48–9
    subjects, 28–30
  chefs, culinary professional and brew masters, 363–82
    brewers and novices assessing beer, 367–8
    data analysis of projective descriptive methods, 365–7
    fast descriptive methods and no formal sensory training, 363–5
    general discussion and recommendations, 377–9
    partial napping of beer, 368–71

sensory testing (cont.)
    project mapping and Napping, 382
    spice blends and pastes, 371–7
  flavourists, 383–98
  methods, 30–7
    affective, 34–7
    descriptive, 33–4
    discrimination, 30–3
    new product development
      children, 473–81
      older people, 485–504
    organisation, 475–6
      legal issues, 475–6
      practical recommendations, 476
sensory testing practice
  open-ended questions, 247–64
    analysing data of getting valuable outcome from different applications, 255–61
    future trends and social media, 261–4
    general pros and cons, 249–51
    processing the answers from raw to clean data, 252–5
    when appropriate, 251–2
sensory toolkit, 9–10
similarity-based method, 364
simplified labelled affective magnitude (SLAM), 60
single Malt Scotch whisky (SMSW), 100–1
single nucleotide polymorphism (SNP), 56
skin conductance responses (SCR), 74
SMACOF, 163
Sniffin' Sticks test, 491, 498
social media, 261–4
SORT-CC method, 166
sorted Napping, 206–10
  analysing data, 207–9
  example for sensory profile technique, 209–10
  Napping as sensory profile technique, 197–213
    analysis of data using R statistical software, 210–12
    projective tests, 197–206
    some behavioural considerations, 206–7
sorting methodology, 188–9
sorting tasks, 191, 499–500
spatial representation
  cars, 177–80

Spectrum, 335
Spectrum Analysis, 41, 48
Spectrum Descriptive Analysis (SDA), 57
spelling correction, 252
spice blends and pastes, 371–7
spider plots, 317
spontaneous responses
  combined from different modalities, 259
standardised TDS curves, 279, 281–3
static condition, 428
STATIS model, 165
statistical methodology, 412
stimulus-response relation, 55
stoplist, 252
strengths, 20
stress, 163
structured experimental design data, 521–4
subject attribute performance, 47
subjects
  Flash Profile, 122–4
  segmentation, 143–5
subjects randomisation, 230
success factors
  guarantee correct application of method, 357
  importance of coordinator and teams, 357
  quality of terms of evaluation, 357
swallowing disorders, 490
symbolic operation stage, 473

tasting protocol
  panel training, 275
    protocol for sensory evaluation of one sip of wine, 275
taxonomic sorting, 158
TDS curves
  difference, 287–8
  parameters, 285–6
  significance limit, 282–4
team tasting, 14
  analysis, 360–1
  improvement in food industry, 345–61
  management and needs, 349–55
    comparative, 349–50, 352–3
    descriptive, 349
    ranking, 353–5

Index   555

  usage depending on project need, 349
  prospects, 361
technical tasting sessions, 387
temporal dominance of sensations (TDS),
    4, 85, 480
  application in cars, 297
  experiment and panel training, 272–8
  future trends, 299–302
  method validation based on wine
      samples, 295–7
  overview, 270–1
  sensory profiling technique, 269–304
    applications, 295–9
    data analysis and comparison
        between products, 287–90
    data analysis and representation of
        product space, 284–7
    data analysis and representation of
        sequence, 279–84
    panel performance, 291–5
temporal measurements, 480
temporal methods, 271
temporal order of sensations (TOS),
    300–1
text mining software packages, 255
Texture Profile, 48
theoretical framework, 188–9
threats, 21–2
time-intensity (TI), 271
TimeSens, 278
trajectory mapping
  TDS, 286–7
TRANSREG algorithm, 270–1
triangle method, 31
Twitter, 261–2

2-way ANOVA model, 290
2-way best-worst scaling, 102–3

Ultra-Flash Profiles (UFP), 249, 366, 370

variable ratio (VR) reinforcement
    schedule, 59
verbal-based methods, 364
verbal self-report, 76–7
  application in sensory and consumer
      field, 78–84
    conceptual profile of Cadbury's
        Bournville Deeply Dark
        Chocolate, 82
    emotion lexicon for EsSense Profile,
        79
    list of emotion terms within six
        dimensions in original and
        modified GEOS, 80
  emotion lexicon, 77–8
verbalisation process, 159
verbalisation task, 154–5, 172, 364
vin de garde, 156
visual analogue scale, 55
vocabulary development, 100

Ward criteria, 168, 323
water, 215
  taste of first example, 215
    product space, , 220
weaknesses, 20–1
Williams Latin square design, 230
working memory, 490

XLSTAT, 403

http://dx.doi.org/
DOI: 10.1533/9781782422587.1.3

Abstract
This chapter analyses the recent evolution in the use of rapid sensory profiling techniques with respect to various situations where descriptive sensory information is needed in research, in product development, as well as in market research and consumer science. The most frequent objectives and constraints are reviewed in order to provide guidelines for users of rapid sensory methods. Implications of methodological changes in descriptive analysis are discussed. The chapter describes how the diversification of the sensory toolkit may impact on sensory activities, and why sensory scientists and their stakeholders should benefit from earlier and more integrated sensory measurements. Eventually, a SWOT analysis of rapid profiling methods is provided.

Keywords: Integrated sensory measures, Sensory diversity, Profiling with consumers, Professional sensory expertise, Sensory toolkit.

http://dx.doi.org/
DOI: 10.1533/9781782422587.1.27

Abstract
The growth of sensory evaluation as a product information resource has been significant over the past four decades. With increased competition, companies look for sensory information that can be obtained quickly without compromising the quality of results. This author describes the basic principles and how they are best used to advantage. Particular attention is given to the subject selecting and qualifying process and its effect on test time and the quality of the results. Additional attention is given to descriptive analysis, a methodology that is used in almost every sensory program but in different ways, reflecting how researchers understand the process. The author describes the process by which the quantitative descriptive analysis (QDA) method was developed, taking into account the time requirements in the context of a rapid method.

Keywords: Qualifying subjects, Test time, Descriptive analysis, QDA, Subject skill, Data analysis.

http://dx.doi.org/
DOI: 10.1533/9781782422587.1.53

Abstract
Sensory scientists typically focus most of their effort on measuring the sensations that products elicit, or measuring the pleasure provided by products. Far less effort is given to measuring consumer behavior with regard to intake or use. Implicitly or explicitly, sensations are assumed to influence liking, acceptance, and preference, which in turn influence use and intake. That is, we all assume formulation ultimately influences use; however, measurement error at each stage complicates efforts to build causal models of the chain from formulation to use. This chapter highlights sources of variability at various points, including biological differences, as well as measurement error in the putative path from formulation to use.

Keywords: Psychophysics, Hedonics, Acceptability, Scaling, Intake, Ingestive behavior.

http://dx.doi.org/
DOI: 10.1533/9781782422587.1.71

Abstract
This chapter begins with a general definition of emotion and discussion of the importance of measuring emotion in sensory and consumer research. General approaches that are used to measure emotions are briefly discussed, including autonomic measures and brain imaging techniques as well as verbal and visual self-report measures. The authors review some of the verbal self-report emotion lexicons that are reported in the current literature and their applications in the sensory and consumer fields. The authors also discuss how consumer emotional responses can be related to the output of sensory descriptive analysis. Finally, some unresolved issues in verbal self-report emotion measurement and topics for future research are discussed.

Keywords: Emotion, Sensory properties, Consumer, Hedonic, Verbal self-report measure.

http://dx.doi.org/
DOI: 10.1533/9781782422587.1.91

Abstract
This chapter describes the importance of conceptual content in defining the very nature of brands, packaging and products, and how the conceptual profiles of these three product touch points should be aligned in order to achieve brand–product consonance, something of fundamental importance in new product success. The psychological basis of conceptualisation is described briefly, along with the role played by conceptual content in delivering reward, which is the primary motivator of product choice. Procedures for conceptual profiling are described in detail. Three case studies are presented to demonstrate the added insight delivered by conceptual profiling in the brand, pack and product development processes.

Keywords: Conceptualisation, Conceptual profiling, Reward, Brand–product consonance/dissonance, Fit-to-brand, Conceptual segmentation.

http://dx.doi.org/
DOI: 10.1533/9781782422587.2.121

Abstract
Flash Profile is a simple sensory descriptive method based on free choice of attributes, a comparative evaluation of the samples, and quantification by the means of ranks for each attribute. It aims at providing quick access to the relative positioning of a set of products. This chapter first presents the principle of Flash Profile, before the main methodological points are illustrated through a simple example of description of dark chocolates. Further methodological issues are then discussed and additional guidelines are provided, together with a presentation of the advantages and limitations of Flash Profile. The last part of the chapter presents the evolution in the use of Flash Profile, and examples of potential applications in R&D and market research are given.

Keywords: Sensory positioning, Experts, Consumers, Professionals, Generalized Procrustes analysis (GPA).

http://dx.doi.org/
DOI: 10.1533/9781782422587.2.153

Abstract
The free sorting task is an efficient technique for assessing the perception of a set of products by a panel of subjects. This holistic and non-verbal task makes it possible to perform sensory evaluation with consumers and to study differences in subjects' perception. This chapter describes different variants of sorting tasks, highlighting their advantages and limitations and presents a review of statistical methods of analysis of free sorting data. A case study in the automotive industry illustrates the use of the free sorting task for assessing the consumers' perceptions of car body style. It shows that the free sorting task is indeed an effective tool to be used in product development.

Keywords: Sorting task, Categorization, Verbalization, Multidimensional scaling (MDS).

http://dx.doi.org/
DOI: 10.1533/9781782422587.2.187

Abstract
Free multiple sorting is an old method that has been recently introduced into sensory science. The method is distinguished most importantly from "single" free sorting, as a whole product space is retrieved from each assessor through *ad libitum* categorization of samples rather than a single category measure. Perceptually, the assessor is less restricted than in "single" free sorting, and free multiple sorting is still considered a rapid approach. For the analyst, global sample spaces are easily obtained, although the descriptive data may take longer time to process. The chapter gives an overview of free multiple sorting, the theory behind it and a practical guide to its implementation.

Keywords: Free multiple sorting, Sorting, Rapid methods, Descriptive methods, Sensory evaluation.

http://dx.doi.org/
DOI: 10.1533/9781782422587.2.197

Abstract
Can (sorted) Napping be considered a rapid method? Can it be considered as an alternative to descriptive analysis? The main objective of this chapter is to shed new light on Napping and sorted Napping in order to provide some clues for answering these two questions. In particular, we will stress the intrinsic nature of the data collected when using Napping, and on the link between Napping and sorted Napping. We will see how the information provided by Napping is unique, and how different it can be from that provided by descriptive analysis. We will see that, at a subject level, Napping is certainly a rapid method, but in order to be used as an alternative to descriptive analysis, a relatively high number of subjects need to be considered.

Keywords: Projective test, Napping, Verbalization, Sorted Napping, Sensory profile.

http://dx.doi.org/
DOI: 10.1533/9781782422587.2.215

Abstract

Polarized sensory positioning (PSP) was originally developed for easy definition of the sensory characteristics of drinking water without presenting too many samples. The method was then revealed to be applicable to various kinds of products (cosmetics, aromas, etc.) and appeared to be a good alternative to classical sensory profiling techniques. PSP is based on comparison with a constant set of references (poles), and provides a new type of sensory data requiring dedicated data analysis. This chapter presents the PSP methodology and its application to the taste of water. Data collection, results reliability and results stability are discussed. Finally, more than a sensory methodology, PSP looks like a philosophy in sensory data aggregation. The applications could be numerous, and the method should be adapted to new problems.

Keywords: Polarized sensory positioning (PSP), Products comparison, Data aggregation, Multidimensional scaling (MDS) unfolding, Drinking water.

http://dx.doi.org/
DOI: 10.1533/9781782422587.2.227

Abstract
Check-all-that-apply (CATA) questions are versatile multiple choice questions which are being increasingly used for product sensory characterization with consumers. The methodology has been reported to be a simple and reliable approach for sensory product characterization of a wide range of products, providing similar results to descriptive analysis with trained assessors. However, several methodological issues related to how CATA questions are implemented can affect the results. This chapter presents the method and discusses recommendations for implementation and data analysis.

Keywords: Check-all-that-apply questions, CATA, Sensory characterization, Consumer research, Check lists.

http://dx.doi.org/
DOI: 10.1533/9781782422587.2.247

Abstract
Why use open-ended questions? This chapter provides an up-to-date overview on the use of open-ended questions in novel rapid sensory methodologies and the potential applications in which they could provide unique benefits. Next, the step-by-step process is described (from task performance to analysis) with its pros and cons, with important considerations to follow. Examples with data from real, recently published case studies are provided to illustrate all the points raised in the chapter. Finally, the future trends of open-ended questions are discussed: Are social media currently useful for sensory science? How can the benefits of open-ended questions be boosted?

Keywords: Free-elicited descriptions, Word association, Spontaneous responses, Idiosyncratic data, Consumers.

http://dx.doi.org/
DOI: 10.1533/9781782422587.2.269

Abstract
Temporal dominance of sensations (TDS) is a relatively recent method in the sensory field aiming at collecting the sequence of the dominant sensations perceived by the subject during the tasting of a product. For instance, while tasting a piece of chocolate from intake to swallowing, the subject can successively perceive the attributes crunchy, cocoa, fat, cocoa again and melting as the dominant sensations. Since the first communication presenting the method in 2003, many developments have been proposed to define adequate protocols, propose dedicated statistical analyses and interpretations, and even extend the TDS approach to a larger scope. This chapter first introduces the TDS method and its specificities, and describes how to set up a TDS experiment before exploring the analysis part and providing some examples of applications.

Keywords: Temporal dominance of sensations, Dominance, Panel training, TDS data analysis, Panel performance.

http://dx.doi.org/
DOI: 10.1533/9781782422587.2.307

Abstract
The Ideal Profile Method (IPM) is a descriptive analysis in which consumers are asked to rate products on both their perceived and ideal intensities against a list of attributes. In addition, overall liking is rated. The information gathered from the IPM is rich, and directly actionable for product development. However, these data need special attention, since they are provided by consumers who are rating a fictive product. In order to fully appreciate the data, the Ideal Profile Analysis (IPA) has been set up. This methodology helps in (1) evaluating the consistency of the ideal data, (2) assessing homogeneous groups of consumers/products, (3) defining the sensory profile of the ideal product of reference, and (4) optimizing the products according to that reference.

Keywords: Ideal Profile Method, Just About Right scale, Liking, Optimization, Product development.

http://dx.doi.org/
DOI: 10.1533/9781782422587.3.335

Abstract
Flash Profile is an interesting methodology to start developing sensory analysis in a highly competitive industrial environment where resources are limited. In this chapter, we will explain why/how we decided to implement Flash Profile for descriptive sensory analysis for new product development, and how this methodology can be easily and quickly implemented to answer a diversity of sensory issues: benchmarking against competition, development of sensory vocabulary, recipe fine-tuning during the product development process. We will examine the advantages and the limitations of the use of Flash Profile in an industrial context, and explain how we can adapt this methodology to new sensory challenges.

Keywords: Comparative free choice profiling (FCP), Flash Profile, Descriptive sensory analysis, Preference Mapping.

http://dx.doi.org/
DOI: 10.1533/9781782422587.3.345

Abstract
Increasingly, we need to evaluate our products through projects inside project teams. This type of evaluation, which we call "team tasting", needs to integrate a trade-off between an easy set-up that is adapted to our constraints, and the preservation of a sound and objective sensory description. Therefore, first we define the four main aims of the evaluation: describe, follow, rank and compare. On the basis of these four goals, three methods are customized and adapted for Danone. Taking account of the judges' low numbers and poor training, we use an approach based on consensus or robust statistical tests. This approach has a very positive impact on the integration of sensory reactions into such projects, in two regards: the project team is more involved, and the most precise tools can be used for the most important steps of the project.

Keywords: Sensory characterization, Team tasting, Describe, Compare, Rank, Internal testing.

http://dx.doi.org/
DOI: 10.1533/9781782422587.3.363

Abstract
We present two case studies with fast sensory methods to capture and document the core of sensory properties in an inexpensive manner outside the safe environment of the sensory laboratory. One is in the development lab with prototypes, with brewers, and one is with chefs working in a more explorative fashion to chart the properties of a large range of spice blends.

For users of fast sensory methods who are not sensory scientists, the main challenge is the data analysis. The necessary statistical skills are often lacking with these professionals. The chapter includes recommendations for further development of fast sensory methods and their data analysis to accommodate the working environment and skills available in development labs.

Keywords: Projective mapping, Realistic product development settings, Brewers, Chefs, Culinary professionals.

http://dx.doi.org/
DOI: 10.1533/9781782422587.3.383

Abstract
Among various types of information, flavourists now have access to sensory insights. The translation of sensory insights into flavours is only possible if the communication between flavourists and sensory experts is optimal. But various challenges appear as the backgrounds, fields of expertise and ways of working for flavourists and sensory experts are very different, from the way they evaluate the food products to the ways they analyse and express their flavour perceptions. Givaudan has implemented strategies to overcome these challenges and profit from the complementarity of both types of experts in order to provide the right flavour in the right application for the right market.

Keywords: Sensory experts, Flavourists, Flavour perception and evaluation, Complementarity, Communication, Adapted descriptive methodologies.

http://dx.doi.org/
DOI: 10.1533/9781782422587.3.401

Abstract
Fine fragrances are a crucial spearhead for the beauty industry. However, perfumes are not easy to describe and their evaluation usually requires a high level of expertise. Perfume creators are the best placed to describe fine fragrances, due to their acknowledged expertise. On the other hand, their expert vocabulary is sometimes difficult for consumers to understand.

In our first study, Flash Profile, with its flexibility and rapidity, allowed us to demonstrate a strong similarity between the fragrance perceptions of experts and consumers. Furthermore, it appeared that consumer' perception is not only olfactory but also, and maybe even more so, evocative. Thus, we upgraded the methodology in a second intercultural consumers study by widening the attributes generation with projective categories such as Emotions, Mental Images and Lifestyles.

Keywords: Flash Profile, Projective categories, Fragrance, Perfume creators, Consumers.

http://dx.doi.org/
DOI: 10.1533/9781782422587.3.427

Abstract

This chapter is the result of more than 15 years of experience of sensory evaluation in the car industry. After explaining the context of automotive sensory evaluation and the need for sensory science in the car industry, we will develop three examples of studies, chosen to highlight the interest of rapid sensory methods including Flash Profiling. Two of these examples concern dynamic aspects of the car. Therefore, the sensory description is obtained during the task of driving, which is a specificity of the car industry. We will discuss the pros and cons of Flash Profiling, in particular with respect to product development taking consumers' perceptions into account.

Keywords: Flash Profiling, Automotive sensations, Comfort, Dynamic perception, Consumer study, Preference Mapping.

http://dx.doi.org/
DOI: 10.1533/9781782422587.3.455

Abstract
Mobile research, i.e. data collection via smartphones and tablets, dominates the current agenda. The sensory community appreciates the external validity given by a natural tasting context and instantaneous consumer reports. We summarize advantages and limitations in mobile research, and present a sensory-app case study exploring a consumer journey in its natural environment. Sensorial attributes of goods from the food and cosmetics segment are effective in specific, phases during such a journey. To capture this phase specificity, instantaneous (i.e. online) data collection is crucial. The case study explores a consumer journey of AXE hair wax in Germany. The results showed a phase specificity of sensorial drivers of liking. Mobile research as a paradigm allows rapid immediate assessments during sensorial experiences.

Keywords: Mobile research, Smartphone app, Sensorial products, Consumer journey, Sensorial codes.

http://dx.doi.org/
DOI: 10.1533/9781782422587.4.473

Abstract
This chapter aims at briefly reviewing applications of sensory descriptive testing methods with children, as well as their limits in relation to the cognitive development of children. Children's sensory perceptions and preferences may differ from those of adults, which may make working directly with children a must for successful product development, at least at some stages of product development. However, most reported studies with children have characterized their preferences. Some studies have used discrimination tests and ranking tasks to gain knowledge about differences among products, or about sensory abilities development. A much smaller number of studies have involved children in sensory panels – probably because of both the limited analytical abilities in children, and the higher cost related to developing sensory evaluation with this population.

Keywords: Children, Teenagers, Sensory evaluation, Consumer testing, Descriptive techniques, Discrimination, Ranking, Intensity scaling.

http://dx.doi.org/
DOI: 10.1533/9781782422587.4.485

Abstract
Very few studies today validate the use of various sensory tools among elderly subjects, although the population is ageing rapidly. This chapter deals with applied sensory assessment in the food sector. It first presents heterogeneity factors in this population: physical and psychological health and dependency, decrease of chemosensory capacities, changes in oral-motor skills and modifications of cognitive abilities. Then, a study aiming at comparing the capacity of young and more or less dependent elderly subjects to use a discrete scale (discriminatory power and repeatability) in a monadic sequential presentation is described. Finally, recommendations are given about the cases where a panel of elderly people should be necessary and the precautions to undertake during a sensory test among elderly subjects.

Keywords: Elderly, Sensory descriptive analysis, Hedonic scaling, Repeatability, Discrimination.

http://dx.doi.org/
DOI: 10.1533/9781782422587.4.509

Abstract
We introduce Empathy and Experiment™, joining in-depth qualitative methods (listening, observing) with experimental design of ideas (conjoint measurement) to understand consumers' food values, determine which "elements" (communication points) drive behaviors, uncover mind-set segments, and create a typing tool to identify those population segments. The goal is to generate knowledge about people, their food habits, and getting them to try new products and improve nutrition. We show how to create a rapid deep understanding of the consumer mind (empathy) and discovery of what to say. This short streamlined process takes weeks. Empathy and Experiment™ is used to increase interest in Golden Rice, containing vitamin A precursor.

Keywords: Conjoint analysis, Experimental design, Consumer research, New food products, Nutrition, Population differences.

Printed in the United States
By Bookmasters